T0191623

Advances in Intelligent Systems and Computing

Volume 433

Series editor

Janusz Kacprzyk, Polish Academy of Sciences, Warsaw, Poland
e-mail: kacprzyk@ibspan.waw.pl

About this Series

The series "Advances in Intelligent Systems and Computing" contains publications on theory, applications, and design methods of Intelligent Systems and Intelligent Computing. Virtually all disciplines such as engineering, natural sciences, computer and information science, ICT, economics, business, e-commerce, environment, healthcare, life science are covered. The list of topics spans all the areas of modern intelligent systems and computing.

The publications within "Advances in Intelligent Systems and Computing" are primarily textbooks and proceedings of important conferences, symposia and congresses. They cover significant recent developments in the field, both of a foundational and applicable character. An important characteristic feature of the series is the short publication time and world-wide distribution. This permits a rapid and broad dissemination of research results.

More information about this series at http://www.springer.com/series/11156

Suresh Chandra Satapathy
Jyotsna Kumar Mandal · Siba K. Udgata
Vikrant Bhateja
Editors

Information Systems Design and Intelligent Applications

Proceedings of Third International
Conference INDIA 2016, Volume 1

 Springer

Editors
Suresh Chandra Satapathy
Department of Computer Science
  and Engineering
Anil Neerukonda Institute of Technology
  and Sciences
Visakhapatnam
India

Jyotsna Kumar Mandal
Kalyani University
Nadia, West Bengal
India

Siba K. Udgata
University of Hyderabad
Hyderabad
India

Vikrant Bhateja
Department of Electronics and
  Communication Engineering
Shri Ramswaroop Memorial Group
  of Professional Colleges
Lucknow, Uttar Pradesh
India

ISSN 2194-5357   ISSN 2194-5365   (electronic)
Advances in Intelligent Systems and Computing
ISBN 978-81-322-2753-3   ISBN 978-81-322-2755-7   (eBook)
DOI 10.1007/978-81-322-2755-7

Library of Congress Control Number: 2015960416

Printed on acid-free paper

This Springer imprint is published by SpringerNature
The registered company is Springer (India) Pvt. Ltd.

Preface

The papers in this volume were presented at the INDIA 2016: Third International Conference on Information System Design and Intelligent Applications. This conference was organized by the Department of CSE of Anil Neerukonda Institute of Technology and Sciences (ANITS) and ANITS CSI Student Branch with technical support of CSI, Division-V (Education and Research) during 8–9 January 2016. The conference was hosted in the ANITS campus. The objective of this international conference was to provide opportunities for researchers, academicians, industry personas and students to interact and exchange ideas, experience and expertise in the current trends and strategies for Information and Intelligent Techniques. Research submissions in various advanced technology areas were received and after a rigorous peer-review process with the help of programme committee members and external reviewers, 215 papers in three separate volumes (Volume I: 75, Volume II: 75, Volume III: 65) were accepted with an acceptance ratio of 0.38. The conference featured seven special sessions in various cutting edge technologies, which were conducted by eminent professors. Many distinguished personalities like Dr. Ashok Deshpande, Founding Chair: Berkeley Initiative in Soft Computing (BISC)—UC Berkeley CA; Guest Faculty, University of California Berkeley; Visiting Professor, University of New South Wales Canberra and Indian Institute of Technology Bombay, Mumbai, India, Dr. Parag Kulkarni, Pune; Dr. Aynur Ünal, Strategic Adviser and Visiting Full Professor, Department of Mechanical Engineering, IIT Guwahati; Dr. Goutam Sanyal, NIT, Durgapur; Dr. Naeem Hannoon, Universiti Teknologi MARA, Shah Alam, Malaysia; Dr. Rajib Mall, Indian Institute of Technology Kharagpur, India; Dr. B. Majhi, NIT-Rourkela; Dr. Vipin Tyagi, Jaypee University of Engineering and Technology, Guna; Prof. Bipin V. Mehta, President CSI; Dr. Durgesh Kumar Mishra, Chairman, Div-IV, CSI; Dr. Manas Kumar Sanyal, University of Kalyani; Prof. Amit Joshi, Sabar Institute, Gujarat; Dr. J.V.R. Murthy, JNTU, Kakinada; Dr. P.V.G.D. Prasad Reddy, CoE, Andhra University; Dr. K. Srujan Raju, CMR Technical Campus, Hyderabad; Dr. Swagatam Das, ISI Kolkata; Dr. B.K. Panigrahi, IIT Delhi; Dr. V. Suma, Dayananda Sagar Institute, Bangalore; Dr. P.S. Avadhani,

Vice-Principal, CoE(A), Andhra University and Chairman of CSI, Vizag Chapter, and many more graced the occasion as distinguished speaker, session chairs, panelist for panel discussions, etc., during the conference days.

Our sincere thanks to Dr. Neerukonda B.R. Prasad, Chairman, Shri V. Thapovardhan, Secretary and Correspondent, Dr. R. Govardhan Rao, Director (Admin) and Prof. V.S.R.K. Prasad, Principal of ANITS for their excellent support and encouragement to organize this conference of such magnitude.

Thanks are due to all special session chairs, track managers and distinguished reviewers for their timely technical support. Our entire organizing committee, staff of CSE department and student volunteers deserve a big pat for their tireless efforts to make the event a grand success. Special thanks to our Programme Chairs for carrying out an immaculate job. We place our special thanks here to our publication chairs, who did a great job to make the conference widely visible.

Lastly, our heartfelt thanks to all authors without whom the conference would never have happened. Their technical contributions made our proceedings rich and praiseworthy. We hope that readers will find the chapters useful and interesting.

Our sincere thanks to all sponsors, press, print and electronic media for their excellent coverage of the conference.

November 2015   Suresh Chandra Satapathy
                                                                             Jyotsna Kumar Mandal
                                                                                    Siba K. Udgata
                                                                                 Vikrant Bhateja

Organizing Committee

Chief Patrons

Dr. Neerukonda B.R. Prasad, Chairman, ANITS
Shri V. Thapovardhan, Secretary and Correspondent, ANITS, Visakhapatnam

Patrons

Prof. V.S.R.K. Prasad, Principal, ANITS, Visakhapatnam
Prof. R. Govardhan Rao, Director-Admin, ANITS, Visakhapatnam

Honorary Chairs

Dr. Bipin V. Mehta, President CSI, India
Dr. Anirban Basu, Vice-President, CSI, India

Advisory Committee

Prof. P.S. Avadhani, Chairman, CSI Vizag Chapter, Vice Principal, AU College of
Engineering
Shri D.N. Rao, Vice Chairman and Chairman (Elect), CSI Vizag Chapter, Director
(Operations), RINL, Vizag Steel Plant
Shri Y. Madhusudana Rao, Secretary, CSI Vizag Chapter, AGM (IT), Vizag Steel
Plant
Shri Y. Satyanarayana, Treasurer, CSI Vizag Chapter, AGM (IT), Vizag Steel Plant

Organizing Chair

Dr. Suresh Chandra Satapathy, ANITS, Visakhapatnam

Organizing Members

All faculty and staff of Department of CSE, ANITS
Students Volunteers of ANITS CSI Student Branch

Program Chair

Dr. Manas Kumar Sanayal, University of Kalyani, West Bengal
Prof. Pritee Parwekar, ANITS

Publication Chair

Prof. Vikrant Bhateja, SRMGPC, Lucknow

Publication Co-chair

Mr. Amit Joshi, CSI Udaipur Chapter

Publicity Committee

Chair: Dr. K. Srujan Raju, CMR Technical Campus, Hyderabad
Co-chair: Dr. Venu Madhav Kuthadi,
Department of Applied Information Systems
Faculty of Management
University of Johannesburg
Auckland Park, Johannesburg, RSA

Special Session Chairs

Dr. Mahesh Chandra, BIT Mesra, India, Dr. Asutosh Kar, BITS, Hyderabad: "Modern Adaptive Filtering Algorithms and Applications for Biomedical Signal Processing Designs"
Dr. Vipin Tyagi, JIIT, Guna: "Cyber Security and Digital Forensics"
Dr. Anuja Arora, Dr. Parmeet, Dr. Shikha Mehta, JIIT, Noida-62: "Recent Trends in Data Intensive Computing and Applications"
Dr. Suma, Dayananda Sagar Institute, Bangalore: "Software Engineering and its Applications"
Hari Mohan Pandey, Ankit Chaudhary: "Patricia Ryser-Welch, Jagdish Raheja", "Hybrid Intelligence and Applications"
Hardeep Singh, Punjab: "ICT, IT Security & Prospective in Science, Engineering & Management"
Dr. Divakar Yadav, Dr. Vimal Kumar, JIIT, Noida-62: "Recent Trends in Information Retrieval"

Track Managers

Track #1: Image Processing, Machine Learning and Pattern Recognition—Dr. Steven L. Fernandez
Track #2: Data Engineering—Dr. Sireesha Rodda
Track #3: Software Engineering—Dr. Kavita Choudhary
Track #4: Intelligent Signal Processing and Soft Computing—Dr. Sayan Chakraborty

Technical Review Committee

Akhil Jose Aei, Vimaljyothi Engineering College (VJEC), Kannur, Kerala, India.
Alvaro Suárez Sarmiento, University of Las Palmas de Gran Canaria.
Aarti Singh, MMICTBM, M.M. University, Mullana, India.
Agnieszka Boltuc, University of Bialystok, Poland.
Anandi Giri, YMT college of Management, Navi Mumbai, India.
Anil Gulabrao Khairnar, North Maharashtra University, Jalgaon, India.
Anita Kumari, Lovely Professional University, Jalandhar, Punjab
Anita M. Thengade, MIT COE Pune, India.
Arvind Pandey, MMMUT, Gorakhpur (U.P.), India.
Banani Saha, University of Calcutta, India.
Bharathi Malakreddy, JNTU Hyderabad, India.
Bineet Kumar Joshi, ICFAI University, Dehradun, India.
Chhayarani Ram Kinkar, ICFAI, Hyderabad, India.
Chirag Arora, KIET, Ghaziabad (U.P.), India.
C. Lakshmi Devasena, IFHE University, Hyderabad, India.
S.G. Charan, Alcatel-Lucent India Limited, Bangalore, India
Dac-Nhuong Le, VNU University, Hanoi, Vietnam.
Emmanuel C. Manasseh, Tanzania Communications Regulatory Authority (TCRA)
Fernando Bobillo Ortega, University of Zaragoza, Spain.
Frede Blaabjerg, Department of Energy Technology, Aalborg University, Denmark.
Foued Melakessou, University of Luxembourg, Luxembourg
G.S. Chandra Prasad, Matrusri Engineering College, Saidabad, Hyderabad
Gustavo Fernandez, Austrian Institute of Technology, Vienna, Austria
Igor N. Belyh, St. Petersburg State Polytechnical University
Jignesh G. Bhatt, Dharmsinh Desai University, Gujarat, India.
Jyoti Yadav, Netaji Subhas Institute of Technology, New Delhi, India.
K. Kalimathu, SRM University, India.
Kamlesh Verma, IRDE, DRDO, Dehradun, India.
Karim Hashim Kraidi, The University of Al-Mustansiriya, Baghdad, Iraq
Krishnendu Guha, University of Calcutta, India

Contents

About the Editors

Dr. Suresh Chandra Satapathy is currently working as Professor and Head, Department of Computer Science and Engineering, Anil Neerukonda Institute of Technology and Sciences (ANITS), Visakhapatnam, Andhra Pradesh, India. He obtained his Ph.D. in Computer Science Engineering from JNTUH, Hyderabad and his Master's degree in Computer Science and Engineering from National Institute of Technology (NIT), Rourkela, Odisha. He has more than 27 years of teaching and research experience. His research interests include machine learning, data mining, swarm intelligence studies and their applications to engineering. He has more than 98 publications to his credit in various reputed international journals and conference proceedings. He has edited many volumes from Springer AISC and LNCS in the past and he is also the editorial board member of a few international journals. He is a senior member of IEEE and Life Member of Computer society of India. Currently, he is the National Chairman of Division-V (Education and Research) of Computer Society of India.

Dr. Jyotsna Kumar Mandal has an M.Sc. in Physics from Jadavpur University in 1986, M.Tech. in Computer Science from University of Calcutta. He was awarded the Ph.D. in Computer Science & Engineering by Jadavpur University in 2000. Presently, he is working as Professor of Computer Science & Engineering and former Dean, Faculty of Engineering, Technology and Management, Kalyani University, Kalyani, Nadia, West Bengal for two consecutive terms. He started his career as lecturer at NERIST, Arunachal Pradesh in September, 1988. He has teaching and research experience of 28 years. His areas of research include coding theory, data and network security, remote sensing and GIS-based applications, data compression, error correction, visual cryptography, steganography, security in MANET, wireless networks and unify computing. He has produced 11 Ph.D. degrees of which three have been submitted (2015) and eight are ongoing. He has supervised 3 M.Phil. and 30 M.Tech. theses. He is life member of Computer Society of India since 1992, CRSI since 2009, ACM since 2012, IEEE since 2013 and Fellow member of IETE since 2012, Executive member of CSI Kolkata Chapter. He has delivered invited lectures and acted as programme chair of many international conferences and also edited nine volumes of proceedings from Springer AISC series, CSI 2012 from McGraw-Hill, CIMTA 2013 from Procedia Technology, Elsevier. He is reviewer of various international journals and conferences. He has over 355 articles and 5 books published to his credit.

Dr. Siba K. Udgata is a Professor of School of Computer and Information Sciences, University of Hyderabad, India. He is presently heading Centre for Modelling, Simulation and Design (CMSD), a high-performance computing facility at University of Hyderabad. He obtained his Master's followed by Ph.D. in Computer Science (mobile computing and wireless communication). His main research interests include wireless communication, mobile computing, wireless sensor networks and intelligent algorithms. He was a United Nations Fellow and worked in the United Nations University/International Institute for Software Technology (UNU/IIST), Macau, as research fellow in the year 2001. Dr. Udgata is working as principal investigator in many Government of India funded research projects, mainly for development of wireless sensor network applications and application of swarm intelligence techniques. He has published extensively in refereed international journals and conferences in India and abroad. He was also on the editorial board of many Springer LNCS/LNAI and Springer AISC Proceedings.

Prof. Vikrant Bhateja is Associate Professor, Department of Electronics and Communication Engineering, Shri Ramswaroop Memorial Group of Professional Colleges (SRMGPC), Lucknow, and also the Head (Academics & Quality Control) in the same college. His areas of research include digital image and video processing, computer vision, medical imaging, machine learning, pattern analysis and recognition, neural networks, soft computing and bio-inspired computing techniques. He has more than 90 quality publications in various international journals and conference proceedings. Professor Vikrant has been on TPC and chaired various sessions from the above domain in international conferences of IEEE and Springer. He has been the track chair and served in the core-technical/editorial teams for international conferences: FICTA 2014, CSI 2014 and INDIA 2015 under Springer-ASIC Series and INDIACom-2015, ICACCI-2015 under IEEE. He is associate editor in International Journal of Convergence Computing (IJConvC) and also serves on the editorial board of International Journal of Image Mining (IJIM) under Inderscience Publishers. At present, he is guest editor for two special issues floated in International Journal of Rough Sets and Data Analysis (IJRSDA) and International Journal of System Dynamics Applications (IJSDA) under IGI Global publications.

Study and Analysis of Subthreshold Leakage Current in Sub-65 nm NMOSFET

Krishna Kumar, Pratyush Dwivedi and Aminul Islam

Abstract As the technology scales down, the subthreshold leakage increases exponentially which leads to a dramatic increase in static power consumption especially in nanoscale devices. Consequently, it is very important to understand and estimate this leakage current so that various leakage minimization techniques can be devised. So in this paper we attempt to estimate the subthreshold leakage current in an NMOSFET at 16, 22, 32 and 45 nm technology nodes. Various factors which affect the subthreshold leakage such as temperature, drain induced barrier lowering (DIBL) and other short channel effects have also been explored. All the measurements are carried out using extensive simulation on HSPICE circuit simulator at various technology nodes.

Keywords Subthreshold leakage · Scaling · Threshold voltage · Short channel

1 Introduction

MOSFETs have been scaled down aggressively since 1970s as per scaling rules followed by the industry (guided by Dennard's scaling theory) in order to achieve higher packing density, reduced power consumption and increased performance [1]. However, leakage currents have also increased drastically with continuous technology scaling. Scaling MOSFET below 32 nm is very much challenging task. Hence, new device structures like SOI and multiple gate transistors are being investigated. Scaling below 32 nm such as 22 and 14 nm requires new kind of

K. Kumar (✉) · P. Dwivedi · A. Islam
Birla Institute of Technology, Mesra, Ranchi 835215, Jharkhand, India
e-mail: be1006713@bitmesra.ac.in

P. Dwivedi
e-mail: be1024713@bitmesra.ac.in

A. Islam
e-mail: aminulislam@bitmesra.ac.in

© Springer India 2016
S.C. Satapathy et al. (eds.), *Information Systems Design and Intelligent Applications*, Advances in Intelligent Systems and Computing 433,
DOI 10.1007/978-81-322-2755-7_1

1

device structure such as tri-gate or multi-gate [2]. Various leakage currents contribute to total leakage in a MOSFET but there are three major types of leakage currents—the subthreshold, the gate tunneling and band to band tunneling [3]. As a consequence of scaling the MOSFETs, their supply voltages have to be reduced to prevent the high field effects and reduce power consumption. Therefore, to maintain reliable operation, the threshold voltage (V_{TH}) has to be scaled down. For short channel MOSFETs, the drain voltage starts decreasing the barrier between the source (S) and the channel. This is called drain induced barrier lowering (DIBL) effect. The threshold voltage of MOSFET also decreases with decrease in the channel-length which is called V_{TH} roll off due to short channel effect (SCE). All these effects cause a substantial increase in the subthreshold leakage current. Since subthreshold current (I_{SUB}) is the diffusion component of drain to source current, temperature also affects it significantly.

The rest of the paper is structured as follows. Section 2 describes the subthreshold current in detail. Section 3 explains how the various factors namely temperature and SCE (such as DIBL) affect the subthreshold leakage current. Section 4 concludes the paper.

2 Subthreshold Current (I_{SUB})

Depending on the gate (G), source (S) and drain (D) voltages, an NMOSFET can be biased in three regions-subthreshold region, linear region and saturation region. For an enhancement type NMOSFET, the V_{GS} (G to S voltage) at which the concentration of electrons in the channel is equal to the concentration of holes in the p⁻ substrate is known as threshold voltage (V_{TH}). An NMOSFET is said to be biased in subthreshold region when the gate-source voltage (V_{GS}) < (V_{TH}). Ideally an MOSFET should be off in that condition but a drain to source current does flow which is called the subthreshold leakage current (I_{SUB}). According to International Technology Roadmap (ITRS), I_{SUB} is defined as the D-to-S current when the D is biased to V_{DD} and the gate, substrate and source terminals are all biased to ground (GND). The expression for the drain to source current in the subthreshold region is given by [4]

$$I_{SUB} = I_0(m-1)e^{\frac{V_{GS}-V_{TH}}{mV_T}}\left(1 - e^{-\frac{V_{DS}}{V_T}}\right) \qquad (1)$$

where,

$$I_0 = \mu_0 C_{ox}\frac{W}{L}V_T^2 \qquad (2)$$

Fig. 1 Log (I_{DS}) versus V_{GS}
at 22 nm technology node

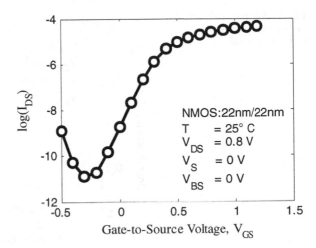

Gate-to-Source Voltage, V_{GS}

and

$$V_T = \frac{K_B T}{q}.$$ (3)

The factor m in the above equation equals $\Delta V_{GS}/\Delta \Psi_S$, which comes out to be $1 + C_{dm}/C_{ox}$. Here, Ψ_S represents surface potential and C_{dm} represents depletion layer capacitance.

In the subthreshold region, the drain to source current is dominated by the diffusion current [5]. This subthreshold region can be identified as the linear region in the semi log plot of I_{DS} versus V_{DS} as shown in Fig. 1.

The reciprocal of the slope of the linear region in the plot, shown in Fig. 1 is known as the subthreshold slope. Ideally, it should be small so that the MOSFET can be turned off effectively in subthreshold region. However, for 22 nm technology node, it comes out be $\cong$ 92 mV/decade. This shows that subthreshold leakage is indeed a major problem in sub-65 nm technology regime. An important conclusion that can be drawn from Fig. 2 is that, whereas for long channel MOSFETs I_{SUB} is almost independent of V_{DS}, it starts increasing with V_{DS} in short channel MOSFETs. It can also be observed that as the device-dimensions scale down, the subthreshold leakage (I_{SUB}) increases exponentially.

Fig. 2 Subthreshold current versus V_{DS} at different technology node

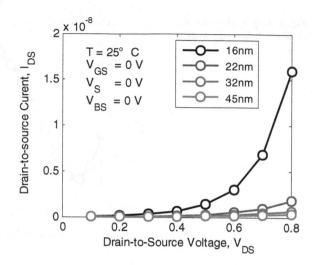

3 Factors Affecting Subthreshold Leakage

3.1 Body Effect

The threshold voltage for a MOSFET is given by

$$V_{TH} = V_{T_0} + \gamma(\sqrt{2\varphi_F - V_{BS}} - \sqrt{2\varphi_F}) \tag{4}$$

where, V_{TH} = threshold voltage, V_{BS} = substrate potential w.r.t. source, V_{T0} = threshold voltage at $V_{BS} = 0$ V and $2\varphi_F$ = surface potential.

It can be seen that, if source/body junction is more reverse-biased, i.e. V_{BS} is negative the V_{TH} increases. This is called body effect. This correspondingly causes a decrease in the subthreshold current. Figure 3 shows the semi log plot of I_{DS} versus V_{GS} for different V_{BS} at 22 nm technology node. Virtually no change can be observed in the SS (subthreshold slope) at different body biases while a downward shift can be observed in the weak inversion region indicating a decrease in the subthreshold current.

3.2 Mechanism for DIBL

In a long-channel MOSFET, the drain (D) and source (S) junctions are quite apart and so the drain bias does not affect the potential barrier between the S (source) and the C (channel). This potential barrier is largely controlled by the gate bias. However, in short-channel MOSFETs, the S and D are so close to each other that the drain-source potential has a strong effect on drain-to-source current (I_{DS}). This closeness results in considerable field-penetration from D to S. Due to this

Fig. 3 Log (I_{DS}) versus V_{GS} at different V_{BS} at 22 nm technology node

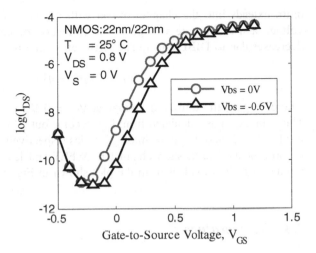

field-penetration the potential barrier (PB) at the S is lowered. The degree of the PB lowering depends upon the drain voltage (also on channel-length (L) and channel-doping concentration (NCH)) [6]. This phenomenon is known as so called drain-induced barrier lowering (DIBL). Since the barrier height is reduced, more number of electrons can move into the channel resulting in higher I_{DS} at a given gate voltage. DIBL can be modeled as the upward shift in the semi log plot of I_{DS} versus V_{GS} for different V_{DS}, as shown in Fig. 4 [7].

The physics of DIBL can be explained from a different angle by considering the depletion layer around the drain diffusion region. Under the zero V_{DS} condition, the depletion layers around S/D diffusion regions are very thin. As V_{DS} increases the depletion layer widens around drain diffusion region.

Equation (4) is not suitable for predicting V_{TH} at scaled technology where impact of DIBL is pronounced. In short-channel devices, the depletion region around the

Fig. 4 Log (I_{DS}) versus V_{GS} for different V_{DS} at 22 nm technology node

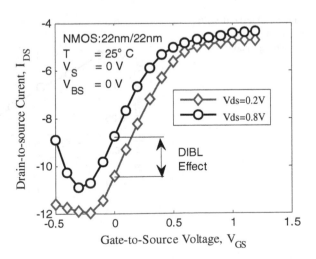

drain extends into the channel region with increase in V_{DS}. Therefore, the drain voltage assists the gate voltage in the formation of inversion layer. Hence, V_{TH} decreases due to DIBL effect. The DIBL effect can be simply modeled as:

$$V_{TH} = V_{TH0} - \eta V_{DS} \tag{5}$$

where, V_{TH0} is the threshold voltage at $V_{DS} = 0$ V, and η is the DIBL coefficient. The values of η as calculated from Fig. 5 come out to be 0.27, 0.23, 0.18, and 0.01 for 16, 22, 32, and 45 nm technology nodes respectively. It shows that DIBL effect is more noticeable in short-channel MOSFET and leads to decrease in threshold voltage, which is evident from the plot shown in Fig. 5.

3.3 Temperature

Since VLSI circuits operate at elevated temperature due to heat generated in the circuit, the analysis of variation of subthreshold leakage with temperature becomes imperative. In subthreshold region major component of I_{DS} is diffusion current (and not the drift current). The diffusion current density for one dimensional flow of electrons is given by the equation

$$J_n(x) = qD_n \frac{dn(x)}{dx} \tag{6}$$

where, $J_n(x)$ = diffusion current density due to electrons, D_n = electron diffusion coefficient, $dn(x)/dx$ = concentration gradient of electron.

The equation contains a temperature dependent parameter: Electron diffusion coefficient (D_n). Moreover, the subthreshold leakage depends on the threshold

Fig. 5 Threshold voltage versus V_{DS} at different technology nodes

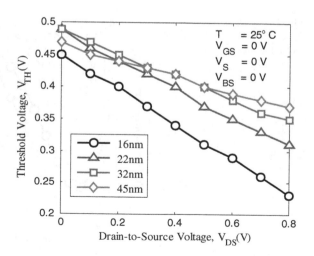

voltage which is also temperature dependent. Hence, the thermal behavior of subthreshold current can be explained on the basis of following two factors.

3.3.1 Electron Diffusion Coefficient (D_n)

With the increase in temperature, the kinetic energy of the electrons increases and hence, their ability to diffuse along the channel also increases, which leads to an increased diffusion coefficient. Consequently, the diffusion current density rises and thus the subthreshold current increases substantially.

3.3.2 Threshold Voltage

The threshold voltage equation is given by [8]

$$V_T = V_{FB} + 2\varphi_F + \gamma\sqrt{2\varphi_F} \tag{7}$$

where, V_T = threshold voltage, $2\varphi_F$ = surface potential, γ = body effect coefficient. The term V_{FB} is the flat band voltage.

In the threshold voltage equation, there are two temperature dependent terms: surface potential ($2\varphi_F$) and flat band voltage (V_{FB}). So, the temperature dependence of threshold voltage can be explained on the basis of following two factors.

Surface Potential ($2\varphi_F$)

The energy gap between the intrinsic fermi level (which is almost at the middle of the band gap (E_i)) and the fermi level of a doped semiconductor is given by the term $q\varphi_F$. As temperature increases, carrier-concentration increases in semiconductor due to thermally generated electron-hole pairs. Therefore, a p-type semiconductor becomes more p-type due to more concentration of holes. Thus, the fermi level shifts towards valence band and hence φ_F increases.

Flat Band Voltage (V_{FB})

The flat band voltage (V_{FB}) as given in (8) is actually the difference between the work functions of n^+ doped poly gate and p-type body

$$\varphi_{MS} = \varphi_M - \varphi_S \tag{8}$$

where, φ_M = work function of n+ polysilicon, φ_S = work function of p-type substrate. φ_{MS} always comes out to be negative for this case. With increase in

Fig. 6 Subthreshold leakage
versus temperature at 22 nm
technology node

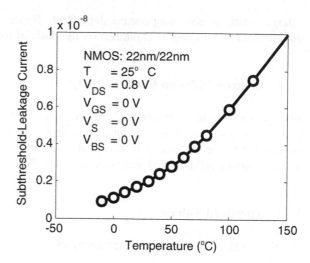

Fig. 6 Subthreshold leakage versus temperature at 22 nm technology node

temperature, the fermi level of the p-type substrate shifts towards the valence band which increases φ_S (since $\varphi_S = 2\varphi_F$) and hence, φ_{MS} becomes more negative.

So, with rise in temperature, $\varphi_{MS} = V_{FB}$ becomes more negative i.e. it decreases. Therefore, the terms φ_F and V_{FB} have conflicting dependence on temperature. However, the rate of decrease of flat band voltage with temperature is more than the rate of increase of φ_F with temperature [9]. Hence, the threshold voltage has a negative temperature coefficient. So, the subthreshold current increases with temperature as shown in Fig. 6.

3.4 Short-Channel Effect (SCE)

In short channel devices, the distance between the source and drain become comparable to the width of the depletion region formed between the diffusion regions and the substrate. Due to this, the electric field pattern becomes two dimensional in short channel devices [6]. For very small geometries, the threshold voltage decreases with channel length. This is called short-channel effect (SCE). It can be explained on the basis of "charge sharing" model shown in Fig. 7 [7]. Since electric field is perpendicular to equipotential contours, it can be inferred that the depletion charges which are in the approximately triangular regions near the source and drain junctions, have their field lines terminated not on the gate but instead on the source and drain junctions. These depletion charges are called "shared" charges since they are electrically shared with the source and drain and should not be included in the equation of threshold voltage. Therefore, the net effective depletion charge (Q_D) which is to be included in the Eq. (9) decreases and is equal to charge contained in the trapezoidal region.

Fig. 7 Charge sharing model
of an NMOSFET

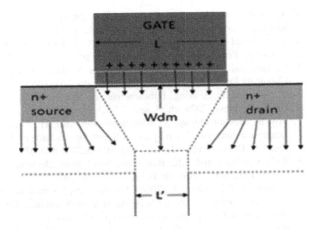

It simply means that the total bulk charge which the gate has to invert decreases and hence the threshold voltage decreases as given by (9) [10].

$$V_T = \varphi_{MS} - \frac{Q_i}{C_i} - \frac{Q_D}{C_i} + 2\varphi_F \qquad (9)$$

where, $\varphi_{MS} - Q_i/C_i$ = flat band voltage, Q_i = effective positive charge at the interface, Q_D = effective depletion charge at the interface, C_i = capacitance of the insulator, $2\varphi_F$ = surface potential. Due to this V_{TH} roll off, the subthreshold leakage increases.

4 Conclusion

This paper carries out thorough investigation on subthreshold leakage current of nanometer MOSFETs and concludes that the leakage current has drastically increased with continuous technology scaling. This increased leakage current has resulted in increased leakage power. Due to this leakage power, very limited room for dynamic power is left since total power for a given system is predefined.

References

1. R. Dennard, et al., "Design of ion-implanted MOSFETs with very small physical dimensions," IEEE Journal of Solid State Circuits, vol. SC-9, no. 5 pp. 256–268, Oct. 1974.
2. Warnock J, IBM Syst. & Technol. Group, T.J. Watson Res. Center, Yorktown Heights, NY, USA "Circuit Design Challenges at 14 nm Technology Node", Design Automation Conference (DAC) 2011 48th ACM/EDAC/IEEE, pp. 464–467.

3. Agarwal, A; Mukhopadhyay, S.; Ray Chowdhury, A.; Roy, K., "Leakage Power Analysis and Reduction for Nanoscale Circuits", IEEE Micro, vol. 26, no. 2, pp. 68–80, Mar. – April. 2006.
4. Y. Taur and T. H. Ning, Fundamentals of Modern VLSI Devices, New York: Cambridge University Press, 2009, ch. 3, sec. 3.1.3.2, pp. 165.
5. Y. Taur and T. H. Ning, Fundamentals of Modern VLSI Devices, New York: Cambridge University Press, 2009, ch. 3, sec. 3.1.3.1, pp. 164–165.
6. Y. Taur and T. H. Ning, Fundamentals of Modern VLSI Devices, New York: Cambridge University Press, 2009, ch. 3, sec. 3.2.1.1, pp. 177–179.
7. Roy K, Mukhopadhyay, S., Mahmoodi-Meimand, H., "Leakage Current Mechanisms and Leakage Reduction Techniques in Deep-Submicrometer CMOS Circuits," IEEE Trans. Electron. Devices, vol. 91, no. 2, pp. 305–327, Feb 2003.
8. B.G. Streetman and S.K. Baneerjee, Solid State Electronic Devices, India: PHI Learning Private Ltd., 6th edition, ch. 6, sec. 6.4.3, pp. 281.
9. Filanovsky IM, Allam A (2001) Mutual compensation of mobility and threshold voltage temperature effects with applications in CMOS circuits. IEEE Trans Circuits and Syst I: Fundamental Theory and Applications 48:876–884.
10. B.G. Streetman and S.K. Baneerjee, Solid State Electronic Devices, India: PHI Learning Private Ltd., 6th edition, ch. 6, sec. 6.4.4, pp. 284.

Implementations of Secure Reconfigurable Cryptoprocessor a Survey

Rajitha Natti and Sridevi Rangu

Abstract One among the several challenges in the area of applied cryptography is not just devising a secure cryptographic algorithm but also to manage with its secure and efficient implementation in the hardware and software platforms. Cryptographic algorithms have widespread use for every conceivable purpose. Hence, secure implementation of the algorithm is essential in order to thwart the side channel attacks. Also, most of the cryptographic algorithms rely on modular arithmetic, algebraic operations and mathematical functions and hence are computation intensive. Consequently, these algorithms may be isolated to be implemented on a secure and separate cryptographic unit.

Keywords Trust · FPGA security · Cryptographic processor · Reconfigurable cryptosystems

1 Introduction

There is an alarming need for securing wide area of applications of cryptography that we use in our daily life besides military, defense, banking, finance sectors and many more. To cater to this need innumerable products/services have been developed which are predominantly based on encryption. Encryption in turn relies on the security of the algorithm and the key used. The different encryption algorithms proposed so far have been subjected to various forms of attacks. While it is not possible to devise an algorithm that works perfectly well and sustains all forms of attacks, cryptographers strive to develop one that is resistant to attacks and that

R. Natti (✉) · S. Rangu
Department of Computer Science & Engineering, Jawaharlal Nehru
Technological University, Hyderabad, India
e-mail: rajitha2k2@yahoo.co.in

S. Rangu
e-mail: sridevirangu@jntuh.ac.in

© Springer India 2016
S.C. Satapathy et al. (eds.), *Information Systems Design and Intelligent Applications*, Advances in Intelligent Systems and Computing 433,
DOI 10.1007/978-81-322-2755-7_2

performs well. The task is not just to propose a new algorithm but to create an environment that improves the performance of the algorithm and that protects the keys from attacks.

A cryptoprocessor is a specialized processor that executes cryptographic algorithms within the hardware to accelerate encryption algorithms, to offer better data, key protection. Commercial examples of cryptoprocessors include IBM 4758, SafeNet security processor, Atmel Crypto Authentication devices. We present the different forms of attacks possible on cryptoprocessors in Sect. 2. The implementation of cryptoprocessors on reconfigurable platform is considered in Sect. 3.

The following are the different architectures of cryptographic computing [1].

- Customized General Purpose Processor: The processor is extended or customized to implement the cryptographic algorithms efficiently. Typical commercially available solutions are CryptoBlaze from Xilinx or the AES New Instructions (AES-NI) incorporated in the new Intel processors. [2, 3] focus on instruction set extensions for cryptographic applications and embedded systems respectively.
- Cryptographic processor (cryptoprocessor): It is a programmable device with a dedicated instruction set to implement the cryptographic algorithm efficiently. [4–6] are examples of cryptoprocessor implementations.
- Cryptographic coprocessor (crypto coprocessor): It is a logic device dedicated to the execution of cryptographic functions. Unlike the cryptoprocessor it cannot be programmed, but can be configured, controlled and parameterized.
- Cryptographic array (crypto-array): It is a coarse grained reconfigurable architecture for cryptographic computing (Fig. 1).

1.1 Motivation

A number of reconfigurable systems have been developed since the 1990s, as they offer strong protection of IP, better protection of key data against vulnerabilities, by integrating functions in software layers into hardware i.e., to System on Chip

Fig. 1 Architecture of cryptoprocessor [1]

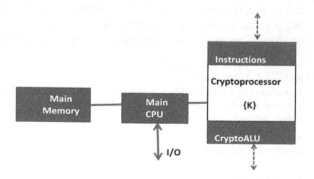

(SoC) or field programmable gate array (FPGA) or application specific integrated circuit (ASIC) depending on functionality. Reconfigurable systems have the ability to hardwire the keys i.e., they can be 'zeroized' and can be unseen by world. They are tamper resistant and offer trust.

Reconfigurable systems are more adaptable to applications at low production cost. Video games and all microprocessor unit embedded systems like home appliances and those found in vehicles are examples. They have their use in telecom applications to handle high speed data stream, in rapid prototyping and emulation, in image and signal processing and many more.

1.2 Novelty and Significance to State of the Art

As software security cannot prevent against hardware attacks, secure processor architectures have been proposed. We identify problems ignored by the current cryptoprocessors like security of interconnects; secure booting, secure key management with focus on lightweight cryptography and post quantum code based cryptography.

The state of art reveals that cryptoprocessor have been implemented with respect to applications [3] or to address specific issue for instance to overcome DPA attack [7] or Cryptoprocessor for SDR etc. or cryptoprocessor for a particular algorithm [8, 9].

1.3 Applications of Cryptoprocessors

Numerous applications of cryptoprocessor exist. Cryptoprocessors can be used in Automated Teller Machine Security, E-commerce applications, smart cards, wireless communication devices, resource constrained devices such as sensors, RFID tags, smart phones, smart cameras, digital rights management, trusted computing, prepayment metering systems, pay per use, banking, military and defense applications.

1.4 Importance of Security in Hardware Processes

The secure development of software processes has led to the need for inclusion of security as an aspect to software development lifecycle. This is done as a part of the training phase at the beginning of the SDLC as per Microsoft Security Development Lifecycle.

Similarly, there is a need to incorporate security activities into all phases of development lifecycle for hardware processes. Cryptographic hardware is

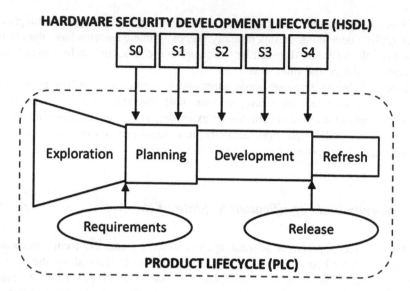

Fig. 2 Incorporating hardware SDL checkpoints into the PLC [10]

vulnerable to both software and hardware attacks and hence security needs to be considered as an integral part of product lifecycle.

Security Assurance is one of the important parameters for many embedded applications which are lightweight and resource constrained. The emergence of system on chip, network on chip applications has further fuelled the need for security assurance. Hence the software development lifecycle has been adapted to hardware in [10] as depicted in Fig. 2.

2 Attacks on Cryptoprocessor

The different forms of hardware attacks of algorithmic implementations on cryptographic devices in literature have been identified as given below

(i) *Side Channel Attack*: A study of the literature reveals that a major amount of research has been expended during the last decade on side channel attacks and countermeasures. Side channel attacks can happen in one of the following ways:

    (a) *Timing Analysis*: Time required by the device to perform encryption/decryption can be used to get additional data to perform an attack.

    (b) *Electromagnetic analysis*: It is based on the electromagnetic radiation from the circuit that executes the encryption/decryption algorithm.

(c) *Power Analysis*: Power consumed by the device implementing the algorithms can be used to perform the attack. It can be of the form Simple Power Analysis or Differential Power Analysis.

Side channel attacks and countermeasures can be found in [11–14]. Side channel attack on bit stream encryption of Altera Stratix II and Stratix III FPGA family in the form of black box attack can be found in [11]. To combat IP theft and physical cloning bit stream encryption is used.

(ii) *Fault Injection Attacks*: involves inserting fault deliberately into the device and to observe erroneous output.

(iii) *Counterfeiting*: is to use your name illegally on a clone.

(iv) *Insert Trojan Horse*: is a common method used to capture passwords.

(v) *Cold boot attack*: is a technique to extract disk encryption keys [15].

(vi) *Cloning*: in which your design is copied without knowing how it works.

(vii) *Reverse Engineering*: is finding out how the design works.

(viii) *Steal IP*: IP is stolen either with the intention to sell it to others or to reverse engineer.

Another classification of attacks on cryptoprocessor as mentioned in [16] is as follows:

2.1 Invasive

Invasive attack gives direct access to internal components of the cryptographic device. The attack can be performed by manual micro probing, glitch, laser cutting, ion beam manipulation etc.

2.2 Local Non Invasive

This form of attack involves close observation to operation on the device. The side channel attacks listed above may be considered as an example of such an attack.

2.3 Remote Attacks

Remote attacks involve manipulation of device interfaces. Unlike the previous attacks these attacks do not need physical access. API analysis, protocol analysis, cryptanalysis are examples of such an attack. While API analysis is concerned with cryptographic processor, cryptanalysis involves finding out the flaws in the algorithms primitives.

3   Implementations of Cryptosystems

There is a growing interest in devising cryptographic hardware that is resistant to the attacks mentioned in Sect. 2, side channel attacks in particular. Security in the digital world is primarily fulfilled by using cryptography. Numerous optimizations have been proposed and implemented for enhancing the performance and efficiency of the cryptographic algorithms that serve the innumerable applications in various fields. We present few such algorithms which have been implemented on FPGA and also on ASIC in certain cases. The significant consideration of most of them is time area product, besides analysis related to side channel resistance, amount of hardware resources utilized etc.

3.1   Symmetric Key Algorithm Implementations

We now discuss few implementations of symmetric key cryptographic algorithms on FPGA. Cryptoraptor [17] considers high performance implementation of set of symmetric key algorithm. The architecture comprises of processing elements (PE) linked by connection row (CR). The PE has independent functional units for arithmetic, shift, logical, table look permutation and operations. Multiplication is limitation due to the limited addressing structure of table look-up unit (TLU). It also lacks support for varying modulo in modular arithmetic operations.

Implementation of AES, RC5 and RC6 block cipher algorithms is considered in [18] in which they discuss area analysis and power consumptions.

3.2   Implementations of Asymmetric Cryptographic Algorithms

Many implementations of the asymmetric cryptographic algorithms exist with optimizations to address the needs of embedded system applications. Tim Erhan Guneysu in [19] investigates High Performance Computing implementation of symmetric AES block cipher, ECC and RSA on FPGA (Table 1).

Table 1 Characteristic of ECC and RSA crypto blocks [23]

Feature	ECC (146 bits)	RSA (1024 bits)
Frequency (MHz)	50	28
Logic size (slices)	3,036	4,595
Execution time	7.28 ms (scalar multiplication)	58.9 ms (decryption with 1024-bit key)

3.3 Implementations of Hash Functions

Hash functions are used for authentication, for providing data integrity and along with public key algorithms as digital signatures. MD5, SHA1, SHA-512 are prominent hash digest algorithms. BLAKE is one of the candidates of SHA3 and Keccak is SHA3 finalist which is based on sponge structure (Table 2).

3.4 Implementations of Lightweight Cryptography

For the fast growing applications of ubiquitous computing, new lightweight cryptographic design approaches are emerging which are investigated in [20].

FPGA implementation on low cost Spartan III of ultra light weight cryptographic algorithm Hummingbird is considered in [21]. Hummingbird has its application in RFID tags, wireless control and communication devices and resource constraint devices (Table 3).

3.5 A Glance on Code Based Cryptography and Its Implementations

Encryption with Coding Theory by Claude Shannon as basis is used in McEliece and Niederreiter which are considered as candidates for post quantum cryptosystems. McEliece is based on binary Goppa Codes which are fast to decode. McEliece and Niederreiter differ in the description of the codes. While the former cannot be used to generate signatures the later can be used (Table 4).

Table 3 Implementation results of PRESENT-128 [20]

Cipher	Block size	FPGA device	Maximum frequency (MHz)	Throughput (Mbps)	Total equiv. slices	Efficiency (Mbps/slice)
PRESENT-128	64	Spartan-III XCS400-5	254	508	202	2.51

Table 2 Comparison of hardware implementation of Hash functions [24]

Algorithm	Technology	Area	Frequency (MHz)	Throughput (giga bits per second)
Blake-512 [25]	FPGA Virtex 5	108 slices	358	0.3
Keccak-1600 [26]	FPGA Stratix III	4684 LUT	206	8.5

Table 4 McEliece
decryption implementations
[27]

Property	Spartan-3an	Virtex-5
Slices	2979	1385
BRAMs	5	5
Clock frequency	92 MHz	190 MHz
Clock cycles	94,249	94,249
Decryption latency	1.02 ms	0.50 ms
Security	80 bits	80 bits

4 Conclusion and Open Problems

A study on existing approaches to software engineering for hardware processes is considered in this paper besides investigating the general characteristics, implementations and uses of reconfigurable cryptoprocessors.

One of the challenges is the remote attacks (in the form of API attack) on cryptoprocessor which may be passive or active and which unlike the physical or invasive attacks doesn't need any contact with the implementation unit.

We need to focus on the formal tools for verification and validation of hardware design processes. The metrics that can be used for assessing the performance is still limited to throughput which again is platform, algorithm and application dependent and not generic. Hence the comparison of the algorithm performance cannot be judged very easily. There is a lot of scope on the adaptation of the process development lifecycle with respect to hardware or embedded software.

Wollinger et al. [22] discuss on the architectures of programmable routing in FPGA in the form of hierarchical and island style. FPGA security resistance to invasive and non-invasive attacks is still under experimentation as new attacks are devised before existing attacks are solved.

Much of the work on cryptoprocessors is specific to the application domain or to address a particular form of attack and is not generic to cater to many applications unless customized.

Key management in general is not considered as part of the cryptoprocessor implementation. Several designs of cryptoprocessors are proposed and implemented but still fully functional cryptoprocessor designs addressing integrity, key generation, key management, privacy of both symmetric and asymmetric cryptosystems is still a challenge.

Acknowledgments This work has been carried out as a part of Ph D under TEQIP-II.

References

1. Lilian Bossuet, Architectures of Flexible Symmetric Key CryptoEngines – A Survey: From Hardware Coprocessor to Multi-Crypto-Processor System on Chip, ACM Computing Surveys, Vol 45, No. 4, Article 41, August 2013.
2. Sandro Bartolini, Instruction Set Extensions for Cryptographic Applications, Springer Cryptographic Engineering, 2009.
3. Stefan Tillich, Instruction Set extensions for support of Cryptography on Embedded Systems, Ph D thesis, Graz University of Technology Nov 2008.
4. Lubos Gaspar, Cryptoprocessor –Architecture, Programming and Evaluation of the Security, Ph D Thesis, November 2012.
5. Sujoy Sinha Roy et al, Compact Ring-LWE Cryptoprocessor, Springer LNCS Vol 8731, 2014.
6. Michael Grand et al, Design and Implementation of a Multi-Core Crypto-Processor for Software Defined Radios, Springer LNCS Vol 6578 2011.
7. Kotaro Okamoto et al, A Hierarchical Formal Approach to Verifying Side-channel Resistant Cryptographic Processors in Hardware-Oriented Security and Trust (HOST), 2014 IEEE International Symposium.
8. Santosh Ghosh et al, BLAKE-512-Based 128-Bit CCA2 Secure Timing Attack Resistant McEliece Cryptoprocessor, IEEE Transactions on Computers 2014.
9. Hans Eberle et al, A Public-key Cryptographic Processor for RSA and ECC, IEEE proceeding 2004.
10. Hareesh Khattri et al, HSDL: A Security Development Lifecycle for Hardware Technologies 2012 IEEE International Symposium on Hardware-Oriented Security & Trust.
11. Pawel Swierczynski et al, Physical Security Evaluation of the Bitstream Encryption Mechanism of Altera Stratix II and Stratix III FPGAs, ACM Transactions on Reconfigurable Technology and Systems, Vol. 7, No. 4, Article 7, Publication date: December 2014.
12. Power Kotaro Okamoto, A Hierarchical Formal Approach to Verifying Side-channel Resistant Cryptographic Processors, IEEE, 2014.
13. Amir Moradi, Side-Channel Leakage through Static Power Should We Care In Practice.
14. Jen-Wei Lee, Efficient Power-Analysis-Resistant Dual-Field Elliptic Curve Cryptographic Processor Using Heterogeneous Dual-Processing-Element Architecture, IEEE Transactions On Very Large Scale Integration Systems, Vol. 22, No. 1, January 2014.
15. J. Alex Halderman et al, Lest We Remember: Cold Boot Attacks on Encryption Keys, Proc. 2008 USENIX Security Symposium.
16. MoezBen MBarka, Cryptoprocessor Application & Attacks Survey, May 2008.
17. Gokhan Sayiler, Cryptoraptor: High Throughput Reconfigurable Cryptographic Processor, IEEE 2014.
18. Rajesh Kannan et al, Reconfigurable Cryptoprocessor for Multiple Crypto Algorithm, IEEE Symposium 2011.
19. Tim Erhan Guneysu, Thesis, Cryptography and Cryptanalysis of Reconfigurable Devices Bochum, 2009.
20. Axer York Poschmann, Ph D Thesis, Lightweight Cryptography Feb 2009.
21. Xin Xin Fan et al, FPGA Implementation of Humming bird cryptographic algorithm, IEEE International Symposium on Hardware-Oriented Security and Trust (HOST), 2010.
22. Thomas Wollinger et al, Security on FPGAs: State of Art Implementations and Attacks, ACM Transactions on Embedded Computing Systems, Vol. 3, No. 3, August 2004.
23. HoWon Kim et al, Design and Implementation of Private and Public Key Cryptoprocessor and its Application to a Security System.
24. Zhije Shi et al, Hardware Implementation of Hash Function, Springer LLC 2012.
25. Bertoni, The Keccak sponge function family: Hardware performance 2010.
26. Beuchat et al, Compact Implementation of BLAKE on FPGA, 2010.
27. Santosh Gosh et al, A speed area optimized embedded Coprocessor for Mc Eliece Cryptosystem, IEEE Conference 2012.

Mitigating and Patching System Vulnerabilities Using Ansible: A Comparative Study of Various Configuration Management Tools for IAAS Cloud

Sanjeev Thakur, Subhash Chand Gupta, Nishant Singh
and Soloman Geddam

Abstract In an organization, Configuration management is an essentially important technique for assuring that the desired configuration are intact all the time.
Configuration keeps an eye on management and consistency of a software product's
versions, update etc. Currently many system administrators manage and maintain
their systems using a collection of batch scripts. Ansible replaces this plodding and
makes application deployment over cloud very simple. In this paper we will
understand and exploit Ansible's true potential. We will write playbooks for
patching system vulnerabilities and also understand the advantages of using
Ansible.

Keywords YML · Configuration management · Chef · Puppet · Ansible

1 Introduction

A Large number of IT organizations depend on golden images, manual procedures
and custom scripts to perform repetitive tasks. Since these IT Organizations are
spanned across the globe with large teams, these techniques are difficult to maintain,

S. Thakur (✉) · S.C. Gupta · N. Singh · S. Geddam
Amity University, Noida, India
e-mail: sthakur3@amity.edu

S.C. Gupta
e-mail: scgupta@amity.edu

N. Singh
e-mail: nishant704@gmail.com

S. Geddam
e-mail: geddamsolomon@gmail.com

© Springer India 2016
S.C. Satapathy et al. (eds.), *Information Systems Design and Intelligent
Applications*, Advances in Intelligent Systems and Computing 433,
DOI 10.1007/978-81-322-2755-7_3

track and keep up. This would result in diminished profit and low productivity. To overcome this issue, configuration management came into the rescue of IT organizations around the globe. There are numerous meanings of configuration management and variety of conclusions on what it truly is. Configuration management is a procedure of anticipating, recognizing and confirming the change in a product. Configuration management is built with the intentions to remove error and confusions caused by any kind of mismanagement of version upgrades, deployment, maintenance etc. It deals with holding the inexorable change under control [1].

Configuration management is a discriminating foundation of IT automation, it provides tools which permit you to deal with the packages, configure files and many different settings that prepare servers to do their allotted tasks. If automated configuration management is not implied in such scenarios then one has to do all the above mentioned tasks probably manually or with batch scripts. It's not only time consuming but also inclined to human mistake. Automating configuration management reduces a ton of manual work and makes more noteworthy reliability along with consistency. Configuration management is additionally about rolling out any improvements to the system in a sorted out way so that your servers are adjusted deliberately and effectively, while representing connections between system elements [2].

A system vulnerability is considered to be as a major security hole in the deployment process. Administrators around the world keep a sharp eye on the state of the system but the stock machine really needs to be secured first even before the context of the system changes afterwards. Before automation came into light, the administrators faced a huge challenge of writing the batch scripts and then running them manually on each system. Moreover this task is also prone to errors since the administrator might just forget to patch a few open ports or just forget a script to run. To help administrators with the initial patching of servers we use a Ansible Playbook which securely patches all the systems in one go.

2 Need for a Configuration Management Tool

The first question that arises while dealing with configuration management is the fact that why automate the whole process. To begin with, administrators have been overseeing the system deployment and configurations with the help of shell scripts or some per-configured system image. These methodologies do provide us with a standardized technique of building the system but at the same time disallow us from keeping the system at desired state. Of-course we can reach the fancied state with an extraordinary script, however this script won't let you know when design float happens because of little manual changes or mechanized overhauls. Moreover this script is also not able to reset the server to the right state. Cloud moreover is considered to be an expanding framework and if you are using scripts to manage your systems then it only increases the already complicated tasks. Writing manual custom scripts give rise to a lot of problems which include [3].

Human mistakes, Delicacy in custom made scripts, Custom made scripts might not be easy to understand, Unsalable, Unpredictable nature of script.

Automated configuration management takes out a considerable measure of manual work, and makes more noteworthy steadfastness and consistency. The administrator simply needs to notify the configurations he wants and the configuration management system roles them on to the servers and bring them back to the desirable states. Operations including upgrades, monitoring, status checking are all performed at a swift speed and even in a case of failure the administrator is notified specifically. One can rapidly check that the staging environment are arranged the same as production environment. Clearly, the benefits of automation are foreseen since it makes the whole process highly efficient and more productive.

3 Kick Starting Configuration Management

When it comes down towards the bare question of how to actually begin with configuration management, the answer is starting 'small'. One needs to begin with the things that are repetitive in nature and also more inclined toward errors. To dig deeper into automated configuration management, choosing a small problem in a system controlled by you is highly proactive in nature since it is isolated from the outside environment and saves time [4].

A numerous amount of system administrator start from automation of SSH, DNS, NTP, Squid etc. i.e. the regular practices which consume a huge amount of time. Apart from these tasks, application deployment, applying fixes and upgrades can be carried out too. Evaluation of capabilities and levels of integrations offered by a configuration management tool needs to be performed before implementing it for the business, and this must be done in such a way that it fulfills the intended purpose. The technology must also be chosen with ability to scale with the growing demands. The configuration management tool should be able to meet the following needs: Assist the cardinal configuration management along with composite tasks, Backings current use cases along with the technical roadmap, Should be able to provide pre-configured solutions and Inserts right in the automation tool series.

4 Ansible

Ansible is a fundamentally basic automation engine that automates configuration management, provisioning in cloud, intra-service orchestration, application deployment, and many other IT needs. It is designed for multi-tier deployments. Ansible models the IT framework by portraying how the majority of systems relate with each other, instead of simply dealing with only one server individually. Ansible is agentless and has no extra custom made security infrastructure which makes it very easy to deploy. Moreover it uses YAML for writing playbooks which permits the depiction of automation jobs in a manner that corresponds to simple English. It also supports idempotency and makes changes noticeable. Ansible is designed with the aim of being less difficult and simpler to comprehend. Since it utilizes an agentless structural engineering, the need of bootstrapping machines with a client vanishes. Apart from this, it uses a straightforward easy to understand file formatting which can be understood both by system administrators as well as developers [5]. Initially tools like "SmartFrog" where used in practices but it didn't supported continuous delivery, later more configuration management tools rolled up like Puppet and chef. Ansible on the other hand is the advanced ad hoc execution engine and is considered to be the 3rd generation configuration management tool. Ansible uses modules to be pushed on the nodes which are being managed. The modules are basically small programs which are composed with resource models of the coveted state of node. After this Ansible executes these modules using a SSH connection by default and automatically removes them when the tasks get completed. The library which contains modules can dwell on any machines and no daemons, servers or databases are needed for this purpose. The basic architectural working is shown as in Fig. 1.

There are three key concepts on which Ansible rely upon which includes Inventory files, Playbooks and Modules. Moreover Ansible also lets the flexibility to call inventory files from Rackspace, Amazon EC2 instances etc. Ansible also allows the placement of this inventory file in anywhere as you like [6].

After the inventory files, comes the Ansible Playbooks. Playbooks permit the organization of configurations and management in a very straightforward, intelligible ".yaml" files. Playbooks consists of tasks which are known as plays written in YAML files.

Ansible provides abstraction by implementing modules. In numerous ways, this is the place the genuine power in Ansible untruths. Ansible has the capacity to make

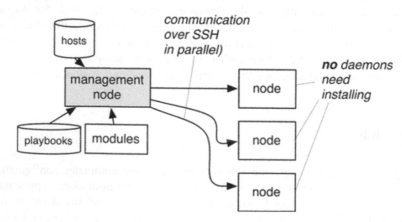

Fig. 1 Ansible mode of operation

framework administration idempotent by abstraction of states and commands in modules. This is an imperative idea that makes configuration management tools significantly more capable and protected than something like a run of the mill shell script. It is sufficiently testing to compose a shell script that can arrange a framework (or bunches of frameworks) to a particular state. It is greatly difficult to compose one that can be run over and again against the same frameworks and not break things or have unintended symptoms. At the point when utilizing idempotent modules, Ansible can securely be run against the same frameworks over and over without fizzling or rolling out any improvements that it doesn't have to make [7].

5 Comparison with Other Configuration Management Tool

Configuration management tools are designed with the aim to simplify configuration and management of hundreds or even thousands of servers placed around the world. There are a variety of tools available in the IT industry which perform configuration management for such large server farms both on premises or in the cloud. We study some of the best 3rd generation tools for configuration management tools and give our reasons why Ansible is the best tool out there.

5.1 Puppet

Puppet is considered to be the most intact when it comes to modules, actions and user interfaces. Puppet consists of simple and nice language which describe the nodes. Puppet speaks to the entire picture of server farm arrangement, enveloping pretty much every working framework and offering profound instruments for the primary Operating systems. Puppet works on the basis of client and server basis. Puppet supports both GUI as well as CLI. Puppet provides a dashboard in GUI which can be accessed via web and this gives the flexibility to orchestrate from anywhere. Command line interface (CLI) is quite is clear and permits the module to be downloaded and establishment by means of the 'puppet' command. At that point, alterations made on the configuration files are obliged to customize the module for the correct task. Moreover Puppet do gets slow from time to time whenever the number of clients increase, the common defect of client server system [8, 9]. Many administrators prefer running it in the masterless mode where the manifests are put inside git or bit-bucket (version control systems) and later rolled out on every hosts they manage. This can be scheduled as a cron job too.

5.2 Chef

Chef is quite alike to Puppet in most of its ideas. It constitutes of a Central server and agents which reside on the nodes being managed. This clearly states that Chef unlike Ansible is not agent-less and requires agents to be explicitly installed over the nodes. Moreover, Chef requires a workstation to oblige the central server. Installation of the agents can be done by the help of 'knife' which basically employees SSH connection. Authentication is done using certificates. Chef configure revolves mostly around the popular version control system, Git. Chef provides a web interface, however it does not provides the capacity to change configuration. It lacks far behind Puppet's web interface, yet takes into account the node organization and inventory control. Chef also has a huge collection of modules and scripts which are termed as 'recipes' which depend vigorously on Ruby. This is a reason Chef is very appropriate in development-centric infrastructures [10].

5.3 Salt

The Command line interface of Salt is quite closely related with Ansible's CLI and makes use of the push method to communicate with client. But similar to Puppet it follows a basic client and server architecture. Git or Package managers can be used to install this system. The nodes are referred as minions. Salt support multiple level of servers which helps to increase the redundancy. The master-level server can control the slave servers which directly control minions. Also, the minions can directly contact the master servers which allows direct correspondence between master and slave. On the downside, Salt is still very buggy and under improvement. The web interface lacks a lot of feature as compared with other systems [11] (Table 1).

Table 1 Various configuration management tool and their attributes

Configuration management tools	Agent based	Architecture	License	Language
Ansible	No	Rapid deployment	GPL	Python
Puppet	Yes	Client-server	Apache	Ruby
Chef	Yes	Client-server	Apache	Ruby
Salt	Both	Client-server	Apache	Python

6 Patching System Vulnerabilities Using Ansible

Any prepared system administrators can let you know that as you develop and include more servers and designers, client organization unavoidably turns into a weight. Keeping up customary access gives in the earth of a quickly developing startup is a daunting task. Whenever a system administrator is asked to setup new servers for the purpose of deployment, the first task that he does is to patch up the sever to get rid of any kind of vulnerability. But imagine doing this on hundreds or even thousands of servers using scripts is a very tedious task and also quite prone to errors. So the first few minutes on the server could seem like a nightmare. This task can be done in the first 5 min while being on a control node and the idea is to automate. Ansible allows all this in just a few minutes as desired. In this section we explore the shear strength of Ansible by deploying a full playbook onto a EC2 Ubuntu server placed in cloud. Ansible recommends the use of a flexible directory structure which involves sub divisions into roles and modules. A system administrator want to perform a serious of tasks which could take a lot of time if done using the shell scripts individually. A typical system administrator logs in the server and sets the root password first. Then creates a user account followed by securing the SSH. Then he usually adds the SSH keys used by his team onto the servers. When done with this, the administrator sets up automatic security updates while locking the firewall down. Some system administrator also configure a log service to deliver the server logs on daily basis.

In this paper we are using an EC2 Ubuntu 14.04 image from the Amazon web services cloud for deploying it on cloud. We will be connecting to this server from a local control node running Ubuntu on-premise. Since Ansible works upon SSH and doesn't require any other agent on the host machine, the first thing which needs to be done is to generate a ssh keygen for the purpose of authentication. The public key is stored on the EC2 Server placed in cloud inside the secure ssh directory. The playbook structure is simple to understand as shown in Fig. 2.

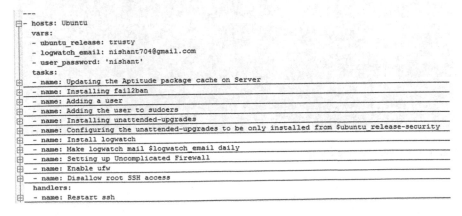

```
---
- hosts: Ubuntu
  vars:
    - ubuntu_release: trusty
    - logwatch_email: nishant704@gmail.com
    - user_password: 'nishant'
  tasks:
    - name: Updating the Aptitude package cache on Server
    - name: Installing fail2ban
    - name: Adding a user
    - name: Adding the user to sudoers
    - name: Installing unattended-upgrades
    - name: Configuring the unattended-upgrades to be only installed from $ubuntu_release-security
    - name: Install logwatch
    - name: Make logwatch mail $logwatch_email daily
    - name: Setting up Uncomplicated Firewall
    - name: Enable ufw
    - name: Disallow root SSH access
  handlers:
    - name: Restart ssh
```

Fig. 2 The playbook structure defining all the roles

The tasks defined in Fig. 2 are self-describing and understandable even to a novice programmer. We write the below playbook for the implementation purpose of this research. The playbook is defined as consisting of tasks or more specifically plays which are executed one step at a time. This makes the playbook secure and consistent with its overall workflow.

The playbook is executed on the control node, this node can be even accessed from a remote machine over ssh channel. Our playbook patches most of the common vulnerabilities which exists on a stock machine. The output from the successful run of this playbook is depicted in Fig. 3.

```
ok: [52.74.37.13]

TASK: [Updating the Aptitude package cache on Server] *************************
ok: [52.74.37.13]

TASK: [Installing fail2ban] **************************************************
changed: [52.74.37.13]

TASK: [Adding a user] ********************************************************
changed: [52.74.37.13]

TASK: [Adding the user to sudoers] ********************************************
changed: [52.74.37.13]

TASK: [Installing unattended-upgrades] ****************************************
ok: [52.74.37.13]

TASK: [Configuring the unattended-upgrades to be only installed from $ubuntu_release-security] ***
ok: [52.74.37.13]

TASK: [Install logwatch] *****************************************************
changed: [52.74.37.13]

TASK: [Make logwatch mail $logwatch_email daily] ******************************
changed: [52.74.37.13]

TASK: [Setting up Uncomplicated Firewall] *************************************
changed: [52.74.37.13] => (item=22/tcp)
changed: [52.74.37.13] => (item=443/tcp)
changed: [52.74.37.13] => (item=60023/udp)

TASK: [Enable ufw] **********************************************************
changed: [52.74.37.13]

TASK: [Disallow root SSH access] *********************************************
changed: [52.74.37.13]

NOTIFIED: [Restart ssh] ******************************************************
changed: [52.74.37.13]

PLAY RECAP ******************************************************************
52.74.37.13                : ok=13   changed=9    unreachable=0    failed=0
```

Fig. 3 Playbook resulting changes

7 Conclusion

This paper introduces Ansible as one of the most upcoming configuration management tool in the IT industry. We give our reasons to show why Ansible is prominently a better solution than other tools which involve more complexity. With the major emphasis laid on the simplicity, Ansible supports a wide variety of rapid deployments and quick fixes offerings. Some of the major advantages we noticed about Ansible included idempotency, declarative nature, re-usability, extensibility, immediate rollouts, staggered push and workflow push, cloud integration and many more. We wrote a playbook which showed the agility of Ansible over cloud platform. This gives a clear idea of the working and syntax of Ansible's playbook automation. Some of the minor drawbacks which are still being perfected is the constant demand for a windows support. The windows support is currently being incorporated upon and apart from this, a compliance report feature of many pull based system can also be integrated with some extra efforts.

References

1. Anne Mette Jonassen Hass.: "Configuration Management Principles and Practice", 1st edition, Addison Wesley, 2002, pp 1–25.
2. Addendum zu Kapitel.: "Software Configuration Management", University of Zurich, Pearson Education, 2004.
3. "Automated Configuration management", Puppet Labs, 2015.
4. "Guide to Software configuration management", ESA Board for Software Standardizations and Control, 1st issue, March 1995.
5. Ansible documentation, 2015, Available: http://docs.ansible.com/.
6. Lorin Hochstein, "Ansible: Up and Running", O'Relilly Media, 2015, pp 1–17.
7. René Moser, "Ansible Cookbook 2014", 2014, pp 1–25.
8. "Puppet documentation", Puppet Labs, 2013 Available at: https://downloads.puppetlabs.com/docs/puppetmanual.pdf.
9. "An Introduction to Puppet Enterprise", Puppet Labs, 2013 Available at: http://downloads.puppetlabs.com/training/materials/intro/Introduction_Student_Guide-v3-1-2.pdf.
10. Andy Hawkins.: "Introduction to Chef", Available at: http://www.netways.de/uploads/media/Andy_Hawkins_Introduction_intoChef_05.pdf.
11. Joël Perras.: "An Introduction to Configuration Management with SaltStack", Available at: https://speakerdeck.com/jperras/an-introduction-to-configuration-management-with-saltstack.

Adaptive Fractal Image Compression Based on Adaptive Thresholding in DCT Domain

Preedhi Garg, Richa Gupta and Rajesh K. Tyagi

Abstract The encoding procedure for fractal image compression consumes up huge amount of time due to enormous number of search operations through domain pool. Sequential search continues till a best harmonious block is found for a range block. Domain pool reduction, an adaptive fractal image encoding method is one solution to overcome this computational complexity. The use of variance domain selection in DCT domain with lesser operations for domain-range comparison is fruitful. In this paper, variances of range and domain blocks are compared and only domain blocks under a threshold value are taken. To further optimize the performance, adaptive thresholding in quadtree partitioning is applied in ally with variance domain selection approach. Experimental results show that executing the proposed method on grayscale images shortens the encoding time and improves the PSNR.

Keywords Fractal image compression · No search · DCT domain · Variance domain selection · Adaptive quadtree partitioning

1 Introduction

Encoding images compactly has a vital role in minimizing the space for mass storage of digital images and in fostering communication channel for faster transmissions Fractal image encoding uses important property of fractals known as

P. Garg (✉) · R. Gupta
CSE Department, Amity University, Noida, Uttar Pradesh, India
e-mail: preedhigarg@gmail.com

R. Gupta
e-mail: richagupta@amity.edu

R.K. Tyagi
IT Department, KIET, Ghaziabad, Uttar Pradesh, India
e-mail: profrajeshkumartyagi@gmail.com

© Springer India 2016
S.C. Satapathy et al. (eds.), *Information Systems Design and Intelligent Applications*, Advances in Intelligent Systems and Computing 433,
DOI 10.1007/978-81-322-2755-7_4

self-similarity within images [1]. This novel idea of fractal image compression ensures high compression ratio, fast decompression. Fractal encoded images include beneficial properties which make this technique more powerful and demanding. The decoded image is resolution independent because of multiresolution property of fractals [2].

The idea originally introduced in 1988 by Barnsley et al. was based on Iterated Function System which encodes the image with the aid of affine transformations [3]. His idea was based on the principle that whole image is a fractal [4].

Further in 1989, Jacquin [5] utilized the self-similarity characteristic within images and achieved high compression ratio based on Partitioned Iteration Function System (PIFS). Fractal Image compression shows unsymmetrical behavior [6] for encoding and decoding procedures. Fractal image compression was firstly applied in spatial domain where each and every original pixel of the image got manipulated [7]. Spatial domain works by considering each single pixel of range cell and finding the best possible match for the pixel in domain cell. In 1994, Monro and Woolley [8] implemented no searching approach in spatial domain to vanquish the enormous encoding time of fractal image encoding. Non search technique attempts to search for an optimal position for a domain block on range block [9].

In full searching, the domain pool is created using a sliding window which moves in horizontal and vertical directions with integral step equals to 1, starting from top left corner and ending at bottom right corner of the original image [10]. A fast no search technique is introduced by Furao and Hasegawa [11] to eliminate search over domain pool by fixing location of range cell on domain cell. A faster technique was applied by combining no search algorithm with adaptive quadtree thresholding.

The traditional quadtree portioning was suggested by Fisher [12] which divides the range block into four sub squares when it is not perfectly matched with domain block. Adaptive quadtree portioning incorporates the use of adaptive tolerance at each level of decomposition.

In 1996, Thao [13] in his work has described the use of DCT domain for its capability of concision of information. The basic operating principle of DCT is to disintegrate the image into segments which provides fewer transform coefficient with better signal approximation [14]. DCT domain allows rapid application of symmetrical operations and RMS distance measure is more accurate with DCT coefficients [15, 16]. Mehdi and Hamid [17] proposed a knack skill of performing non search fractal image compression under frequency domain using the concept of DCT. To wave off the long encoding time, DCT was applied to range and domain blocks and then no search algorithm are executed.

Chong and Minghong [18] brought in the concept of adaptive search of removing the incompetent domain cells so as to foster the speed of image compression. Hasan and Xingqian [19] introduced the various adaptive approaches which include reducing the complexity as well as the number of similarity computation to reduce the encoding time.

In this paper, a new approach of adaptive fractal image compression using variance domain selection is put forward with no need of doing any search.

The proposed algorithm will also take into consideration the effect of using adaptive thresholding in quadtree partitioning. The combined algorithm will save the search time and raise the compression ratio with image quality as much as possible. Comparative results will be shown for propose algorithm with the baseline method which includes only non search fractal image compression in DCT domain.

Following section throws light on searchless algorithm in DCT domain. Section 3 puts forward the concept of quadtree partitioning and also the affect of using adaptive thresholding for the same. This section is organized to talk about zero means intensity level approach, reduced domain size method, range exclusion method and reduced domain pool method also known as variance domain selection method. Finally, Sect. 4 outlines the proposed algorithm with detailed steps describing the working of the algorithm. Proposed method will be taking up variance domain selection approach with adaptive thresholding. Experimental results and conclusion follows the Sect. 4.

2 Searchless Encoding in DCT Domain

The information compacting capability of discrete cosine transform acts as beneficial advantage for fast and easy compression of the images [15]. The DCT of a discrete 2D function is computed using Eq. (1).

$$F(u,v) = \alpha(u)\alpha(v) \sum_{x=0}^{M-1} \sum_{y=0}^{N-1} f(x,y)\cos\left[(2x+1)\frac{\pi u}{2M}\right]\cos\left[(2y+1)\frac{\pi u}{2N}\right] \quad (1)$$

And inverse 2D transform is calculated using Eq. (2)

$$f(x,y) = \sum_{x=0}^{M-1} \sum_{y=0}^{N-1} \alpha(u)\alpha(v)F(u,v)\cos\left[(2x+1)\frac{\pi u}{2M}\right]\cos\left[(2y+1)\frac{\pi u}{2N}\right] \quad (2)$$

where x, u = 0,1,…, M − 1 and y, v = 0,1,…, N − 1 and $\alpha(u) = \begin{cases} \frac{1}{\sqrt{2}}, & u = 0 \\ 1, & u \neq 0 \end{cases}$

If DCT Eqs. (1) and (2) are applied on block of size B × B then it will give B × B dc coefficients. The frequency transform of whole image turn the small coefficients of edges into zero and consumes up great amount of encoding time. As long as cell size increases, the count of DCT coefficients also keep on increasing which affects the encoding time [17].

The no search fractal image encoding technique based on DCT domain is portrayed in the Procedure 2.1.

Procedure 2.1 No search fractal image encoding in DCT domain procedure

- Divide the original image into range blocks of size B × B and into domain blocks of size 2B × 2B.
- Apply DCT transform to range and domain cells.
- Calculate mean of range block and of four neighboring pixels of domain blocks.
- Choose a value for threshold TH.
- For each range block

    - Set the range block on the domain block at position ($row_R - B/2$, $col_R - B/2$)
    - Calculate error using Eq. (3)

$$E(R, D) = \sum_{i=1}^{n} (sd_i + o - r_i) \qquad (3)$$

    - where R and D are the mean value of range and domain block respectively, s is scaling factor and o is contrast scaling.
    - If $E(R, D) < TH$ then store the dc coefficient which is $F_r(1, 1)$.
    - Store the best scaling offset by considering $F_r(1, 1) = F_d(1, 1) = 0$.
    - Otherwise perform the quadtree partitioning on range cells.

The dc coefficient of range cell, $F_r(1, 1)$ and domain cell, $F_d(1, 1)$ are computed using Eq. (1). Setting the range cell on the domain cell at position ($row_R - B/2$, $col_R - B/2$) as it gives good reconstruction fidelity [17].

3   Adaptive Thresholding

High flexibility and low overhead of segmenting the range blocks is being provided by quadtree partitioning. The conventional theory of Yuval Fisher [12] suggests the pre-fixing of threshold value by expert experiences for range-domain cell matching.

The working of quadtree partitioning is based on tree structure. The root node represents the original image and its 4 child nodes depict the sub blocks called range blocks [20]. And each child node correspondingly has 4 child nodes which act as sub range blocks. The maximum depth of the tree and maximum allowable fixed threshold is aforementioned this procedure. Partitioning algorithm iterates until maximum depth of the tree is achieved [20]. Adaptive thresholding technique introduced by Liangbin and Lifeng [19] considers the features of input image and sets the adaptive threshold value accordingly. Another adaptive thresholding method introduced by Furao et al. [11] suggests adaptive RMS tolerance. A linear equation is used to change the tolerance at each decomposition level which is shown in Eq. (4) where n denotes the nth decomposition level.

$$T_{n+1} = 2T_n + 1 \qquad (4)$$

The outlined Procedure 3.1 depicts the summary of the adaptive threshold in quadtree encoding algorithm.

Procedure 3.1 Adaptive thresholding in quadtree encoding procedure

- Read an original image as input of size 512×512.
- Generate range blocks and domain blocks of size $B \times B$ and $2B \times 2B$ respectively.
- Set an initial value for the adaptive tolerance.
- For each range block

  - Match the range block with domain block and calculate RMS for current range block and domain block by using Eq. (3).
  - If RMS < Adaptive Tolerance then save the IFS parameters.
  - Otherwise apply quadtree partitioning to range block and repeat this step.

4 Proposed Algorithm

The fundamental Image compression with the help of Domain Pool Reduction FIC based on no search algorithm in DCT domain includes an encoding procedure as well as decoding procedure. The encoding procedure aims at reducing the encoding time while maintaining the image quality and takes as input the original image to be compressed and present the IFS code. On the other hand, the decoding procedure decodes the IFS code (input) into the decoded image.

Algorithm 5.1 Searchless fractal image compression using variance domain selection approach in DCT domain algorithm

- Load the original image of size $M \times N$ into an array.
- Generate all the non overlapping range blocks of size $B \times B$.
- Generate all the domain blocks of size $2B \times 2B$.
- Choose an initial threshold value for quadtree partitioning.
- Choose a standard threshold for variance domain selection.
- Set the adaptive thresholding equation for each decomposition level of quadtree partitioning.
- For each range block

  - Compute its mean value r and variance Vr value using Eqs. (1) and (2)
  - Calculate the variance for each domain block in domain pool and store it in Vd.

- Verify the condition using Eq. (3), if condition is true then keep the domain block and

  Perform no search in DCT domain using Procedure 2.1
  Execute Procedure 3.1 if domain block does not match.

- Otherwise discard the domain block

- Generate the IFS code with the help of Huffman coding using scaling parameter s and dc coefficient of the range block.
- At decoding side, load the IFS code and decode it using Huffman decoding and generate the decoded image.

5 Experimental Results

To accentuate the encoding time with constant quality of the decoded image, the proposed algorithm is compared with baseline method i.e. without variance domain approach and adaptive thresholding. The results are also shown for domain pool reduction method with fixed threshold for quadtree partitioning. The values are being encoded using 2 bits for scaling parameter and 8, 7, 7, 7 bits for dc coefficient at each level of decomposition of quadtree partitioning respectively. The maximal size of the range block is taken to be 16×16 whereas 4×4 is taken to be minimal size.

For all the algorithms, the decoded image is obtained with the same number of recursions of the encoded affine transformations. The simulation of the proposed method is done using MATLAB 7 on computer with A8 AMD processor and Windows 7 operating system. The tested images are the grey scale images of size 512×512 8 bpp. The suggested algorithm is 1.2 times faster than baseline algorithm with approximately same PSNR. The initial value for the adaptive thresholding is taken to be $t = 1$ and is varied using Eq. (4). Also the standard threshold value for variance domain selection can be 10–50 [11]. In the experiment performed the value is taken to be $SD_{TH} = 25$. Table 1 depicts the comparison values for proposed algorithm and baseline algorithm. It can be seen in the table that encoding time has been decreased and also compression ratio have been increased significantly. The results for the baseline algorithm are being quoted from [17].

Table 2 shows the result for proposed method with fixed threshold having $SD_{TH} = 25$. The encoding time is about 2 times faster than the baseline algorithm with higher compression ratios. For this case, one can obtain higher performance but the image quality falls down significantly.

Table 1 Comparison of proposed algorithm and baseline algorithm	Baseline algorithm			Proposed algorithm		
	CR	ET (s)	PSNR	CR	ET (s)	PSNR
Lena	41.15	18	29.44	19.88	17.22	30.82
Baboon	5.87	157	23.17	7.45	41.41	22.33
Peppers	9	26	32.12	14.95	21.18	29.84

Table 2 Performance parameters for propose algorithm with fixed threshold

	Proposed algorithm with fixed threshold		
	CR	ET (s)	PSNR
Lena	25.54	12.95	29.13
Baboon	8.64	35.88	21.07
Peppers	23.62	15.23	28.16

Table 3 Effect of different value of initial adaptive threshold on Lena image using proposed method

T (initial value)	CR	ET (s)	PSNR
−1	5.33	64.99	31.79
−0.5	6.32	50.27	31.78
0	11.00	31.58	31.56
0.5	15.26	23.39	31.21
1	19.08	18.88	30.82
2	28.01	13.42	29.88
3	36.56	10.16	29.07

The initial value for the adaptive thresholding in quadtree partitioning can be varied to get different values for performance analysis parameters. The values are being shown in Table 3 where initial value is varied and results are being noted.

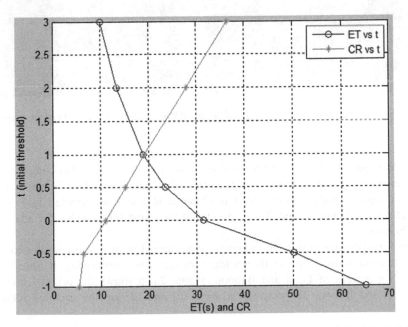

Fig. 1 Graph showing change in compression ratio and encoding time wrt t (initial threshold) for Lena image

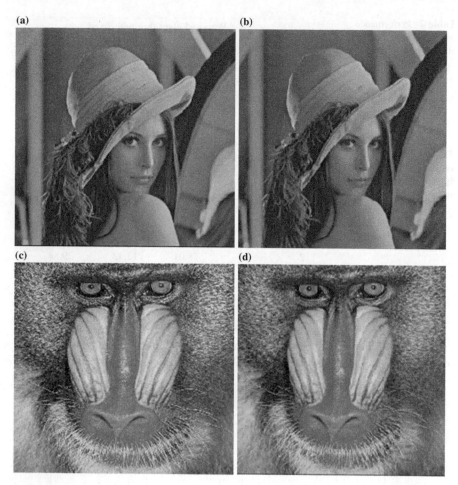

Fig. 2 Experimental results of proposed method: **a** and **b** are original images and **c**, **d** are decoded images

With decreasing value for t, PSNR and encoding time increases, also compression ratio decreases. We need to set such an initial value that it gives good image quality with minimal encoding time which is obtained at t = 1 as depicted in Fig. 1. Therefore, for Tables 1 and 2 we have taken the same initial value for adaptive thresholding.

The experimental results for adaptive non search fractal image compression in DCT domain with adaptive thresholding in quadtree partitioning is shown in Fig. 2 for images of Lena and Baboon. Figure 2a, b shows the original images for Lena and Baboon respectively whereas (c) and (d) are the decoded images for both.

6 Conclusion

The most distinctive benefit of using the algorithm that is put forward in this paper is the combination of reduced domain pool and non search mode with adaptive decomposition. As results have proved, by directly assigning less number of domain blocks to each range cell for matching, the search time can be shortened. This communication comes up with the implementation of adaptive no search fractal image compression using adaptive thresholding in DCT domain so as to improve performance. The selection of initial value for adaptive thresholding has also been discussed in this paper. Degradation in image quality has been observed with fixed threshold in quadtree partitioning as compared with adaptive threshold. Moreover, the proposed algorithm improves the compression ratio with the PSNR as much as similar. To further vanquish the encoding time with same image quality, the new algorithm could be implemented using Fast DCT and Real DCT.

References

1. Galabov, Miroslav. "Fractal image compression." Proceedings of International Conference on Computer Systems and Technologies (CompSysTech'03). 2003.
2. Baviskar, Amol G., and S. S. Pawale. "Efficient Domain Search for Fractal Image Compression Using Feature Extraction Technique." *Advances in Computer Science, Engineering & Applications.* Springer Berlin Heidelberg, 2012. 353–365.
3. Barnsley, Michael F., and Alan D. Sloan. "A better way to compress images."*Byte* 13.1 (1988): 215–223.
4. Barnsley, Michael F. *Fractals everywhere*. Academic press, 2014.
5. Jacquin, Arnaud E. "Fractal image coding: a review." *Proceedings of the IEEE*81.10 (1993): 1451–1465.
6. Wohlberg, Brendt, and Gerhard De Jager. "A review of the fractal image coding literature." *Image Processing, IEEE Transactions on* 8.12 (1999): 1716–1729.
7. Truong, T. K., et al. "Fast fractal image compression using spatial correlation."*Chaos, Solitons & Fractals* 22.5 (2004): 1071–1076.
8. Monro, Donald M., and Stuart J. Woolley. "Fractal image compression without searching." *Acoustics, Speech, and Signal Processing, 1994. ICASSP-94., 1994 IEEE International Conference on.* IEEE, 1994.
9. Zhou, Yiming, Chao Zhang, and Zengke Zhang. "Fast Fractal Image Encoding Using an Improved Search Scheme." *Tsinghua Science & Technology* 12.5 (2007): 602–606.
10. Furao, Shen, and Osamu Hasegawa. "A fast no search fractal image coding method." *Signal Processing: Image Communication* 19.5 (2004): 393–404.
11. Fisher, Yuval. *Fractal image compression with quadtrees*. Springer New York, 1995.
12. Thao, Nguyen T. "A hybrid fractal-DCT coding scheme for image compression."*Image Processing, 1996. Proceedings., International Conference on.* Vol. 1. IEEE, 1996.
13. Rawat, Chandan, and Sukadev Meher. "A Hybrid Image Compression Scheme Using DCT and Fractal Image Compression." *Int. Arab J. Inf. Technol.* 10.6 (2013): 553–562.
14. Fu, Chong, and Zhi-Liang Zhu. "A DCT-based fractal image compression method." *Chaos-Fractals Theories and Applications, 2009. IWCFTA'09. International Workshop on.* IEEE, 2009.

15. Curtis, K. M., G. Neil, and V. Fotopoulos. "A hybrid fractal/DCT image compression method." *Digital Signal Processing, 2002. DSP 2002. 2002 14th International Conference on.* Vol. 2. IEEE, 2002.
16. Salarian, Mehdi, and Hamid Hassanpour. "A new fast no search fractal image compression in DCT domain." *Machine Vision, 2007. ICMV 2007. International Conference on.* IEEE, 2007.
17. Tong, Chong Sze, and Minghong Pi. "Fast fractal image encoding based on adaptive search." *Image Processing, IEEE Transactions on* 10.9 (2001): 1269–1277.
18. Mohammed Hasan, Taha, and Xingqian Wu. "An Adaptive Fractal Image Compression." (2013).
19. Zhu, Shiping, Liang Yu, and Kamel Belloulata. "An improved fractal image coding algorithm based on adaptive threshold for quadtree partition." *Seventh International Symposium on Instrumentation and Control Technology.* International Society for Optics and Photonics, 2008.
20. Zhang, Liangbin, and Lifeng Xi. "A Novel Fractal Image Coding Based on Quadtree Partition of the Adaptive Threshold Value." *Theoretical Advances and Applications of Fuzzy Logic and Soft Computing.* Springer Berlin Heidelberg, 2007. 504–512.

Improved Resource Exploitation by Combining Hadoop Map Reduce Framework with VirtualBox

Ramanpal Kaur, Harjeet Kaur and Archu Dhamija

Abstract MapReduce is a framework for processing huge volumes of data in parallel, on large groups of nodes. Processing enormous data requires fast coordination and allocation of resources. Emphasis is on achieving maximum performance with optimal resources. This paper portraits a technique for accomplishing better resource utilization. The main objective of the work is to incorporate virtualization in Hadoop MapReduce framework and measuring the performance enhancement. In order to realize this master node is setup on physical machine and slave nodes are setup in a common physical machine as virtual machines (VM), by cloning of Hadoop configured VM images. To further enhance the performance Hadoop virtual cluster are configured to use capacity scheduler.

Keywords Hadoop · MapReduce · HDFS · Virtualization · Cluster

1 Introduction

Everyday 2.5 quintillion bytes of data is produced in structured, semi structured and unstructured formats [1]. With the explosion this Big Data, companies are now focusing on controlling data, to support the decisions making for their future growth. The intricacy of today's analytics requirements is beating the existing computing capabilities of legacy systems. Apache Hadoop software library framework provides simple programming model that permits distributed processing of large datasets within clusters [2]. It is capable of scaling up from single server to thousands of physical machines, where each machine offers its own local compu-

R. Kaur (✉) · H. Kaur · A. Dhamija
Lovely Professional University, Phagwara, India
e-mail: ramanpalbrar@gmail.com

H. Kaur
e-mail: harjeet.kaur@lpu.co.in

A. Dhamija
e-mail: er.archudhamija@gmail.com

© Springer India 2016
S.C. Satapathy et al. (eds.), *Information Systems Design and Intelligent Applications*, Advances in Intelligent Systems and Computing 433,
DOI 10.1007/978-81-322-2755-7_5

41

tation and storage resources. Hadoop library has been designed to identify and control failure at the application layer only instead of depending on hardware for high-availability. It is self-managing and can easily handle hardware failure. HDFS which is the distributed file system of Hadoop for cheap and reliable data storage and MapReduce Engine which provides high performance parallel data processing, are the two major components of Hadoop. In addition Hadoop can scale up or down its deployment with no changes to codebase.

With its distributed processing, Hadoop can handle large volumes of structured and unstructured data more efficiently than the traditional enterprise data warehouse. While first deployed by the main web properties Hadoop is now widely used across many sectors including entertainment, healthcare, government and market research. Being an open source and ability to run on commodity hardware, the cost saving is visible from initial stages to further growth of organizational data.

1.1 MapReduce

High scalability allows Hadoop Map Reduce to compute large volume of data in a moderate time. This is achieved by allocating the job in small tasks that are divided among a large pool of computers. Map refers to data collection phase where data is divided into smaller parts and processed in parallel. Key/Value pair is passed as input to Mapper to produce intermediate Key/Value pair. This intermediate Key/Value pair is passed on to Reducer, used for data transformation. This is the data processing stage as it combines the results of map functions into one final result. Map uses the data of one domain and gives the output back in another domain of data.

$$MAP(K1, V1) \rightarrow LIST(K2, V2)$$

Intermediary keys and values are listed for each block of data. MapReduce collects the data of same key. The reduce function is applied in parallel to each output of map phase.

$$REDUCE(K2, LIST(V2)) \rightarrow LIST(V3)$$

Figure 1a, b example of word count demonstrate the concept of map reduce.

The rest of the paper is structured as follows. Next section describes the ongoing work in this field. Main emphasis in Sect. 3 is placed on exhibiting the design and implementation details of the incorporation of virtualization in Hadoop MapReduce. Section 4 supports the whole discussion with experimental results (in Sect. 5) to prove the effectiveness of the technique used and finally Sect. 6 concludes the paper with future enhancements.

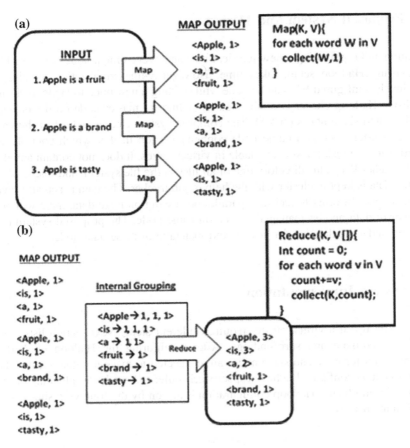

Fig. 1 a Map output of word count. b Reduce output of word count

2 Previous Work

In order to process huge volumes of data efficiently, fast allocation of resources is required. If whole setup is established from physical machines, achieving quick and proper allocation along with fast coordination of nodes, is a big problem, as computations involved are huge. Although Hadoop MapReduce works very well, a lot of work is going on to further improve the performance. Adding and configuring new physical machine with increasing number of requests takes significant amount of time. David de Nadal Bou, in his work to provide support for managing Hadoop cluster dynamically, used a modified internal Hadoop scheduler that adapts the available resources in the cluster for running jobs according to time restrictions [3]. With the adapted internal scheduler a flexible and self-adaptive service for Hadoop environments was developed. Virtualized Hadoop performance was presented in performance study by Jeff Buell, clearly stating the role of virtual machine for performance improvement [4].

3 Proposed System and Design

Creating and configuring a new physical machine every time a node is to be added in existing Hadoop setup, takes time. However with virtualization, cloning of previously configured VM can be done while adding a new node to Hadoop cluster. A virtual Hadoop cluster is established by setting up master node on the physical machine and slave nodes on VM. The complete system can be viewed as shown in Fig. 2. Master node is a name node containing a job tracker which controls and coordinates the task trackers available in virtual slaves. It does not contain any data of files rather keeps the directory tree of all files in the file system, and tracks where the file data is kept in cluster with the help of job tracker. The data is actually saved in Data node. In order to facilitate parallel processing on huge data, there will be n number of data anodes managed by central name node. The proposed system uses cloning and virtualization for adding and maintaining these data nodes.

4 System Implementation

The mount up of a virtual cluster requires different steps. Next section lists them, with the requisite and surroundings. After configuring the Hadoop on Virtual machine, master node and slave nodes are connected to make a virtual cluster. The master node is configured to have capacity scheduler to have better planning of jobs in this virtual cluster. Hadoop configuration is driven by the following site-specific configuration files:

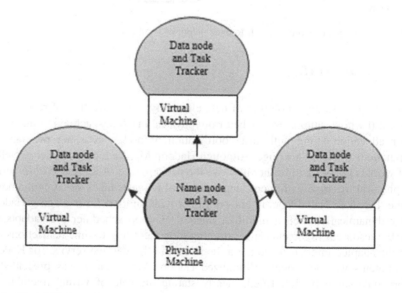

Fig. 2 Virtual Hadoop cluster

- conf/core-site.xml
- conf/hdfs-site.xml
- conf/mapred-site.xml

4.1 Configuring Hadoop

Hadoop 2.4 is used for experiment with JDK1.8. Hadoop is installed and a cluster of multi node setup is made [5]. The whole setup is established on a dedicated Hadoop user in Ubunto. Secure shell (SSH) access is required in Hadoop as the nodes attached exist either on different systems or VMs. SSH key is produced for all devices in the cluster. The Hadoop daemons, Namenode and Datanode along with Jobtracker and Tasktracker are configures to execute MapReduce job on the cluster setup. In order to install ssh and setup certificate, following set of commands are used

- sudo apt-get install openssh-server
- ssh-keygen -t rsa -P "
- cat ~/.ssh/id_rsa.pub≫ ~/.ssh/authorized_keys
- ssh localhost

Hadoop configuration is complete after setting up of Hadoop environment variable and modification of conf/core-site.xml, to include a base for other temporary directories along with the name of default file system.

4.1.1 Configuring HDFS

Configuring HDFS includes specification of path on local file system where namespaces are saved in NameNode along with transaction logs. Paths on the local file system of the data node where blocks are stored, is also mentioned. All specifications are done in conf/hdfs-site.xml.

4.1.2 Configuring Map Reduce

Alterations are made in conf/mapred-site.xml to describe parameters associated with JobTracker and TaskTracker. IP address and port number for JobTracker is specified. The path of the data files stored by MapReduce framework on HDFS is also specified in this file. These are stored as input to various jobs.

4.1.3 Configuring Capacity Scheduler

The master node by default makes use of FCFS scheduler. In process to have better resource utilization, conf/mapred-site.xml is modified again to have capacity scheduler for JobTracker. CLASSPATH in conf/hadoop-env.sh is also updated by specifying the capacity-scheduler jar. Then the file conf/mapred-site.xml is modified to include a new property, mapred.jobtracker.taskScheduler where scheduler is set to capacity scheduler.

4.2 Building Hadoop Cluster

Oracle Virtual-box has been used for creating three virtual machines with the base memory 512 MB and a maximum capacity of 8 GB. Dynamically allocated virtual disk file are used to ensure flexibility in storage management. This has major role in performance enhancement [6].

4.2.1 Configuration of Virtual Machine

Virtual machines are setup to establish a network connection. Later on different VMs are created with the help of cloning of VM. VirtualBox 11.0.0 is installed for demonstrating 64 bit Ubuntu.

Setting up of Network Connection

Host machine and VM are required to be SSH connected through network. For this vtun, uml and Bridge utils need to be installed on the Host machine. Network address translation Adapter exists in the system by default; bridge and tap devices are created in Host machine. When theses setting are done successfully on each machine then Ssh is configured on both the machines. In each machine the setup of the bridge and tap device is done automatically at the booting time. Each VM is attached to corresponding one tap device to have same connection on multiple. Multiple tap devices are created for this.

Installing Hadoop on VM and Cloning of VM

Installation of Java on one VM is done followed by SSH installation and configuration. Hadoop source file is copied to this VM. As soon as correct working of Hadoop on VM is assured, it is added as a slave node in the cluster. In order to speed up the process of creating clusters, rest of the slaves are cloned instead of configuring a new machine. Cloned machines will be having same name and IP address as the machine from which it has been cloned has. Changes are made in / etc./hosts/and /etc./hostname files for allocating distinct names and IP Addresses.

Entire setup is complete with the establishment of successful connection between all nodes of the cluster and running MapReduce program.

5 Results and Discussions

Performance is recorded by running MapReduce program of word count on both physical and virtual cluster. Size of input data is mentioned in Table 1. Execution time is observed and comparison is presented in Table 2 for physical and virtual cluster. Charts shown in Fig. 3a, b show the results in different perspective.

For recording the performance, two types of setup are established. One with a group of physical machines and other one by connecting one physical machine with two virtual machines. One has been configured as master node and other two as slave nodes. Each virtual machine has been given with 1024 MB RAM and 2 processors with 8 GB HDD. MapReduce program of word count is executed on three different data sets. Observations are recorded in Table 2.

Table 2 also contains the execution time taken to run the same program on same datasets but using virtual cluster. Virtual machines has been allocated the same configuration as physical one. Results obtained are analysed in two different ways as shown in Fig. 3a, b.

Chart in Fig. 3a compares the execution time for program of word count. While Fig. 3b describes performance in terms of input size. Figure 3a clearly depicts the success of having virtual machine as slave over physical machines. Bars showing the time taken by physical slave machine is always exceeding the second bar. Addition of third virtual machine on this configuration lead to slow working of processor. That can easily be overcome by increasing the capabilities of master machine.

Table 1 Size of input dataset

Name	Size in MB
Dataset 1	2.5
Dataset 2	3.8
Dataset 3	6.2

Table 2 Execution time of word count in physical cluster and virtual cluster

Name	Physical cluster (time in s)	Virtual cluster (time in s)
Dataset 1	32	26
Dataset 2	54	39
Dataset 3	58	42

Fig. 3 **a** Comparison of execution time. **b** Input size with respect to time

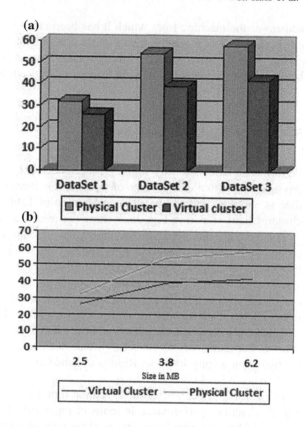

Figure 3b shows the time variation with input size while using virtual machine and physical machine. Execution time is increasing with the size of input data.

Adding more number of nodes to cluster can decrease execution time significantly. It all depends upon the capacity of machine on which this setup is made.

The data set used for this experimental setup is small. Whereas the capabilities of Hadoop allows it to process huge volumes of data.

6 Conclusion and Future Work

An experimental setup to have better resource utilization by combining MapReduce framework with VirtualBox, is made. Integration of virtualization with multimode setup of Hadoop, for executing programs of MapReduce reduced the execution time significantly. In future work can be extended to include analysis of cluster size and number with respect to time. Scalability issue can be addressed by including multiple master in single cluster.

References

1. http://www-01.ibm.com/software/data/bigdata/what-is-big-data.html.
2. Hadoop – The Apache Software Foundation, (Dec 2011). (http://hadoop.apache.org).
3. Davi'd D. Nadal, Support for Managing Dynamically Hadoop Clusters, (Sept 2010). (http://upcommons.upc.edu/pfc/bitstream/2099.1/9920/1/63274.pdf
4. J. Buell, A benchmarking Case Study of Virtualized Hadoop Performance on VMware vSphere 5, (Oct 2011). http://www.vmware.com/files/pdf/techpaper/VMWHadoop-Performance-vSphere5.pdf).
5. M.G. Noll, Running Hadoop on Ubuntu Linux MultiNode Cluster, (Aug 2007). (http://www.michael-noll.com/tutorials/running-hadoop-on-ubuntu-linux-multi-nodecluster/).
6. Ravindra, Building a Hadoop Cluster using VirtualBox, (Oct 2010). (http://xebee.xebia.in/2010/10/21/building-a-hadoop-cluster-using-virtualbox/).

Density Based Outlier Detection Technique

Raghav Gupta and Kavita Pandey

Abstract Outlier Detection has become an emerging branch of research in the field of data mining. Detecting outliers from a pattern is a popular problem. Detection of Outliers could be very beneficial, knowledgeable, interesting and useful and can be very destructive if remain unexplored. We have proposed a novel density based approach which uses a statistical measure i.e. standard deviation to identify that a data point is an outlier or not. In the current days there are large variety of different solutions has been efficiently researched. The selection of these solutions is sometimes hard as there is no one particular solution that is better than the others, but each solution is suitable under some specific type of datasets. Therefore, when choosing an outlier detection method to adapt to a new problem it is important to look on the particularities of the specific dataset that the method will be applied. To test the validity of the proposed approach, it has been applied to Wisconsin Breast Cancer dataset and Iris dataset.

Keywords Outlier · Density · Outlier · Outlier detection technique

1 Introduction

An outlier is an observation which is significantly deviated from other observations of the datasets due to abnormal/suspicious behavior of the system, generating data values. Outlier detection refers to the problem of finding those unusual non-conforming/inconsistent values in the datasets i.e. values which are not according to the usual pattern/trend of the other values. From a very long period, a large number of outlier detection techniques has been applied to detect and, remove

R. Gupta (✉) · K. Pandey
Department of Computer Science, Jaypee Institute of Information Technology, Noida, India
e-mail: rgupta1992@gmail.com

K. Pandey
e-mail: kavita.pandey@jiit.ac.in

© Springer India 2016
S.C. Satapathy et al. (eds.), *Information Systems Design and Intelligent Applications*, Advances in Intelligent Systems and Computing 433,
DOI 10.1007/978-81-322-2755-7_6

noisy observations from datasets. The introduction of outlier values is the result of mechanical failure, deviations in system behavior, deceitful behavior, human error, instrumentation error or simply through natural deviations in populations. Their detection can find faults in a system and credit card frauds before they escalate with potentially catastrophic consequences. These day amount of data is increasing at a very drastic rate. There is a need of mining the data to get knowledge from this data. This huge data may consist of unexpected, doubtful, unconventional, noisy values which are known as Outliers. These inconsistent values have different name according to area of application domain i.e. outliers, anomalies, exceptions, strident observations, novelties or noise. They could be very useful and interesting if they are detected and can be very distractive if left undiscovered. Outlier detection have some very meaningful real world applications e.g. Fraud Detection, Insider Trading, Medical Health Anomaly Detection, Fault Diagnosis, Social Welfare, Environmental Monitoring etc. Sometimes outlier detection can be useful in security purposes for example air safety. Lets' take an example where outlier detection served helpful in security. One of the high jacked airplanes of 9/11 had a particular anomaly. It has five passengers which (i) were non-US citizens (ii) had links to a particular country (iii) had purchased a one-way ticket (iv) had paid in cash (v) had no luggage. Of course one or two such passengers in an airplane is a normal observation but 5 in a particular plane is an anomaly.

Outlier detection is an important aspect in the domain of data mining and knowledge discovery, which aims to identify abnormalities in observations in a large dataset. In the Fig. 1, the observations points are spread uniformly in the dimension. The observation 'o' is an outlier. As the number of points that lie in its radius 'r' are very few as compared to the other points. In other words the locality of point 'o' is quite sparse whereas the locality of other points is dense.

Many good techniques methodologies exists for outlier detection. Generally density based approaches are more preferred. These approaches have large number of computations. In this paper we proposed a new approach based on density based approach to reduce number of computations. This paper is divided into five sections. Section 2 presents related work about density based algorithms. Section 3 presents our version of algorithm, proposed work. In Sect. 4 results are presented and the Sect. 5 presents conclusion and future work.

Fig. 1 Visualization of outliers [5]

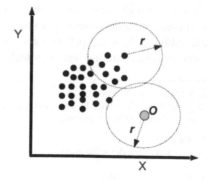

2 Related Work

The traditional definition of an outlier is given by Hawkins [1] which defines an outlier is an observation that deviates so many other observations as to arouse suspicion it has been generated by a different mechanism. Breunig et al. introduced the concept of local outlier, a kind of density-based outlier. He assigns each data a local outlier factor (LOF) which tell about an observation that given point is an outlier depending on their neighbourhood data points. These type of approaches are closely linked with the k-nearest neighbor. These methods selects a definite counts of nearest neighbors and based on that decides data point is an outlier or not. These approaches select those points that are within a fixed range/radius of the test data point, i.e. those points that are within an imaginary hypersphere around the test observation data points. If the number of points are equal or less than the selected percentage of the total number of data observations, that data observation is considered to be an outlier. Techniques that require user set attributes are generally not preferred in an unsupervised scenario, but this approach is interesting since its percentage can easily be compared to a p-value. One issue with these techniques is that they have a difficulty with data sets that contain regions of varying density. Breunig et al. [2] solves this, by giving each data point an outlier score based on the ratio of the average local density where the data point resides and the data point itself, which is called the local outlier factor (LOF). This method has been extended both to improve the efficiency and to relate to different structures in the data.

Jiang et al. [3] presented a clustering based outlier detection which uses the concept of density based outlier detection. This is a two-step approach where first clustering is done using one pass algorithm then on the basis of outlier factor score outlier is detected. Tao [4] has also presented an outlier detection method which also based on clustering. In the proposed approach DBSCAN clustering algorithm is applied to core objects after that LOF is applied on non-core objects. This technique reduces the computation of LOF algorithm as number of data observation points is reduced. This reduction reduces computation time so as it increases the efficiency of LOF method.

Mohemmed et al. [5] proposed an approach based on Particle Swarm Optimization which automatically selects the initialization parameters. Buthong et al. [6] presented a new approach based on LOF is introduced for parameter free outlier identification algorithm to calculate the Ordered Distance Outlier Factor. A new formulation is introduced to determine outlier score for each object point by using the difference of ordered distances and later on this value is used to compute an outlier score. The score of every iteration is used to give a degree of outlier to every instance and compare it with the Local Outlier Factor. The algorithm work as first the distance matrix is calculated using the computing the distanced between every iterations. Then all the distances are ordered row-wise to construct ordered distance matrix and the difference of the ordered distance is calculated between the distance d_i and the next distance d_j, in this way difference ordered matrix is generated. Then ordered outlier factor (OOF) is computed for every data point then are arranged in ascending order and the first k instances are the outlier. The complexity of the algorithm is $O(n^2)$.

Dai et al. [7], concentrated on privacy preserving Local Outlier Factor. A novel algorithm is proposed for privacy preserving k-distance neighbors search. A privacy preserving kDN search protocol has been proposed. The proposed algorithm consists of two parts: computation and communication. The Table 1 shows the general approaches to estimate the outlier detection. Various advancements have been done on the basis of these approaches and still going on.

3 Proposed Work

Assume a circle of radius 'r' whose center is the data point or we can say that an imaginary circle is drawn of radius 'r' at data point as a center. For an observation to be an outlier it should contain a lesser number of data points in it. The

Table 1 Comparison of various approaches to outlier detection

Approach	Assumption	Problem	Solution
Depth based	Outliers are objects on outer layers	Inefficient in 2D/3D spaces	Points having a depth $\leq$ k are reported as outliers
Deviation based	Outliers are the outermost points of the data set	Naïve solution is in O (2^n) for n data objects	The variance of the set is minimized when removing the outliers
Distance based	Normal data objects have a dense neighborhood	Approaches are local	Outliers are far apart from their neighbors, i.e., have a less dense neighborhood
Density based	The density around an outlier is considerably different to the density around its neighbors	Exponential runtime w. r.t. data dimensionality	Compare the density around a point with the density around its local Neighbors
High dime-nsional	Normal points are in the center of the data distribution	Data is very sparse, almost all points are outliers, concept of neighborhood becomes meaningless	Use more robust distance functions and find full Use more robust distance functions and find full-dimensional outliers dimensional outliers
Statistical based	All the data observations follow a same characteristics statistical measures	In a multi valued attribute dataset every attribute has own characteristic trend	Use dimension reduction techniques to reduce attributes. Then calculate the statistical parameters.

Fig. 2 Principal component analysis

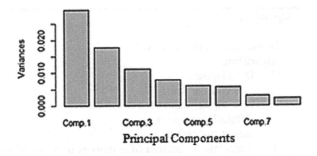

selection of 'r' is done for the whole dataset using the statistical mean of each attribute individually. This mean is summed up to calculate the Manhattan distance which forms the radius for every data point. Further the number of points in the radius are calculated to get outlier factor. We compute a fitness ratio $\frac{k}{r}$ where 'k' is the number of points in neighborhood within the radius 'r'. If the fitness values comes below a threshold value, the point is marked as outlier. When estimating the neighboring points, consider only the points which are not marked outliers. The fact behind doing this, the point which is estimated as outlier already, no point of considering it again. This reduces the number of computational steps. The proposed algorithm is presented in the box below. This algorithm is suitable for multi-dimensional numerical datasets in which data observations lies uniformly.

For experimental work we have installed the RStudio and R-language project on the Windows system. Firstly we created a dummy dataset of 20 observations with the 4 outlier values. This is just to check the feasibility of the proposed approach. A generalize radius of dataset is identified using standard deviation and the radius is used to estimate number of points that lie in neighbourhood within the radius, if it is found below a certain threshold the point is in sparse region, it can be outlier. Then we applied the approach to Iris Dataset which has four attributes and Breast Cancer dataset which has 10 attributes. Because of large number of dimensions, the dimension reduction of the Breast Cancer dataset is done. We have applied the Principal Component Analysis on the Breast Cancer Dataset. The result of this analysis is shown in Fig. 2 which depicts that the majority of the information lies up to component 3. As a result we have considered the first four attributes of the datasets.

Algorithm

Input: dataset, threshold value
Algorithm:
1. D<- Dataset;
2. Add column flag to D;
3. r<-compute_radius(D);
4. While(D!=empty){
5. Select 'm' objects from D;
6. Calculate 'n' number of neighbors in radius 'r' whose outlier flag is zero;
7. Evaluate only those observations with flag value= =0;
8. Check fitness of 'n/r':
9. If(n/r < threshold){
10. Mark observation flag value as outlier;
11. }
Output: Observations with flag value 1 are outliers

We checked whether the data point is the outlier or not for those points, if yes, marked the point as outlier, again selecting the points until all the points are evaluated. Note that, the points which are marked outliers will not me use again for further analysis. We have selected Wisconsin Breast Cancer Dataset and Iris Dataset and applied our approach on them. Firstly we did a dimensionality reduction using Principal.

Component Analysis. Then we computed the radius 'r' for the remaining attributes of the datasets. Then randomly m objects are selected. The fitness ratio is calculated and is compared with the threshold value. If value is lesser, the point is an outlier. Finally we got outliers. The time complexity of the proposed approach in $O(n^2)$.

4 Results

The results of our proposed approach are compared with the traditional density based approach. The results shows that with the proposed algorithm there is a reduction in the iterations in comparison to the traditional density based outlier detection algorithm. The results are tabulated in Table 2.

In Fig. 3a we have compared graphically the decrease in the number of iterations of the proposed approach with traditional density based approach. The Table 3 gives the analysis the number of outliers detected by both the approaches i.e. traditional density based approach and proposed approach. Both the approaches detects same number of outliers. We can find the correctness of the proposed approach with the traditional approach. In Fig. 3b, the graphical representation of Table 2 is presented, we can compare from the graph that there is not a major

Table 2 Comparison of results

Dataset	No. of observation points	Total number of iterations		Percentage reduced (%)
		In density based approach	In proposed approach	
Dummy	20	400	320	20
Iris	100	10,000	9100	9
Breast cancer	300	90,000	87,500	3

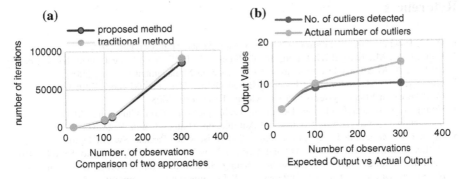

Fig. 3 a Graphical representation of results. **b** Analysis of output results

Table 3 Analysis of outputs

Datasets	No. of observation points	Total number of outliers detected		Actual number of outliers
		By proposed approach	By traditional approach	
Dummy	20	4	4	4
Iris	100	9	9	10
Breast cancer	300	10	10	15

difference in the actual number of outliers and the number of outliers detected. It can be conclude from the results the proposed approach shows significant improvement. Also if datasets of 'n' points and having 'm' outliers then complexity will me $O(n*(n - m))$ and in worst case the complexity will be $O(n^2)$.

5 Conclusion and Future Work

Outlier detection is a very emerging field of data mining. This field has number of applications in the current scenario because data size is increasing day by day. In this paper, we have presented a new approach based on density based outlier. The new approach is applied for the validation to different datasets of different dimensions, densities and we get the better results in terms of number of computations. For the feasibility it is first applied to dummy dataset and then applied to standard datasets to find its effectiveness.

The future work can be done in two directions. First is the proposed work can be made scalable to high dimensionality datasets and second is the selection of threshold value for evaluation of fitness can be automated.

References

1. D. M. Hawkins, Identification of Outliers. Chapman and Hall, London, 1980.
2. Breunig, M. M.; Kriegel, H.-P.; Ng, R. T.; Sander, J., LOF: Identifying Density based Local Outliers, ACM SIGMOD International Conference on Management of Data, 2000.
3. Sheng-yi Jiang, Qing-bo An, Clustering-Based Outlier Detection Method, Fifth International Conference on Fuzzy Systems and Knowledge Discovery, 2008.
4. Huang Tao, Research Outlier Detection Technique Based on Clustering Algorithm, 7th International Conference on Control and Automation, Hainan, China, Page No. 12– 14, 2014.
5. Ammar W Mohemmed, Mengjie Zhang, Will N Browne, Particle Swarm Optimisation for Outlier Detection, GECCO'10, Portland, Oregon, New York, NY, USA, Page No. 83–84, 2010.
6. Nattorn Buthong, Arthorn Luangsodsai, Krung Sinapirom saran, "Outlier Detection Score Based on Ordered Distance Difference", IEEE International Computer Sci- -ence and Engineering Conference, 2013.
7. Zaisheng Dai, Liusheng Huang, Youwen Zhu, Wei Yang, Privacy Preserving Density-Based Outlier Detection, IEEE International Conference on Communication and Mobile Computing, Shenzhen, China, Page No. 80–85, 2010.

Systematic Evaluation of Seed Germination Models: A Comparative Analysis

Lalita Chaudhary, Rajesh, Vikas Deep and Preeti Chawla

Abstract Agriculture field has recently adopted the various techniques to boost the production and monitoring of seed germination in a professional manner. Temperature is the key factor in amending the germination of non-dormant seeds. This paper presents the various computational developed prediction models for calculating seed germination growth rate. Also in this paper their merits and demerits are depicted to analyze them. These are helpful in monitoring and controlling the seed germination process to enable higher quality storage and produce good yield of crops.

Keywords Seed germination · Germination rate · Prediction model · Temperature

1 Introduction

There are numerous factors which affects the germination of non-dormant seeds. These factors include seed size, water content, gene control, temperature, depth of planting, seed dormancy, soil moisture, humidity, etc [1–3]. Temperature is the amidst factor that intensely influences the germination rate of non-dormant seeds

L. Chaudhary (✉) · V. Deep · P. Chawla
Amity University, Noida 201301, India
e-mail: lalita.chaudhary19@gmail.com

V. Deep
e-mail: vdeep.amity@gmail.com

P. Chawla
e-mail: preetichawla06@gmail.com

Rajesh
CSIR-Central Scientific Instrumentation Organisation, Chandigarh 160030, India
e-mail: calltorajesh@gmail.com

© Springer India 2016

S.C. Satapathy et al. (eds.), *Information Systems Design and Intelligent Applications*, Advances in Intelligent Systems and Computing 433,
DOI 10.1007/978-81-322-2755-7_7

59

[4]. The relationship between environmental temperature and seed germination rate is complex. But the enhanced research field in agriculture has embraced numerous methods and techniques to improve the production and monitoring of seed germination. Analyzing the cause affecting the crop manually is quite impossible. To speed up the process of detection of any harm caused to the seed germination or yield of crop we need to apply computer technology which also will help in controlling temperature, water content, and moisture etc requirement accordingly to the crop or seed. Further the advancement in technologies, artificial neural network enables the development of various prediction methods that helps in improving germination rate of non dormant seeds. The artificial neural network prediction models provide a mechanism that organize, monitor and control the seed germination process in an effective way. This paper enlightens few computational seed germination prediction models which are influenced by temperature with few other germination parameters. Further a comparison among them is depicted below. Temperature is the basis for models used in predicting germination rate for non-dormant seeds [5].

There is an ongoing research to augment the tools and technique used for measuring various physical parameters required for germinating seeds. We use computational prediction models for a variety of purposes in diverse applications. With respect to agriculture we use it for monitoring crops, automating the agriculture process and controlling the seed germination process to enable higher quality storage and produce good yield of crops.

2   Germination Rate and Analysis

In the field of agriculture and gardening, the germination rate portrays how many non-dormant seeds are likely to germinate in a specified period. In seed physiology, germination rate is the reciprocal of time taken by the seed for the process of germination to accomplish starting from time of sowing. Predominantly, germination rate increases or upsurge with raise in temperature up to the optimum temperature (T0) and declines at temperatures exceeding it [6]. Plant growth rate increases with rises in temperature from the base (Tb) to the optimum temperatures and declines at temperatures between the optimum and the ceiling ones (Tm) [7]. A seed germinates in an ample range of temperature, but their maximum germination noticeably changes at the upper and lower thresholds of this range. The bound of temperature within which germination is at its maximum varies depending on species and seed quality [8] (Fig. 1).

Different models used in this paper for quantifying germination rate against temperature. The advantages of these functions are that they used parameters are referring to biological concepts such as fundamental temperatures, native germination rate, and emergence [9–12]. Also they are computationally and manually executable. Yousry A. El-Kassaby describes a mathematical cumulative seed germination method based on four methods which are cumulative germination

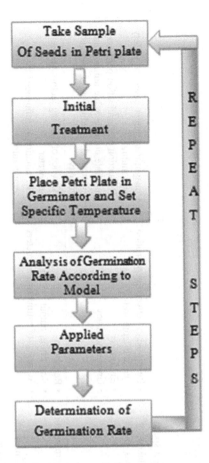

Fig. 1 Block diagram for germination rate prediction model

percentage at time $x(y_o)$, germination capacity (%) (α), shape and steepness of the germination course (b) and germination speed (c) [13]. Zahra Heidari compared 3 temperature functions for calculating germination rate and showed that temperature has significant effect on germination rate [14]. Ghasem Parmoon used a non linear regression method to show the response of seed germination by using a temperature function. Stuart P. Hardegree concludes that the characterizations of seed germination response for mostly models are based on cardinal temp. Wei-Quing Wang presented a mathematic model that enumerates the effects of temperature along with storage time and on germination rate of aspen seeds. Mehmet Basbag illustrated that the germination rate was concerned by different temperature for different duration (Table 1).

Table 1 Comparison of prediction models for seed germination rate [13–17]

S. no.	Germination model	Parameters used	Merits	Demerits
1	Yousry A. El-Kassaby $$y = y_o + \frac{ax^b}{c^b + x^b}$$ $$TMGR = b\sqrt{\frac{c^b(b-1)}{b+1}}$$	• y_o = cumulative germination percentage at time x • α = germination capacity (%) • b = shape and steepness of the germination Course • c. germination speed	• Stratification treatment was effective • Raise with the various parameters	• Inadequacy of measures due to lack of association with time over germination course • Speed of compacting the germination affects curve shape and steepness
2	By Zahra Heidari $$GR = \frac{1}{D_{50}}$$	• D_{50} = time taken for 50 % of seeds to germinate	• Relationship between germination rate and temp well • Uniformity • Greater accuracy	• Insignificant effect based on variety and mutual • Reduction of germination percentage and rate at undesirable temperature • Attributed to reduction of enzymatic activity
3	By Ghasem Parmoon, Seyed Amir Moosavi, Hamed Akbari, Ali Ebadi $$G = \frac{G_x}{1 + \exp[a\,(t-b)]} \quad GR = \frac{f(T)}{f_o}$$	• G_x = max germination percentage • t = time • $\frac{1}{f_c}$ = inherent max rate of germination at optimal temp • $f(T)$ = t function range between 0 and 1	• Reliable • Best fitting output • Accuracy of estimated Parameters	• Absence of limiting factors like lights and water • Germination is influenced by temperature

(continued)

Table 1 (continued)

S. no.	Germination model	Parameters used	Merits	Demerits
4	By Stuart P. Hardegree $$R = a\,\exp\left[\frac{-\ln(2)}{\ln(d)^2}\ln(x) + 1\right]$$ $$x = \frac{(T-b)(d^2-1)}{cd}$$	• T = temperature • a - d = equation specific coefficient	• Useful in generating indices for comparing relative • Better fit model • Germination rate response to temperature	• All species show same general pattern in total germination percentage • Possible source of error is drop in germination percentage at high and low temperature • Increase in temperature leads to degradation of seed • Low temperature results in very large error in prediction of field emergence in early spring
5	By Wei-Quing Wang, Hong-Yan Cheng, Song-Quan Sang $$\frac{1}{t_g} = \frac{(T-T_b)}{\theta_T(g)}$$ $$\frac{1}{t_g} = \frac{(T-T_m(g))}{\theta_{Tm}}$$	• $1/t_g$ = germination rate • $\theta_T(g)$ = thermal time to germination of given percentage g • T_b = base • $T_m(g)$ = minimum temp • θ_{Tm} = thermal time	• Improved model accuracy • Higher likelihood to correct in describing the effect of temperature • More accurately predict the germination	• Poor fitting to germination data • Regression did not fit the data at 5 °C • Sensitivity at low temperature • Lose viability during storage under natural condition after dispersion
6	Mehmet Basbag, Ozlem Toncer and Sema Basbag $$GR = \frac{O(n*t)}{T}$$	• n = no. of days for each counting of germinated seeds • t = no. of germinated seeds in each counting day • T = total no. of germinated seeds	• Dry heat temp and duration to breaking dormancy were enhance germination percentage • Dry heat improved the germination	• Germination were affected by different temperature and duration

3   Future Works and Conclusion

This paper describes the various prediction models used for calculating seed germination rate using various parameters. The paper also describes the merits and demerits of various seed germination rate predication models, which are very helpful in the field of agriculture to monitor and control the germination growth of seed. Further the seed germination prediction model could also be applied to develop a user interface that collects all the variations and represents it on a graphical system on the basis of which we can monitor and control the seed germination process to enable higher quality storage and produce good yield of crops.

Acknowledgments These authors gratefully acknowledge the support extended by CSIR-CSIO, Chandigarh. Thanks and appreciation to all the helpful people at CSIO, Chandigarh for their support. Finally the authors thank the reviewers and the editors for their constructive remarks.

References

1. Shafii B, Price WJ (2001) Estimation of cardinal temperatures in germination data analysis. J Agric Bio Environ Statis 6(3):356–366.
2. Rajasekaran LR, Stiles A, Caldwell CD (2002) Stand establishment in processing carrots-Effects of various temperature regimes on germination and the role of salicylates in promoting germination at low temperatures. Can J Plant Sci 82(2):443–450.
3. Windauer L, Altuna A, Arnold BR (2007) Hydrotime analysis of Lesquerella fendleri seed germination responses to priming treatments. Ind Crop Prod 25(1):70–74.
4. Kamkar B, Ahmadi M, Soltani A, Zeinali E (2008) Evaluating non-linear regression models to describe response of wheat emergence rate to temperature. Seed Sci Technol 2: 53–57.
5. Bewley JD, Black M (1994) Seeds: physiology of Development and Germination. (2nd ed), Plenum press, New York, USA.
6. Kebreab E, Murdoch AJ (2000) The effect of water stress on the temperature range for germination of Orobanches aegyptiaca seeds. Seed Sci Res 10(2):127–133.
7. Soltani A, Robertson MJ, Torabi B, Yousefi-Daz M, Sarparast R (2006) Modeling seedling emergence in chickpea as influenced by temperature and sowing depth. Agr For Meteorol 138 (1–4):156–167.
8. Ellis RH, Roberts EH (1981) The quantification of aging and survival in orthodox seeds. Seed Sci Technol 9(2):373–409.
9. Mwale SS, Azam-Ali SN, Clark JA, Bradley RG, Chatha MR (1994) Effect of temperature on the germination of sunflower. Seed Sci Technol 22(3):565–571.
10. Ellis RH, Covell S, Roberts EH, Summerfield RJ (1986) The influence of temperature on seed germination rate in grain legumes. II. Intraspecific variation in chickpea (Cicer arietinum L.) at constant temperatures. J Exp Bot 37(10):1503–1515.
11. Jame YW, Cutforth HW (2004) Simulating the effects of temperature and seeding depth on germination and emergence of spring wheat. Agr For Meteorol 124(3–4):207–218.
12. Hardegree SP (2006) Predicting germination response to temperature. I. Cardinal temperature models and subpopulation-specific regression. Ann Bot 97(6):1115–1125.
13. Yousry A. El-Kassaby Seed Germination: Mathematical Representation and Parameters Extraction FOR. SCI. 54(2):220–227.

14. Soltani A, Robertson MJ, Torabi B, Yousefi-Daz M, Sarparast R (2006) Modeling seedling emergence in chickpea as influenced by temperature and sowing depth. Agr For Meteorol 138 (1–4):156–167.

15. KESHET, L. 2006. Calculus course notes. Dept. of Mathematics, Univ. of British Columbia, Vancouver, BC, Canada. Available online at www.ugrad.math.ubc.ca/coursedoc/math102/ keshet.notes/index.html; last accessed Feb. 5, 2007.

16. THOMSON, A.J., AND Y.A. EL-KASSABY. 1993. Interpretation of seed-germination parameters. New For. 7:123–132.

17. Heidari Z, Kamkar B, Masoud Sinaki J (2014) Influence of Temperature on Seed Germination Response of Fennel. Adv Plants Agric Res 1(5): 00032. doi:10.15406/apar.2014.01.00032.

Business Modeling Using Agile

Lalita Chaudhary, Vikas Deep, Vishakha Puniyani, Vikram Verma
and Tajinder Kumar

Abstract The selection of methodology used is of great importance in business process modeling. It has a great impact on satisfaction of the customer. The aim of this paper is to fulfill the gaps of the existing model using agile methodology. There also exist several aspects in the business modeling as proposed by various scholars such as market-oriented aspects, value aspects, product oriented aspects, actors in business aspects etc. There were various shortcomings in the existing business agility. Keeping those short comings in mind the model is being proposed and hence various advantages are being extracted.

Keywords Business model · Agile structure · Quality

1 Introduction

Business process modeling is an event of showing how processes of an enterprise be improved and how they are being analyzed. Business process modeling in this concern may or may not require information technology. But business process

L. Chaudhary (✉)
G.L. Bajaj Institute of Technology and Management, Greater Noida 201306, U.P., India
e-mail: lalita.chaudhary19@gmail.com

V. Deep · V. Puniyani
Amity University, Noida 201301, U.P., India
e-mail: vdeep.amity@gmail.com

V. Puniyani
e-mail: puniyani.vishakha@gmail.com

V. Verma · T. Kumar
Seth Jai Parkash Mukund Lal Institute of Engg and Technology,
Radaur 135133, Haryana, India
e-mail: vikram_it@gmail.com

T. Kumar
e-mail: tajinder_114@gmail.com

© Springer India 2016
S.C. Satapathy et al. (eds.), *Information Systems Design and Intelligent Applications*, Advances in Intelligent Systems and Computing 433,
DOI 10.1007/978-81-322-2755-7_8

Fig. 1 Aspects in business modelling

modeling can be improved using agile methodology. Business process modeling provides a generic classification scheme for all artifacts that can be used to describe the Enterprise.

Agile methodology is based on iterative and incremental development where requirement and solutions evolve through collaborations between self organizing and cross functional terms. It promotes various approaches in iterative environment. Agile approach aims to improve system modeling by combining best practices in context of a particular project.

Business process modeling can be simply defined as an activity that represents and organizes enterprise working process. These days due to the demanding nature of customers over quality an important issue of agility comes across where the business entities need to fulfill customer's requirements at a steady pace keeping quality in mind. There also exist several aspects in the business modeling as proposed by various scholars such as market-oriented aspects, value aspects, product oriented aspects, actors in business aspects etc. (Fig. 1).

All these aspects are necessary to be considered so as to provide a quality to the customers related to product in a specific business model. Although many business enterprises take advantage of the automated business processes but still what they lack is agility due to operating in highly turbulent environment due to rapid globalization.

2 Literature Survey

It was to investigate the business process agility and to develop execution. Taking into consideration a systematic and comprehensive analysis of enterprise agility different view point architecture was developed. The structure which is already studied has two phase architecture. When first phase is considered every view point is considered in terms of agility. Wherein second phase has a combinational view point where viewpoints are combined and tested in terms of agility. It was an event driven approach and was applied to promote agility [1–10].

There were various obstacles in the existing system due to which another model was being proposed:

- The lack of understanding of various assets and capital investment in an organization.
- The lack of understanding of various assets and capital investment in an organization.
- Less flexibility in budget.
- Unprioritized business agility.
- Team conflicts based on priorities and business models.
- Long processes occupying business models should be avoided.
- Some obstacles also include a very high initial cost required to start up a project.
- Inefficiency amongst the collaboration of stakeholders.

3 Proposed Model

Enterprise business agility is modeled to improve the organizational structure where fulfilling the demand of the customers is a challenging task where quick delivery with assured quality is a priority. The market oriented business model has various aspects where the economic, social and environmental factors need to be considered to satisfy the customers. When we come to agility where we require great outcomes there exist a planned structure which needs to be followed and various conditions or drivers for business agility that need to be considered. The structure which is improved is described in the figure.

3.1 Innovate Business Model in Agile Environment

In this step we plan the prerequisites to be delivered to customers. But to prove the agile requirement we need to have a continuous interaction with the customer. So reviews and feedback come out to be an important step where we can improve the quality. Also the customer requirement may change over time so the model of business agility should be free to adapt and implement new changes (Fig. 2).

3.1.1 Adapt Process to Satisfy Agility Business Model

As the technological advances, the enterprise business model should overthrow old processes and should be ready to adapt new practices, tool and techniques to ensure fast deliver.

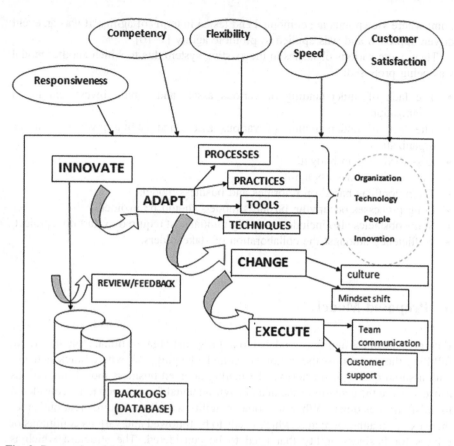

Fig. 2 Proposed model for business modeling using agile

3.1.2 Changes in Business Modelling Agility

In an enterprise business model we also need to consider the culture of people or stakeholders in an organization and a mindset shift to adopt agility to overcome obstacles in fast and quality delivery to the customers.

3.1.3 Execution

When all the data is reported in accordance with the customer we need to execute the finalized implementation as a final product that is of assured quality the customer is supported further, still any changes need to be accommodated.

3.2 Drivers to Ensure Agile Capabilities

- **Responsiveness** In an enterprise business model if the stakeholder needs to be quick in responding to new changes to ensure a quick delivery.
- **Competency** As there exists a competition amongst various business enterprises so agile business model ensures to provide an efficient structure to satisfy the customers.
- **Flexibility** Flexibility can be coined as a major driver in an agile business model. It needs to be applied in a variety of component factors such as budget designed by a business model, i.e. budget should be flexible as per the requirements of the customers
- **Flexibility in restructuring requests** customers can change the requirements so agile business model ensures flexibility.
- **Speed** A rapid delivery of the request of customer requirements is a major concern and important concept in agile business model. Long processes of business that take too long to be implemented should be avoided.

3.3 Advantages Ofenterprise Business Modelling Agility

- **Improved quality of application or product developed** Since the customer interaction is implemented for fast and quality delivery in business agility. So we get an enhanced application at the end.
- **Increased technical investment return in business agility** Using the business agility improves the end productivity of an organ which ultimately benefits all the stakeholders involved.
- **Benifiting the existing system by rapidly including emerging technology** As discussed in figure, in the adaption phase we are open to accept new practices,tools,processes that can help us to benefit our existing system and improve the quality.
- **Modular structure** As in figure we can reduce the development time and reduce the interdependency by specifying specific tasks in various phases so that when various teams wish to use only specific ones are able to do so.
- **Improved quality** Since we ensure that customers are involved whenever it is required in the long term improves the quality at the end.

4 Conclusion

In this paper a business model is being proposed which uses agility. It helps in overcoming certain obstacles of the existing system. The proposed model helps in extracting various advantages like improvement in quality, increasing technical investment, improved quality of development.

References

1. Dynamic Resource Allocation Using Virtual Machines for Cloud Zhen Xiao,Senior Member, IEEE, Weijia Song, and Qi Chen.
2. Multi-perspective Process Variability: A Case for Smart Green Buildings Aitor Murguzur Information Technologies Area.
3. Business process modeling and Organizational agility cutting through complexity with business agility.
4. Strategic IT alignment with business strategy: service oriented approach Achieving Business process agility in engineering change management with agent technology.
5. An Integrated model for the design of agile supply chains Agile manufacturing: Framework for research and development, A. Gunasekaran.
6. Enterprise agile: Business modeling Scottwambler.
7. Recommended practice for architecture description of software intensive systems IEEE 2002.
8. A holistic approach for enterprise agility for enterprise agility Alexopoulou nancy, Kanellis panagotis, Martokas Nancy alexopoulou.
9. Economist intelligence unit 2009 sponsered by EMC IBM white paper, Abdi, M dept. of computer & inf. Sci., Univ. Teknol. PETRONAS Giovanni rimassa, Birgit Burmeister Professor martin Christopher.
10. The Internet of Things: Challenges & Security Issues, L. Chaudhary, G.S. mathur, P. Upadhyay, 10th IEEE ICET 2014.

An Efficient Hybrid Encryption Technique Based on DES and RSA for Textual Data

Smita Chourasia and Kedar Nath Singh

Abstract The data security in almost every field is a challenging concern all around the globe. The application area may be as wide in the area of banking, internet, network and mobile data etc. The main focus of this paper is to secure the text data and provide a comparison with different parameters. DES and RSA are being used for comparison. A hybrid approach has been proposed in this paper based on the combination of DES and RSA algorithm. The comparison is done on the basis of size, length, number of keys and the time of encryption and decryption. The overall results suggest the hybrid encryption approach for the encryption and decryption process.

Keywords DES · RSA · Hybrid approach · Cryptography

1 Introduction

Data security is the main importance of today's age as the data has been communicated and shared through web, network, banking and sensor etc. [1–5]. There are several research work are carried in this direction to go through in the main communication stream with different control strategy.

In [6] there are three essential strategies for secured data security and accessibility are cryptography, steganography and watermarking. Among these three, the first one, cryptography [7–9], manages the improvement of systems for changing over data in the middle of clear and incoherent structures amid data trade. Steganography [10, 11], then again, is a method for concealing and separating data to be passed on utilizing a transporter signal [6]. The third one, watermarking [12, 13], is a method for creating legitimate strategies for concealing restrictive data in the perceptual information. In

S. Chourasia (✉) · K.N. Singh
TIT Science, Bhopal, India
e-mail: gazalojha11w@gmail.com

© Springer India 2016
S.C. Satapathy et al. (eds.), *Information Systems Design and Intelligent Applications*, Advances in Intelligent Systems and Computing 433,
DOI 10.1007/978-81-322-2755-7_9

73

[14] authors have recommend that the large portion of the information, the estimations of the neighboring information are firmly related (i.e. the estimation of any given pixel can be sensibly anticipated from the estimations of its neighbors [15, 16]. So to accomplish the higher relationship quality and expanding the entropy worth can make more impact [17]. Distinctive cryptography calculations are, undoubtedly used nowadays for securing correspondence channels using open key like DES, AES and Blowfish calculations [18]. An open key exchange depends on upon a key which is created through time and logical strategy to scramble the data encryption depends on upon the measure of key used [19]. It is clear from the above discussion that the data security is very crucial concern. So concerning this view a hybrid security approach has been presented. This method is also compared with simple DES and RSA.

2 Related Works

In 2011, Matalgah et al. [20] motivated by system coding hypothesis a productive half and half encryption-coding calculation that obliges utilizing customary encryption just for the first little measure of information. In their proposed calculation, all whatever remains of the data will then be transmitted safely over the remote channel, utilizing system coding, without a requirement for utilizing customary encryption. The same with the distinctive methodology has been proposed in [21, 22]. In 2011, Wai Zin et al. [23] watch that because of expanding the advancements security frameworks are extremely well known in numerous territories. The security of data can be attained to by utilizing encryption and steganography. In 2011, Sandeep Bhowmik et al. [24] recommend that the adequacy of the insurance through encryption relies on upon the calculation connected and in addition on the nature of the "key" utilized. It determines the way in which the "key" is to be picked. This work concentrates on an absolutely new approach towards the "key" era for encryption algorithms. In 2011; Rohollah Karimi et al. [25] explore shortcomings in existing Geoencryption frameworks and propose a few answers for increment the wellbeing and dependability in these frameworks. For this reason they show another geoencryption convention that will permit portable hubs to impart to one another securely by confine disentangling a message in the particular area and time period. In 2012, Rajavel et al. [26] proposed another cryptographic calculation in light of mix of hybridization and revolution of shapes. Hybridization was performed utilizing enchantment 3D shapes with m number of n request enchantment square for the creating crossover blocks. In 2012, Fanfara et al. [27] recommend that Communication security is one of numerous informatics parts which have gigantic advancement. Delicate information is progressively utilized as a part of correspondence and that is the motivation behind why security prerequisite is all the more convenient and essential. In 2012, Lili Yu et al. [28] recommend that calculation security is extraordinarily enhanced, through examining a few renowned information encryption calculations, and enhancing some information encryption calculations, and organizing encryption calculations in some request. In 2012,

Seung-Hoon Cho et al. [29] proposes an ongoing information stockpiling framework that is made out of the pressure of the flight and voice information in view of DPCM, the encryption of the compacted information utilizing AES encryption calculation and the re-plan of the encoded information by rearranging system.

3   Method

This paper presents an efficient scheme for text data encryption. In this paper an efficient framework has been presented for data security. In these framework two separate frames is designed separately for client and server. The server has the privilege to register the user and the register user only can see the data after proper authorization. The server can upload the data also. There are three phases in the process completion which are explained subsequently. In the first phase DES algorithm is applied which is working as per the steps suggested in algorithm 1. The data is first prepared by the admin and then the cipher text is send to the client. The authorized client can view the data by applying the appropriate key. Here there is only one key is needed. Based on the process the key length, number of keys, encryption and decryption time along with the size of the file is registered in the log details. In the second phase RSA algorithm is applied which is working as per the steps suggested in algorithm 2. The data is first prepared by the admin and then the cipher text is send to the client. The authorized client can view the data by applying the appropriate key. Here there are there are three keys are needed namely public, private and modulus key. Based on the process the key length, number of keys, encryption and decryption time along with the size of the file is registered in the log details. In the third phase Hybrid algorithm is applied which is working as per the steps suggested in algorithm 3. The data is first prepared by the admin and then the cipher text is send to the client. The authorized client can view the data by applying the appropriate key. Here there are there are three keys are needed namely public, private, modulus, string and integer key. Based on the process the key length, number of keys, encryption and decryption time along with the size of the file is registered in the log details.

Algorithm 1: DES [30, 31]

Step 1   DES utilizes 16 rounds. Each round of DES is a Feistel figure.
Step 2   The heart of DES is the DES capacity. The DES capacity applies a 48-bit key to the furthest right 32 bits to produce a 32-bit yield.
Step 3   Expansion P-box
         Since RI − 1 is a 32-bit information and KI is a 48-bit key, we initially need to grow RI − 1 to 48 bits.
Step 4   Although the relationship between the information and yield can be characterized scientifically, DES uses Table to characterize this P-box.

Step 5   After the extension stage, DES utilizes the XOR operation on the extended right area and the round key. Note that both the right segment and the key are 48-bits long. Likewise take note of that the round key is utilized just in this operation.

Step 6   S-Boxes

The S-boxes do the genuine disarray. DES utilizes 8 S-boxes, each with a 6-bit information and a 4-bit.

Step 7   Apply s-box standard.

Step 8   We can make each of the 16 adjusts the same by including one swapper to the 16th round and include an additional swapper after that (two swappers wipe out the impact of one another). The round-key generator makes sixteen 48-bit keys out of a 56-bit key.

Algorithm 2: RSA [32–34]

The Rivest-Shamir-Adleman (RSA) calculation is a standout amongst the most mainstream and secure open key encryption routines. The calculation profits by the way that there is no proficient approach to component huge (100–200 digit) numbers.

Utilizing an encryption key (e,n), the calculation is as per the following:

1. Represent the message as a number somewhere around 0 and (n − 1). Huge messages can be separated into various pieces. Every square would then be spoken to by a whole number in the same reach.
2. Encrypt the message by raising it to the eth power modulo n. The outcome is a ciphertext message C.
3. To unscramble ciphertext message C, raise it to another force d modulo n.
4. The encryption key (e,n) is made open. The unscrambling key (d,n) is kept private by the client.

The most effective method to Determine Appropriate Values for e, d, and n

1. Choose two huge (100+ digit) prime numbers. Indicate these numbers as p and q.
2. Set n equivalent to p * q.
3. Choose any huge whole number, d, such that GCD(d, ((p − 1) * (q − 1))) = 1
4. Find e such that e * d = 1 (mod ((p − 1) * (q − 1)))

Algorithm 3: Hybrid Algorithm

This hybrid algorithm is the combination of DES and RSA.

Step 1   Receive the initial text message

Step 1   KEYLENTGH = 256;

Step 2   The key pair is generated by the Key pair scheme based on RSA.

Step 3   It initializes the key length based on the same mechanism adopted in the other.

Step 4   We can make each of the 16 adjusts the same by including one swapper to the 16th round and include an additional swapper after that (two swappers wipe out the impact of one another). The round-key generator makes sixteen 48-bit keys out of a 56-bit key.

Step 5   The key pair is generated by the key pair method.

Step 6   The method used for finding the public key is getPublic() method.

Step 7   The method used for finding the private key is getPrivate();

Step 8   The instance of RSA is use for finding the modulus and integer key.

Step 9   It is then parsed to generate the cipher text.

4   Results Evaluation

The results based on the log generated are discussed in this section. We have compared the results based on four different parameter. First is the number of key as the number of key increases it enhances the security. So in the first case hybrid algorithm approaches good security as five consecutives keys are used. In the second case the key length Hybrid and RSA performs good as shown in Figs. 3 and 4. In third case of variations in the file size again hybrid algorithm out performs as the changes are negligible. In case of time DES outperforms and then hybrid gives good performance. So based on the overall evaluation hybrid is good in most of the cases but in terms of time DES performs better (Figs.1 and 2).

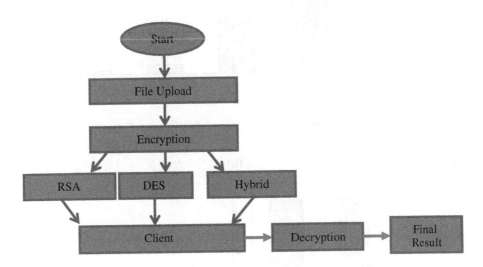

Fig. 1 Flowchart

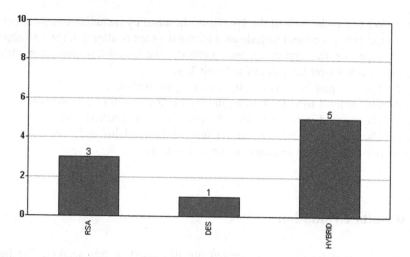

Fig. 2 Number of keys used in RSA, DES and hybrid

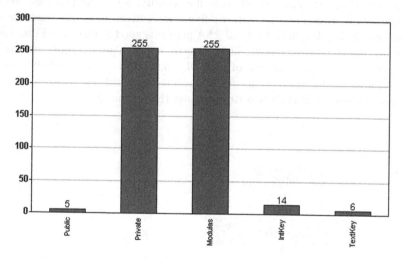

Fig. 3 Length of keys used in hybrid

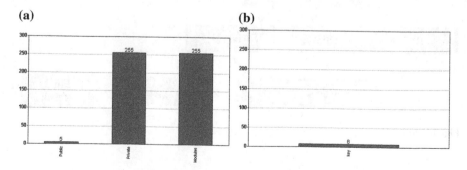

Fig. 4 a Length of Keys used in RSA **b** Length of Keys used in DES

5 Conclusions

In this paper an efficient hybrid encryption technique has been proposed. The proposed scheme has been compared with the DES and RSA algorithm. The parameters used for the comparison are length of key, number of key, size of file and time of encryption and decryption. In case of number of key hybrid algorithm approaches good security as five consecutives keys are used. In the second case the key length Hybrid and RSA performs well. In third case of variations in the file size again hybrid algorithm out performs as the changes are negligible. In case of time DES outperforms and then hybrid gives good performance.

References

1. Piyush Rajan Satapathy, "Performance Measurment of AES Crypto Algorithm in Microcode Environment of IXP2400 Platform," University of California, Riverside Projects, spring 2005.
2. Mozaffari-Kermani, Mehran Reyhani-Masoleh, Arash, "Concurrent Structure-Independent Fault Detection Schemes for the Advanced Encryption Standard," IEEE Transactions on Computers, Vol. 59, No. 5, May, 2010.
3. Bhavesh Joshi and Anil Khandelwal," Rivest Cipher based Data Encryption and Clustering in Wireless Communication", International Journal of Advanced Technology and Engineering Exploration (IJATEE), Volume-2, Issue-2, January-2015, pp. 17–24.
4. Farhadian, A. Aref, M.R," Efficient Method for simplifying and approximating the S-Boxes Based on Power Functions," Information Security IET, Vol.3 No. 3, Sept, 2009.
5. Sirwan Geramiparvar and Nasser Modiri, "Security as a Serious Challenge for E-Banking: a Review of Emmental Malware", International Journal of Advanced Computer Research (IJACR), Volume-5, Issue-18, March-2015, pp. 62–67.
6. A. Mitra, Y V. Subba Rao, and S. R. M. Prasnna, "A new image encryption approach using combinational permutation techniques," Journal of computer Science, vol. 1, no. 1, p. 127, 2006.
7. A. J. Elbirt and C. Paar, "An Instruction-Level Distributed Processor for Symmetric-Key Cryptography," IEEE Trans. Parallel and distributed systems, vol. 16, no. 5, pp. 468–480, May 2005.
8. Nath, Asoke, et al. "Multi Way Feedback Encryption Standard Ver-2 (MWFES-2)." International Journal of Advanced Computer Research (IJACR) 3.1 (2013).
9. W. Stallings, Cryptography and Network Security. Englewood Cliffs, NJ: Prentice Hall, 2003.
10. Satish Bhalshankar and Avinash K. Gulve, "Audio Steganography: LSB Technique Using a Pyramid Structure and Range of Bytes", International Journal of Advanced Computer Research (IJACR), Volume-5, Issue-20, September-2015, pp. 233–248.
11. S. Trivedi and R. Chandramouli, "Secret Key Estimation in Sequential Steganography," IEEE Trans. Signal Processing, vol. 53, no. 2, pp. 746–757, Feb. 2005.
12. Y. Wu, "On the Security of an SVD-Based Ownership Watermarking," IEEE Trans. Multimedia, vol. 7, no. 4, pp. 624–627, Aug. 2005.
13. Y. T. Wu and F. Y. Shih, "An adjusted-purpose digital watermarking technique," Pattern Recognition 37, pp. 2349–2359, 2004.
14. Mohammad Ali Bani Younes and Aman Jantan," Image Encryption Using Block-Based Transformation Algorithm", IAENG International Journal of Computer Science, 35:1, IJCS_35_1_03.

15. S. P. Nana'vati., P. K. panigrahi. "Wavelets:applications to image compression- I,". joined of the scientific and engineering computing, vol. 9, no. 3, 2004, pp. 4– 10.
16. C. Ratael, gonzales, e. Richard, and woods, "Digital image processing," 2nd ed, Prentice hall, 2002.
17. AL. Vitali, A. Borneo, M. Fumagalli and R. Rinaldo,"Video over IP using standard-compatible multiple description coding," Journal of Zhejiang University-Science A, vol. 7, no. 5,2006, pp. 668– 676.
18. S. Masadeh W. Salameh. End to end keyless self-encrypting/decrypting streaming cipher. In: Information Technology & National Security Conference. 2007.
19. A. Nadeem MYJ. A performance comparison of data encryption algorithms. In: First International Conference on Information and Communication Technologies. 2005:84– 89.
20. Matalgah, Mustafa M., Magableh, A.M., "Simple encryption algorithm with improved performance in wireless communications", IEEE 2011.
21. Asoke Nath, Debdeep Basu, Surajit Bhowmik, Ankita Bose, Saptarshi Chatterjee, "Multi Way Feedback Encryption Standard Ver-2(MWFES-2)", International Journal of Advanced Computer Research (IJACR), Volume-3, Issue-13, December-2013, pp. 28–34.
22. Sagar Chouksey, Rashi Agrawal, Dushyant Verma,Tarun Metta, "Data Authentication Using Cryptography", International Journal of Advanced Computer Research (IJACR), Volume-3, Issue-10, June-2013, pp. 183–186.
23. Wai Wai Zin and Than Naing Soe," Implementation and Analysis of Three Steganographic Approaches", IEEE 2011.
24. Sandeep Bhowmik and Sriyankar Acharyya," Image Cryptography: The Genetic Algorithm Approach", IEEE 2011.
25. Rohollah Karimi and Mohammad Kalantari, "Enhancing security and confidentiality in location-based data encryption algorithms", IEEE 2011.
26. Rajavel, D., Shantharajah, S.P., "Cubical key generation and encryption algorithm based on hybrid cube's rotation", IEEE 2012.
27. P. Fanfara, E. Danková and M. Dufala, "Usage of Asymmetric Encryption Algorithms to Enhance the Security of Sensitive Data in Secure Communication", 10th IEEE Jubilee International Symposium on Applied Machine Intelligence and Informatics, SAMI 2012.
28. Lili Yu, Zhijuan Wang and Weifeng Wang, "The Application of Hybrid Encryption Algorithm in Software Security", 2012 Fourth International Conference on Computational Intelligence and Communication Networks.
29. Seung-Hoon Cho, Chan-Bok Jeong, Seok-Wun Ha, Yong Ho Moon," A Flight Data Storage System with Efficient Compression and Enhanced Security", IEEE 2012.
30. Shikha Joshi, Pallavi Jain," A Secure Data Sharing and Communication with Multiple Cloud Environments with Java API", International Journal of Advanced Computer Research (IJACR) Volume 2 Number 2 June 2012.
31. Saket Gupta, "Secure and Automated Communication in Client and Server Environment", International Journal of Advanced Computer Research (IJACR), Volume-3, Issue-13, December-2013, pp. 263–271.
32. Ashutosh Kumar Dubey, Animesh Kumar Dubey, Mayank Namdev, Shiv Shakti Shrivastava," Cloud-User Security Based on RSA and MD5 Algorithm for Resource Attestation and Sharing in Java Environment", CONSEG 2012.
33. Khandelwal, Anil, et al. "Establishing Secure Event Detection with Key Pair in Heterogeneous Wireless Sensor Network." Advanced Materials Research. Vol. 433. 2012.
34. Wuling Ren; Zhiqian Miao, "A Hybrid Encryption Algorithm Based on DES and RSA in Bluetooth Communication," Modeling, Simulation and Visualization Methods (WMSVM), Second International Conference on, vol., no., pp. 221, 225, 15–16 May 2010.

Application of Machine Learning on Process Metrics for Defect Prediction in Mobile Application

Arvinder Kaur, Kamaldeep Kaur and Harguneet Kaur

Abstract This paper studied process metrics in detail for predicting defects in an open source mobile applications in continuation with our previous study (Moser et al. Software Engineering, 2008). Advanced modeling techniques have been applied on a vast dataset of mobile applications for proving that process metrics are better predictor of defects than code metrics for mobile applications. Mean absolute error, Correlation Coefficient and root mean squared error are determined using different machine learning techniques. In each case it was concluded that process metrics as predictors are significantly better than code metrics as predictors for bug prediction. It is shown that process metrics based defect prediction models are better for mobile applications in all regression based techniques, machine learning techniques and neuro-fuzzy modelling. Therefore separate model has been created based on only process metrics with large dataset of mobile application.

Keywords Code metrics · Process metrics · Software metrics · Defect prediction

1 Introduction

Process [1] metrics [2, 3] are quantitative measures which are collected across the project over long period of time that has the capability to improve the efficiency of software process. Process metrics can be used to predict the bugs before formal testing which enhance the quality of the software [4]. Bug prediction greatly reduces the testing cost and usage of resources. Process metrics are used as

A. Kaur (✉) · K. Kaur · H. Kaur
USICT, GGS, Indraprastha University, Sec-16C, Dwarka, Delhi, India
e-mail: arvinderkaurtakkar@yahoo.com

K. Kaur
e-mail: kdkaur99@gmail.com

H. Kaur
e-mail: harguneet.sethi@gmail.com

© Springer India 2016

81

S.C. Satapathy et al. (eds.), *Information Systems Design and Intelligent Applications*, Advances in Intelligent Systems and Computing 433,
DOI 10.1007/978-81-322-2755-7_10

indicators that lead to long term process improvement. Different process metrics are recorded from the comments posted by developers in open source repositories.

Earlier researchers started predicting bugs using code metrics [5–7] but at current state of the art many researchers have worked on code and process metrics both on different datasets but very few research has been done on defect prediction of mobile applications. In our recent research we observed that process metrics are better predictors for bug prediction than code metrics for mobile applications with small dataset. We continue this research with large dataset and more emphasis is on building bug prediction models using process metrics as predictors. The main aim of this research is to extract process metrics from the change history [8] of source codes. Various modelling techniques have been applied to predict fault prone modules. Large dataset of mobile applications has been taken in this research to produce more reliable results. Code metrics are also compared with process metrics with different dataset and different modelling techniques. To the best of our knowledge, this current study may be the first attempt where neuro-fuzzy modelling is applied on process and code metrics both for mobile applications.

The remainder of this paper is divided into the following parts: In Sect. 2 Background presents a brief summary of the relevant researches regarding to the source code metrics and process metrics for defect prediction. Section 3 presents metrics which gives a brief summary of source code metrics in 3.1 and process metrics in 3.2 that are being used in this research. Section 4 gives the description of research methodology and how the data is collected. Analysis has been presented in Sect. 5. Section 6 discusses the limitations. Conclusion and future work are presented in Sect. 7.

2 Background

Already many researchers have done work on code and process metrics both [9–11] for defect prediction, but there is hardly research has been done on defect prediction of mobile applications with large scale dataset of process metrics.

Ruchika Malhotra et al. [12] addressed four issues in their paper: (a) machine learning techniques are compared using non popular large data sets. (b) performance of defect prediction models is measured using inappropriate performance measures. (c) less use of statistical tests. (d) models are validated using same data set from which it is trained. In their research dataset used is MMS application package of Android but in our research we have used different application packages of mobile operating system-Android and in our study process metrics are used as predictors to predict the fault prone modules.

Cagatay Catal [13] provided the systematic review of software fault prediction studies which emphasized on methods, metrics and datasets. Their review is conducted with the help of 74 fault prediction papers where it has been concluded that more public datasets have been used and usage of machine learning algorithms are increased since 2005. Dominant metrics are the method level metrics in fault prediction research area. But in our study we have used a mobile application-Android as a dataset and process metrics for predicting the defects in softwares which help in software process improvement.

Emad Shihab et al. [14] used the statistical approach for building explainable regression models. Experiment is performed on Eclipse open source project in which it is observed that 4 out of 34 code and process metrics effects the likelihood of finding a post release defect. They quantified the impact on predicting the defects. It was also observed that simple models showed comparable performance over complex PCA-based models.

Foyzur Rahman et al. [15] analysed both code and process metrics form different views for building prediction models. They build many prediction models across 85 releases of 12 open source large projects and compared the stability, performance and portability of different set of metrics. It was then concluded that code metrics are less effective predictors than process metrics. The reason concluded for this observation was that code metrics do not change release to release which lead to stagnation in prediction models. This results in the same files to be repeatedly predicted as defective. But in our research we analysed code and different set of process metrics across mobile applications.

Wahyudin et al. [16] investigated defect prediction with large dataset of widely used Open Source Softwares based on product and process metrics and combination of both. The results show that process metrics had strong correlation for predicting defect growth between releases. Combination of both process and product metrics provide better results for predicting defects which improved the accuracy and identify modules for improving process and product.

Johari et al. [17] computed class level object oriented metrics using ckjm-1.9: A tool for calculating Chidamber and Kemerer Java Metrics and Metrics 1.3.6, which is an eclipse plugin. Dataset is taken from SCM (software configuration management) repository which contains the details of revisions made to open source software. Each revision has the details of classes which have a bug. Relationship between the number of revision made to a class and the measure of software metrics extracted from ckjm is studied. Thus buggy areas are identified which can predict the maintainability of the software.

In this current paper, we study three defect prediction models: Code metric model, Process metric model and combined model of code and process metrics both for predicting defects in Android camera application using regression, machine learning techniques and neuro-fuzzy modelling. We also study Code metric model, Process metric model and combined model of code and process metrics both for predicting defects in Android alarmclock application using neuro-fuzzy modelling.

3 Metrics

Two different types of metrics have been studied: Code and Process metrics which are extracted from dataset to predict the defective modules in system. This step is expected to enhance the quality and improve the performance of the software system.

3.1 Code Metrics

Code metrics are the metrics which are also known as product or source code metrics that are calculated from ckjm tool [18]. Code metrics are calculated by ckjm tool [19]. Metrics which are calculated by ckjm for each class are tabulated in Table 1. Description of each metric is given in our previous paper [1].

Table 1 Ckjm metrics description

Metric	Description	Metric	Description
WMC	Weighted methods per class	DAM	Data access metric
DIT	Depth of Inheritance tree	MOA	Measure of aggregation
NOC	Number of children	MFA	It is the ratio of number of methods inherited by a class to the number of methods accessible by member methods of the class
CBO	Coupling between object classes		
RFC	Response for a class	CAM IC	Cohesion among methods of class Inheritance coupling
LCOM	Lack of cohesion in methods	CBM	Coupling between methods
Ca	Afferent couplings		
Ce	Efferent couplings	AMC	Average method complexity
NPM	Number of public methods	CC	McCabe's cyclomatic complexity
LCOM3	Lack of cohesion in methods. LCOM3 varies between 0 and 2	AC count	It is the average of the McCabe's cyclomatic complexity which is calculated for each method
LOC	Lines of code	MAX count	It is the maximum value of CC among all the methods

3.2 Process Metrics

In our study process metrics are gathered from the repository where developers of particular application have posted the comments for all the bugs and commits occurred. Process metrics improve the performance of software and they are extracted manually. These metrics gets changed from revision to revision so gives better results so that performance can be compared for different revisions. Process metrics are collected over long period of time across all revisions. Following metrics are recorded from the comments by different authors for the particular revision of the dataset given in Table 2:

Table 2 Process metrics

LOC_ADDED	Number of lines added. It is the sum of lines added for all the code added, comments added and blanks added in a class for a specific revision of dataset
LOC_DELETED	Number of lines deleted. It is the sum of lines deleted for all the code deleted, comments deleted and blanks deleted in a class for a specific revision of dataset
CODE_ADDED	Number of source code lines added.This metric refers to the source code lines added in a class due to bug.It is the sum of the source code lines added for all the changes in a class
CODE_DELETED	Number of source code lines deleted. This metric refers to the source code lines deleted in a class due to bug. It is the sum of the source code lines deleted for all the changes in a class
COMMENT_ADDED	Number of comment lines added. This metric refers to the comment lines added in a class due to bug. It is the sum of the comments added for all the changes in a class
COMMENT_DELETED	Number of comment lines deleted. This metric refers to the comment lines deleted in a class due to bug. It is the sum of the comments deleted for all the changes in a class
BLANKS_ADDED	Number of blanks added. This metric refers to the blanks added in a class due to bug. It is the sum of the blanks added for all the changes in a class
BLANKS_DELETED	Number of blanks removed. This metric refers to the blanks removed in a class due to bug. It is the sum of the blanks removed for all the changes in a class
NA	Number of Authors. This metric refers to the number of distinct authors that checked or changed a particular file or classdue to the bug into the repository for the particular revision of dataset
NR	Number of Revisions. This metric calculates the number of times the class is referred for each bug in a specific revision
NB	Number of Bugs. This is the most important metric which is recorded as number of times a particular class is involved for bug-fixes. This metric is considered as the predictor variable for defet prediction during modelling

4   Methodology and Data Collection

In our study, mainly process metrics of a mobile application have been studied to predict faulty modules for the Android release 1.6 camera application package which has been extracted from the Android platform. Code metrics are also compared with process metrics of android camera package. Neuro-fuzzy modelling is applied on both alarm clock and camera application package of Android and then results are compared to find which set of metrics is better. In this paper camera is a huge software package for which we have extracted process metrics for defect prediction. The research methodology is shown in Fig. 1, which demonstrates how to build defect prediction models using process and code metrics. At first, fact finding deals with the study of related work in this field. Once we are thorough with the understanding of our topic, then we understand the metric extraction tools, statistical tools, machine learning tools and open source software in planning phase. Next in data extraction phase, dataset is extracted from the open source repository github then it is processed to extract 20 static code metrics listed in Table 1 using the tool-CKJM [18]. 11 Process metrics listed in Table 2 are recorded from the comments posted by the developers. The comments due to bugs are extracted using the filters from the openhub repository [20]. Next the data analysis phase deals with predicting the number of bugs both with the code and process metric sets using different types of modelling techniques and comparing the results that which one is the better metric set and creating the separate prediction models based on process metrics for camera package of Android. Lastly in conclusion phase, final results and model are presented which are build during the result Only Java files are considered as Android is mainly built with java programming language. In this study we also analyze the relationship between process metrics and number of defects in each class files. Table 3 gives the detail description of Camera package for Android with total number of bugs noted down.

Fig. 1 Steps for predicting defects

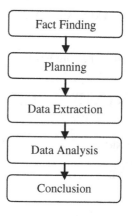

Table 3 Summary of the android camera package

Dataset	#Java classes	Commits	Contributors	Number of bugs
Camera release 1.6	40	8032	114	1560

Once source code metrics and process metrics are calculated, modelling is performed using Weka tool, SPSS 20 and MATLAB. Different modelling techniques have been applied to the metrics for predicting the fault prone modules like Regression, Machine Learning techniques and Neuro-Fuzzy modelling. Table 4 gives the brief overview of all the techniques used.

In our current paper we have given our emphasis on process metrics. First we applied neuro-fuzzy modelling to Android Alarmclock Package then we compared code metrics with process metrics taking huge dataset of Android Camera Package. At last we have taken all the buggy classes and created the model using Process

Table 4 Modelling techniques used

Modelling technique	Description	Tool
Regression analysis	The process of constructing a mathematical model or function that can be used to predict or determine one variable by another variable or variables	IBM SPSS version 20
RBFNetwork	The RBF network is one hidden layer neural network with several forms of radial basis activation functions. The most common one is the Gaussian function	Weka 3
Simple linear regression	It is the lowest square estimator of a linear regression model with a single dependent variable	Weka 3
SMOreg	SMOreg implements the support vector machine for regression. The parameters can be learned using various algorithms. The algorithm is selected by setting the RegOptimizer	Weka 3
IBk	It is a k-nearest-neighbour classifier that uses the same distance metric	Weka 3
Kstar	KStar (K*) is an instance-based learner which uses entropy as a measure, and results are presented which compare favorably with several machine learning algorithms	Weka 3
LWL	Locally Weighted Learning is a class of function approximation techniques, where a prediction is done by using an approximated local model around the current point of interest	Weka 3
MLP	Multi Layer Perceptron Classifier is a network consisting of a layer of input nodes, one or more hidden layers of computation nodes (Perceptrons) and an output layer of computation nodes connected through synaptic links	Weka 3
Neuro-fuzzy modelling	Neuro fuzzy logic is implemented using matlab R2011a. To demonstrate the fuzzy logic control capabilites of MATLAB, simulation is performed	MATLAB R2011a

metrics of Android Camera Package. In our previous research we have taken small dataset of Android Alarmclock package and compared the models based on three predictors: Correlation Coefficient, Mean Absolute Error and Root Mean squared Error. But in this study we have applied an advanced technique neuro-fuzzy modelling in process metrics based defect prediction of mobile application. Regression Analysis is also done using different tool. Based on our study we concluded process metrics are better measures for predicting fault prone modules, that is why we have created separate models using only process metrics. During modelling two types of variables are considered: Independent variables and Dependent variables. All the code and process metrics which are extracted from the dataset are taken as independent variables. Here we are predicting buggy areas so number of bugs which is extracted from the large dataset for each class is taken as dependent variable. The aim of this study is to investigate advanced modelling techniques in process metrics based defect prediction, using large dataset of mobile application-Android camera package. We have used three different tools for analysing code and process metrics: SPSS, Weka, MATLAB. We analyse the following:

- Firstly we compared code and process metrics based defect prediction models of Android Alarmclock package using neuro-fuzzy modeling technique
- Secondly we compared code and process metrics of large dataset of Android Camera Package using various modelling techniques.
- Thirdly we have created models using only process metrics of Android Camera Package taking all the buggy classes and applied various modelling algorithms.

5 Analysis

In analysis at first step, Neuro-Fuzzy modelling technique is applied to the dataset of Android Alarmclock Package. MATLAB is used for performing neuro-fuzzy modelling. Mean absolute error is calculated from the code written in MATLAB and number of bugs which are extracted manually from the openhub repository. Three models are built using only code metrics and process metrics separately and then the combination of both set of metrics. First Fuzzy logic is performed to calculate mean absolute error from the output. Then neuro-fuzzy modelling is performed. Mean absolute error $= \sum |$ Number of Bugs-Predicted number of bugs/Number of classes (5.1) meanabserror1 is calculated for fuzzy logic and Meanabserror2 is for neuro-fuzzy modelling. Predicted number of bugs is calculated from fuzzy logic and number of bugs is the process metric for each class which is extracted manually from the comments posted by various developers. Fuzzy modelling is proved slightly better than neuro-fuzzy modelling. Mean absolute error should attain less value with various modelling techniques. Three models are built using neuro-fuzzy modelling on Android Alarmclock package.

Table 5 Mean absolute error for process metric (fuzzy-alarmclock)

No. of bugs	Meanabserror1	Meanabserror2	No. of bugs	Meanabserror1	Meanabserror2
1	1.003743	1.25491	8	6.098037	5.503654
1	0	2.34E-08	5	1.599141	0.260016
1	0.652597	0.594572	1	2E-15	3.34E-08
6	2	2	0	0.633817	0.969809
2	2	2	4	1.088268	1.805571
1	1.878505	0.799555	3	0.411765	0.269997
0	1.84466	2.132034	3	0.362069	0.257186
0	1.301703	3.808601	5	2.504899	3.13229
–	–	–	–	1.4612	1.549262

Mean absolute error for process metrics, code metrics and combined set of metrics using fuzzy modelling is given in Tables 5, 6 and 7 respectively.

It is observed that mean absolute error of Android alarmclock package is:

Code metrics: 1.825229

Process metrics: 1.4612

Combined metrics: 1.649135557

Thus process metrics are proved better predictors using fuzzy logic than code metrics. Next we take large dataset of mobile application-Android Camera Package where various modelling algorithms are used to analyse the performance.

Table 6 Mean absolute error for combined metric (fuzzy-alarmclock)

No. of bugs	Meanabserror1	Meanabserror2
1	1.363636	1.268418
1	3.727272	3.740019
1	3	3
6	2	2
2	1.621951	1.612161
1	0.460526	0.471106
0	2.31E-19	2.15E-19
0	0.921053	0.956796
8	5	5
5	1	1
1	0.80154	0.798332
0	4	4
4	2.307692	2.283809
3	1	1
3	1	1
5	1	1
–	1.825229	1.820665

Table 7 Mean absolute error for combined metric (fuzzy-alarmclock)

No. of bugs	meanabserror1	meanabserror2
1	0.5	0.144701
1	0.965116279	0.909708
1	3	3
6	2	2
2	2	2
1	3	3
0	5.29963E-10	7.4E-10
0	0.921052632	0.960458
8	5	5
5	1	1
1	1	1
0	4	4
4	0	0
3	1	1
3	1	1
5	1	1
–	1.649135557	1.625929

Linear regression is applied using SPSS 20 tool which looks at the predicting powers of the established constructs on the dependent variable and inspect if the metrics are significant predictor or not. In our study bug prediction models are built from regression analysis where measures Coefficient of Correlation (R) and The coefficient of determination (R^2) are determined. R^2 value should be close to 1. Table 8 presents the bug prediction models that are being developed by using Regression analysis on static metric dataset of Android_package_camera.

Next machine learning algorithms which have been used for predicting defects on the basis of accuracy measures: Correlation Coefficient, Mean Absolute Error and Root Mean Squared error using Weka tool on Android Camera Package version 1.6. Here also we built three models, two with code and process metrics each and third with combination of both for predicting fault prone modules.

Table 9 displays the value of correlation coefficient of process metrics, code metrics and combination of two For Android Camera Package. Accurate models should attain high value of correlation coefficient. It is visible from the graph of

Table 8 Bug prediction models using regression of android camera package

Datasets	Metrics	R	R square	Predictors
Android_Camera	Code	1.000	1.000	MAX,DAM,DIT,NOC,Ca,MOA,CAM,AC, CBM,LCOM3,AMC
Android_camera	Process	1.000	1.000	NA, BLANK_DEL, COMMENT_DEL, BLANK_ADD, NR, CODE_ADD, COMMENT_ADD

Table 9 Correlation coefficient of camera package

Technique	Linear regression model	Multilayer	RBFNetw	Simple Li	SMOreg	Ibk	Kstar	LWL
Process	−0.9735	0.9883	0.9683	0.9999	0.9802	0.9822	0.9833	0.9633
Code	0.5371	0.8953	0.4461	0.8634	0.8177	0.9743	0.9442	0.9682
Combined	0.939	0.9635	0.9126	0.9999	0.9537	0.9743	0.9787	0.9705

Fig. 2 that in all the machine learning algorithms, process metrics have higher value than code metrics for Correlation coefficient. The combination of both also attain almost same value of process metrics so there is no need to calculate code metrics. With process metrics we may attain the satisfactory value.

Mean absolute error is also calculated from Weka tool in which process metrics performs significantly better than code metrics and combination of two as shown graphically in Fig. 3. Modelling techniques should have low value of mean absolute error. Table 10 displays the techniques and process metrics and code metrics with different values of mean absolute error for camera Package. Root mean squared error is measured for process metrics, code metrics and combination of two by various modelling algorithm whose value should be 1 as shown in Table 11 and graphically in Fig. 4.

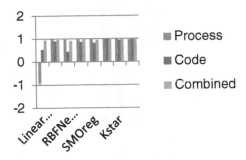

Fig. 2 Graphical representation of correlation coefficient of camera package

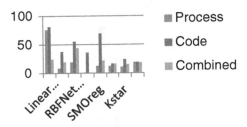

Fig. 3 Graphical representation of mean error of camera package

Table 10 Mean absolute error of camera package

Technique	Linear Regression model	Multilayer	RBFNetw	Simple Li	SMOreg	Ibk	Kstar	LWL
Process	75.9727	8.5548	19.1204	0.9459	12.6379	12.3333	10.9392	19.1754
Code	81.1727	37.5159	55.2858	36.051	69.2144	16.5833	23.9778	19.0327
Combined	24.5904	19.4133	44.0396	0.9459	21.7356	16.5833	15.4259	18.2606

Table 11 Root mean squared error of camera package absolute

Technique	Linear regression model	Multilayer	RBFNetw	Staple Li	SMOreg	Ibk	Kstar	LWL
Process	102.945	14.4655	23.4761	1.2461	18.861	17.8699	17.1841	25.5811
Code	97.5271	47.2336	85.6917	48.1813	81.411	22.5111	31.3647	23.6477
Combined	33.2646	25.4229	75.8565	1.2461	28.6358	22.5111	21.8115	23.0833

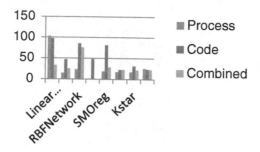

Fig. 4 Graphical representation for root mean squared error of camera package

From the above and with our previous research [1] we can conclude that whether its the large dataset or small dataset of mobile application process metrics are the better predictor to predict the faulty modules than code metrics with machine learning modelling techniques. We have also observed that in comparison to various machine learning algorithms, simple linear regression outperforms all other with satisfactory results in defect prediction of mobile applications. Neuro-Fuzzy modelling is also applied to Android Camera Package to calculate the same mean absolute error using MATLAB for code metrics, process metrics and combination of both in Tables 12, 13 and 14 respectively.

It is observed that mean absolute error of Android camera package using neuro fuzzy logic is: Code metrics: 78.04695 Process metrics: 8.845541667 Combined metrics: 78.84153 Thus process metrics are proved better predictors using fuzzy logic than code metrics.

Next we use only the process metrics of large dataset of mobile application: Android Camera Package containing all the buggy classes in which developers have found the bug since 1.6 version of Android.11 process metrics have been

Table 12 Code Metrics

No. of bugs	Meanabserror1	Meanabserror2
58	44.9143	45.5476
1	138	138
18	9.227	9.1695
23	116	116
5	0.3714	0.4323
11	11.5692	11.3395
6	0.4815	0.4878
1	138	138
253	114	114
4	135	135
277	150	150
60	79	79
–	78.04695	78.08139

Table 13 Process Metrics

No. of bugs	Meanabserror1	Meanabserror2
58	26.5263	31.6465
1	0	0
18	7.5277	6.107
23	12.0512	16.9027
5	1.1895	0.9762
11	3.4001	0.4072
6	0.0128	8.0208
1	0	0
253	21.6908	26.4557
4	3	15.1429
277	21.8731	28.5984
60	8.875	0.0152
–	8.845541667	11.18938

recorded in this study from openhub repository for this particular application. We have created bug prediction models using only the process metrics by applying various modelling algorithms including Regression Analysis, Machine Learning Algorithms and Neuro-Fuzzy Modelling. For process metrics R square value is found to be .999 which is the acceptable value during regression analysis. Machine Learning Algorithms are applied to process metrics consisting of 40 classes of mobile application: Android Camera Package. Bug prediction models are built on the basis of three predictors: Correlate Coefficient, Root Mean Squared Error, Mean Absolute Error. It is found that Simple Linear Regression model is the best one to predict bugs as it estimates the desired value of all the predictors: Correlation coefficient: 0.9994, Mean absolute error: 1.2327 and Root mean squared error: 1.9306

Table 14 Combined metrics of Code and Process

No. of bugs	Meanabserror1	Meanabserror2
58	52.49351	49.09192
1	138	138
18	9.226994	0.586197
23	116	116
5	4.012987	3.290809
11	9.883375	6.378347
6	0.481481	4.152978
1	138	138
253	114	114
4	135	135
277	150	150
60	79	79
–	78.84153	77.79169

Neuro Fuzzy Modelling is applied only on process metrics of 40 faulty classes of camera package using MATLAB. Mean Absolute Error (MAE) is calculated which is found to be 9.71 with large dataset which is comparable to previous calculated MAE with comparable small dataset.

6 Limitations

Not many researchers have done their work on mobile applications for bug prediction so there might be error in data collection while extracting code and process metrics from it. We have done our analysis by taking small and large both types of datasets for mobile application as to strengthen our point towards bug prediction.

Earlier we have used simple techniques on Android Alarmclock Package for version 1.6. In this study we have used more advanced techniques to compare the results obtained. So we have tried to eliminate bias in selection of techniques and datasets. When we obtained code metrics, we have to create jar file of the source code of Android Alarmclock Package and Android Camera Package in Android Environment. This may have excluded some class files of packages. So there might be some classes which may have bugs but missed out which can be threat to the validity of results. The datasets which are used in this study are based on Android package camera and Android alarmclock package are written in Java language. The models build by this study likely to be valid for Object oriented languages, for example C++ or Java, however further research can establish its usefulness in predicting bugs in other paradigms. Only java class files are used to calculate metrics for the android package. Other files are excluded in our study. Some unknown errors might creep in while mapping between source files and defects, extraction of code and process metrics. In addition process metrics are extracted from Android github and openhub repository which may contain some errors. Defect prone areas are observed from the Android package of single version 1.6. Other versions of same package may be included for getting more confidence about the results.

7 Conclusions and Future Work

The aim of this study was to find the metrics that are significant in predicting bugs and to build bug prediction models on large dataset of mobile application. Defect predictors are identified using a set of 20 code metrics and 11 process metrics extracted from openhub repository. Android Camera package of version 1.6 is used to build bug prediction models based on process metrics. It has been observed that whether its a small dataset of mobile application or large process metrics are better predictors than code metrics. Prediction models are built to identify the class files which are defective or which are not. We presented a new set of process metrics that are significant predictors of faulty areas other than Chidamber and Kemerer metric suite. Future work will focus on following issues for defect prediction: (1) Other than object oriented java programming language class files other files may also be included to extract code and process metrics for bug prediction. (2) More versions of Android application should be carried out with other packages in order to calculate faulty areas. (3) The results of the gathered metrics should thoroughly be verified using other metrics gathering sources/tools.

References

1. Arvinder Kaur, Kamaldeep Kaur, Harguneet Kaur " A Comparative Study of the Accuracy of Code and Process Metrics for Defect Prediction of Mobile Applications".
2. Jureczko, Marian, and Lech Madeyski. "A review of process metrics in defect prediction studies." Metody Informatyki Stosowanej 5 (2011): 133–145.
3. Madeyski, Lech, and Marian Jureczko. "Which process metrics can significantly improve defect prediction models? An empirical study." *Software Quality Journal* (2014): 1–30.
4. D'Ambros, Marco, Michele Lanza, and Romain Robbes. "An extensive comparison of bug prediction approaches." *Mining Software Repositories (MSR), 2010 7th IEEE Working Conference on.* IEEE, 2010.
5. Jureczko, Marian, and Diomidis Spinellis. "Using object-oriented design metrics to predict software defects." Models and Methods of System Dependability. Oficyna Wydawnicza Politechniki Wrocławskiej (2010): 69–81.
6. O'Keeffe, Mark, and Mel O. Cinnéide. "Search-based software maintenance." *Software Maintenance and Reengineering, 2006. CSMR 2006. Proceedings of the 10th European Conference on.* IEEE, 2006: pp. 10.
7. Yogesh Singh, Arvinder Kaur, and Ruchika Malhotra. "Empirical validation of object-oriented metrics for predicting fault proneness models." *Software quality journal* 18.1 (2010): 3–35.
8. Graves, T. L., Karr, A. F., Marron, J. S., Siy, H. 2000. Predicting fault incidence using software change history. IEEE Transactions on Software Engineering, 26(7): 653–661 (July 2000).
9. Jureczko, Marian. "Significance of different software metrics in defect prediction." Software Engineering: An International Journal 1.1 (2011): 86–95.
10. Malhotra, Ruchika, Nakul Pritam, and Yogesh Singh. "On the applicability of evolutionary computation for software defect prediction." *Advances in Computing, Communications and Informatics (ICACCI, 2014 International Conference on.* IEEE, 2014, 2249–2257.
11. Moser, Raimund, Witold Pedrycz, and Giancarlo Succi. "A comparative analysis of the efficiency of change metrics and static code attributes for defect prediction." *Software Engineering, 2008,* 181–190.
12. Malhotra, Ruchika, and Rajeev Raje. "An empirical comparison of machine learning techniques for software defect prediction." *Proceedings of the 8th International Conference on Bioinspired Information and Communications Technologies.* ICST (Institute for Computer Sciences, Social-Informatics and Telecommunications Engineering), 2014, 320–327.
13. Catal, Cagatay, and Banu Diri. "A systematic review of software fault prediction studies." *Expert systems with applications* 36.4 (2009): 7346–7354.
14. Shihab, Emad, et al. "Understanding the impact of code and process metrics on post-release defects: a case study on the eclipse project." *Proceedings of the 2010 ACM-IEEE International Symposium on Empirical Software Engineering and Measurement.* ACM, 2010, p. 4.

15. Rahman, Foyzur, and Premkumar Devanbu. "How, and why, process metrics are better." *Proceedings of the 2013 International Conference on Software Engineering.* IEEE Press, 2013, pp. 432–441.
16. Wahyudin, Dindin, et al. "Defect Prediction using Combined Product and Project Metrics-A Case Study from the Open Source" Apache" MyFaces Project Family." *Software Engineering and Advanced Applications, 2008. SEAA'08. 34th Euromicro Conference.* IEEE, 2008, pp 207–215.
17. Johari, Kalpana, and Arvinder Kaur. "Validation of object oriented metrics using open source software system: an empirical study." ACM SIGSOFT Software Engineering Notes 37.1 (2012): 1–4.
18. S. Chidamber, and C. Kemerer, "A metrics suite for object oriented design", IEEE Transactions on Software Engineering, vol. 20, no. 6, pp. 476-493, 1994.
19. ckjm—Chidamber and Kemerer Java Metrics -http://www.spinellis.gr/sw/ckjm/.
20. https://www.openhub.net/.

Automatic Insurance and Pollution Challan Generator System in India

Harleen, Shweta Rana and Naveen Garg

Abstract In today's time there is paper wastage of issuing the pollution and insurance certificate. Traffic police used to check the documents which cause traffic jam on the road. This paper proposes use of Information technology to save the usage of paper; corruption and it reduce possibility of misuse of power by proper monitoring by introducing smart chips in the number plates of the vehicles which will provide necessary details required for monitoring. Everything will be done automatically just by fetching the number plate of the vehicle this will reduce the problem of traffic jam as well as consumption of paper.

Keywords Insurance · Pollution · Challan · AIPCS

1 Introduction

The traffic police were established as early as 1929 in the British Raj in some states of the country. The Traffic Police are a specialized unit of the Police responsible for overseeing and enforcing traffic safety compliance on city roads as well as managing the flow of traffic. As time passed, government changed its system and converted the paper challan system into the electronic challan system.

Traditional challan system that is spot challan, in this system challan is issued on the spot if there is any traffic rule violation [1].

E-challan introduced in 2012, it is an electronic format of the challan. E-Challans are taken by photo evidence only. It is an electronic receipt for challan payment.

Harleen (✉) · S. Rana · N. Garg
Department of Information Technology, Amity University, Noida, India
e-mail: harleenaujla18@gmail.com

S. Rana
e-mail: shweta11rana@gmail.com

N. Garg
e-mail: ngarg1@amity.edu

© Springer India 2016
S.C. Satapathy et al. (eds.), *Information Systems Design and Intelligent Applications*, Advances in Intelligent Systems and Computing 433,
DOI 10.1007/978-81-322-2755-7_11

This challan can be paid directly by cash, at any E-seva Centre or by any other payment mode as specified on the challan [1].

1.1 Handheld Machine [2]

In this system, the entire challan—right from driver's name, driving license number, nature of offence, section of the Motor Vehicle Act invoked and the fine amount—is written manually and consumes a lot of time. With the help of machine, in small period of time more cases could be booked and the waiting time for the offender would get reduced drastically (Fig. 1).

1.2 Manual System [2]

In manual system, when the challan is issued, the fine is paid on the spot or over a period of time, but the previous record is not available on that time means instantly.

Traffic violations will hereafter be booked by Multi-Purpose Handheld System (VIOLET) dedicated for Spot Fining System. The software runs on any android-based tablet or cellular phone and is integrated to a Bluetooth printer to dispatch receipts [3]. These include details in which the device was used, at what location the fine has been collected, details of the overall fine collected per day, information about which officer had used the system. These could be forwarded or printed out for official records.

In our system we basically focus on the insurance and pollution challan issued by the traffic police. In the traditional challan system, the major problem is that we don't have any records of the previous challans paid or not, paper wastage, unable to prove the vehicle is being stolen, corruption, misuse of power and consume a lot of time. E-challan works on GPRS/GSM. In our proposed system, we can establish

(a)   **(b)**

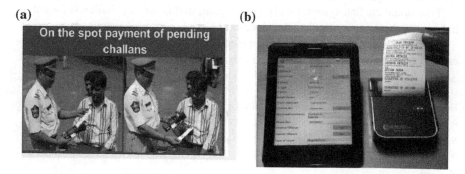

Fig. 1 **a** Spot Printout of Pending E-challan against your vehicle [1]. **b** A hand-held machine to issues e-challan [2]

a website where we can check the history, if the vehicle has any traffic offences registered against it. Traffic police using these sites know about the details of fine, user charges, when the challan is issued and vehicle is not having the updated pollution as well as an insurance check. It helps the traffic police to maintain a database of traffic violations and keep track of habitual offenders. Insurance and pollution registration certificates are mandatory to carry. The limit of insurance is 1 year and the pollution is almost for 3 months after that we renew these two certificates. We establish software in which we can check all the details online. When the vehicle is registered then all the details are submitted. This website will give details of offence description, fine amount, user charges, and the total fine amount.

It will provide you with the details of your challan which you can pay easily. This will obviously reduce corruption as every challan will have an entry and audit can be made to verify details.

2   Challenges Faced by Current System of Issuing Challan

There are some serious challenges faced by the police department in enforcing traffic laws which are being discussed below:

2.1 **Corruption** This is the most serious issue faced by the police as the enforcing law officer takes bribe and does not discharge his duty honestly. It encourages the serial offenders as they feel like they can get away easily after violating traffic rules which results in heavy casualties on the road.

2.2 **No Record of Previous Violations at the Go** The law enforcement officer on the field having direct contact with the public do not have any real time data relating to the previous violations committed by the offender. This problem handicaps the law enforcement officer in evaluating the seriousness of the violations causing unnecessary leniency in some cases.

2.3 **Misuse of Power** The law enforcement officers misuse their powers to gain undue financial favors from the violators to let them go which causes various serious problems in the nearby future and loss of human life.

3   Proposed System

The Camera Placed on roadside will detect the number plate of Car. From Car number plate, it will fetch all information of car i.e. Pollution_Expiry_date, Insurance_Expiry_date, Insurance company and other required details. These dates will be compared from Today_date automatically and Challan will be generated. A Challan will be generated from this system, if expiry_date is less than today_date. Amount of Challan will be taken by the system and Challan can be sent to registered address which has been fetched from Car number plate. In this process, no manual system is required. For Automation of Challan for Insurance and Pollution, proposed system require following steps:

3.1 Automatic License Plate Reader

First step is to detect car number plate. An Automatic License Plate Reader (ALPR) technique is available for scanning number plates and collecting toll tax via this technique. ALPR is an image processing technology used to identify vehicles by these license plates. It is a special form of OCR (Optical Character Recognition) where algorithms are employed to transform the pixels of the digital image into the text of the number plate. Systems commonly use infrared lightning to allow the camera to take the picture at any time of day.

This technique is already available of scanning number plates and collecting toll tax. AIPCS (Automatic Insurance Pollution Challan System) will detect car number plate and fetch the details from registration number of car (Figs. 2 and 3).

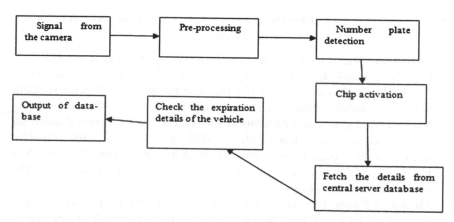

Fig. 2 How the automatic generator system work

Fig. 3 Automatic license plate recognition [4]

3.1.1   Details Behind Car Number Plate

Every Car has its registration number which is pasted on front and rear side of car. This registration number will have following details which will be helpful in generating Challan:

Car Registration number, Name, Address, Mobile number, mail id of Car Owner, Pollution details and Insurance details.

Insurance Details: Insurance policy number, Number of Insurance company, Expiry date and Type of Insurance.
Pollution details: Pollution_Id, Pollution_Expiry_date and Region of Pollution.

After scanning of car number plate, administrator has all details. So, next step is to organize data of Pollution and Insurance.

3.2   Pollution

It is mandatory for the owner of the vehicle to carry the pollution control certificate. Pollution check is prepared in the pollution booth Centre; these booths are available at many petrol pumps. Computerized facilities are provided for checking the pollution level and issues of pollution control certificate. Pollution under control certificate contain each information regarding vehicle. If the pollution check is not issued regarding that vehicle then it is against the rules related to the traffic police. It is most common violation which goes undetected as it does not have proper database/records. Till now, after pollution check, we receive hard copy of pollution certificate which we have to produce at the time of checking by Police.

The proposed system will generate online pollution certificate which we will be saved at administrator database and customer will receive confirmation of Pollution check on email id and mobile number via message. By using this technique, no hard copy of pollution certificate is required which will save paper.

The details of the pollution check of the vehicles are recorded to the system and uploaded to the central server directly such as vehicle registration number, model, category, date of mfg., fuel, date, time, valid up to, and address from where pollution check is done.

This data will be saved in two tables i.e. Customer_table and Administrator_table. Customer_table and administrator_table will have details as shown in Tables 1 and 2 respectively.

Table 1 Customer_table of pollution

Vehicle registration number	Vehicle company	Category (W)	Date of mfg.	Fuel	Current date	Time	Valid up to
DL2CAP5408	Maruti	4	19/Jul/2012	Petrol	19/Jul/2015	12:16:58 PM	18/Oct/2015
DL2CAP4509	Hyundai	4	16/Mar/2013	Petrol	16/Mar/2015	13:45:56 PM	15/June/2015
DL2CAP6798	Honda	4	13/Apr/2015	Diesel	13/May/2015	11:34:45 AM	12/Aug/2015

Vehicle registration number	Category (W)	Expiry_Date
DL2CAP5408	4	18/Oct/2015
DL2CAP6798	4	12/Aug/2015
DL2CAP4509	4	15/Jun/2015

Table 2 Administrator_table of pollution

Customer_table will help to check customer profile at instant

3.3 Insurance

Insurance paper is prepared by the Insurance Companies such as National Insurance Company Limited, HDFC, and Reliance etc. Insurance Certificate includes each information of vehicle.

If the vehicle does not contain the Insurance Certificate then it is against the rules related to the traffic police. It is the violation which is done by the vehicle owner due to their negligence. Information related to all the vehicles by all the insurance companies will be uploaded on the central server database which will be accessible for the enforcement agency at all times. Fields such as Vehicle Registration Number and Place of Registration, Policy number, Issue Date, Period of insurance and Name of the Company and Vehicle Owner address are uploaded to the central server by the company.

A central database is created which is shown in Tables 3 and 4.

Each pollution booth and various insurance companies are connected to the central server, when the camera rays fall on the chip which is embedded in the number plate, and then it fetches each information related to that vehicle. Then information is passed on to the administration. Through the administration table, it will check all the details related to the vehicle. Administrator_table check the current date and expiry date of the vehicle for pollution as well as for insurance. If expiry date is greater than the current date then it issues the challan and send the letter to the address of the vehicle owner and alternatively it alert the customer about the pollution check and insurance challan through messages. If there is issued any challan related to pollution and insurance send message to the registered mobile number, tell the customer about where the challan is taken out, on which date and time.

3.4 Expiry Check Table

Vehicle registration number is unique number for each vehicle. Insurance_Expiry_date and Pollution_Expiry_date are checked with respect to the present_date. Pollution check and Insurance check are carried out, if it is expired, or not.

Table 3 Customer_table for insurance

Vehicle registration number	Policy number	Issue date	Start date	Expiry date	Model of vehicle	Company name	Address of vehicle holder	Type of insurance version
DL2CAP5408	351010311461354375 96	18-Sep-2014	27-Sep-2014	26-SEP-2015	ALTO	Maruti	H. no-34, Rohini, Delhi	Third party insurance
DL2CAP4509	586429631257452586 14	20-Nov-2014	24-Nov-2014	23-Nov-2015	I10	Hyundai	B-55, Kriti Nagar, Delhi	First party insurance
DL2CAP6798	354896231578412521 22	16-Jan-2015	19-Jan-2014	18-Jan-2015	Swift	Maruti	H. no-65, D block, Noida	Third party insurance

Table 4 Administartor_table for insurance

Vehicle registration number	Issue date	Start date	Expiry date
DL2CAP5408	18-Sep-2014	27-Sep-2014	26-Sep-2015
DL2CAP4509	20-Nov-2014	24-Nov-2014	23-Nov-2015
DL2CAP6798	16-Jan-2015	19-Jan-2014	18-Jan-2015

3.5 Challan Table

After comparison of Current_date with Pollution_Expiry_Date and Insurance_Expiry_Date, we received the list of all vechicles whose pollution or Insurance has been expired. As discussed in Table 5 that Pollution of vehicle number DL2CAP4509 and Insurance of Vehicle number DL2CAP6798 has been expired. So, Administrator will get table of all violated vehicles along with their expiry date as shown in Table 6.

Now, software will generate the challan related to that policy and send it to the particular address, it will ease the workload and reduce the wastage of paper. Help the traffic police maintain a database of traffic violations and keep track of habitual offenders. By using this approach, traffic police catch the frequent offenders in a more efficient and structured manner (Fig. 4).

There are different possibility of challan either for pollution or insurance.

1. If the pollution check and the insurance check date are less than the current date, then there will be no case of challan.
2. If the pollution check is greater than the current date and the insurance check is less than the current date, then the challan will be issued only for the pollution not for insurance.
3. If the pollution check is less than current date but insurance check greater than current date, then the challan will be issued only for the insurance.
4. If the pollution check and the insurance check are greater than the current date, then challan will be issued for both.

Challan Letter is generated for the owner of the vehicle if he/she doesn't have their updated pollution as well as insurance check (Fig. 5).

Vehicle owner can submit the challan by any mode of payment and administrator will confirm the submission of challan payment.

Table 5 Expiry_check table

Vehicle registration number	Insurance expiry date	Pollution expiry date	Current date	Insurance expired	Pollution expired
DL2CAP5408	26/Sep/2015	18/Oct/2015	20/Jul/2015	No	No
DL2CAP4509	23/Nov/2015	15/June/2015	20/Jul/2015	No	Yes
DL2CAP6798	18-Jan-2015	12/Aug/2015	20-Jul-2015	Yes	No

Table 6 Challan status enquiry table

Vehicle registration number	Address of vehicle holder	Pollution expired	Pollution_expiry_date	Insurance expired	Insurance_expiry_date	Place of challan
DL2CAP4509	H. no-20, B block, Golf course, Noida	Yes	15/June/2015	No	–	Shalimar Bagh
DL2CAP6	H. no-65, D block, Noida	No	–	Yes	18-Jan-2015	Pari Chowk, Noida

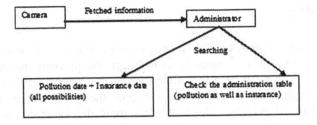

Fig. 4 Function performed at administrator end

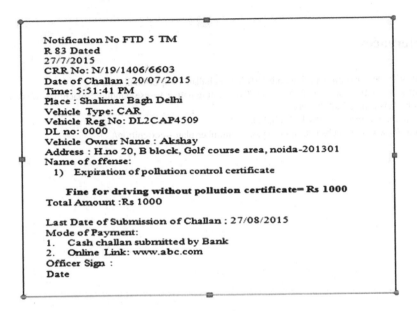

Fig. 5 Challan format

4 Advantage

4.1 **Transparency and Accountability** The proposed system will definitely bring transparency in the system as all the transactions will be made through use of technology which will lead to more accuracy in processing resulting into higher accountability.

4.2 **Save Paper** In the modern times every effort should be made to protect the environment and use of technology that leads to protection of trees and forests by being paperless should be encouraged.

4.3 **Information for Pollution and Insurance Renewal** A message and email will be sent to registered mobile number and mail id respectively before 10 days of expiry of Insurance and pollution. By this process, every person will take care of renewal of Insurance and pollution.

5   Conclusion

The system which is mentioned above once implemented will help to obey the rule and regulation developed by ministry of transport and prevents the surroundings from pollution. There can be a few difficulties in this automated system such as initial cost of implementing the whole system. Once the system is setup and work in progress the cost will decrease. This will help in managing the information of every vehicle and manage the pollution and insurance details of every vehicle and provide the transparency in the system.

References

1. http://hyderabad.trafficpolice.co.in/2014/01/echallan-vs-spot-challan.html.
2. http://www.thehindu.com/news/cities/Coimbatore/police-to-use-handheld-machines-to-issue-echallans/article5209064.ece.
3. http://en.wikipedia.org/wiki/E-challan.
4. http://www.neurallabs.net/en/ocr-systems/number-plate-recognition/.

Effectively Implementation of KNN-Search on Multidimensional Data Using Quadtree

B.D. Jitkar, S.D. Raut and S.T. Patil

Abstract In modern systems like location based network systems, radio frequency identification based network systems where we use different techniques to collect or capture the data. This represents them in position with the Longitude and Latitude which we call as Multidimensional Data. So study of multidimensional data is attracting more attention in research field. In this paper we are focusing mainly on effective searching for nearest neighbor search. Data is collected using some type of capturing/scanning device. This captured region in divided into quad format two times and an indexing is applied for fast and accurate searching of objects. By considering different facts we propose a new technique of nearest neighbor for effective cost model. Comprehensive experiments shows that quad region technique gives better performance in nearest neighbor search on multidimensional objects.

Keywords Multidimensional data · Quadtree · KNN search

1 Introduction

Computer Technology is growing at a pace very fast in all the directions like hardware, software, networking, support system. Because of which computer is playing very major role in every business. In every business data plays a major and important role for further improvement in the business. As number of days passes, data goes on increasing. This huge data is stored using many advanced techniques.

B.D. Jitkar (✉) · S.T. Patil
Department of Computer Science & Engineering, D.Y. Patil COET, Kolhapur, India
e-mail: bjitkar@rediffmail.com

S.T. Patil
e-mail: patilsangram@hotmal.com

S.D. Raut
Department of Computer Science & Engineering, Orchid COE, Solapur, India
e-mail: suhasraut@gmail.com

© Springer India 2016

S.C. Satapathy et al. (eds.), *Information Systems Design and Intelligent Applications*, Advances in Intelligent Systems and Computing 433,
DOI 10.1007/978-81-322-2755-7_12

111

This data can be collected either personally or by appointing agency or through any sensing device. This indicates that we can collect data through multiple ways which is required for business enhancement or working or processing. The collected data is single dimensional data as it can be identified by only one attribute of record. The applications which are location based collects the data using some type of sensing devices or radio frequency devices [1]. This data contains data with its location. So for identifying the data with attribute we also need its location which raises the need for multidimensional data management system or applications. Many real time applications such as environment monitoring, transportation system, sensor network, medical diagnosis etc. have multidimensional data [2]. These all applications either scans or sensors or generates data at frightening rate. It is very much important, in such applications where the explosive growth of information or data is taking place, how we store the huge data, and access the information or data which are most important part for user or application. There are many data base management systems which provide efficient way to store the data. But for the fast result it is not only important to store the data but also effectively retrieve the data at faster rate. This can be possibly partially achieved by using fast storage device. If data goes on increasing it might be possible that query takes more time for execution or display required result. Indexing is one of the most effective and ever-present tool for reducing query execution time in traditional database system. All database systems support variety of index structures like B-Tree, hash indexing, R-Trees [1]. Different database system use different index structure that works on single dimensional data. For multidimensional data we have designed effective indexing structure for storing as well as fast and accurate retrieving of the data.

The queries like point query, range query, top-k query and probabilistic nearest neighbors query can be executed on multidimensional data. Searching of nearest neighbor or range search is main problem in many applications like location based services (LBS), global position system (GPS), and sensor data analysis [1]. Recursively dividing region into Quad format (two times) is the new technology which can be used for effective and accurate searching. This is very flexible technology and here we adaptively build collections of the objects so that overall cost of the range query, nearest neighbor search can be minimized regarding the cost model using Quad region format. For range query and nearest neighbor search on uncertain objects this index structure will help effectively for accurate and fast retrieval of data.

2 Related Work

In many places or applications Radio frequency identification (RFID) networks, Location Based System or sensor networks are widely used [3]. The nearest neighbor and range searching is fundamental problem in these types of applications. In many location based system location of the objects with their details are obtained by sensor devices or some type of readers. Resent research involves filtering and verification paradigm which evolves filtering cost in terms of I/O or CPU cost [4]. Considering this cost an effective index structure must be used for effectively indexing the objects in the database to get fast and accurate response for nearest neighbor and range query for multidimensional objects. This indexing must be in such a way that range query and nearest neighbor query must work effectively and accurately [5].

3 Proposed Work

There are exactly four leaf nodes for every internal node of a Quadtree structure. In this structure a two-dimensional space is portioned recursively by subdividing it into four quadrants or regions two times [5]. This structure is a novel index structure which can be used to effectively organize the multidimensional objects n quad region format. Also arbitrary PDFs are also calculated for every region using formula (Fig. 1)

$$\frac{\sum Cell\_Objects}{\sum Total\_Objects}$$

where every divided region is called as cell.

Quadtree can support nearest neighbor query, range query, uncertain range query where the search region is multidimensional [5]. Comprehensive experiments show the efficiency of the Quadtree technique in many types of queries such as uncertain range query and K Nearest Neighbor query with different factors like number of records, different theta values, different K values etc.

Fig. 1 System architecture

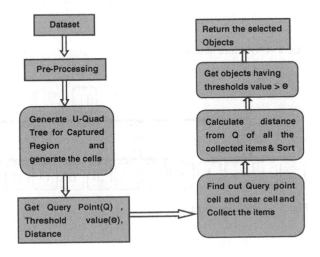

For effectively and flexibly processing of multidimensional objects a new technique is constructed which significantly improve the performance of the K Nearest Neighbor. In this new technique filtering and selecting multidimensional object is applied only on selected objects from selected cells/regions instead of going through all the multidimensional objects. Therefore the overall cost is minimized.

3.1 Quad Tree

A quad tree is a region partitioning tree data structure in which a region is recursively subdivided into 2 sub regions (cells). Many applications widely use this quad tree technique because of its simplicity and regularity. In this paper, we focus on two-dimensional region and all techniques suggested are immediately applied on given region/spaces. In our work, we say a Quadtree is optimal as the cost model if is minimized [5]. In using quad tree, focus is on efficient construction of Quadtree to reduce/minimize the overall cost of the nearest neighbor and range search query. The region is divided into 16 cells and is numbered using Hilbert code as shown in Fig. 2 (Fig. 3).

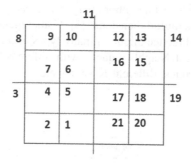

Fig. 2 U-Quadtree structure numbering

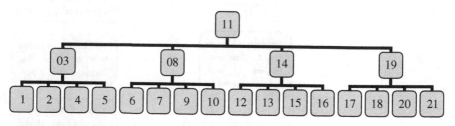

Fig. 3 U-Quadtree structure

3.2 Range Query

In this type of query instead of searching in all the regions or cells we get the lower and upper bounds followed by filtering-and-verification of objects. Also PDFs are calculated for every cell for selected region. This technique extracts all nonempty cells which are contained or overlapped by range query. This gives the lower and upper bounds of the appearance probabilities of the objects and the cost to compute lower and upper bounds of the appearance probabilities. Since the selected region of search is restricted to select bounded area rather than full area the cost is minimized.

3.3 K Nearest Neighbor Search

Here a new KNN algorithm based on the Quadtree is used. The essential idea is to identify the promising objects based on the summaries on different factors like threshold and distance of the objects from query point Q. The computational cost is still expensive if simply we try to calculate the distance using formula (1) of each object from query point Q.

For example consider following Fig. 4.

We are considering Q as Query Point. Then from Q we are calculating distance of every object (p) using formula

$$Distance = ABS\sqrt{(Q.x - p.x)^2 + (Q.y - p.y)^2} \tag{1}$$

After calculating distance using above formula, rearrange the objects according to distance of p from Q. as shown below (Fig. 5).

Fig. 4 KNN search

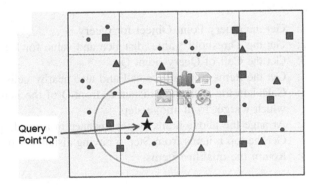

Query Point "Q"

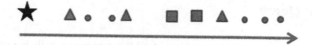

Fig. 5 Distance wise arranged objects

With distance with threshold value per cell is also mentioned for searching criteria if required. To calculate the threshold value of objects of same type or every cell formula is used as

$$\frac{\sum Cell\_Objects}{\sum Total\_Objects} \tag{2}$$

Now from arranged objects select TOP K objects as per requirement and return the objects.

In the following, a new KNN algorithm is considered which incrementally extends the search region from the query point Q to prune objects based on their summaries following the collecting and filtering paradigm.

In algorithm, 3rd step gets the cell to which query point Q belongs to. Then all objects including current cell and nearby cells are collected together in 4th step. In 5th step PDFs and distance is calculated of all the objects collected in previous step and those objects are arranged in ascending order on distance n 6th step. In 7th step we selected the required objects which are returned in 8th step.

Algorithm for KNN Search

Input: UQ: the UQuad tree set region from U sets.

Q: Query Point
θ: Threshold Value
k: Number of items
d: distance for range

Output: Set of items with expected threshold and within distance from query point Q.

1: Get the Query Point Object for Query
2: Get the Threshold Value, distance and value for k for Search
3: Get the Cell of Query point Q.
4: Get the items from above cell and also nearby cells.
5: Calculate the PDFs and distances from Q of the items
    which we get from above step.
6: Arrange the above items as per distance in ascending order.
7: Get the top k items from Step 6 having distance < d and if required PDF >= θ.
8: Return the qualified items.

Fig. 6 Time complexity for
different THETA (Θ) values

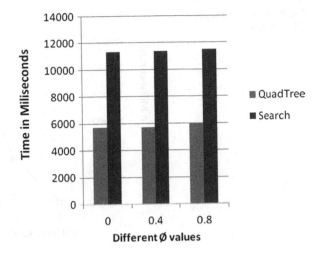

Fig. 7 Time complexity for
different number of records

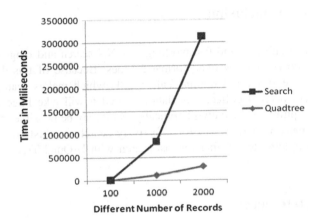

The cost required is comparatively less because we are searching into limited region instead of full region.

The experimental results are shown below for different values of K, theta (Θ), records (Fig. 6).

Above experimental result shown for different theta values which are processed for regular search and Quadtree structure search (Fig. 7).

Above experimental result shown for different number of records and same theta, K, distance values which are processed for regular search and Quadtree structure search (Fig. 8).

Above experimental result shown for different number of K values and same number of records, theta, distance values which are processed for regular search and U-Quadtree structure search.

Fig. 8 Time complexity for different K values

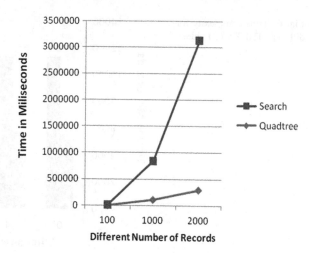

4 Conclusion

For effective and fast working of KNN search and range query a region is divided into recursively quad format 2 times. Because of this divide format searching is done in limited regions called cells which makes searching accurate and faster. However if collectively data is stored it will take huge memory space as data is captured reclusively periodically. So it may be possible that U-Quad tree technique many take time. So for effectively searching clustering of data can be possibly implemented further for fast search with U-Quad Tree.

References

1. Chih-Wu Chung, Ching-Hung Pan, Chuan-Ming Liu, *"An Effective Index for Uncertain Data"*, 2014, IEEE DOI 10.1109/IS3C.2014.132.
2. Rui Zhu, Bin Wang, Guoren Wang, *"Indexing Uncertain Data for Supporting Range Queries"*, WAIM 2014, LNCS 8484, pp.72–83, Springer 2014.
3. James Bornholt, Todd Mytkowicz, Kathryn S. McKinley, *"A First-Order Type for Uncertain Data"*, ACM 978-4503-2305-5/14/03, March 2014.
4. Navya E. K., M. Madhan Kumar, *"Processing Uncertain Database Using U-Skyline Mechanism"*, ISSN Vol. 2, Issue 4, April 2014.
5. Ying Zhang, Wenjie Zhang, Qianlu Lin, Xuemin Lin, *"Effectiveli Indexing the Multidimensional Uncertain Objects"*, IEEE transactions on knowledge and data engineering, Vol 26, No. 3, March 2014.

Edge Detectors Based Telegraph Total Variational Model for Image Filtering

Subit K. Jain and Rajendra K. Ray

Abstract For the existing issues of edge blur and uncertainty of parameter selection during image filtering, a novel telegraph total variational PDE model based on edge detector is proposed. We propose image structure tensor as an edge detector to control smoothing process and keep more detail features. The proposed model takes advantages of both telegraph and total variational model, which is edge preserving and robust to noise. Experimental results illustrate the effectiveness of the proposed model and demonstrate that our algorithm competes favorably with state of the-art approaches in terms of producing better denoising results.

Keywords Image denoising · Telegraph equation · Variational model · Partial differential equations (PDEs)

1 Introduction

From last two decades, nonlinear partial differential equations (PDEs) based methods have been widely employed for image denoising and edge detection in digital images, due to its well established theory in literature [1–5]. Most of the nonlinear diffusion models are generalized from the following model, in which the diffusion is controlled by a variable coefficient [6].

$$\begin{cases} I_t & = \nabla(c(|\nabla I|^2)\nabla I) & in\ \Omega \times (0, +\infty) \\ \frac{\partial I}{\partial n} & = 0 & in\ \partial\Omega \times (0, +\infty) \\ I(x, y, 0) & = I_0(x, y) & in\ \Omega \end{cases} \tag{1}$$

S.K. Jain (✉) · R.K. Ray
Indian Institute of Technology, Mandi, India
e-mail: jain.subit@gmail.com

R.K. Ray
e-mail: rajendra.ray@gmail.com

© Springer India 2016
S.C. Satapathy et al. (eds.), *Information Systems Design and Intelligent Applications*, Advances in Intelligent Systems and Computing 433,
DOI 10.1007/978-81-322-2755-7_13

119

Subsequently, Rudin, Osher and Fatemi proposed a total variational model which smooths the original image and obtains the resultant image by minimizing the energy function [7].

To our best knowledge, parabolic equations acquire the center stage in the field of image denoising. But, the hyperbolic PDEs could improve the quality of the detected edges and so enhance the image better than parabolic PDEs [8]. In their model (the TDE model), proposed by Ratner and Zeevi [8], the image is viewed as an elastic sheet placed in a damping environment,

$$\left\{ \begin{array}{lll} I_{tt} + \gamma I_t & = \nabla(c(|\nabla I|^2)\nabla I) & in \; \Omega \times (0, +\infty) \\ \frac{\partial I}{\partial n} & = 0 & in \; \partial\Omega \times (0, +\infty) \\ I(x,y,0) & = I_0(x,y), \quad I_t(x,y,0) = 0 & in \; \Omega \end{array} \right\} \qquad (2)$$

where c and γ are the elasticity and damping coefficients, respectively. Note that the TDE model is a hyperbolic equation which interpolates between the diffusion equation and the wave equation. It has been proved that using this family of equations enables better edge preservation and enhancement when compared with diffusion-based approaches. For different elasticity coefficients, the TDE model can be treated as the improved versions of the corresponding nonlinear diffusion models.

Inspired by the work of Catté et al. [9], Cao et al. [10] presented the following improved telegraph diffusion (ITDE) model applied to image restoration,

$$I_{tt} + \gamma I_t = \nabla(c(|\nabla I_\sigma|^2)\nabla I) \quad in \; \Omega \times (0, +\infty) \qquad (3)$$

With $I_\sigma = G_\sigma * I : I_\sigma$ is the smoothed version of image I convolved by a Gaussian smoothing kernel G_σ.

The variational methods are proven tools to stabilize the PDE and makes it less dependent on the number of iterations [4]. Inspired by the ideas of [7, 8, 11], we propose a novel telegraph total variational model using the telegraph equation and total variational model along with an edge detector function. We first propose the Telegraph Total Variation (TTV) model and then we add an edge detector function in the propose model which is named here as edge detector function based Telegraph Total Variation (ETTV) model. The paper is organized as follows: The Sect. 2 deals with the proposed model. Section 3 shows an appropriate numerical realization. Numerical Experiments are carried out and studied in Sect. 4. Finally, the paper is concluded in Sect. 5.

2 Telegraph Total Variational Model

Noise removal and feature preservation are the main task of image filtering. Hence, image denoising process consists two major objectives, (1) Minimization of high frequency components and (2) Enhance and preserve the features of image.

Motivated from the work of [7, 8], it is natural to investigate a model inherits the advantages of the Telegraph equation and Total variation model. Our proposed Telegraph Total Variation (TTV) model as follows:

$$\left\{\begin{array}{ll} I_{tt} + \gamma I_t & = \nabla\left(\frac{\nabla I}{|\nabla I|}\right) - \lambda(I - I_0) & \text{in } \Omega \times (0, +\infty) \\ \frac{\partial I}{\partial n} & = 0 & \text{in } \partial\Omega \times (0, +\infty) \\ I(x, y, 0) & = I_0(x, y), \quad I_t(x, y, 0) = 0 & \text{in } \Omega \end{array}\right\} \quad (4)$$

Using the information obtained by magnitude of the gradient, the diffusion function lower the diffusion near the edges. This model may not be good enough in the case of edge preservation when there is a loss of edge connectivity. Hence, to increase the edge preservation, we introduce a structure tensor function [12] in the proposed scheme (ETTV model), which yields

$$\left\{\begin{array}{ll} I_{tt} + \gamma I_t & = \nabla\left(\alpha(I)\frac{\nabla I}{|\nabla I|}\right) - \lambda(I - I_0) & \text{in } \Omega \times (0, +\infty) \\ \frac{\partial I}{\partial n} & = 0 & \text{in } \partial\Omega \times (0, +\infty) \\ I(x, y, 0) & = I_0(x, y), \quad I_t(x, y, 0) = 0 & \text{in } \Omega \end{array}\right\} \quad (5)$$

where, $\alpha(I) = 1 - F(I)$ provides a pixel wise edge characterization using the image local structure measure function, $F(I) = \delta_1 * \delta_2 + |\delta_1 - \delta_2|^2$, where δ_1 and δ_2 are the eigenvalues of Hessian matrix of the image I.

3   Numerical Scheme

To solve the proposed model, here we construct an explicit numerical method. The explicit schemes are commonly used in the literature and considered as the simplest option but the shortcoming with this is that it needs small time steps for stability [13, 14]. In this work, we use an explicit scheme with small time step to validate our model.

Let h represents the spatial step grid size and τ is the time step size. Denote $I_{i,j}^n = I(x_i, y_j, t_n)$ where $x_i = ih, y_j = jh$ and $t_n = n\tau$. Since the diffusion term is approximated by central differences, we use the following notations,

$$\frac{\partial I}{\partial t} = \frac{I_{i,j}^{n+1} - I_{i,j}^n}{\tau}, \frac{\partial^2 I}{\partial t^2} = \frac{I_{i,j}^{n+1} - 2I_{i,j}^n + I_{i,j}^{n-1}}{\tau^2},$$

$$\frac{\partial I}{\partial x} = \frac{I_{i+h,j}^n - I_{i-h,j}^n}{2h}, \frac{\partial I}{\partial y} = \frac{I_{i,j+h}^n - I_{i,j-h}^n}{2h},$$

$$\frac{\partial^2 I}{\partial x^2} = \frac{I_{i+h,j}^n - 2I_{i,j}^n + I_{i-h,j}^n}{h^2}, \frac{\partial^2 I}{\partial y^2} = \frac{I_{i,j+h}^n - 2I_{i,j}^n + I_{i,j-h}^n}{h^2},$$

Hence, we can write the proposed model as the discrete form as follows:

$$(1+\gamma\tau)I_{i,j}^{n+1} - (2+\gamma\tau)I_{i,j}^{n} + I_{i,j}^{n-1} = \tau^2\left(\nabla(\alpha(I_{i,j}^{n})\frac{\nabla I_{i,j}^{n}}{|\nabla I_{i,j}^{n}|}) - \lambda(I_{i,j}^{n} - I_0)\right)$$

$$\Rightarrow (1+\gamma\tau)I_{i,j}^{n+1} = (2+\gamma\tau)I_{i,j}^{n} - I_{i,j}^{n-1} + \tau^2\left(\nabla(\alpha(I_{i,j}^{n})\frac{\nabla I_{i,j}^{n}}{|\nabla I_{i,j}^{n}|}) - \lambda(I_{i,j}^{n} - I_0)\right),$$

where image local structure measure function can be calculated as,

$$F(I_{i,j}^{n}) = \delta_1^n * \delta_2^n + |\delta_1^n - \delta_2^n|^2.$$

4 Experimental Results

In this section, we evaluate the experimental results obtained by applying both the proposed models (Eqs. 4 and 5) on some standard test images. The performance of the proposed models are compared with those obtained by PM model [6], TDE model [8] and ITDE model [10].

We use the Mean Structural Similarity Index (MSSIM) [15] and Peak Signal to Noise Ratio (PSNR) [16], to evaluate the performance of our models. The formulas used for calculating PSNR and SSIM are as follows,

$$\text{SSIM}(x,y) = \frac{(2\mu_x\mu_y + c_1)(2\sigma_{xy} + c_2)}{(\mu_x^2 + \mu_y^2 + c_1)(\sigma_x^2 + \sigma_y^2 + c_2)}$$

$$\text{PSNR} = 10\log_{10}\left(\frac{\max(I_0)^2}{\frac{1}{rc}\sum_{i-1}^{r}\sum_{j=1}^{c}(I_t - I_0)^2}\right).$$

Here $\mu_x, \mu_y, \sigma_x^2, \sigma_y^2, \sigma_{xy}$ is the average, variance and covariance of x and y, respectively. c_1 and c_2 are the variables to stabilize the division with weak denominator. For both quality measures, a high value suggests that the filtered image is closer to the noise free image. Also for objectively evaluate the performance of edge detector, we have used Figure of Merit (FOM) [17] parameter. Apart from this, some other approaches have also been proposed for defining the performance of edge detector [18].

For experiment, we have used the set of standard test images of size 256×256. Images are degraded with white Gaussian noise of zero mean and different standard deviations (e.g. $\sigma \in (20, 40)$). In these experiments we have considered the parameters values as follows: the time step size $\Delta t = 0.25$, the grid size $h = 1$ and threshold parameter $k = 10$.

Figure 1a, b show the noise free "Lena" image and its noisy image with Gaussian noise of mean 0.0 and s.d. 20, respectively. Figure 1c–g depict the enhanced images by using PM, TDE, ITDE, proposed TTV and ETTV model,

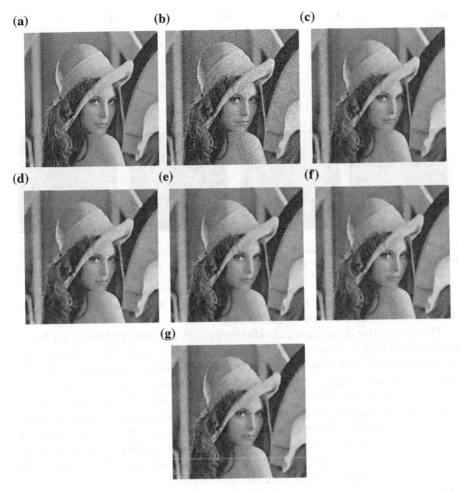

Fig. 1 **a** Original Lena image. **b** Lena image with Gaussian noise of mean 0.0 and s.d. 20, Denoising of image using **c** PM model. **d** TDE model. **e** ITDE model. **f** TTV model. **g** ETTV model

respectively. We observed that the proposed ETTV model has an impressive smoothing effect which is visually comparable to the existing methods along with proposed TTV model, in restoring and edge-preservation. It confirms the usefulness of the telegraph model combined with TV model as a better denoising model, along with structure tensor to detect the true edges more efficiently.

Similarly, Fig. 2 shows the denoising performance of the proposed methods on a "House" image, degraded with white Gaussian noise of zero mean and $\sigma = 40$. Figure 2a–d show the denoised images and corresponding edge maps using different approaches discussed above. All these studies confirm that our proposed

(a)   (b)   (c)   (d)

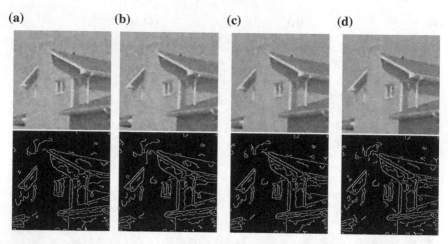

Fig. 2 Denoised House image and corresponding edge map using **a** TDE model, FOM = 0.5079. **b** ITDE model, FOM = 0.5160. **c** TTV model, FOM = 0.5266. **d** ETTV model, FOM = 0.5358

ETTV model can achieve both image enhancement as well as noise reduction, better than any other models discussed here.

The quantitative comparisons of different models, in terms of MSSIM and PSNR values, for different noise levels are shown in Table 1. From this table, one can observe that proposed TTV model produces better results than existing PM, TPM and ITPM models, whereas proposed ETTV model produces best results among all the models in all the cases. Also, proposed ETTV model efficiently preserves the finer details as compared to the standard TDE model as it inherits the advantage of the standard diffusion method and interpolates between the parabolic and the hyperbolic PDE models. So, it can better preserve and describe the textures of the image. That is why, the proposed ETTV model obtains higher MSSIM and PSNR than the other methods, for the different noise levels.

5 Conclusion

In this paper, we proposed couple of new models, i.e., TTV and ETTV models, for better preservation of edges and small details in an image, based on the combination of telegraph equation and variational model along with a structure tensor based edge detector. Numerous numerical studies are carried out with different noise levels to validate our models. Our numerical studies reveal that both TTV and ETTV models are better than the existing PM, TPM and ITPM models, in general, and ETTV model is better than the TTV model, in particulate, for denoising ability in terms of visual quality and quantitative quality measures (e.g. MSSIM and PSNR).

Table 1 PSNR and MSSIM of the test images for different restorations

Images	Measure	σ = 20					σ = 30					σ = 40				
		PM model [6]	TPM model [8]	ITPM model [10]	TTV model	ETTV model	PM model [6]	TPM model [8]	ITPM model [10]	TTV model	ETTV model	PM model [6]	TPM model [8]	ITPM model [10]	TTV model	ETTV model
Lena	MSSIM	0.8566	0.8569	0.8569	0.8590	**0.8607**	0.8107	0.8110	0.8110	0.8136	**0.8156**	0.7790	0.7792	0.7801	0.7825	**0.7848**
	PSNR	29.79	29.80	29.81	29.91	**29.97**	27.96	27.97	27.97	28.05	**28.10**	26.82	26.83	26.94	26.95	**27.00**
Barb	MSSIM	0.7908	0.7921	0.7921	0.7977	**0.7991**	0.7418	0.7423	0.7427	0.7469	**0.7494**	0.7029	0.7033	0.7069	0.7082	**0.7105**
	PSNR	27.26	27.30	27.30	27.51	**27.55**	25.79	25.80	25.87	25.98	**26.02**	24.82	24.83	24.96	24.98	**25.01**
Cameraman	MSSIM	0.8202	0.8211	0.8211	0.8303	**0.8325**	0.7718	0.7723	0.7723	0.7766	**0.7800**	0.7331	0.7334	0.7335	0.7341	**0.7389**
	PSNR	28.50	28.56	28.56	28.48	**28.70**	26.42	26.45	26.45	26.33	**26.53**	24.90	24.92	24.92	24.77	**24.96**
House	MSSIM	0.7901	0.7908	0.7908	0.8023	**0.8038**	0.7573	0.7576	0.7578	0.7704	**0.7726**	0.7344	0.7347	0.7353	0.7485	**0.7506**
	PSNR	31.05	31.06	31.07	31.29	**31.38**	29.42	29.42	29.44	29.63	**29.72**	28.29	28.30	28.41	28.55	**28.62**

6 Future Work

In future, we will compare the proposed system with some statistical approaches (e.g. Bayesian model) along with time complexity and computational complexity of proposed system. Also to improve convergence speed of proposed models, Advanced numerical schemes and better algorithms can be used. This will be a good future scope of the present work.

References

1. Alvarez, L., Guichard, F., Lions, P.L., Morel, J.M.: Axioms and fundamental equations of image processing. Archive for rational mechanics and analysis **123**(3) (1993) 199–257.
2. Weickert, J.: Anisotropic diffusion in image processing. Volume 1. Teubner Stuttgart (1998).
3. Gonzalez, R.C., Woods, R.E.: Digital image processing (2002).
4. Aubert, G., Kornprobst, P.: Mathematical problems in image processing: partial differential equations and the calculus of variations. Volume 147. Springer (2006).
5. Nolen, J.: Partial differential equations and diffusion processes. Technical report, Technical report, Stanford University. Department of Mathematics (2009).
6. Perona, P., Malik, J.: Scale-space and edge detection using anisotropic diffusion. Pattern Analysis and Machine Intelligence, IEEE Transactions on **12**(7) (1990) 629–639.
7. Rudin, L.I., Osher, S., Fatemi, E.: Nonlinear total variation based noise removal algorithms. Physica D: Nonlinear Phenomena **60**(1) (1992) 259–268.
8. Ratner, V., Zeevi, Y.Y.: Image enhancement using elastic manifolds. In: Image Analysis and Processing, 2007. ICIAP 2007. 14th International Conference on, IEEE (2007) 769–774.
9. Catté, F., Lions, P.L., Morel, J.M., Coll, T.: Image selective smoothing and edge detection by nonlinear diffusion. SIAM Journal on Numerical analysis **29**(1) (1992) 182–193.
10. Cao, Y., Yin, J., Liu, Q., Li, M.: A class of nonlinear parabolic-hyperbolic equations applied to image restoration. Nonlinear Analysis: Real World Applications **11**(1) (2010) 253–261.
11. Prasath, V.S., Singh, A.: Edge detectors based anisotropic diffusion for enhancement of digital images. In: Computer Vision, Graphics & Image Processing, 2008. ICVGIP'08. Sixth Indian Conference on, IEEE (2008) 33–38.
12. Li, C., Liu, C., Wang, Y.: Texture image denoising algorithm based on structure tensor and total variation. In: Intelligent Networking and Collaborative Systems (INCoS), 2013 5th International Conference on, IEEE (2013) 685–690.
13. Thomas, J.W.: Numerical partial differential equations: finite difference methods. Volume 22. Springer (1995).
14. Jain, S.K., Ray, R.K., Bhavsar, A.: Iterative solvers for image denoising with diffusion models: A comparative study. Computers & Mathematics with Applications **70**(3) (2015) 191–211.
15. Wang, Z., Bovik, A.C., Sheikh, H.R., Simoncelli, E.P.: Image quality assessment: from error visibility to structural similarity. Image Processing, IEEE Transactions on **13**(4) (2004) 600–612.
16. Wang, Z., Bovik, A.C.: Mean squared error: love it or leave it? a new look at signal fidelity measures. Signal Processing Magazine, IEEE **26**(1) (2009) 98–117.
17. Abdou, I.E., Pratt, W.K.: Quantitative design and evaluation of enhancement/thresholding edge detectors. Proceedings of the IEEE **67**(5) (1979) 753–763.
18. Bhateja, V., Devi, S.: A reconstruction based measure for assessment of mammogram edge-maps. In: Proceedings of the International Conference on Frontiers of Intelligent Computing: Theory and Applications (FICTA), Springer (2013) 741–746.

Cloud Based K-Means Clustering Running as a MapReduce Job for Big Data Healthcare Analytics Using Apache Mahout

Sreekanth Rallapalli, R.R. Gondkar
and Golajapu Venu Madhava Rao

Abstract Increase in data volume and need for analytics has led towards innovation of big data. To speed up the query responses models like NoSQL has emerged. Virtualized platforms using commodity hardware and implementing Hadoop on it helps small and midsized companies use cloud environment. This will help organizations to decrease the cost for data processing and analytics. As health care generating volumes and variety of data it is required to build parallel algorithms that can support petabytes of data using hadoop and MapReduce parallel processing. K-means clustering is one of the methods for parallel algorithm. In order to build an accurate system large data sets need to be considered. Memory requirement increases with large data sets and algorithms become slow. Mahout scalable algorithms developed works better with huge data sets and improve the performance of the system. Mahout is an open source and can be used to solve problems arising with huge data sets. This paper proposes cloud based K-means clustering running as a MapReduce job. We use health care data on cloud for clustering. We then compare the results with various measures to conclude the best measure to find number of vectors in a given cluster.

Keywords Big data · Clustering · Hadoop · K-means · Mahout · NoSQL

S. Rallapalli (✉)
R&D Centre, Bharathiyar University, Coimbatore, India
e-mail: rsreekanth1@yahoo.com

R.R. Gondkar
AIT, Bangalore, India
e-mail: rrgondkar@gmail.com

G.V. Madhava Rao
Botho University, Gabarone, Botswana
e-mail: venumadhavaraog@gmail.com

© Springer India 2016
S.C. Satapathy et al. (eds.), *Information Systems Design and Intelligent Applications*, Advances in Intelligent Systems and Computing 433,
DOI 10.1007/978-81-322-2755-7_14

127

1 Introduction

Clustering algorithms are used to solve the most of cluster related issues which are focused for small data sets. With increase in volume of data over the years, data sets now a day are very huge and cannot be accommodated into the memory. There are many efficient clustering algorithms [1] in operational which can solve the data intensive problems. The most commonly used algorithm for clustering is K-means. It is simple to solve the problems. In order to solve the large data set problems, technologies based on open source are useful. Mahout on Hadoop [2] platform can be promising for solving large data set problems. Cloud [3] based K-means clustering with mahout shows good results when run on Amazon EC2 [4]. Further sections in this paper focus on various aspects of clustering algorithm and structured as follow: Sect. 2 presents some fundamental aspects of clustering and various algorithms for Mahout. Section 3 discuss about Proposed Mahout K-means clustering algorithm on cloud for healthcare data. Section 4 discuss about MapReduce job for K-means clustering. Section 5 presents the results and analysis. Conclusion is presented in Sect. 6.

2 Fundamental Aspects

2.1 Clustering in Mahout

A collection of an algorithm, a notion of similarity and dissimilarity, a condition to stop are the things which are required for clustering. Similar items can be grouped together for certain groups. Clustering is all about grouping the items near to particular set. Implementation of clustering is discussed as many ways in Mahout. K-means, fuzzy K-means and canopy clustering are few implementations. For health care records analysis we use K-means clustering. Hadoop library [5] consists of binary file format. Mahout clustering also takes input as binary format. Create an input file and then follow the three steps to input to algorithm. Once the input is ready we can use the clustering K-means algorithm.

2.2 Clustering Algorithms in Mahout

There are many clustering algorithms which work for the data sets. But few algorithms do not fit for the large data sets. K-means is one of the fundamental algorithm which works for all types of data sets. The other algorithms are fuzzy K-means, canopy clustering, dirichlet and latent dirichlet analysis.

K-means algorithm initially starts with a set of k centroid points [6]. The algorithm process in multiple rounds until maximum limit of set criterion is reached. In mahout to run a clustering data we need to implement KMeansClusterer class or KMeansDriver class. In-memory clustering of data is done by KMeansClusterer. To implement K-means as a MapReduce job KMeansDriver act as an initial point. These methods can run on Apache Hadoop cluster.

Canopy clustering has an ability to quickly generate clusters. Optimal number of clusters k, can be specified which is required by K-means. This algorithm can be executed using CanopyClusterer or CanopyDriver class [6].

Fuzzy K-means algorithm generates overlapping clusters from the data set [7]. It discovers the soft clusters. The classes used for this algorithm is FuzzyKMeans-Clusterer and FuzzyKMeansDriver.

Dirichlet algorithm is based on Bayesian clustering algorithm [6]. This algorithm start with a set of data points called as Model Distribution. In memory clustering is based on DirichletClusterer class and DirichletDriver as a MapReduce job.

Latent Dirichlet algorithm is a model of dirichlet clustering. This algorithm reads all the data in a mapper phase in parallel and finds the probability of each occurrence in the document [6]. In memory implementation in this algorithm is not available, where as MapReduce implementation can be achieved by LDADriver class.

Cloud based K-means clustering shows best performance for computation in a datacenter on DISA Virtual Machine (DVM) [8]. Experiments conducted on different data size and number of nodes proved that for K-means clustering. Mahout K-means clustering works for large data sets and is scalable. Consistency is maintained by mahout K-means clustering between several experiments conducted [4]. Clustering patient health records via sparse subspace representation is given in [9].

3 Proposed Mahout K-Means Clustering Algorithm on Cloud for Healthcare Data

In this section we take a set of health care data and discuss how Mahout K-means clustering can be promising. Table 1 shows the code and description for various Ill health's. Table 2 shows the patient id and their ill health data. We then take a data set for running K-means clustering using the class.

Table 1 Code description for ill health

Ill health code	Description of ill health
HBP	High blood pressure
DHBP	Diabetic with high blood pressure
DHBPHS	Diabetic, high blood pressure with heart stroke
DINS	Diabetic with insulin dependent
DKD	Diabetic with kidney related disorder

Table 2 Patient ID and their ill health data set

Patient ID	Ill health code data set
P111	HBP, DHBP
P112	HBP, DHBP, DHBPHS
P113	HBP, DKD
P114	DINS, DHBPHS
P115	DHBP, DKD

In the first stage K-means clustering start with random data as the centroid (Fig. 1). The map stage assigns each point to the cluster near it (Fig. 2).

As discussed in Sect. 2.2 K-means clustering algorithm in-memory clustering of the data is achieved by KMeansClusterer. The algorithm given below will take the points in vector format around the given data of health codes. This program can execute on apache hadoop cluster on cloud.

Fig. 1 K-means clustering with random points

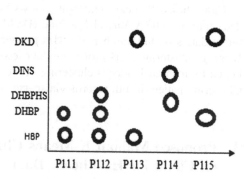

Fig. 2 Map stage

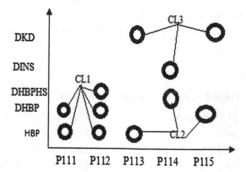

Algorithm 1 Computing K-means for healthcare data

Input: Data is in List<vector> format.
Output: K-means for patient data

---

1. Begin with n clusters, each containing one object and we will number the clusters 1 through n.

2. Compute the between-cluster distance D(p,q) as the between-object distance of the two objects in p and q respectively, p, q =1, 2, ..., n. Let the square matrix D = (D(p, q)). If the objects are represented by quantitative vectors we can use Euclidean distance.

3. Next, like pairs of p and q clusters is identified, such that the distance, D(p, q), is minimum among all the pair wise distances.

4. Join p and q to a new cluster t and compute the between-cluster distance D(t, k) for any existing cluster $k \neq p, q$. Once the distances are obtained, corresponding rows and columns are eliminated to the old cluster p and q in the D matrix, because p and q do not exist anymore. Then insert a row and column in D which matches with cluster t.

5. Repeat Step 3 a total of n − 1 times until there is only one cluster left.

---

4   MapReduce Job for K-Means Clustering

In Mahout KMeansDriver class can be instantiated for MapReduce [10]. The RunJob method is the entry point to MapReduce Job. The input parameters for the algorithm are the hadoop configuration, the sequence file which contains the input vectors. Initial cluster centers are contained in sequence file. The K-means clustering for MapReduce job can be given as

KmeansDriver.runJob(hadoopconf,InputVectorFilesDirPath,
ClusterCenterFilesDirPath,   OutPutDir,   new   EuclideanDistanceMeasure(),
ConvergenceThreshold, numIterations, True,False)

Figure 3 shows K-means clustering for MapReduce [11] Job. The input file is taken as sequence file containing vectors. The data chunks from each vector are then sent to parallel mappers. The mappers then send the data to reducer's stage where the data is reduced to all points of clusters from mapper.

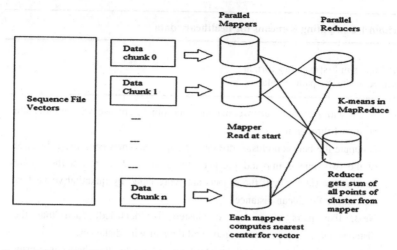

Fig. 3 MapReduce for K-means clustering

5 Results and Analysis

In this section we find out the distance measure between two health care data codes
using various methods. The results between two points are then compared for the
number of iterations each method takes to do the K-means algorithm processing.
The Euclidean distance measure between the points in the health care data graph is
given by

$$D = \sqrt{x_1 - y_1)^2 + (x_2 - y_2)^2 + \cdots + (x_n - y_n)^2} \tag{1}$$

where x_1, y_1 are the points to be considered in Health data codes.

The squared Euclidean distance measures is given by

$$D = (x_1 - y_1)^2 + (x_2 - y_2)^2 + \cdots + (x_n - y_n)^2 \tag{2}$$

The Manhattan distance measure is given by the equation and shown in Fig. 4.

$$D = |x_1 - y_1| + |x_2 - y_2| + \cdots + |x_n - y_n| \tag{3}$$

The cosine distance measure is given by

$$D = 1 - \frac{(x_1y_1 + x_2y_2 + \cdots + x_ny_n)}{(\sqrt{(x_1^2 + x_2^2 + \cdots + x_n^2)}(\sqrt{(y_1^2 + y_2^2 + \cdots + y_n^2)})} \tag{4}$$

Tanimoto distance measure is given by

Fig. 4 Distance measure between two healthcare records

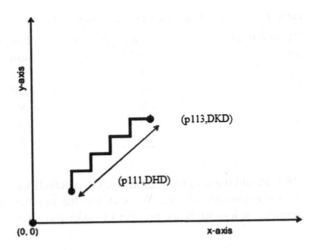

$$D = 1$$

$$-\frac{(x_1y_1 + x_2y_2 + \cdots + x_ny_n)}{\sqrt{(x_1^2 + x_2^2 + \cdots + x_n^2)} + \sqrt{(y_1^2 + y_2^2 + \cdots + y_n^2)} - (x_1y_1 + x_2y_2 + \cdots + x_ny_n)}$$

$$(5)$$

A Weighted Distance Measure class allows giving weights to different dimensions in order to increase or decrease the effect of a dimension.

The cosine distance measure takes one iteration whereas squared Euclidean distance measure takes five iterations and remaining measure takes three iterations for the algorithm processing.

Cluster 1: HBP, DHBP, DHBPHS—Patients with High Blood Pressure and Diabetes are prone to Heart strokes.

Cluster 2: HBP, DHBP, DINS—Patients with High Blood Pressure and Diabetes also will be Insulin Dependent in future.

Cluster 3: DINS, DKD—Patients with Diabetes and Insulin based are likely to get any type of Kidney disorders in near future.

5.1 Experimental Results

For the proposed algorithm we intended to take patient data set which contains 500,000 records of patients with more than 100 different parameters. These records are then supplied to cloud based hadoop cluster running k-means as MapReduce job. The tests are performed on Amazon Elastic Cloud Computing (EC2). The results obtained are mentioned in the Table 3. The block size of the HDFS was

Table 3 k-means as MapReduce Job on Multinode clusters

Patient data set file (%)	Number of nodes	K-means (s)	Iterations
10	2	40.0	50
30	5	35.0	30
50	10	25.2	20
70	20	19.5	10
90	30	10.5	05
100	40	5.0	01

128 MB, and the data replication is set to a default value of 3. The k-means can be run on multimode cluster. We can see the number of iterations taken by the algorithm is less when the number of nodes increases.

6 Conclusion

In this paper we have discussed how Mahout can be a promising tool for algorithm of K-means. It is also a good tool for visualizing the algorithm. DisplayKMeans class shows cluster movement position after every iteration. KMeansClusterer method is the best method for clustering the health records. In-memory clustering using K-means works with large data sets which have many records. Number of hadoop clusters need to be mentioned while running. The MapReduce algorithm helps the large data sets to run on multiple machines with each mapper getting a subset of the points. Mapper job will compute the cluster which is near to it by reading the input points. K-means clustering MapReduce algorithm is designed to run on Hadoop cluster.

References

1. T Kanungo, D. Mount, N. Netanyahu, C. Piatko, R. Silverman, and A. Wu, "An efficient K-means clustering algorithm: Analysis and implementation", Pattern Analysis and Machine Intelligence, IEEE Transactions, Vol 24, No 7, pp. 881–892, 2002.
2. White, T: Hadoop the definitive guide, O'Reilly Media, 2009.
3. Fredrik Farnstorm, J: Scalability for clustering algorithms revisited—SIGKDD Explorations, 2002, 2, pp. 51–57.
4. Rui Maximo Esteves, Chunming Rong, Rui Pais: K-means clustering in the cloud—a Mahout test, IEEE 2011 Workshops of international conference on Advanced information networking and application, pp. 514–519.
5. http://hadoop.apache.org/docs/r2.7.0/hadoop-project-dist/hadoopcommon/NativeLibraries. html.
6. Jain, A.K. and R.C Dubes, 1998: Algorithms for Clustering Data, Prentince Hall, New Jersy.

7. Dweepna Garg, Kushboo Trivedi, Fuzzy k-mean clustering in MapReduce on cloud based Hadoop, 2014 IEEE International Conference on Advanced Communication Control and Computing Technologies (ICACCCT).
8. Lin Gu, Zhonghua sheng, Zhiqiang Ma, Xiang Gao, Charles Zhang, Yaohui Jin: K Means of cloud computing: MapReduce, DVM, and windows Azure, Fourth International Conference on Cloud Computing, GRIDs, and Virtualization (cloud computing 2013). May 27–June 1, 2013, Valencia, Spain.
9. Budhaditya Saha, Dinh Phung, Duc-son Pham, Svetha Venkatesh, Clustering Patient Medical Records via sparse subspace representation from http://link.springer.com/chapter/10.1007/978-3-642-37456-2_11.
10. Sean Owen, Robin Anil, Ted Dunning, Ellen Friedman, Mahout in Action by Manning Shelter Island.
11. J. Dean and S. Ghemawat, "MapReduce simplified data processing on large clusters", In Proc. Of the 6th Symposium on OS design and implementation (OSDI'04), Berkely, CA, USA, 2004, pp. 137–149.

Symbolic Decision Tree for Interval Data—An Approach Towards Predictive Streaming and Rendering of 3D Models

V. Vani and S. Mohan

Abstract 3D content streaming and rendering system has attracted a significant attention from both academia and industry. However, these systems struggle to provide comparable quality to that of locally stored and rendered 3D data. Since the rendered 3D content on to the client machine is controlled by the users, their interactions have a strong impact on the performance of 3D content streaming and rendering system. Thus, considering user behaviours in these systems could bring significant performance improvements. To achieve this, we propose a symbolic decision tree that captures all attributes that are part of user interactions. The symbolic decision trees are built by pre-processing the attribute values gathered when the user interacts with the 3D dynamic object. We validate our constructed symbolic tree through another set of interactions over the 3D dynamic object by the same user. The validation shows that our symbolic decision tree model can learn the user interactions and is able to predict several interactions with very limited set of summarized symbolic interval data and thus could help in optimizing the 3D content streaming and rendering system to achieve better performance.

Keywords Symbolic data analysis · Symbolic decision tree · 3D streaming · Predictive models · User interactions · 3D rendering · Decision tree

V. Vani (✉) · S. Mohan
College of Computer and Information Systems, Al Yamamah University,
Riyadh, Kingdom of Saudi Arabia
e-mail: v_vani@yu.edu.sa

S. Mohan
e-mail: m_gounder@yu.edu.sa

© Springer India 2016
S.C. Satapathy et al. (eds.), *Information Systems Design and Intelligent Applications*, Advances in Intelligent Systems and Computing 433,
DOI 10.1007/978-81-322-2755-7_15

137

1 Introduction

In recent times, 3D modeling and rendering has gained attention over the internet and most of the multiuser virtual environment renders the entire world once it is fully downloaded from the server [1]. Therefore, to get the first response from the server, the clients/users ought to wait till the requested region of the model is completely downloaded with all necessary refinements to get rendered. Due to the increased complexity of the 3D model, even with the high bandwidth user has to wait for a longer time to view the model. To reduce the waiting time of the user, prediction based 3D streaming and rendering [2–5] technique is made available to the users. On the other hand, for the last few decades, the Decision tree based predictors [6] are considered to be one of the most popular approaches for representing predictors or classifiers. Researchers from various disciplines have used this decision tree for possible data classification and prediction. In this paper, instead of considering the large set of user interaction data available data as such, it is proposed to convert those data with necessary preprocessing to symbolic data with the summarized data set. Further, these summarized data sets can be used to construct the symbolic decision tree [7–10]. Also, an attempt is made to consider Symbolic Decision Tree for necessary classification and prediction of the profiled user interactions. The user interaction profiles collected offline helps us to conduct experiments and estimate the misclassification rate and prediction accuracy.

In the proposed work, the user interaction patterns are profiled by fixing the number of camera views to six to construct predictive models based on Symbolic Decision Tree. The considered six views are Top, Bottom, Left, Right, Front and Back with respect to 3D object. These six viewpoints are considered to be sufficient to visualize the projected object from different angles. However, in addition to the camera views, the eye position (x, y, z), eye orientation (x, y, z), object position (x, y, z), current key press as well as next key press are recorded to train the predictors. The user interaction patterns with the above mentioned parameters are profiled by considering different users across various 3D models with static or dynamic behaviors. The predictors are trained based on the collated user interaction patterns. These trained predictors are used to predict the probable next key press, and thus to stream and render the appropriate portions of 3D models more efficiently to enrich the user experience.

2 Decision Tree

One of the classical soft computing approaches proposed by Han et al. [6] is a decision tree classifier which is referred in this paper to predict the next key press of the user during his/her interactions with the chosen 3D model. The decision tree classifier is a recursive partitioning algorithm which divides the nodes into sub nodes based on the discrete function. As a result of this division/partition, the leave node of the decision tree is assigned to one class that specifies most appropriate target value or the probability value.

Figure 1a describes a sample CART [6] decision tree that reasons the next user key press. Given this classifier, the analyst can predict the response of a user (by sorting it down the tree), and understand the behavioral characteristics of the users while interacting with the 3D objects. In Fig. 1a, there are 11 nodes and each node is labeled with its corresponding values. Leaf nodes represent the possible 6 target classes ranging from 1 to 6 which represent the next key press and CK represents the Current Key and UZ represents camera orientation (Up Vector: UX, UY, UZ) in the Z direction. Figure 1b gives the corresponding decision tree rules for classification/prediction.

3 Proposed Work

3.1 An Approach to Construct Symbolic Decision Tree

Input: Profiled User Interactions of Dynamic 3D Cow Model Training Data Set (93 tuples)
   Process:

1. Extract the user interaction data set and normalize the continuous data to fall in the range between -1 and $+1$ using Min-Max Normalization [6] algorithm.
2. Partition the tuples in the data set based on the Current Key(CK), View Point (VP) and Next Key (NK).

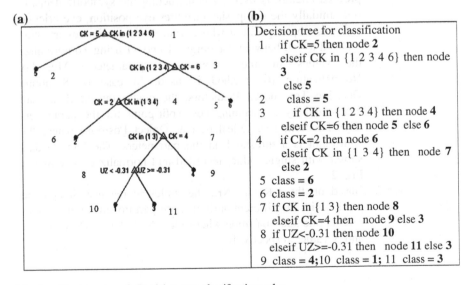

(a)

(b)

Decision tree for classification
1   if CK=5 then node **2**
    elseif CK in {1 2 3 4 6} then node **3**
    else **5**
2   class = **5**
3   if CK in {1 2 3 4} then node **4**
      elseif CK=6 then node **5** else **6**
4   if CK=2 then node **6**
    elseif CK in {1 3 4} then node **7**
      else **2**
5   class = **6**
6   class = **2**
7   if CK in {1 3} then node **8**
    elseif CK=4 then node **9** else **3**
8   if UZ<-0.31 then node **10**
    elseif UZ>=-0.31 then node **11** else **3**
9   class = **4**;10 class = **1**; 11 class = **3**

Fig. 1 **a** Decision tree. **b** Decision tree classification rules

3. Covert the normalized and partitioned data, into interval data with minimum and max value for each attribute and reduce the number of tuples and make it as symbolic data.
4. Construct the histogram using standard set of bins for all the interval data pertaining to the Partition ij to determine the splitting attribute
5. Compute the Information of each attribute using histogram distribution as well as for the partitioned data Dij
6. Compute Information Gain between Info(Dij) and InfoA(Dij)
7. Choose the attribute with maximum Information Gain as the splitting attribute
8. Based on the splitting attribute Divide the chosen partition Dij to reach the possible output class
9. If the possible output class is not reached or else number of tuples in the partition is more than 1 then repeat the steps 3 to 7
10. Repeat the steps 3 to 8 by choosing the next partition from the available 'n' partitions

Output: Constructed Symbolic Decision Tree with Set of Decision Rules for Prediction.

3.2 Illustration

The illustration of the step by step approach to construct the Symbolic Decision Tree is specified in this section.

Step 1: The user interaction over 3D dynamic cow model that was profiled already is used for constructing the symbolic decision tree. Initially the composite attributes eye position, eye orientation and object position whose values are continuous are normalized to fall between the range −1 and +1 using To normalize the values in these continuous parameters, Min-Max Normalization [6] method is used. In case of Symbolic Decision Tree based Predictors, the continuous attributes are discretized by determining best split-point for the parameters (Eye Position, Eye Orientation & Object Position) where the split-point is a threshold on the parameters. The Sample data subset with single values recorded and normalized is shown in Fig. 2.

Step 2: Based on the CK and NK, the tuples in the user interaction profile with 10 different attributes are partitioned. Figure 2 represents one such partition where the CK, VP and NK with 3, 1 and 3 values respectively.

Current Key	View Point	Eye Position			Eye Orientation			Object Pos	Next Key
		X	Y	Z	X	Y	Z	X	
0	1	3.7761	-13.15	4	0	0	1	0.77612	3
3	1	3.3579	-13.15	4.357	-0.099833	0	0.995	23.584	3
3	1	2.9062	-13.15	4.6705	-0.19867	0	0.98007	13.581	3
3	1	2.4254	-13.15	4.9373	-0.29552	0	0.95534	4.5788	3
3	1	1.9204	-13.15	5.1547	-0.38942	0	0.92106	-0.91464	3
3	1	1.3962	-13.15	5.3207	-0.47943	0	0.87758	-6.4229	3

Current Key	View Point	Eye Position			Eye Orientation			Object Pos	Next Key
		X	Y	Z	X	Y	Z	X	
0	1	0.17	-0.63	-0.31	0.06	0.22	1.00	0.00	3
3	1	0.15	-0.63	-0.27	-0.09	0.22	0.98	0.77	3
3	1	0.13	-0.63	-0.24	-0.23	0.22	0.90	0.44	3
3	1	0.11	-0.63	-0.21	-0.38	0.22	0.78	0.13	3
3	1	0.09	-0.63	-0.19	-0.52	0.22	0.61	-0.05	3
3	1	0.07	-0.63	0.17	-0.65	0.22	0.40	-0.24	3

Fig. 2 Un-normalized and normalized data set

Partitioned Tuples with Single Value Attributes									
Current Key	View Point	Eye Position			Eye Orientation			Object Pos	Next Key
		X	Y	Z	X	Y	Z	X	
3	1	0.15	-0.63	-0.27	-0.09	0.22	0.98	0.77	3
3	1	0.13	-0.63	-0.24	-0.23	0.22	0.90	0.44	3
3	1	0.11	-0.63	-0.21	-0.38	0.22	0.78	0.13	3
3	1	0.09	-0.63	-0.19	-0.52	0.22	0.61	-0.05	3
3	1	0.07	-0.63	-0.17	-0.65	0.22	0.40	-0.24	3

Partitioned Tuples with Interval Values represented by min and max																	Next Key
C K	VP	Eye Position						Eye Orientation						Object Pos			
		Xmin	Xmax	Ymin	Ymax	Zmin	Zmax	Xmin	Xmax	Ymin	Ymax	Zmin	Zmax	Xmin	Xmax		
3	1	0.07	0.15	0.63	0.63	0.27	0.17	0.65	0.09	0.22	0.22	0.4	0.98	-0.2	0.77		3

Fig. 3 Conversion of single data set to interval data

Step 3: The partitioned tuples with single data are converted into interval data by considering minimum and maximum values under each continuous attributes. Figure 3 shows normalized single data set converted into interval data. In this case, 5 tuples are reduced to 1 tuple with the interval data.

Step 4: The interval data are assigned to bins by using following algorithm to determine the splitting attribute based on the information gain [8]. Figure 4 shows the sample histogram distribution for the considered training set with 93 tuples.

Algorithm to distribute the partioned interval data into histogram bins.

1. For every interval data set in the partition, the appropriate bin is determined. In the below algorithm, let us assume eye position Xmin and Xmax interval data is chosen with 93 rows in it.

2. The number of bins is considered to be 10 and are initialized and transformed as follows $x = [0, 0, 0, 0, 0, 0, 0, 0, 0, 0]$; $x = x'$;

3. Total number of tuples is assigned with the value $n = 93$;

4. For every tuple in the set
   for $i = 1:n$
   bincount = (xmin(i)−xmax(i))/0.2; binpos = (xmin(i)−1)/0.2; binpos = binpos + 1;
   for $j = 0$:bincount x(binpos) = x(binpos) + 1; binpos = binpos + 1 end
   end

5. Bar graph with the updated bin value is displayed

Steps 5, 6 and 7: The $Info_A$ (D) and Info (D) are calculated to compute Gain (D) where A indicates information based on the Attribute and D indicates information for whole partition.

$$Gain(A) = Info\ (D) - InfoA\ (D) \tag{1}$$

The attribute with the maximum gain is chosen as the splitting attribute for the available data. From the given distribution in Fig. 4, Info $_{Eye\ Position\ X}$(D), Info $_{Eye\ Position\ Y}$(D), Info $_{Eye\ Position\ Z}$(D), Info $_{Eye\ Orientation\ X}$(D), Info $_{Eye\ Orientation\ Y}$(D), Info $_{Eye\ Orientation\ Z}$(D) and Info $_{Eye\ Position\ X}$(D) are computed along with Info(D), $Info_{CK}$ (D) and $Info_{VP}$ (D). From the Table 1, its observed that $Info_{VP}$ (D) has the maximum gain and it is considered as the splitting attribute to continue with the next iteration with the available data. The Steps 1 through 7 is repeated to construct the decision tree for the considered 3D dynamic cow model training data set with 93 tuples. Figure 5 shows the complete symbolic decision tree for the chosen model.

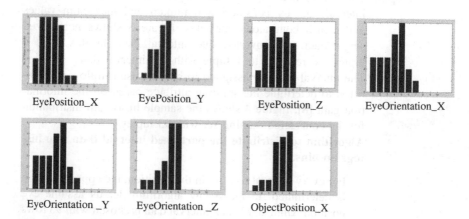

EyePosition_X   EyePosition_Y   EyePosition_Z   EyeOrientation_X

EyeOrientation _Y   EyeOrientation _Z   ObjectPosition_X

Fig. 4 Histogram distribution for interval data (3D dynamic cow model: training set)

Table 1 Information gain

Info		Gain
Info (D)	2.40855	
Info$_{CK}$ (D)	2.40855	0.00
Info$_{VP}$ (D)	1.32218	1.09
Info$_{Eye\ Position\ X}$ (D)	2.49197	−0.08
Info$_{Eye\ Position\ Y}$ (D)	2.08272	0.33
Info$_{Eye\ Position\ Z}$ (D)	2.24548	0.16
Info$_{Eye\ Orientation\ X}$ (D)	2.5013	−0.09
Info$_{Eye\ Orientation\ Y}$ (D)	2.44062	−0.03
Info$_{Eye\ Orientation\ Z}$ (D)	2.1635	0.25
Info$_{Object\ Position\ X}$ (D)	2.19909	0.21

Fig. 5 Symbolic decision tree for dynamic 3D cow model

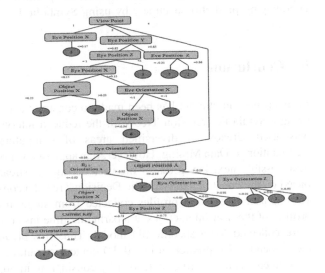

4 Results and Inferences

For the experimentation purpose, single user interactions over dynamic 3D cow model at two different instant of time is considered to generate training and testing data sets. Figure 5 shows the constructed symbolic decision tree for dynamic 3D cow model. There were 93 training tuples and 30 testing tuples in the original data set. These tuples are reduced to 22 and 13 respectively during the conversion process of single data to interval data. This indicates that when the number of interactions in the same direction is high, the reduction is the number of tuples will also be high. These 22 training tuples are considered to construct the symbolic decision tree as given in Fig. 5. In order to validate and estimate the prediction accuracy of the constructed symbolic decision tree, 13 testing tuples are considered. From the prediction result, it is found that the constructed symbolic decision tree could classify and predict 10 testing tuples correctly out of given 13 tuples. This

Table 2 Comparison of CART and symbolic decision tree performance

Model	Number of training tuple	Number of testing tuple	Predictor	Misclassification in %	Accuracy in %
3D Dynamic cow model	93	30	Classification and regression tree	60	40
	22	13	Symbolic decision tree	15.38	84.62

gives the misclassification rate of 15.38 %. Also, Table 2 shows the comparison of the misclassification and accuracy rate for the same dynamic 3D cow model across Classification and Regression Tree (CART) and Symbolic Decision Tree. For the considered 3D model user interaction profiles, we could achieve quite encouraging prediction accuracy by using Symbolic Decision Tree over CART.

5 Conclusion

In this paper, an attempt has been made to construct and validate the predictors by using a Symbolic Decision Tree with the reduced interval data set. This predictor was used already in diversified areas of computing ranging from Pattern Recognition to Data Mining. However, this approach is first attempted to predict the best possible next key press of the user during 3D streaming and rendering. It is based on the constructed Symbolic Decision Tree. To construct/train the predictors, various user interaction profiles collected from single user are considered. These profiles of the user interactions at a particular time instance or over a period of time were collected for a specific 3D model. Thus this prediction approaches can be recommended for prediction based 3D content streaming and rendering. Also, the experiments can be further extended by considering multiple user interactions over various static and dynamic 3D models and the prediction accuracy can be estimated.

Acknowledgments We thank the management of Al Yamamah University, KSA for supporting financially to publish our research work.

References

1. Blanc, A.L., Bunt, J., Petch, J. and Kwok, Y.: The virtual learning space: an interactive 3D environment. In Proceedings of the Tenth international Conference on 3D Web Technology, Web3D '05. ACM, New York, NY, pp. 93–102, (2005).
2. Vani, V., Pradeep Kumar, R. and Mohan, S.: Profiling User Interactions of 3D Complex Meshes for Predictive Streaming & Rendering. Lecture Notes in Electrical Engineering, Vol. 221(1), pp. 457–467, Springer ISSN 18761100, (2013).

3. Vani, V., Pradeep Kumar, R. and Mohan, S.: Predictive Modeling of User Interaction Patterns for 3D Mesh Streaming. International Journal of Information Technology and Web Engineering, Vol. 7(4), pp. 1–19, DOI:10.4018/jitwe.2012100101, (2012).
4. Vani, V., Mohan S.: Neural Network based Predictors for 3D content streaming and rendering. 18th ICSEC 2014, KKU, Thailand, IEEE pp. 452–457, ISBN 978-1-4799-4965-6, (2014).
5. Vani V., Pradeep Kumar R., Mohan S.: 3D Mesh Streaming based on Predictive Modeling, Journal of Computer Science 8 (7): 1123–1133, 2012, ISSN 1549-3636, Science Publications (2012).
6. Han, J., Kamber, M. and Pei, J.: Data Mining: Concepts and Techniques. Morgan Kaufmann Publishers Inc., p. 744, chapter 6, ISBN: 9780123814791, (2011).
7. Diday E.: An Introduction to Symbolic Data and Analysis and the SODAS software. Journal of Symbolic Data Analysis, Vol. 0, No. 0, ISSN 1723-5081, (2002).
8. Mballo, C. and Diday, E.: Decision trees on interval valued variables. Journal of Symbolic Data Analysis, Vol. 3, No. 1, ISSN 1723-5081, (2005).
9. Perner P., Belikova T.B., and Yashunskaya N.I.: Knowledge Acquisition by Decision Tree Induction for Interpretation of Digital Images in Radiology. Advances in Structural and Syntactical Pattern Recognition, P. Perner, P. Wang, and A. RosenFeld (Eds.), Springer Verlang LNC 1121, (1996).
10. Seck D., Billard L., Diday.E. and Afonso F.: A Decision Tree for Interval-valued Data with Modal Dependent Variable. COMSTAT, (2010).

Predictive 3D Content Streaming Based on Decision Tree Classifier Approach

S. Mohan and V. Vani

Abstract 3D content streaming and rendering system has attracted a significant attention from both academia and industry. However, these systems struggle to provide comparable quality to that of locally stored and rendered 3D data. Since the rendered 3D content on to the client machine is controlled by the users, their interactions have a strong impact on the performance of 3D content streaming and rendering system. Thus, considering user behaviours in these systems could bring significant performance improvements. Towards the end, we propose a decision tree that captures all parameters making part of user interactions. The decision trees are built from the information found while interacting with various types of 3D content by different set of users. In this, the 3D content could be static or dynamic 3D object/scene. We validate our model through another set of interactions over the 3D contents by same set of users. The validation shows that our model can learn the user interactions and is able to predict several interactions helping thus in optimizing these systems for better performance. We also propose various approaches based on traces collected from the same/different users to accelerate the learning process of the decision tree.

Keywords 3D streaming · Decision tree · Machine learning · Predictive model · Progressive meshing · User interactions profiling

S. Mohan (✉) · V. Vani
College of Computer and Information Systems, Al Yamamah University,
Riyadh, Kingdom of Saudi Arabia
e-mail: m_gounder@yu.edu.sa

V. Vani
e-mail: v_vani@yu.edu.sa

© Springer India 2016
S.C. Satapathy et al. (eds.), *Information Systems Design and Intelligent Applications*, Advances in Intelligent Systems and Computing 433,
DOI 10.1007/978-81-322-2755-7_16

147

1   Introduction

Over the past few years, constant research towards audio and video streaming standards have improved the user experience in accessing audio and video contents over the network [1]. This has been supplemented with various wireless technologies from 3G to 6G which has created a wide business opportunity for audio/video streaming like in YouTube. With current users adapted to broadband networks and 3D acceleration cards, delivery of huge 3D data streams could be made viable through a considerable amount of research on 3D streaming and rendering over the network. This also will ensure that there is no plug-ins required to be installed on the client side in contrast to existing application like Second Life. In 3D streaming, the streaming sequence is based on the visibility or interest area of each user, making it individually unique among the users. On the other hand in audio and video, streaming is sequential though we can forward and go to the specific frame. Frames are well organized and it is not view dependent [2, 3]. Enriching the user experience through 3D streaming over the network is evident as it would provide the initial response quickly to the client without the need for a complete download of the 3D data representing the 3D scene. 3D streaming is the process of delivering 3D content in real-time to the users over a network [4, 5]. 3D streaming is carried out in such a way that the interactivity and visual qualities of the content may match as closely as if they were stored locally.

This proposed study, attempts to create a client-server framework to stream exactly the visible portion of the 3D content that is being viewed by the user and render it appropriately on a client machine. This reduces remarkable amount of mesh transmissions from the server to the client because most of the time the users do not view the 3D scene in entirety. The user moves while interacting with 3D objects are profiled for necessary predictions. The well-known soft computing approach called Decision Tree based Predictors is chosen for necessary classification and prediction of the profiled user interactions. The user profiles collected offline helps us to determine the close match with the move and reduce the amount of data being brought from the server to the client [6–8].

2   An Approach to Discretize 3D Parameters

In the proposed work, an attempt is made to construct predictive models based on well proven soft computing approach namely Decision Tree based predictor. In order to construct a decision tree, the user interaction patterns are profiled by fixing the number of camera views to six. The considered six views are Top, Bottom, Left, Right, Front and Back with respect to 3D object. These six viewpoints are considered to be sufficient to visualize the projected object from different angles. However, in addition to the camera views, the eye position (x, y, z) EP, eye orientation (x, y, z) EO, object position (x, y, z) OP as well as current key press are recorded to train the predictor. The user interaction patterns with the above mentioned parameters are profiled by considering different users across various 3D models with static or dynamic behaviors. Also, progressive user interactions at different instant of time on the same 3D model are considered. The predictors are trained based on the collated user interaction patterns. These trained predictors are used to predict the probable next key press, and thus to stream and render the appropriate portions of 3D models more efficiently to enrich the user experience.

3 Approaches to Construct and Validate Predictors

To construct and validate the predictors, three approaches are considered in the proposed work based on User and 3D model perspectives. In these two perspectives, the single user-multiple 3D models and multi user-single 3D model approaches are considered to construct and validate the predictors. These two approaches along with the progressive single user-single 3D model approaches are fair enough to conclude whether the predicted move is 3D model dependent or user dependent.

Progressive Single User-Single 3D Model (PSU-S3DM): In this approach (Fig. 1), single 3D complex (e.g. "Armadillo") model is considered and user interactions at different instant of time on the model are profiled. Here the validation is centered on the user and the 3D model as well. It is basically a one to one relationship. This approach is used to predict the accurate path of a particular user while interacting with the model over a period of time. The prediction results after training the predictors are tested against another set of user interaction patterns collected from the same user while interacting with the same model. From the result of this approach, the pattern of a particular user on a 3D model can be predicted precisely.

Single User-Multiple 3D Models (SU-M3DM): In this approach (Fig. 2), single user interactions across multiple 3D models are profiled. It is used to predict how a specific user interacts with different 3D models. This leads us to the conclusion that whether the interaction patterns vary with respect to the 3D model or it is same across all the 3D models. The prediction results after training the predictors can be

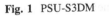 **Fig. 1** PSU-S3DM

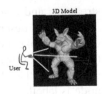

Fig. 2 SU-M3DM

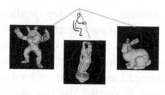

Fig. 3 MU-S3DM

validated against another set of interaction patterns collected from same user while interacting with a new 3D model. From the result of this approach, the pattern of a particular user on a new 3D model can be predicted to certain extent. This type of prediction may not be accurate as the user interaction patterns may sometimes be subjective (depends on the 3D model being viewed). Hence this results in a more user specific patterns.

Multiple Users-Single 3D Model (MU-S3DM): In this approach (Fig. 3), multiple user interaction patterns on a 3D model are profiled. It can be used to predict the possible user interaction pattern on a specific 3D model (e.g. "Armadillo"). The prediction results can be validated against a new user interaction patterns on the same 3D model. From the results of this approach, user interaction patterns on a specific 3D model can be derived more accurately. Hence this results in a more model specific patterns.

4 Decision Tree Based Predictors

The Decision Trees are considered to be one of the most popular approaches for representing classifiers. Researchers from various disciplines such as Statistics, Machine Learning (ML), Pattern Recognition (PR), and Data Mining (DM) have dealt with the issue of growing a decision tree from available data. For the first time, the decision tree based predictor is considered to construct predictive model for 3D streaming and rendering in the proposed work.

Figure 4a describes a sample decision tree that reasons the next key press the user might take. Given this classifier, the analyst can predict the response of a user (by sorting it down the tree), and understand the behavioural characteristics of the users about the 3D object interactions. In Fig. 4a, there are 11 nodes and each node is labelled with its corresponding values. Leaf nodes represent the possible 6

(a)

(b)

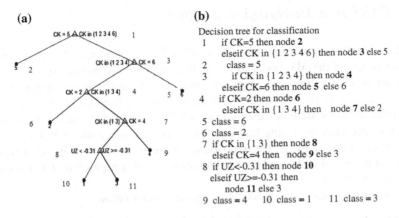

Decision tree for classification
1   if CK=5 then node **2**
     elseif CK in {1 2 3 4 6} then node **3** else 5
2   class = 5
3   if CK in {1 2 3 4} then node **4**
        elseif CK=6 then node **5** else 6
4   if CK=2 then node **6**
     elseif CK in {1 3 4} then   node **7** else 2
5   class = 6
6   class = 2
7   if CK in {1 3} then node **8**
    elseif CK=4 then   node **9** else 3
8   if UZ<0.31 then node **10**
    elseif UZ>=-0.31 then
        node **11** else 3
9   class = 4   10   class = 1   11   class = 3

Fig. 4 a Decision tree. **b** Decision tree classification rules

target classes ranging from 1 to 6 which represent the next key press and CK represents the Current Key and UZ represents camera orientation (Up Vector: UX, UY, UZ) in the Z direction. Figure 4b gives the corresponding decision tree rules for classification/prediction.

In Decision Tree base Predictors, the continuous attributes are discretized by determining "best" split-point for the parameters (EP, EO and OP) where the split-point is a threshold on the parameters [9, 10]. We first sort the values of P in increasing order. Typically, the midpoint between each pair of adjacent values is considered as a possible split-point. Therefore, given v values of P, then $v - 1$ possible splits are evaluated. For example, the midpoint between the values p_i and p_{i+1} of P is

$$\frac{(p_i + p_{i+1})}{2} \tag{1}$$

$$\mathrm{Info}_A(D) = \sum_{j=1}^{v} \frac{|D_j|}{|D|} - \mathrm{Info}(D_j) \tag{2}$$

If the values of P are sorted in advance, then determining the best split for P requires only one pass through the values. For each possible split-point for P, we evaluate $\mathrm{Info}_P(D)$, where the number of partitions is two, that is v = 2 (or j = 1;2) in Eq. (2). The point with the minimum expected information requirement for P is selected as the split point for P. D1 is the set of tuples in D satisfying P ≤ split point, and D2 is the set of tuples in D satisfying P > split point. The process of identifying the split point is repeated for each continuous attributes.

4.1 CART and Misclassification Rate

The Classification and Regression Tree (CART) is constructed by considering training tuples of the 3D model [9, 10]. Once the tree is constructed, the testing tuples are used to validate and compute error rate of the decision tree based predictor. The cost of the tree is the sum over all terminal nodes of the estimated probability of that node times the node's cost. The constructed tree is also used to predict the presence of testing tuples. If the prediction succeeds, then the success count is incremented by 1. The ratio between the success count in other words number of predicted tuples and the total number of testing tuples will determine the misclassification rate.

Steps for computing misclassification rate are given below:

1. During training, train a Decision Tree based predictor from a training set $(I_1, O_1)(I_2, O_2)\ldots, (I_n, O_n)$ where I is Input parameter, O is the output label.
2. During testing, for new test data $(I_{n+1}, I_{n+2}, \ldots, I_{n+m})$, the predictor generates predicted labels $(O'_{n+1}, O'_{n+2}, \ldots, O'_{n+m})$,
3. Calculate the Test set accuracy(acc) with the true test label $(O_{n+1}, O_{n+2}, \ldots, O_{n+m})$

$$acc = \frac{1}{m} \sum_{i=n+1}^{m} 1_{O_i = O'_i} \qquad (3)$$

4. Calculate the test set

$$\text{misclassification (error)rate} = 1 - acc \qquad (4)$$

5 Experiment Analysis

5.1 Experiments

To conduct the experiment and draw inference on the constructed predictors, two 3D static models and three 3D dynamic models are considered. There are 8 input parameters namely, "Current Key Press", "View Point", "Eye Position", "Eye Orientation" and an output parameter "Next Key Press" is considered in case of 3D static models. In case of 3D dynamic model, in addition to the 8 input parameters, object's position which is dynamic (Ox–x direction linear movement is considered) is also considered as given in Table 1 with actual value recorded during interaction and after normalizing it. The Current Key Press and Next Key Press takes any of the following values ranging from 1 to 6 indicating rotation operation R_Θ in any one of the directions: $+\Theta x/-\Theta x$ (2/1), $+\Theta y/-\Theta y$(4/3), $+\Theta z/-\Theta z$ (5/6). While building the

Table 1 Un-normalized and normalized data set

Current Key	View Point	Eye Position			Eye Orientation			Object Pos	Next Key
		X	Y	Z	X	Y	Z	X	
0	1	3.7761	-13.15	4	0	0	1	0.77612	3
3	1	3.3579	-13.15	4.357	-0.09983	0	0.995	23.584	3
3	1	2.9062	-13.15	4.6705	-0.19867	0	0.98007	13.581	3
3	1	2.4254	-13.15	4.9373	-0.29552	0	0.95534	4.5788	3
3	1	1.9204	-13.15	5.1547	-0.38942	0	0.92106	-0.91464	3
3	1	1.3962	-13.15	5.3207	-0.47943	0	0.87758	-6.4229	3
0	1	0.1	-	-	0.06	0	1.00	0.00	3
3	1	0.1	-	-	-	0	0.98	0.77	3
3	1	0.1	-	-	-	0	0.90	0.44	3
3	1	0.1	-	-	-	0	0.78	0.13	3
3	1	0.0	-	-	-	0	0.61	-0.05	3
3	1	0.0	-	-	-	0	0.40	-0.24	3

3D model for streaming and rendering based on the user interactions, a bounding sphere is used to enclose any type of 3D models being visualized. A bounding sphere is used to make the centre of the camera (eye) to always focus on the 3D model's centre. This is required to compute the viewpoints with respect to the centre of the 3D model. These parameters are used to construct the Decision Tree based Predictors The parameters namely Current Key Press, Next Key Press, View Points are discrete attributes whereas the parameters namely EP, EO and OP (in case of dynamic model) are continuous attributes. These continuous attributes are normalized using min-max quantization approach and then discretized by using the Eq. 1.

5.2 Results and Inference

Table 2 shows the 3D static models along with its attributes. The highlighted models have been used for user interactions profiling and further experiment analysis. Initially, profiles of progressive user interactions performed at different instance of time over static 3D Armadillo model are considered. There are 3382 training tuples collected at different instant of time and 180 testing tuples of same user is considered for validation. There are totally 23 pruning levels. Figure 5a

Table 2 Decision tree based predictors performance

Static 3D models	No. of vertices	No. of triangles
Dumptruck[a]	26,438	23,346
Armadillo[b]	1,72,974	3,45,944

[a]Open Scene Graph Standard Model
[b]Stanford Model

(a)   (b)   (c)

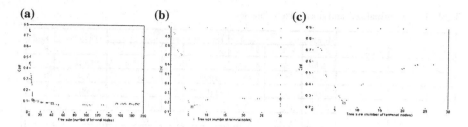

Fig. 5 Error rate of the constructed decision tree

shows the error rate of the tree. The solid line in this figure shows the estimated cost for tree size, the dashed line marks one standard error above the minimum, and the square symbol marks the smallest tree under the dashed line. The constructed decision tree based predictor for progressive single user interactions on single 3D armadillo static model could predict and classify 166 tuples properly out of 180 tuples. Therefore, the achieved misclassification rate is only 0.0778 (i.e., 7.78 %) and 92.22 % testing tuples are classified and predicted exactly by the constructed decision tree.

Secondly, single user interactions over multiple 3D models are considered. Figure 5b shows the error rate of the constructed decision tree for static 3D dumptruck model. There are 517 training tuples and 319 testing tuples considered to construct and validate the decision tree. From the prediction result it is found that the constructed decision tree could classify and predict 240 testing tuples correctly out of given 319 tuples. This gives the misclassification rate of 0.2476 (i.e., 24.76 %).

Figure 5c shows the error rate of the constructed decision tree for static 3D armadillo model. There are 331 training tuples and 340 testing tuples are considered to construct and validate the decision tree. From the prediction result it is found that the constructed decision tree could classify and predict 141 testing tuples correctly out of given 340 tuples. This gives the misclassification rate of 0.5853 (i.e., 58.53 %). The test tuples collected out of same user interactions for the respective models are used to predict the misclassification rate. Also, to affirm that the user interaction pattern is always subjective, test tuples collected when the user interacts with static 3D dumptruck model (horizontal shape) is used to predict the performance of constructed decision tree which is trained using training tuples of static 3D armadillo (vertical shape) model. This gives the misclassification rate of 0.8500 (i.e., 85 %). This result shows that when the models are different, misclassification rate of user interaction pattern is higher. Hence, it is concluded that user interaction pattern varies based on the model that is being viewed.

Finally, multiple user interactions on static 3D armadillo model are considered. Here 54 user profiles are considered to construct the decision tree and 55th user profile is considered to validate the tree and calculate the misclassification rate. There are 25,127 training tuples and 1258 testing tuples are considered to construct and validate the decision tree. From the prediction result it is found that the

Table 3 Decision tree based predictors performance

Approach	Model	No. of training tuples	No. of testing tuples	Misclassification rate in %
SU-S3DM	Armadillo Model	3382	180	7.78
SU-M3DM	Dumptruck Model	517	319	24.76
	Armadillo Model	331	340	58.53
MU-S3DM	Armadillo Model	25,127	1258	7.87

constructed decision tree could classify and predict 1159 testing tuples correctly out of given 1258 tuples. This gives the misclassification rate of 0.0787 (i.e., 7.87 %). This infers that, when there are more user interaction patterns on a particular 3D model, it is more accurate to predict a new user's interaction on the same 3D model. Table 3 shows the decision tree based predictors' consolidated performance results.

6 Conclusion

In this paper, an attempt has been made to construct and validate the predictors by using a soft computing approach namely Decision Tree. This predictor was used already in diversified areas of computing ranging from Pattern Recognition to Data Mining. However, these approaches are first attempted to predict the best possible next key press of the user during 3D streaming and rendering. It is based on the constructed Decision Tree. To construct/train the predictors, various user interaction profiles collected from single user to multiple users are considered. These profiles of the user interactions at a particular time instance or over a period of time were collected for a specific 3D model or set of 3D models. The results of these predictors applied for various approaches reveals that the exact prediction of next user move is possible to a larger extent. Thus this prediction approaches can be recommended for prediction based 3D content streaming and rendering.

Acknowledgments We thank the management of Al Yamamah University, KSA for supporting financially to publish our research work.

References

1. Ullah, I., Doyen, G., Bonnet, G., and Gaïti, D.: A Bayesian approach for user aware peer-to-peer video streaming systems. Signal Processing: Image Communication, Volume 27, Issue 5, pp. 438–456, ISSN 0923-5965, (2012).
2. Chan, M., Hu, S., and Jiang, J. Secure peer-to-peer 3d streaming. Multimedia Tools and Applications, 45(1–3):369–384, (2009).
3. Hu, S-Y., Jiang, J., and Chen, B.: Peer-to-Peer 3D Streaming. IEEE Internet Computing, vol. 14, no. 2, pp. 54–61, (2010).

4. Cheng, W., Ooi, W.T., Mondet, S., Grigoras, R., and Morin, G.: An analytical model for progressive mesh streaming. In Proceedings of the 15th international conference on Multimedia, ACM, New York, NY, USA, pp. 737–746, (2007).
5. Cheng, W., Liu, D., and Ooi, W.T.: Peer-assisted view-dependent progressive mesh streaming. In Proceedings of the 17th ACM international conference on Multimedia (MM '09). ACM, New York, NY, USA, pp. 441–450, (2009).
6. Vani, V., Pradeep Kumar, R. and Mohan, S.: 3D Mesh Streaming based on Predictive Modeling. Journal of Computer Science, 8 (7): 1123–1133, Science Publications, ISSN 1549-3636, (2012).
7. Vani, V., Pradeep Kumar, R. and Mohan, S.: Profiling User Interactions of 3D Complex Meshes for Predictive Streaming & Rendering. ICSIP'12, Proceedings, Lecture Notes in Electrical Engineering, Vol. 221(1), 457–467, Springer ISSN 18761100, (2013).
8. Vani, V., Mohan S,: Neural Network based Predictors for 3D content streaming and rendering. 18th ICSEC 2014, KKU, Thailand, IEEE explore, pp. 452–457, ISBN 978-1-4799-4965-6, (2014).
9. Han, J., Kamber, M. and Pei, J.: Data Mining: Concepts and Techniques. Morgan Kaufmann Publishers Inc., San Francisco, CA, USA, pp 744, chapter 6. ISBN: 978012381479, (2011).
10. Breiman, L., Friedman, J., Olshen, R., and Stone, C.: Classification and Regression Trees. Boca Raton, FL: CRC Press, (1984).

Quantitative Characterization of Radiographic Weld Defect Based on the Ground Truth Radiographs Made on a Stainless Steel Plates

P. Chitra

Abstract This paper presents a new approach for quantification of radiographic defects. This approach is based on calculating the size of the pixel using the known image quality indicator present in the radiographic image. This method is first applied on the ground truth realities of different shapes whose size is known in advance. The proposed method is then validated with the defect (porosity) where the defect is quantified accurately. The image processing techniques applied on the radiographic image are contrast enhancement, noise reduction and image segmentation to quantify the defects present in the radiographic image. The image processing algorithms are validated using image quality parameter Mean Square Error (MSE) and Peak Signal to Noise Ratio (PSNR).

Keywords Defect quantification · Image quality indicator · Image processing · Ground truth radiographs

1 Introduction

The digital information present in the image should be made visible to the viewer. The quality of image formed on radiographic film depends on the exposure factors, source, film types etc. [1–3]. Image enhancement is the first preprocessing step in automatic detection of features in the digital image. An image enhancement step includes contrast enhancement and noise reduction [4].

Once denoised, for retaining the visual information in the image there is a need for preserving the edges of an image. Edges are the representation of discontinuities of the image intensity function. Edge detection algorithm is essential for preserving the discontinuities. The edge detection algorithm is broadly classified into deriva-

P. Chitra (✉)
Faculty of Electrical and Electronics, Sathyabama University,
Chennai 600119, India
e-mail: chitraperumal@gmail.com

© Springer India 2016
S.C. Satapathy et al. (eds.), *Information Systems Design and Intelligent Applications*, Advances in Intelligent Systems and Computing 433,
DOI 10.1007/978-81-322-2755-7_17

tive approach and pattern fitting approach. Derivative approach is done by means of mask and pattern fitting by edge filter [5]. Edges in the image are linked using morphological operation. Dilation and erosion is the primary operator in morphological approach.

Nacereddine et al. [6] suggested the use of geometrical transformation (rotation, translation and Scaling) for quantitative analysis of weld defect images in industrial radiography. The geometric attributes are large in number because of the raw features present in the weld. The quantitative analysis of the weld defect images thus extracted, the major problem remains how to build a set of attributes which characterize the most accurately these defect regions [6]. Vilar et al. [7] Proposed an automatic defect identification and classification procedures for the detection of standard defect such as porosity, slag inclusion and crack. The feature present in the radiographic defect such as area, centroid, major axis, minor axis, eccentricity, orientation, Euler number, equivalent diameter, solidity, extent and position are used for classification of the defects. From the result it was found that the algorithm is able to identify and classify the defects 99 % [7]. Chaudhary et al. [8] used an standard image processing technique (Otsu thresholding) for segmentation of holes of diameter approximately 1, 2, 4, 5 and 10 respectively drilled on a aluminum plate. The holes are assumed as defect in the plate. The holes are quantified in terms of pixels. The percentage error in pixel based segmentation is calculated with respect to the physical measurement. From the result it was found that it 94 % of the drilled holes are correctly segmented and quantified. Later the same technique is applied to the several defects such as chipping, crack and voids present in the radiographic image. The false alarm rate is 8.1 % is obtained by this method [8].

Yazid et al. [9] uses Fuzzy C mean clustering as a first step is to identify and isolate the defects. The method gives the sensitivity of 94.6 %. The drawback of this method is manual selection of threshold value for converting into binary image [9]. Kasban et al. [10] presented a qualitative and quantitative characterization of radiographs using image enhancement and denoising techniques. The parameters considered are PSNR, RMSE, standard deviation and execution time. All these techniques are applied to gamma radiographic images. The image processing technique applied to improve the contrast and denoising the radiographic image are adaptive histogram equalization, Weiner, and wavelet filter. From this analysis it was found that weiner filter gives maximum image quality parameters value [10]. Cogranne and Retraint [11] developed a statistical method for defect detection in radiographic images. The process is carried out with three different phase. First image processing algorithms are applied without knowing prior knowledge about the radiographs. Second the result is compared with the reference radiographs. Using the information obtained from the two phase a statistical model using Bayesian approach is developed for the classification of defects. However a precision calibration is required to match the radiographs with this geometrical model [11]. Zahran et al. [12] Proposed an automatic weld defect detection techniques. The detection process is carried by three stage which includes preprocessing, segmentation and defect identification. Preprocessing includes contrast enhancement and filtering by weiner filter. Segmentation is done by considering the weld region as 1D signal and by creating a database of defect features using Mel-Frequency Cepstral Coefficients. The polynomial coefficients are extracted from the Power Density Spectra of this 1D signal. This parameter is used for training the neural network which is used for automatic identification of the defects [12].

From review of literatures, it was found that manual interpretation of weld radiographs is subjective in nature and is affected by operator fatigue, and hence necessitates, automated interpretation of weld radiographs. Considerable research is carried out in developing automated weld defect detection systems for radiographs [13, 14]. However most of the researchers have concentrated in developing segmentation techniques for defect detection and classification. Very limited research is done in the area of digitizing industrial radiographs and uncertainty in sizing of defects. Hence the study aims to develop an automatic weld defect detection system for defect identification and quantification. This paper presents an innovative method of identify the defect using image processing techniques and quantify the defects by using the Image Quality Indicator (IQI) present in the radiographic image.

2 Experimental Methods and Materials

A stainless steel metal of 3.5 mm thickness with four different shapes (Square, Rectangle, Circle, Notch) of varying length and breadth are prepared by Electrical Discharge Machining(EDM) with different depths. The plate is subjected to X-ray and Gamma ray radiography. For gamma ray, iridium 192 radioactive sources are used in experimental set up. The radiation is exposed on D4 (Extra fine grain film with very high contrast) and D7 (Fine grain film with high contrast) film. Two types of Image Quality Indicator (IQI) are placed on the source side of the material Viz the placard or hole-type and the wire IQI. After radiography the film is digitized by using Array Corporation's 2905 X-Ray film digitizer which is a Laser film digitizer. The film was digitized at two different DPI (100, 300) and in two file format (BMP, TIFF). Radiographs of stainless steel plate are prepared in Scanray Private India limited, Chennai. Thirty two radiographs are acquired with four plates exposed on two films with four different Source to Film Distance using X-ray source. Similarly eight radiographs are acquired with four plates exposed on two films using gamma source (2 Curie). Six radiographs are acquired for a rectangular shape plates exposed to two films with three different control voltage using x-ray source. The different voltages considered for experimental analysis are 110, 130, 150 kV with a source to film distance of 1500, 1000, 700, 400 mm. The dimensions of the stainless steel plates are measured manually using non contacting measuring device HAWK SYSTEM 7. The system is having an accuracy of 2 µm and a resolution of 0.5 µm. Figure 1 shows the radiographs obtained for the different shapes made on the stainless steel plate. The shapes are made with varying size and depth whose representation is shown in Fig. 2.

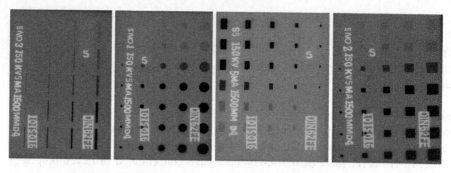

Fig. 1 Radiographic image of different shapes made on a stainless steel plate

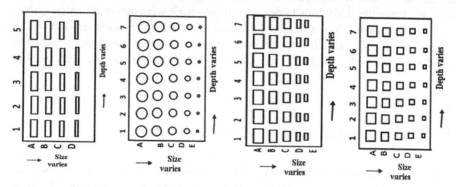

Fig. 2 Representation of shapes on the stainless steel plate

Table 1 shows the measured value using non contacting measuring device for a square shape notch. Similar calculation is also obtained for remaining three shapes. This value is used as a reference value for comparing with the value obtained using image processing algorithm.

Table 1 Square shape notch length and breadth on a stainless steel plate

		1	2	3	4	5	6	7
A	Length	9.74	9.84	9.87	9.90	9.94	10.00	10.00
	Breadth	9.76	9.85	10.00	9.88	10.06	9.85	10.06
B	Length	7.87	7.93	7.87	8.00	8.00	8.10	8.01
	Breadth	7.91	8.01	7.95	8.03	8.05	8.10	8.11
C	Length	5.89	5.96	6.02	5.95	6.15	6.09	6.08
	Breadth	5.93	5.96	6.01	6.06	6.07	6.16	6.04
D	Length	3.99	3.95	3.99	4.05	4.04	4.10	3.98
	Breadth	3.96	3.95	4.00	4.07	4.05	4.09	3.96
E	Length	1.91	1.91	1.96	1.93	1.99	2.04	2.02
	Breadth	1.82	1.91	1.88	2.01	2.00	2.04	1.98

3 Methodology Adapted

Image processing algorithms is applied on the digitized radiographic image. The steps involved in the detection of defects in radiographic image are shown in Fig. 3. Two different image enhancement techniques such as linear and nonlinear contrast enhancement techniques are performed to highlight the region of interest. Adaptive median filter, weiner filter, homomorphic filters are explored to denoise the images [15–17]. Edge detection is performed using different masks. The image processing algorithms are evaluated by using image quality parameter Peak Signal to Noise Ratio (PSNR). The image processing technique applied on the digitized radiographic weld image is shown in Fig. 4. The quality measures of the image processing algorithms are tabulated in Table 2.

Root Mean Square Error is the square root of Mean Square Error (MSE) which is given by

$$MSE = \frac{1}{M*N}\sum_{i=0}^{m-1}\sum_{j=0}^{n-1}[f(i,j) - g(i,j)]^2 \qquad (1)$$

where f(i, j) and g(i, j) be the original and processed radiographic image at the mth row and nth column for M × N size image.

Peak Signal to Noise Ratio is given by

$$PSNR = 10\log_{10}(255^2/MSE) \qquad (2)$$

Based on the image quality parameters measured linear contrast enhancement and adaptive weiner filter gives better result on the radiographic image.

4 Validation of the Method

A Graphical User Interface (GUI) was developed using Matlab 2007 software which includes the entire image processing algorithm which is shown in Fig. 4. Radiographic defect is quantified with respect to Image Quality Indicator (IQI) present in the digital radiographic image. The original radiographic image containing the defect porosity is shown in Fig. 5. The defect to be quantified after applying morphological image processing operator is shown in Fig. 6. The length

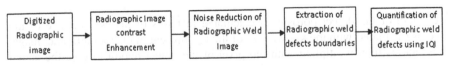

Fig. 3 Steps involved in radiographic weld defect detection

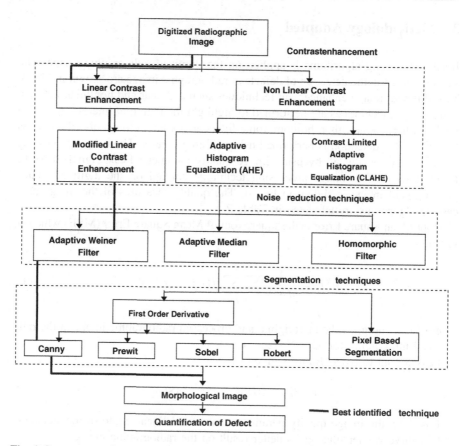

Fig. 4 Image processing steps applied on the radiographic weld image

Table 2 Image quality measures for different image processing technique applied on the radiographic weld image

Techniques applied on the radiographic image			RMSE	PSNR (dB)
Contrast enhancement	Linear contrast enhancement		1.8803	42.6461
	Non-linear contrast enhancement	AHE	0.2205	13.1323
		CLAHE	0.0890	21.0104
Noise reduction	Adaptive median filter		0.9721	48.3770
	Adaptive Weiner filter		0.8970	49.0749
	Homomorphic filter		6.4691	30.0605

and width of the defect porosity is calculated in terms of pixels. The size of the pixel is calculated with the known size of the IQI present in the radiographic image. This ratio is then multiplied with the pixel of the defect region to give the actual

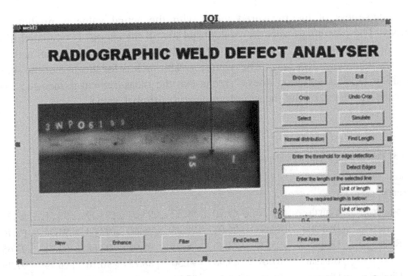

Fig. 5 Radiographic weld defect analyzer with radiographic image containing the defect porosity and image quality indicator

Fig. 6 Segmented defect

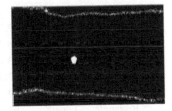

value of the weld defect. From this the width of the defect (porosity) is 1.2 mm (38 pixels) and the length is 1.6 mm (49 pixels). Further the defects are also quantified in terms of features such as major and minor axis.

5 Error Analysis

To know how far the measured value is far from the original (Actual) value, percentage error is calculated. Percentage error is the difference between the measured value and the original value by its original value.

$$\text{Percentage error} = \frac{|\text{Measured value} - \text{Actual vaue}|}{\text{Actual vaue}} \times 100 \qquad (3)$$

The closeness of the measured value with the actual value is predicted by accuracy. The percentage accuracy calculated for all the shapes is tabulated in the Table 3. From the data obtained it is found that the percentage of accuracy decreases when size decreases. The accuracy value is less for the size of the shapes at position D. This variation doesn't mean that the measurement is incorrect. This shows the closeness of the measured value with original value.

Table 3 Percentage accuracy calculated for different shapes

Position			1	2	3	4	5	6	7
Percentage accuracy									
A	Square	Length	99.10	88.70	99.52	93.68	98.39	98.55	98.21
		Breadth	95.09	96.94	98.25	96.66	97.64	91.96	91.46
	Rectangle	Length	95.70	98.86	98.74	98.97	98.68	92.31	92.65
		Breadth	87.01	83.98	94.50	88.75	90.76	96.79	97.75
	Circle	Diameter	96.01	93.11	97.06	89.49	94.39	99.37	98.27
	Line	Length	96.91	95.52	93.16	95.80	92.61	–	–
B	Square	Length	99.40	89.52	97.12	88.71	95.52	98.96	98.75
		Breadth	99.94	97.75	99.50	98.57	98.31	98.90	99.01
	Rectangle	Length	97.35	93.45	94.87	99.92	96.33	89.70	99.10
		Breadth	92.71	90.05	87.37	88.99	95.08	81.62	98.81
	Circle	Diameter	95.23	94.50	90.53	98.16	88.27	96.78	97.79
	Line	Length	93.18	94.58	91.43	92.90	96.81	–	–
C	Square	Length	97.30	99.34	86.18	99.18	93.13	89.57	92.28
		Breadth	89.59	90.18	91.00	91.99	87.68	89.28	91.57
	Rectangle	Length	97.41	94.20	97.24	89.03	86.95	96.92	99.61
		Breadth	96.72	79.82	96.19	91.45	99.95	91.12	89.85
	Circle	Diameter	80.34	91.89	94.10	93.10	95.38	87.16	97.43
	Line	Length	95.94	91.34	94.23	91.50	91.80	–	–
D	Square	Length	97.42	96.66	81.42	94.41	98.71	–	–
		Breadth	89.88	89.46	83.92	99.33	90.00	–	–
	Rectangle	Length	87.69	82.25	89.85	95.49	99.91	94.65	88.64
		Breadth	65.71	74.42	87.30	63.98	63.46	86.12	89.15
	Circle	Diameter	98.40	79.84	94.00	92.79	92.52	–	–
	Line	Length	93.62	92.88	93.41	91.29	93.77	–	–
E	Square	Length	71.46	71.19	74.48	89.81	83.15	79.73	51.65
		Breadth	65.16	71.17	68.98	89.85	63.46	66.39	89.74
	Rectangle	Length	–	–	80.36	79.13	67.20	79.25	67.70
		Breadth	–	–	55.36	92.61	95.70	98.91	98.23
	Circle	Diameter	72.33	63.65	85.21	91.64	91.55	–	94.21

6 Conclusion

To characterize and quantify the defects, image processing algorithms are used. The algorithm includes image enhancement, noise reduction and segmentation techniques for extracting the defects present in the radiographic weld image. The size of the pixel also varies when it is stored in different file formats and resolution due to this the defect size varies for the same image stored in various format. So a calibration based high precision algorithm is developed by using standard Image Quality Indicator (IQI) available in all radiographic weld images. Having known the standard size of IQI, irrespective of the scanner with which it is digitized, the defect is quantified accurately. The experimental results show that the defects are quantified with an accuracy of 96 %.

Acknowledgments I would like to extend my gratitude to Dr. B. Venkatraman, Head, Quality Assurance Division and Radiological Safety division, IGCAR, Kalpakkam who spent so many constructive hours in shaping this research work and provided me with useful inputs. I am also grateful to all faculty members of Scaanray metallurgical services, Chennai for permitting to acquire on-line database and helping me in acquiring radiographs.

References

1. Udupa H., Peter Mah., Dove S.B. and McDavid. W.D.. Evaluation of image quality parameters of representative intraoral digital radiographic systems: Journal of Oral Surgery, Oral Medicine, Oral Pathology, Oral Radiology, Elseiver publication, Vol. 116, issue 6, pp. 774–783 (2013).
2. Strauss L.J. and William I D Rae.: Image quality dependence on image processing software in computed radiography: SA Journal of Radiology, Vol. 16, pp. 44–48. (2012).
3. Baldev Raj and Venkatraman B.: Ionising radiations For Non Destructive Evaluation: Indian Society for radiation Physics, Kalpakkam Chapter, (1989).
4. Gonzalez R.C. and Woods R.E.: Digital Image Processing: Prentice Hall of India (2005).
5. Pratt W.K.: Digital Image Processing: Wiley and Sons, New York (1991).
6. Nacereddine N., Tridi M., Belaifa S., Zelmat M.: Quantitative analysis of weld defect images in industrial radiography based invariant attributes: International Journal of Computer, Information, Systems and Control Engineering, Vol. 1, No 2, pp. 466–469 (2007).
7. Vilar R., Zapata J. and Ruiz R.: An automatic system of classification of weld defects in radiographic images: Journal of NDT&E International, Elsevier publication, Vol. 42, pp. 467–476. (2009).
8. Chaudhary U.K, Iqbal M, Munir Ahmad.: Defect sizing of post-irradiated nuclear fuels using grayscale thresholding in their radiographic images: Journal of Nuclear Engineering and Design, Elsevier publication, Vol. 240, pp. 3455–3461 (2010).
9. Yazid H., Arof H., Ahmad S., Mohamed A.A. and Ahmad F: Discontinuities detection in welded joints based on inverse surface thresholding: Journal of NDT&E International, Elsevier publication, Vol. 44, pp. 563–570 (2011).
10. Kasban H., Zahran O., Arafa H., El-Kordy M., Elaraby S.M.S and Abd El-Samie F.M.: Quantitative and Qualitative Evaluation for Gamma Radiographic Image Enhancement: International Journal of Signal Processing, Image Processing and Pattern Recognition Vol. 5, No. 2, June, 2012, pp. 73–87 (2012).

11. Cogranne R. and Retraint F:Statistical Detection of Defects in Radiographic Images Using an Adaptive Parametric Model: IEEE transaction of signal processing, Elsevier publisher, Vol. 96, pp 173–189 (2013).
12. Zahran O., Kasban H., Kordy M. and SamieF.E.: Automatic weld defect identification from radiographic images: proceeding of NDT&E International, Elseiver publication Vol. 57 (2013) pp. 26–35 (2013).
13. Rajagopalan C., Venkatraman B., Jayakumar T., Kalyanasundaram P. and Baldev Raj: A novel tool for automated evaluation of Radiographic weld images: 16th World Conference on Non Destructive Testing (16th WCNDT–2004), No. 358, www.ndt.net (2004).
14. Yahia N.B, Belhadj T, Brag S, Zghal A: Automatic detection of welding defects using radiography with neural approach: Procedia Engineering Elsevier publication, Vol. 10, pp. 671–679. (2011).
15. Wang. X.: Adaptive multistage median filter: IEEE Transactions on Signal Processing, Vol. No. 40, pp. 1015–1017 (1992).
16. Zhang Jian., Qianga G., Qia L., Xiao-dia C. and Xin-ping L.: Adaptive Wiener filtering for image restoration using wavelet package based on the edge detection: Electronic Imaging and Multimedia Technology IV, Proc. of SPIE, Bellingham, WA, Vol. 5637, pp. 532–539 (2005).
17. Xiaoyin X., Eric L. and Miller.: Entropy Optimized Contrast Stretch to Enhance Remote Sensing Imagery: IEEE, Proceedings of 16th International Conference on Pattern Recognition, Vo. 3, August 11–15, 2002, pp. 915–918 (2002).

Research and Topology of Shunt Active Filters for Quality of Power

Lakshman Naik Popavath, K. Palanisamy and D.P. Kothari

Abstract The utilization of distorting loads has been increasing exponentially, which results power quality issues in electrical power systems. Transmission of the pollution free power is an important task for power engineers. Power quality issues may affect on end user equipment like electronic goods, digital meters etc. which results the spoilage of products. To nullify the power quality concerns the custom power devices are playing a determinant role in the power systems in the present scenario. This paper mainly demonstrates on research and topology of Shunt Active Power Filters (SAF) to magnify the quality of power in sensitive industrial loads, electrical distribution networks, transmission, and power generation systems.

Keywords Power quality (PQ) · Reactive power compensation · Power factor improvement · SAF VSC · SAF topology · STATCOM · Power systems

1 Introduction

Transmission of pollution free power to the end users is one of the major key issues for the electric power engineers, because at present electrical distribution systems have almost all distorting or non-linear loads and reactive loads. Reactive power burden is more in the power system due to the reactive loads, so that the immoderate reactive power demand may increase the feeder losses and it will diminishes the active or true power flow proficiency of the distribution system. In other hands the distorting loads (non-linear loads/sensitive loads) like single phase

L.N. Popavath (✉) · K. Palanisamy · D.P. Kothari
School of Electrical Engineering, VIT-University, Vellore, India
e-mail: lakshman.help1@gmail.com

K. Palanisamy
e-mail: kpalanisamy@vit.ac.in

D.P. Kothari
e-mail: dpkvits@gmail.com

© Springer India 2016
S.C. Satapathy et al. (eds.), *Information Systems Design and Intelligent Applications*, Advances in Intelligent Systems and Computing 433,
DOI 10.1007/978-81-322-2755-7_18

167

and three phase power converters, electronic goods, electric arc furnace, digital meters etc. generates the considerable asymmetric disturbances in the ac mains, which may leads the power quality problems in the electric power systems. Today the most quite recurrent power quality issues are voltage variations, wave distortion (harmonics), transients, voltage flickering, supply frequency variations, DC off-sets, noise and notching etc. [1].

The balanced and unbalanced non-sinusoidal current generates harmonics, extravagant neutral current, reactive power, low power factor and unbalance loading of the ac mains [2]. Poor quality of power may leads to heavy power losses, power quality also degrades the consumer services, system efficiency, productivity, economic conditions and on telecommunication network's performance. At initial stage LC passive filters were played a vital role to nullify the quality issues in the electrical power systems, but due to some determinant drawbacks of passive filters namely large in size, poor dynamic performance, enforcement of fixed compensation and resonance etc. the power engineers concentrated in their research work to develop the fast and dynamic solutions to magnify the quality of power, due to remarkable forward movement in the field of PE devices, active filters have been studied and the huge number of work have been published so that the passive filters are quickly replaced by the Active power filters. Among the different types of active filters the Shunt Active filters (SAF) have been playing a crucial role in the power quality concept. Shunt AF's have proved to provide the efficient and effective solutions for the compensation of electric power related disturbances in electrical distribution systems. Almost all consumer's load related power quality concerns can be solved by the shunt AF's (STATCOM).

The organization of present article is as follows. Section 2 illustrates the power quality concerns, its consequences and standards. Section 3 describes state of art. Sections 4–7 introduces system configuration, the control strategies, advantages and applications of shunt AF's, and conclusion respectively.

2 Power Quality Standards, Issues and Its Consequences

2.1 Power Quality Standards

In national and international level several organizations like IEEE and IEC etc. are working with the power engineers, research organizations to provide familiar platform for participated parties to work together to ensure compatibility for end-use and system equipment. According to the international standards guidelines [3] the consumption of electrical energy by the electric equipment should not produce the contents of harmonics more than the specified values. The current or voltage injections are limited based on the size of load and power systems. The standards are developed by the working group members, which describes to determine the characteristics of quality of power. According to the IEEE standards

519-1992 for below 69 kV systems, the THD value should not be equal or greater than 5 %. The frequencies of harmonic load current should not coincide with system resonances [1]. According to the IEC standards 61000-2-2* for third harmonic with multiples of 3, the compatibility level of harmonic voltage should be less than 5 %.

2.2 Power Quality Issues

One of the most prolific buzzwords in an electric power system concept is *power quality*. In present power scenario with the increase of usage of non-linear loads or sensitive loads drawing non-sinusoidal currents which effects on the end-use equipment [4], so that the demand for high reliable, quality of electric power has been increasing, which may leads to increase the power quality awareness by both utilities and end-users. The main power quality issues are

2.2.1 Voltage Variations

The load variations in the power system mainly results the voltage variations [5]. The real and reactive powers are directly or indirectly related to the load variations. Depending on *rms* voltage variations and time period these are termed as short duration and long duration voltage variations [6, 7]. The variation of voltage is mainly classified as voltage rise/swell, voltage dips/sags, voltage interruptions and long duration voltage variations. The voltage fluctuations are commonly known as voltage flicker issue which represents the dynamic variations [8] caused by varying loads in the power networks.

2.2.2 Wave Distortion/Harmonics

The technical literature was done in 1930 and 1940s on harmonics and it is defined as the integer multiple of system fundamental frequency. Harmonics comes under the wave form distortion concept. Wave form distortions (harmonics) are caused by the sensitive or distortion loads like PE devices, arc furnace and induction furnace in the power networks [1]. The distortion load is one in which the current and voltages do not have directly proportional relationship between each other. Wave form distortion is commonly classified as Notching, Noise, DC-offset, Inter harmonics.

The existence of dc content in the Ac current or voltage in the power system is known as Dc-offset. The main causes for Dc-offset are geomagnetic disturbance and Dc current in Ac network which results saturation of transformers.

2.2.3 Consequences of PQ Issues

Power quality issues like voltage variations, Transients, flickers, interharmonics and harmonics causes over loading, overheating, additional losses, saturation of transformers, data error or data loss and mall-operations of equipment like logic controller, microprocessor based controllers, power meters. Particularly on electric conductor the harmonics causes the over heat, proximity effect and skin effect.

3 State of Art

This section mainly describes the research and development activities of shunt AF's for power quality enhancement. Since at the starting of 1970s active filters have been researched. In 1972 the shunt AF was discovered by *Gyugyi*, at that time the active filters could not realized to the commercial potential, because unavailability of high rated solid state power switches. The solid state technology has been developed with emergence initiation in power semiconductor field. In Japan as per [9] in the time span 1981–1996 nearly five hundreds plus shunt AF's have been developed into commercial application due to continuous active progress. The price, size of shunt AF were reduced and other technological advancements has been improved [9], at present we have 300 kVA rated shunt active filters are available to reduce current dictation factor from 38.4 to 7.4 %. In 1973 the shunt, series active filters and the combination of these power filters have been refined and commercialized for the applications of continuous power supply. Based on voltage source converters (VSC) with capacitive energy storage and current source inverters (CSI) with inductive energy storage concept single phase active filters were developed [10], latterly 3-phase, 3-wire AF's have been developed, as well as shunt active and series active filters are combined with passive filters for some typical configurations [11]. Many more technologies like flicker compensators, static var generators and compensators etc. have been retained in the literature. At initial stage the BJT's, Thyristors and MOSFET's were used for the fabrication of active filters, later GTO's (gate-turn off thyristors), static induction thyristors (SITS's) were employed to develop SAF [12].

With IGBT's initiation shunt AF technology has gained a real boost, at present these are treated as ideal PE devices for the rapid growth of shunt AF's. To provide better control actions to the AF's several control strategies like p-q theory, Id-Iq theory and notch filter methods [13–15] etc. are developed for the development of shunt AF. To measure the performance of active filters, it is very important to develop a good computing instrument, so that these novel concepts have given an exponential growth to instrumentation technology [12]. With the presence of adequate rating of IGBT's, advanced sensor technology, hall-effect sensors and insulation amplifiers at acceptable cost levels the shunt AF's performance were improved. The microelectronics revolution gave the next important discovery in shunt AF's improvement. This improvement allows to use the various control technologies like PI control, Fuzzy logic control and neutral network control etc. to improve steady state and dynamic performance of the AF's. With this improvement shunt active filters have the ability to provide efficient and effective fast corrective action even under dynamically varying distorting loads. Shunt active filter were found to compensate very high order harmonics (up to 25th).

4 System Configurations

Even though several options are available to enlarge the quality of power, the shunt AF's are widely implemented and accepted, because of its flexible nature and robust performance. Basically the shunt active filters are pulse width modulated VSI and CSI. Active filters are mainly classified based on phase numbers, type of converter topology (either CSI or VSI) [16], power ratings and speed of responses [17]. These classifications of shunt AF's explained clearly with neat sketches as follows.

4.1 Converter-Based Classifications

Based on the converter type the shunt AF's are classified as voltage-fed type and current-fed type shunt active filters. Due to its expandability [12, 16, 18], low initial cost, better efficiency voltage-fed PWM power converters are the most desired one compare to current-fed type shunt active filters. The bridge structured current-fed type PWM inverter is shown in Fig. 1. It acts as non-sinusoidal current source to reach the harmonic current requirement of distorting loads [12]. The diode is connected in series with the IGBT will block the reverse voltage (these were restricted frequency of switching). Current source inverters (CSI) have higher losses, so that it requires high rated parallel Ac power capacitor. Figure 2 shows the bridge structured voltage-fed type PWM inverter. It consist of the self-supporting Dc voltage bus with large Dc capacitor. It is more dominant, because it is cheaper, light in weight, enlargeable to multilevel and multistep versions to improve the performance with less switching frequencies [10]. The shunt AF's with this inverter bridge mitigates the harmonics of censorious distorting loads.

Fig. 1 Shunt AF current-fed type

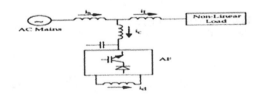

Fig. 2 Shunt AF voltage-fed
type

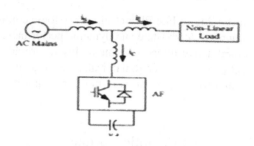

4.2 Phase-Based Classifications

Based on the supply systems and/or load systems phase levels shunt AF's are categorized as 1-phase (2-wire) and 3-phase (3-wire or 4-wire) systems [12].

(B.1) **2-Wire Active Filters**: Many distorting loads, such as domestic gadgets are connected to 1-phase supply system. The two-wire AF's are used in series, shunt and combination of both to nullify the power quality issues. Both current-fed and voltage fed PWM converters will be used to amplify the input characteristics at the supply side. Figure 3 indicates the configuration of SAF with bridge structured current source inverter using inductive storage element. The similar configuration can be obtained with voltage source bridge using capacitive storage element for shunt AF's.

(B.2) **3-Phase, 3-wire AF's**: There are several publications which have appeared on 3-phase, 3-wire shunt AF's. 3-Phase non-linear loads having 3-wires such as ASD's etc. consists of the solid state power converters and laterally these type of loads are incorporated with AF's. The configuration of 3-Phase, 3-wire shunt AF is exhibited in Fig. 4. The 3-Phase active shunt filters are also designed with three 1-phase active filters with isolation transformer for independent phase control, proper voltage matching and reliable compensation with unbalanced systems [10, 12].

(B.3) **3-Phase, 4-Wire AF's**: Three-phase supply system with neutral conductor may supply power to a large number of 1-phase loads, so that the systems may have reactive power burden, excessive neutral current, harmonics and unbalance. To mitigate these issues 4-wire shunt AF's have been attempted. These 4-wire filters have been developed with voltage-fed and current-fed

Fig. 3 1-phase CSI type
shunt AF

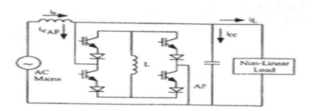

Fig. 4 Three-phase 3-wire
shunt AF

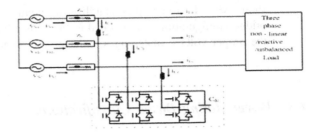

type converters. Figures 5, 6, and 7 indicates the configuration of 3-phase
4-wire AF's [12]. Capacitors midpoint type AF is the first developed configuration for smaller rating applications. The complete neutral current
flows through Dc capacitors having large capacitance values. In switch type

Fig. 5 3-phase 4-wire AF
with midpoint capacitor

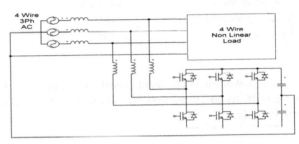

Fig. 6 Four-wire 4-pole
shunt AF

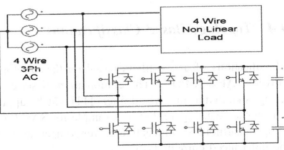

Fig. 7 Three 1-phase 4-wire
bridge type shunt AF

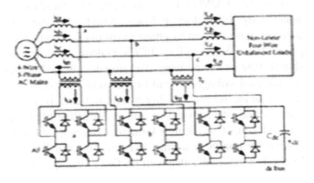

shunt AF's the fourth pole stabilize the neutral of AF. The three 1-phase bridge configuration is illustrated in Fig. 7 is a familiar version will allows the proper voltage matching to enhance the reliability of shunt AF's [12].

4.3 Power Rating-Based Classifications

The factors power rating of compensation level and response speeds are playing a principal role in deciding the feature requirement of the active filters. These two factors have reciprocal relationship. These types of filters are mainly classified based on rating and response speed [17] as follows.

Filter type	Rating	Response time
Low power filters	<100 kVA	100 μs–10 ms
Medium power filters	100 kVA–10 MVA	100 ms–1 s
High power filters	>10 Mva	1–10 s

Low power filters are again classified as 1-phase and 3-phase filters, the response time is 10 μs to 10 ms and 100 ms to 1 s for 1-phase and 3-phase power filters respectively.

4.4 Topology-Based Classifications

Classifications of active filters based on the topology used are shunt, series and universal AF's (UPQC). The combination of series active and passive shunt filtering is known as the hybrid filters [11] which suppress low ordered harmonics at reasonable cost. Figure 2 is an example for SAF (known as STATCOM) used to mitigate harmonics and reactive power compensation, to improve PF and balancing the unbalanced currents [12].

The STATCOM mainly connected at load end to suppress the current harmonics injected by the distorting loads. Shunt AF's can inject the equal and opposite compensating current in phase to nullify harmonics and/ or reactive component at point of common coupling (PCC). Statcom can improve the voltage profile and also stabilize the power systems. To mitigate voltage harmonics, to regulate and balance the terminal voltage of load or line series active filter is used before load (source side) with the Ac mains using matching transformer [12]. Universal active filter is a combination of series active and shunt active filters with Dc-link storage element (either Dc-bus Capacitor or Inductor), universal AF can eliminates both voltage and current harmonics at a time and has the capability to give clean power to critical loads. The main advantages of unified power quality conditioner are, it can balance and regulate the terminal voltage and mitigate the negative sequence currents [10]. The main drawback of universal AF's are its control complexity and high cost.

5 Control Strategies

Control strategy is the heart of active filters. To generate the required compensation current, the control strategies are playing an important role in the designing of shunt AF's. There are several control strategies like instantaneous reactive power theory (p-q theory), synchronous reference theory (Id-Iq theory), perfect harmonic cancellation method (PHC), Unity power factor method (UPF) etc. are there to extract the reference current for SAF's connected to 3-phase source which supplies to the distorting loads. This paper the mainly demonstrates on the p-q theory and Id-Iq theory.

5.1 P-Q Theory

This Control strategy was first proposed by Akagi et al. in 1984 [19]. It is a time domain case, which can be applied for both 3-phase, 3-wire systems and 3-phase 4-wire systems. This control strategy is valid for both steady state and transient state conditions of the systems. Based on α-β transformation the set of voltages and currents are transformed from abc to α-β-0 coordinates [13, 18, 20]. Supply phase voltages are given as

$$Va = VmSin(\omega t), \quad Vb = VmSin\left(\omega t - \frac{2\pi}{3}\right), \quad Vc = VmSin\left(\omega t - \frac{4\pi}{3}\right) \quad (1)$$

Similarly reactive load currents are given by the equations

$$i_{La} = \sum I_{Lan}Sin\{n(\omega t) - \theta_{an}\}, \quad i_{Lb} = \sum I_{Lbn}Sin\left\{n\left(\omega t - \frac{2\pi}{3}\right) - \theta_{bn}\right\}$$

$$i_{Lc} = \sum I_{Lcn}Sin\left\{n\left(\omega t - \frac{4\pi}{3}\right) - \theta_{cn}\right\} \quad (2)$$

$$\begin{bmatrix} V_\alpha \\ V_\beta \end{bmatrix} = \sqrt{\frac{2}{3}}\begin{bmatrix} 1 & -1/2 & -1/2 \\ 0 & \sqrt{3}/2 & -\sqrt{3}/2 \end{bmatrix}\begin{bmatrix} V_a \\ V_b \\ V_c \end{bmatrix}, \quad \begin{bmatrix} i_\alpha \\ i_\beta \end{bmatrix} = \sqrt{\frac{2}{3}}\begin{bmatrix} 1 & -1/2 & -1/2 \\ 0 & \sqrt{3}/2 & -\sqrt{3}/2 \end{bmatrix}\begin{bmatrix} i_a \\ i_b \\ i_c \end{bmatrix} \quad (3)$$

In abc coordinates a, b and c are fixed axes each displaced by $2\pi/3$. The vectors v_a, i_{La} amplitude varies in negative and positive directions with time [13]. The α-β transformation (Clark's transformation) is given as above equations. The real and reactive powers can obtained as

$$p = v_a i_a + v_b i_b + v_c i_c, \quad p = v_\alpha i_\alpha + v_\beta i_\beta, \quad q = -v_\beta i_\alpha + v_\alpha i_\beta \quad (4)$$

$$\begin{bmatrix} p \\ q \end{bmatrix} = \begin{bmatrix} v_\alpha & v_\beta \\ -v_\beta & v_\alpha \end{bmatrix} \begin{bmatrix} i_\alpha \\ i_\beta \end{bmatrix} \quad (5)$$

i_α and i_β can be obtained as

$$\begin{bmatrix} i_\alpha \\ i_\beta \end{bmatrix} = \frac{1}{\Delta} \begin{bmatrix} v_\alpha & -v_\beta \\ v_\beta & v_\alpha \end{bmatrix} \begin{bmatrix} p \\ q \end{bmatrix} \quad (6)$$

where $\Delta = v_\alpha^2 + v_\beta^2$. The powers p and q can decomposed as $p = \bar{p} + \tilde{p}$, $q = \bar{q} + \tilde{q}$. The reference source currents for α-β and abc transformations can be expressed [13] as

$$\begin{bmatrix} i_{s\alpha}^* \\ i_{s\beta}^* \end{bmatrix} = \frac{1}{\Delta} \begin{bmatrix} v_\alpha & -v_\beta \\ v_\beta & v_\alpha \end{bmatrix} \begin{bmatrix} \bar{p} \\ 0 \end{bmatrix} \quad (7)$$

$$\begin{bmatrix} i_{sa}^* \\ i_{sb}^* \\ i_{sc}^* \end{bmatrix} = \sqrt{\frac{2}{3}} \begin{bmatrix} 1/\sqrt{2} & 1 & 0 \\ 1/\sqrt{2} & -1/2 & \sqrt{3}/2 \\ 1/\sqrt{2} & -1/2 & -\sqrt{3}/2 \end{bmatrix} \begin{bmatrix} i_0^* \\ i_{s\alpha}^* \\ i_{s\beta}^* \end{bmatrix} \quad (8)$$

5.2 I_d-I_q Theory

This theory is also known as synchronous reference frame theory (SRF). To generate the final gate or reference signal to STATCOM based on Park's transformation the 3-phase stationary (abc) system can be transform to a rotating coordinate (dq0) system. The d-q reference frame can determined by the angle θ with respective α-β frame, which is used in instantaneous reactive power theory (p-q theory) [21, 22]. The Park's transformation is given as

$$\begin{bmatrix} \mu_d \\ \mu_q \end{bmatrix} = \begin{bmatrix} \cos\theta & \sin\theta \\ -\sin\theta & \cos\theta \end{bmatrix} \begin{bmatrix} \mu_\alpha \\ \mu_\beta \end{bmatrix}, \quad \begin{bmatrix} I_d \\ I_q \end{bmatrix} = \begin{bmatrix} \cos\theta & \sin\theta \\ -\sin\theta & \cos\theta \end{bmatrix} \begin{bmatrix} I_\alpha \\ I_\beta \end{bmatrix} \quad (9)$$

where voltage and currents are

$$\begin{bmatrix} \mu_\alpha \\ \mu_\beta \end{bmatrix} = \begin{bmatrix} V_m \sin\omega t \\ V_m \cos\omega t \end{bmatrix}, \quad \begin{bmatrix} I_\alpha \\ I_\beta \end{bmatrix} = \begin{bmatrix} I_m \sin(\omega t - \delta) \\ I_m \cos(\omega t - \delta) \end{bmatrix} \quad (10)$$

The power equations are given [13] as follows

$$p = \mu_d I_d + \mu_q I_q, \quad q = \mu_q I_d - \mu_d I_q \tag{11}$$

The d-q transformation reduces the 3-quantities to 2-quantities, because for balanced systems the zero component is omitted and when synchronous frame is aligned to voltage then $\mu_q = 0$. So that

$$p = \mu_d I_d, \quad q = -\mu_d I_q \tag{12}$$

I_d-I_q theory is one of the most frequently used control strategy for shunt AF's. Active and reactive power can be controlled by controlling d-q components independently.

5.3 PHC Method

This control method for active power filters is also known as "fundamental positive sequence component based method" [23]. The principal objective of this control method is to compensate the total harmonic currents and local reactive power demand. This control technique also eliminates the imbalance issues in the power system [23]. The source current will be given as

$$i_{Sref} = K \cdot \mu_1^+ \tag{13}$$

The power supplied by the source is

$$P_S = \mu \cdot i_{sref} = \mu \cdot K \cdot \mu_{\alpha 1}^+ = K(\mu_\alpha \mu_{\alpha 1}^+ + \mu_\beta \mu_{\beta 1}^!) \tag{14}$$

where k is calculated as

$$k = \frac{\bar{P} L_{\alpha\beta} + \bar{P} L_0}{\mu_{\alpha 1}^{+2} + \mu_{\beta 1}^{+2}} \tag{15}$$

The reference currents are represented as follows [98].

$$\begin{bmatrix} i_{S0ref} \\ i_{S\alpha ref} \\ i_{S\beta ref} \end{bmatrix} = K \begin{bmatrix} 0 \\ \mu_{\alpha 1} \\ \mu_{\beta 1} \end{bmatrix} = \frac{\bar{P} L_{\alpha\beta} + \bar{P} L_0}{\mu_{\alpha 1}^{+2} + \mu_{\beta 1}^{+2}} \begin{bmatrix} 0 \\ \mu_{\alpha 1}^+ \\ \mu_{\beta 1}^+ \end{bmatrix} \tag{16}$$

5.4 UPF Method

Unity power factor control method for shunt connected filters is also known as "voltage synchronization method", because of the desired in phase relationship between the gauge point voltage and source current vectors [23].

$$i_{Sref} = K.\mu \tag{17}$$

The Power delivered from source is

$$Ps = \mu.i_s = \mu.K.\mu = K(\mu_0^2 + \mu_\alpha^2 + \mu_\beta^2) \tag{18}$$

where the constant k is expressed as

$$k = \frac{\bar{P}L_{\alpha\beta} + \bar{P}L_0}{(\mu_0^2 + \mu_\alpha^2 + \mu_\beta^2)_{dc}} \tag{19}$$

The reference currents were calculated [98] as follows

$$\begin{bmatrix} i_{S0ref} \\ i_{S\alpha ref} \\ i_{S\beta ref} \end{bmatrix} = K \begin{bmatrix} \mu_0 \\ \mu_\alpha \\ \mu_\beta \end{bmatrix} = \frac{\bar{P}L_{\alpha\beta} + \bar{P}L_0}{(\mu_0^2 + \mu_\alpha^2 + \mu_\beta^2)_{dc}} \begin{bmatrix} \mu_0 \\ \mu_\alpha \\ \mu_\beta \end{bmatrix}. \tag{20}$$

6 Advantages and Applications of Shunt Active Filter's

Shunt active filter is an important device to improve the quality of power in the power systems. From last decades onwards many commercial projects were developed on shunt AF's and put into practice because of its less complexity and cost considerations. The Important advantages and applications of shunt active filters are listed as follows.

Advantages: Low cost and less complexity, small in size with high power ratings, rapid response to power disturbances improve transmission reliability and local power quality with voltage support. It has smooth voltage control over a wide range of operating conditions.

Applications/Functions: Current harmonic compensation, Load balancing for unbalanced star and delta systems, Flicker effect compensation for time varying loads, Power factor improvement, Reactive power compensation, are used to reduce the neutral current in 3-phase, 4-wire systems. Shunt active power filters play vital role in grid interconnection of green energy sources at the distribution level to improve the quality of power for both Day and Night time applications [24].

7 Conclusion

This paper represents the importance of research and development of shunt active filters with respect to non-linear loads in the power systems. Statcom is connected at load end to suppress the current harmonics injected by distorting loads; it can also improve the PF and voltage regulation. This paper gave the clear information regarding power quality issues, power quality standards, consequences, system configurations, control strategies, and an important advantages and applications of shunt active filters (Statcom). Over last two decades onwards due to continuous active research progress achievement in static reactive compensation (SAF) technology has produced remarkable or significant benefits to end-users, utility and wind form developers.

References

1. Rogar C Dugan "Electrical Power Systems Quality, Surya Santosa" markF.mc Granaghana, Surya Santoso. IEEE, transactions.
2. J. Nastran, R. Cajhen, M. Seliger, and P. Jereb, "Active Power Filters for Nonlinear AC loads, IEEE Trans. on Power Electronics Volume 9, No. 1, pp. 92–96, Jan(2004).
3. M. F. McGranaghan, "Overview of the Guide for Applying Harmonic Limits on Power Systems—IEEE P519A," Eighth International Conference on Harmonics and Quality of Power, ICHQP Athens, Greece, pp. 462–469, (1998).
4. K. Eichert, T. Mangold, M. Weinhold, "PowerQuality Issues and their Solution", in VII Seminario de Electrónica de Potencia, Valparaíso, Chile, Abril (1999).
5. T.J.E. Miller "Reactive power control in electric systems", IEEE Transactions, ISBN: 978-81-265-2520-1.
6. G.Yalienkaya, M.H.J Bollen, P.A. Crossley, "Characterization of Voltage Sags in Industrial Distribution System", IEEE transactions on industry applications, volume 34, No. 4, July/August, pp. 682–688, (1999).
7. Haque, M.H., "Compensation Of Distribution Systems Voltage sags by DVR and D-STATCOM", Power Tech Proceedings, 2001 IEEE Porto, Volume 1, pp. 10–13, September (2001).
8. Bollen, M.H.J., "Voltage sags in Three Phase Systems", Power Engineering Review, IEEE, Volume 21, Issue: 9, pp. 11–15, September (2001).
9. H. Akagi. New trends in active filters for power conditioning. IEEE Transactions on Industrial Applications, 32(6):1312–1322, November/December (1996).
10. Sandeep G J S M, Sk Rasoolahemmed EEE KL. UNIVERSITY, Asst. Professor, EEE KL. university 2013 Importance of Active Filters for Improvement of Power Quality. International Journal of Engineering Trends and Technology (IJETT)—Volume 4 Issue 4-April. (2013).
11. Z. Chen, F. Blaabjerg, and J.K. Pedersen. A study of parallel operations of active and passive filters IEEE Power Electronics Specialists Conference, 33rd Annual (2):1021–1026, June (2002).
12. B. Singh, K. Al-Haddad, and A. Chandra. A review of active filters for power quality improvement. IEEE Transactions on Industrial Electronics, 46(5):961–971, October (1999).
13. Milands. M. I, Cadavai. E. R, and Gonzalez. F. B, "Comparison of control strategies for shunt active power filters in three phase four wire system," IEEE Trans. Power Electron., vol. 22, no. 1, pp. 229–236, Jan. (2007).

14. Charles. S, Member, IACSIT, G. Bhuvaneswari, Senior Member, IEEE "Comparison of Three Phase Shunt Active Power Filter Algorithms". International Journal of Computer and Electrical Engineering, vol. 2, no. 1, February, (2010).
15. S. Buso, L. Malesani, and P. Mattavelli, "Comparison of current control techniques for active filters applications," IEEE Trans. Ind. Electron., vol. 45, no. 5, pp. 722–729, Oct. (1998).
16. J. Nastran, R. Cajhen, M. Seliger, and P. Jereb, "Active Power Filters for Nonlinear AC loads, IEEE Trans. on Power Electronics Volume 9, No. 1, pp. 92–96, Jan (2004).
17. M.K. Darwish and P. Mehta. Active power filters: A review M. El-Habrouk, IEE Proc.-Elertr. Power Appl., vol. 147, no. 5, September (2000).
18. R. S. Herrera, P. Salmeron, and H. Kim, "Instantaneous reactive power theory applied to active power filter compensation: Different approaches, assessment, and experimental results," IEEE Trans. Ind. Electron., vol. 55, no. 1, pp. 184–196, Jan. (2008).
19. H. Akagi, Y. Kanazawa, and A. Nabae. "Generalized theory of the instantaneous reactive power in threephase circuits." IPEC'83—International Power Electronics Conference, Toyko, Japan, 46(5):1375–1386, (1983).
20. J. Afonso, C. Couto, and J. Martins, "Active filters with control based on the p–q theory," IEEE Ind. Electron. Soc. Newslett., pp. 5–11, Sep.
21. M. Aredes, J. Hafner, and K. Heumann, "Three-phase four-wire shunt active filter control strategies," IEEE Trans. Power Electronics, vol. 12, no. 2, pp. 311–318, (1997).
22. S. Bhattacharya & D. Divan, "Synchronous frame based controller implementation for a hybrid series active filter system," in Proc. 13th IAS Annual meeting, pp. 2531–2540, (1995).
23. María Isabel Milanés Montero, Member, IEEE, Enrique Romero Cadaval, Member, IEEE, and Fermín Barrero González, Member, IEEE "Comparison of Control Strategies for Shunt Active Power Filters in Three-Phase Four-Wire Systems", IEEE Transactions on Power Electronics, vol. 22, no. 1, January (2007).
24. Rajiv K. Varma, Senior Member, IEEE, Shah Arifur Rahman, Member, IEEE, and Tim Vanderheide, Member, IEEE "New Control of PV Solar Farm as STATCOM (PV-STATCOM) for Increasing Grid Power Transmission Limits During Night and Day", IEEE Transactions on Power Delivery, vol. 30, no. 2, April (2015).

Development of 3D High Definition Endoscope System

Dhiraj, Zeba Khanam, Priyanka Soni and Jagdish Lal Raheja

Abstract Recent advances in technology have paved way for 3D endoscope, which has propelled the progress of minimum invasive surgical methods. The conventional two dimensional endoscopy based Minimally Invasive Surgery (MIS) can be performed by experienced surgeons. Inability to perceive depth was the main cause of migration to 3D endoscope. In this paper, a prototype of the stereo endoscopic system is presented. Problems pertaining to the stereo endoscope such as ease of use, inhomogeneous illumination and severe lens distortion are eliminated in the proposed system. Moreover, stereo calibration and rectification have been performed for 3D visualization. Polarization technique is used for depth perception. The proposed system also allows real time HD view to the surgeons.

Keywords Minimally invasive surgery (MIS) · Stereo calibration · Endoscopic lens distortion · Rectification · Stereo endoscope

1 Introduction

Large incisions and cuts are the root cause of the disadvantages faced during the open surgical procedure. Loss of blood, the risk of infection and long recovery time are often associated with the open surgery approach and are a traumatic experience

Z. Khanam · J.L. Raheja
Academy of Scientific and Innovative Research, Chennai, India
e-mail: khanam_zeba@yahoo.co.in

J.L. Raheja
e-mail: jagdish@ceeri.ernet.in

Dhiraj (✉) · Z. Khanam · P. Soni · J.L. Raheja
CSIR-Central Electronics Engineering Research Institute, Pilani, Rajasthan, India
e-mail: dhiraj@ceeri.ernet.in

P. Soni
e-mail: p.soni0109@gmail.com

© Springer India 2016
S.C. Satapathy et al. (eds.), *Information Systems Design and Intelligent Applications*, Advances in Intelligent Systems and Computing 433,
DOI 10.1007/978-81-322-2755-7_19

for the patient. In order to combat the cons, minimally invasive surgery (MIS) was introduced to prevent surgical disorders. Departure from large incisions and cuts was made possible by the use of an endoscopic camera during the surgery. Images acquired by the endoscopic camera helped in visualization of human cavity. The 2D viewing ability of the endoscopic camera made MIS a cumbersome task for the surgeon. Challenges related to the loss of depth perception [1] are the greatest limitation of this surgery. Three dimensionality of human anatomy makes depth an integral part of perception for surgery. This loss of depth allows visual misconception to creep in.

Many solutions have been proposed in the past for the recovery of depth during MIS. Melo [2] proposed a system which used Computer Aided Surgery (CAS) software for decreasing errors of clinical consequences. Medical robot assisted surgery [3] used ultrasound images for 3D reconstruction of the affected region. 3-D Time of Flight endoscope [4] is the current state of art in the time of the flight endoscope system. Stereo Endoscope [5] was also proposed as a solution to the problem. However, the scope of this system was limited to the database of images acquired by Da-Vinci surgical robot with 640 × 480 resolution. Moreover, the focus was on the development of a 3D endoscope as a measurement tool. Rectification of stereo frames was not carried out and instead tracking of the instrument was done. In our paper, we incorporate the step of rectification for the correct perception of 3D.

In this paper, we propose a prototype of a stereo endoscope for MIS. Awaiba NanEye 2C stereo camera was used for 3D endoscope. This miniature CMOS camera has a footprint of 2.2 × 1.0 × 1.7 mm (as shown in Fig. 1b) compared to [1] where 6.5 mm stereo endoscope was reported. Display Screen renders the 3D view of the affected region allowing continuous monitoring during surgery. Surgeons who have used earlier prototypes of 3D Endoscope have reported headaches, dizziness, ocular fatigue and nausea [6]. This can primarily be attributed to active shutter glasses. The weight and flickering caused by the active shutter glasses result in discomfort caused to the surgeon. As an alternative option to active shutter glasses, the prototype uses polarized glasses with passive screen. Issues pertaining to the endoscopic cameras such as inhomogeneous illumination and endoscopic lens distortion have been dealt with while developing the real time vision system.

2   System Overview

The main aim of the system is to retain the depth of images acquired by the endoscope. To fulfil our objectives, we developed a prototype of 3D endoscope for MIS using stereo vision. Stereo vision is inspired from the human vision system where simultaneously two images are captured from slightly different viewpoints [7]. Disparity between two images allows the brain to perceive the depth. Figure 1a shows the assembly of NanEye 2C.

(a)   **(b)**   **(c)**

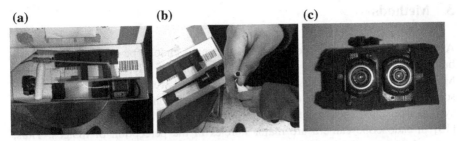

Fig. 1 System setup. **a** Stereo NanEye assembly. **b** Camera head. **c** Stereo rig for HD

With four LEDs coincident with lens (2.5 mm ring), it enables the dynamic illumination control, thus overcoming the problem of inhomogeneous illumination which is a recurrent problem of the endoscope. Figure 2 illustrates the sequence of steps required for the development of the system. The system starts with the capture of left and right frame simultaneously followed with stereo calibration and rectification is performed. Interlace image is generated using left and right rectified frames for 3D visualization. NanEye 2C camera [8] captures pair of stereo images with 250 × 250 resolution at 44 frames per second. The low resolution for Awaiba NanEye 2C proved to a constraint. To overcome this constraint a prototype for HD system was developed using stereo rig (developed using two Logitech HD cameras) as shown in Fig. 1c.

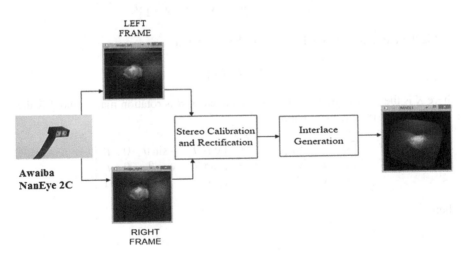

Fig. 2 System flow chart

3 Methods

A real-time vision system developed using stereo endoscope requires stereo calibration and rectification as a necessary condition for inferring 3D information. Many of the endoscopic systems [9–12] developed were restricted to monocular scope. Thus, calibration of endoscopic camera was only carried out. Since, our system uses stereo endoscope, we perform stereo calibration. Severe lens distortions [13] in endoscope camera are major hurdles in the perception of images. Thus, removal of lens distortion has been carried out. Following this, rectification is performed to bring the right and left image planes in frontal parallel configuration.

3.1 Stereo Calibration

Camera calibration allows the establishment of the relationship between an image pixel and 3D world coordinate. Intrinsic and Extrinsic parameters of stereo camera are estimated. For intrinsic camera calibration, the pinhole camera model is used. In mathematical terms [7],

$$x = PX \tag{1}$$

where x is a homogeneous image coordinate matrix and X is a homogeneous world coordinate matrix such that

$$x = [x \quad y \quad 1]^T, X = [X \quad Y \quad Z \quad 1]^T$$

P is the camera matrix which has to be estimated

$$P = K[R|T] \tag{2}$$

where K is the camera matrix (intrinsic parameter), R is rotation matrix and T is the translation matrix (extrinsic parameter).

$$K = \begin{bmatrix} 0 & \alpha_x & x_0 \\ 0 & \alpha_y & y_0 \\ 0 & 0 & 1 \end{bmatrix}, [R|T] = \begin{bmatrix} \cos\theta & -\sin\theta & 0 & t_x \\ \sin\theta & \cos\theta & 0 & t_y \\ 0 & 0 & 0 & 1 \end{bmatrix}$$

where

$$\alpha_x = \frac{f}{d_x}, \alpha_y = \frac{f}{d_y}$$

f focal length
x_0, y_0 align (0, 0)

Fig. 3 Chessboard image.
a Reference. **b** Corner points

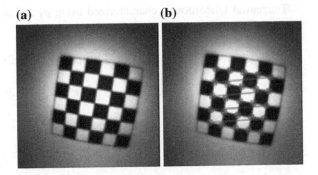

d_x, d_y physical dimension of a pixel
t_x, t_y translation vector
θ rotation angle

Based on geometric camera calibration [14], chessboard is used as reference images. A set of 20 pair of images of a chessboard was used. As seen in Fig. 3a, the chess board used contains 6 × 6 squares, each with side 4 mm. Internal corner points of the chessboard are used as reference points as seen in Fig. 3b. Stereo calibration involves computation of the geometrical relationship between the left and right camera. Thus, rotation and translation vector R and T respectively are calculated for the right camera such that its location becomes relative to the position of the left camera in global space.

3.2 Lens Distortion

The miniature size of the endoscopic camera causes severe lens distortion. Endoscopic cameras are equipped with wide eyed lens (fish eyes lens) in order to obtain a larger field of view [15]. Radial distortion is mainly due to shape of the lens whereas the assembly process of camera is responsible for the tangential distortion. The radial distortion observed in the endoscope lens is more compared to general camera lens. Radial distortion is mainly observed at the periphery of the image and is minimal at the optical center of the image. Based on the Browns method [16], we characterize the radial distortion using Taylor series in Eqs. 3 and 4.

$$x_{corrected} = x\left(1 + k_1 r^2 + k_2 r^4 + k_3 r^6\right) \tag{3}$$

$$y_{corrected} = y\left(1 + k_1 r^2 + k_2 r^4 + k_3 r^6\right) \tag{4}$$

Tangential Distortion is characterized using p_1 and p_2 as additional parameters. We model p_1 and p_2 using Eqs. 5 and 6.

$$x_{corrected} = x + \left[2p_1 y + p_2 \left(r^2 + 2x^2 \right) \right] \tag{5}$$

$$y_{corrected} = y + \left[p_1 \left(r^2 + 2y^2 \right) + 2p_2 x \right] \tag{6}$$

3.3 Rectification

In order to achieve a frontal parallel configuration such that two image planes are row aligned we need to perform stereo rectification. This allows us to align epipolar lines horizontally. We need to re project the image plane such that only horizontal disparity remains, and vertical disparity is removed. Hartley [17] and Bouguet [18] were implemented for stereo rectification.

3.4 Polarization

The passive polarization technique is suitable for 3D viewing during MIS [6]. Left and right rectified images are projected onto a polarized screen, but each image is polarized in different directions. Polarized glasses have different filters for each eye, which allows to see only the right and left eyes images by respective eyes. Input of left and right images to our respective eyes allows our brain to perceive depth.

4 Results

For testing our system, we simulated an environment inside 4 cm wide box as shown in Fig. 4. Different objects of varying sizes from 0.4 mm to 2.5 cm were fixed inside the box. Later on this system was further verified using kidney stone as seen in Fig. 6a.

Stereo rectification was carried using Hartley and Bouguet algorithms. In case of Hartley algorithm, intrinsic parameters are bypassed which leads to left and right camera to have different focal lengths, skewed pixels, different centers of projection, or different principal points. This leads us to have no perception about scale of image. Moreover radial distortion in case of endoscopic camera is more, thus calibration and distortion cannot be bypassed. Bouguet on the other hand gives

(a)   (b)

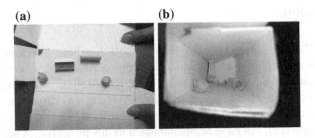

Fig. 4 Simulated environment. **a** Plan view. **b** Assembled box

(a)   (b)

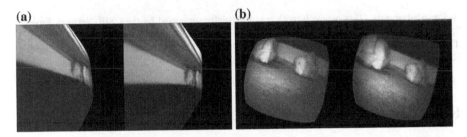

Fig. 5 Side by side view. **a** Hartley. **b** Bouguet

Fig. 6 Interlace generation.
a 3D Kidney stone. **b** 3D HD
view

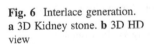

(a)   (b)

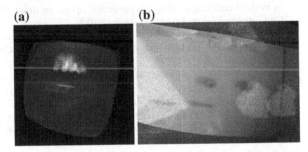

acceptable results as it considers the calibration and distortion coefficients. Figure 5a, b shows the stark difference in the result which clearly led us to reject Hartley and go ahead with Bouguet for our prototype. Figure 6b shows the result obtained using HD stereo rig system. Due to unavailability of HD resolution stereo endoscope camera, Logitech Webcam based stereo rig is used at present. In future, this will be replaced with suitable HD miniature size stereo camera.

5 Conclusion

An efficient 3D endoscopic vision system for MIS has been developed. Essential steps of stereo calibration, rectification and distortion removal has been carried out. Bouguet algorithm has been found suitable for rectification. 3D visualization has been carried out using polarization technique. Severe lens distortion, inhomogeneous illumination and difficult use are few problems which are present in earlier stereo endoscope [13], but have been eliminated in this prototype. A separate stereo Rig for HD view has been developed. This system cannot be directly employed for endoscopic view of human anatomy. In future, we target to develop a full HD system which uses a pair of stereo camera suitable for endoscopic vision.

Acknowledgments This research work was financially supported by the CSIR-Network Project, "Advanced Instrumentation Solutions for Health Care and Agro-based Applications (ASHA)". The authors would like to acknowledge the Director, CSIR-Central Electronics Engineering Research Institute for his valuable guidance and continuous support.

References

1. Brown, S.M., Tabaee, A., Singh, A., Schwartz, T.H., Anand, V.K., et al.: Three-dimensional endoscopic sinus surgery: feasibility and technical aspects. Otolaryngology–Head and Neck Surgery 138(3), 400–402 (2008).
2. Melo, R., Barreto, J., Falcao, G.: Camera calibration and real-time image distortion correction in medical endoscopy using exchangeable optics. In: Joint Workshop on New Technologies for Computer/Robot Assisted Surgery (2013).
3. Du, Q.: Study on medical robot system of minimally invasive surgery. In: Complex Medical Engineering, 2007. CME 2007. IEEE/ICME International Conference on. pp. 76–81. IEEE (2007).
4. Penne, J., Höller, K., Stürmer, M., Schrauder, T., Schneider, A., Engelbrecht, R., Feußner, H., Schmauss, B., Hornegger, J.: Time-of-flight 3-d endoscopy. In: Medical Image Computing and Computer-Assisted Intervention–MICCAI 2009, pp. 467–474. Springer (2009).
5. Field, M., Clarke, D., Strup, S., Seales, W.B.: Stereo endoscopy as a 3-d measurement tool. In: Engineering in Medicine and Biology Society, 2009. EMBC 2009. Annual International Conference of the IEEE. pp. 5748–5751. IEEE (2009).
6. Ohuchida, K., Eishi, N., Ieiri, S., Tomohiko, A., Tetsuo, I.: New advances in three-dimensional endoscopic surgery. J Gastroint Dig Syst 3(152), 2 (2013).
7. Hartley, R., Zisserman, A.: Multiple view geometry in computer vision. Cambridge university press (2003).
8. Overview, N.: Naneye family overview | awaiba cmos image sensors (Oct 2014), http://www.awaiba.com/product/naneye-stereo/.
9. Shahidi, R., Bax, M.R., Maurer Jr, C.R., Johnson, J., Wilkinson, E.P., Wang, B., West, J.B., Citardi, M.J., Manwaring, K.H., Khadem, R., et al.: Implementation, calibration and accuracy testing of an image-enhanced endoscopy system. Medical Imaging, IEEE Transactions on 21 (12), 1524–1535 (2002).
10. Barreto, J., Roquette, J., Sturm, P., Fonseca, F.: Automatic camera calibration applied to medical endoscopy. In: BMVC 2009-20th British Machine Vision Conference. pp. 1–10. The British Machine Vision Association (BMVA) (2009).

11. Melo, R., Barreto, J.P., Falcao, G.: A new solution for camera calibration and real-time image distortion correction in medical endoscopy–initial technical evaluation. Biomedical Engineering, IEEE Transactions on 59(3), 634–644 (2012).
12. Furukawa, R., Aoyama, M., Hiura, S., Aoki, H., Kominami, Y., Sanomura, Y., Yoshida, S., Tanaka, S., Sagawa, R., Kawasaki, H.: Calibration of a 3d endoscopic system based on active stereo method for shape measurement of biological tissues and specimen. In: Engineering in Medicine and Biology Society (EMBC), 2014 36th Annual International Conference of the IEEE. pp. 4991–4994. IEEE (2014).
13. Wengert, C., Reeff, M., Cattin, P.C., Székely, G.: Fully automatic endoscope calibration for intraoperative use. In: Bildverarbeitung für die Medizin 2006, pp. 419–423. Springer (2006).
14. Heikkilä, J.: Geometric camera calibration using circular control points. Pattern Analysis and Machine Intelligence, IEEE Transactions on 22(10), 1066–1077 (2000).
15. HSU, C.H., MIAOU, S.G., CHANG, F.L.: A distortion correction method for endoscope images based on calibration patterns and a simple mathematic model for optical lens. Biomedical Engineering: Applications, Basis and Communications 17(06), 309–318 (2005).
16. Duane, C.B.: Close-range camera calibration. Photogram. Eng. Remote Sens 37, 855–866 (1971).
17. Hartley, R.I.: Theory and practice of projective rectification. International Journal of Computer Vision 35(2), 115–127 (1999).
18. Bouguet, J.: The calibration toolbox for matlab, example 5: Stereo rectification algorithm. code and instructions only), http://www.vision.caltech.edu/bouguetj/calib_doc/htmls/example5.Html.

Excel Solver for Deterministic Inventory Model: INSOLVER

Pratiksha Saxena and Bhavyanidhi Vats

Abstract An excel solver is proposed for optimized result of deterministic inventory models. Seven inventory models including price break models can be solved by using this Excel Solver. This solver is designed by using yes-no algorithm. INSOLVER has benefit of solving even quantity discount inventory models with different price breaks, which makes it unique.

Keywords Excel solver · Inventory models · Deterministic inventory models · Quantity discount models

1 Introduction

Inventory control problems deal with accounting the optimized decisions for cost and quantity. The designing of optimal inventory model is the task to fulfil the demand of the customers and make profit to the organization. A number of models and algorithms have been developed for this purpose. Main objective of inventory model is optimization of ordering quantity and minimization of total cost associated with it. Research and models are developed due to randomly changing demand, effective management of production and inventory control operations in dynamic and stochastic environments.

EOQ model was proposed by T.C.E Chang, on the basis of the fact that the product quality is not always perfect and a relationship was formulated between unit production cost and demand [1]. In case of sporadic demand or infrequent demand the conventional inventory policy is replaced with base stock policy by the Scultz. The base stock inventory policy was evaluated and forecasting procedure for the size and frequency of sporadic demand was presented. It determines the base stock level and the timing of replenishment orders so that holding cost is signifi-

P. Saxena (✉) · B. Vats
Department of Applied Mathematics, Gautam Buddha University, Greater Noida, India
e-mail: pratiksha@gbu.ac.in

© Springer India 2016
S.C. Satapathy et al. (eds.), *Information Systems Design and Intelligent Applications*, Advances in Intelligent Systems and Computing 433,
DOI 10.1007/978-81-322-2755-7_20

cantly reduced with a minor risk of stock outs [2]. Drezener et al. considered the case of substitution in the EOQ model when two products are required. Three cases for no substitution, full substitution and partial substitution between the products are considered. An algorithm is presented to compute the optimal order quantities and observed that full substitution is never optimal; only partial or no substitution may be optimal. The case of three products is a local extension of that problem [3]. Prasad Classified the inventory models so that the selection of an appropriate inventory model can be made easy for the particular set of requirements and inventory conditions. The objective is to differentiate the inventory models according to the salient inventory features for the better understanding of inventory literature and to discover related research areas of this field [4]. A literature review is presented for the dynamic pricing inventory models. Current practices and future directions related to dynamic pricing and its applications in different industries were discussed [5]. In 2005, Haksever and Moussourakis developed a model for optimizing multiple product inventory systems with multiple constraints by using mixed integer linear programming technique [6].

An EOQ Model has been developed by Bose et al. for deteriorating items with linear time dependent demand rate and allowed Shortages. The inflation rate and the time value of money were counted into the model and assumed that the deteriorating rate is constant over the period of time. The model is solved by both analytical and computational method and a sensitive analysis is also done for the different parameters of a system [7]. Wee has developed a deterministic inventory model with the quantity discount, pricing and partial backordering. The deteriorating condition following the Weibull rate of distribution was also accounted. The aim is to maximize profit rather than optimizing or minimizing inventory costs [8]. A study was summarized on the recent trends, advances and categorization of deteriorating inventory models [9]. A model was developed to determine the lot size for the perishable goods under finite production, partial backordering and lost sale. This model also involved the assumption of exponential decay [10]. With the passage of time there is a huge advancement in methods and techniques of inventory control and its optimization. In 2011, Abuizam et al. used simulation technique for the optimization of (s, S) periodic review inventory model with uncertain demand and lead time. In this paper the result is compared by using two software's @Risk simulation and @Risk Optimizer to get the minimum cost of inventory over a period of time [11]. Fuzzy set theory has been applied by De and Rawat for the optimization of fuzzy inventory model under fuzzy demand rate and lead time using exponential fuzzy numbers. The optimal value of the reorder point and safety stock is calculated by the mean of fuzzy numbers and graded mean integration representation method [12]. In 2011, Amir Ahmadi Javid worked on the integration of location; routing and inventory decisions. Objective is to design a multisource distribution network at an optimal cost. Mixed integer linear programming formulation and three phase heuristic approach were used for this purpose. A numerical analysis comparison has been done between these two methods [13]. Different inventory control techniques for efficient inventory management system were analyzed and compared by Tom Jose et al. In this study various methods such as EOQ method, ABC Classification, Safety stock, FSN are used to compare the results [14]. An integrated vendor buyer inventory model with controllable lead time, ordering cost reduction and service level constraint was presented and an algorithm was developed to find the optimal solution of the proposed model [15]. In 2014, Venkateswarlu et al. proposed an inventory model for perishable items by taking into account the assumptions that the demand is quadratic function of time and the rate of deterioration is constant. EOQ model was developed for optimizing the total inventory cost under inflation and the sensitivity analysis was also done by taking a numerical example [16].

The proposed INSOLVER has additional aspects from the previous research work as it comprises more inventory parameters. It also includes quantity discount inventory models with one and two price-breaks. It makes different from the existing excel solvers. By using the comparison of alternative conditions for the best outcome is possible which ultimately results in optimized results.

2 Materials and Methods

Seven deterministic models have been integrated for single item inventory models. Different conditions and assumptions are taken for each model. Models are incorporated with and without shortages, quantity Discounts, constant and varying demand rate in this Excel Solver. The Proposed Excel Solver has been integrated with seven different conditions for better inventory management. The alternative Conditions for different cases can be compared through this model and best one can be selected to minimize total Inventory Costs. The Individual excel sheet has been taken to separate home page (conditions are mentioned), result page and input page for easy access. Selection of the different conditions has been designed with the help of yes-no algorithm.

The model I is considered with constant demand and no Shortage Condition. Assumptions for this model are taken as;

1. The demand is known and constant
2. Supply is instant that means lead time is constant and known
3. Shortages and Quantity Discounts are not allowed
4. Carrying cost per year and Ordering cost per order are known and constant
5. The inventory model is used for single product/item.

The model II is considered with the different rates of demand and with no shortage condition. Assumptions for this model are;

1. Demand is varying from period to period
2. Shortages and Quantity Discounts are not allowed
3. Supply is instant that means lead time is constant
4. Carrying Cost per year and Ordering cost per order are constant and known
5. The inventory model is used for single product/item.

For the model III, gradual supply is involved with no shortage condition. Assumptions for model III are;

1. Demand is constant and known
2. Production of the item is continuous and at a constant rate until the Quantity (Q) is complete that means the supply/replenishment is gradual
3. The rate of production (p) is greater than the rate of consumption (d)
4. Production set up cost is fixed.

Model IV is included with all assumption of Model I and existence of storages allowed. Model V represents the conditions of constant demand, fixed order cycle time with Shortages allowed. Model VI is considered with assumptions of model III and shortages allowed. Model VII is considered with the condition of quantity discounts with one and two price breaks. Assumptions are;

1. Demand is known and constant
2. Shortages are not allowed
3. Replenishment is instantaneous
4. Quantity Discounts are allowed

These seven deterministic inventory models are included in this solver to obtain quick and exact results. Flow chart for INSOLVER is given below.

Algorithm for INSOLVER can be described as;

1. The Model has been developed by using the yes-no algorithm which is based on the binary numbers 0 and 1.
2. 'yes' implies for 1, that means the model will consider that condition and 'no' implies 0 when model does not consider the condition.
3. All the Conditions has been provided on the home page using yes-no algorithm.
4. All the Input parameter has been defined on the input sheet which provides initial and boundary conditions for the problem.

3   Discussion

The proposed inventory model is designed on the MS Excel for the ease of calculating the outputs. The seven different conditions are integrated in the models by putting the formulas for each case. The individual excel sheets are defined for various conditions. Each sheet is designed with the help of yes-no algorithm. The working of this Model is shown by the following steps:

Step 1   On the home page, yes/no conditions are inserted according to the model. For example, to insert shortage allowed condition, yes/no is inserted in the form of binary digit. If model allows shortage condition 1 will be assigned to the row and if shortage is not allowed for the model, 0 will be assigned to the corresponding row. After checking shortage condition for the model,

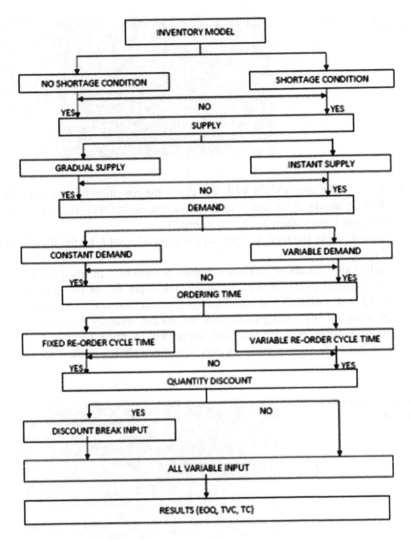

Fig. 1 Flow chart for INSLOVER

it will check other parameters of the model. Figure 1 represents home page with first parameter of chosen allowed shortage model (Fig. 2).

Step 2 After selecting shortage condition of the model, it will check the demand rate for the model. If demand rate is constant, 1 will be assigned to the corresponding row and 0 will be assigned to the third row for variable demand rate. Vice versa can be chosen according the selected model.

Step 3 Supply pattern can also be checked by INSOLVER. It may be gradual or instant and accordingly binary digit can be assigned to corresponding row.

Fig. 2 Home screen for
INSOLVER

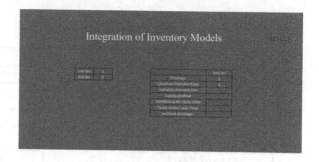

Fig. 2 Home screen for INSOLVER

Step 4 At the fourth step, INSOLVER will check the condition for Ordering Time (Fixed ordering cycle time/Variable ordering cycle time). Home page with assigned binary digits is shown in Fig. 3.

After defining the model, input page will be selected from the INSOLVER (Fig. 4).

Step 5 At this step, values of all input variables according to the model conditions are assigned to the defined variables. The input values for demand, carrying cost, ordering cost, shortage cost, production rate and unit cost will be assigned. According to the selected model, quantity discount and discount scheme will be specified according to price breaks. The input parameters for the different models have been shown in the Fig. 5.

Fig. 3 Home screen for INSOLVER

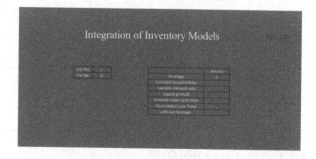

Fig. 4 Home screen with all input parameter conditions

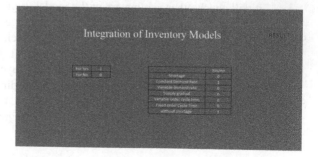

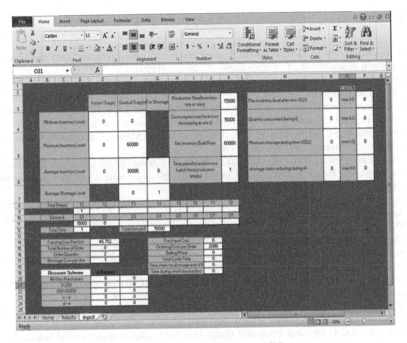

Fig. 5 Input page represents input variables for different conditions

Fig. 6 Result page

	Instant Supply	Gradual Supply	Shortage
Total Carrying Cost	0.25	194.8714661	0.3125
Order Quantity	10000		
Total Ordering Cost	12.5	12500	
Fixed Purchase Cost	100000		
Total Shortage Cost		11248593.93	2.5

	Instant supply(constant /variable	Gradual Supply(with/with out shortage)	Shortage(variable/fixed reorder cycle	DISCOUNT For all	a - b	b - c	c - d	d - e
Total Variable Inventory Cost	12.75	11261288.8	15.3125					
Optimal Quantity to Be Ordered(Q*)	1000	1417.744688	1002.496683	101.015	100	102.598	105.409	108.465
Optimal Interval Between Successive Order	0.1	0.472581563	0.100249688					
Optimal Number of Order	10	8.055456159						
Optimal Total Variable Inventory Cost (TVC*)	100	70.71067812	99.75093361					
Optimal Total Inventory Cost	100100	100070.7107		980990	1001000	950975	900949	850822
Optimal Inventory Level (M*)			997.5093361					
Optimal Shortage level (R*)			4.987546681					

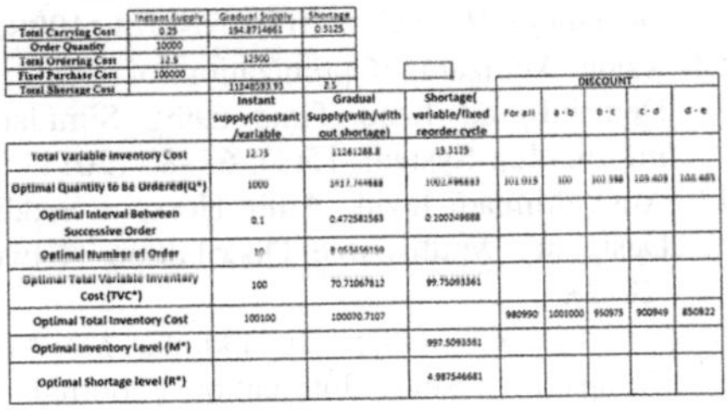

Output variables can be obtained on the result page according to the chosen model and input variables (Fig. 6).

4 Conclusion

Different inventory models are integrated into this MS Excel Solver to obtain output. The different alternative conditions can be combined for minimum inventory cost by using this INSOLVER. The benefit of using this INSOLVER is ease of accessibility of result for less time consumption. The solver can also be applied in different sectors such as manufacturing industries, warehouses, multi-brand and multi product retail stores which shows its wide scope for inventory management.

References

1. T.C.E Cheng,: An EOQ model with demand dependent unit production cost and imperfect production processes. IEEE Transactions, 23, 23–28 (1991).
2. Carl R. Scuhltz,: Forecasting and Inventory Control for sporadic demand under periodic review. The Journal of the Operations Research Society, 38(5), 453–458 (1987).
3. Zvi Drzener, HareshGurnani and Barry A. Pasternack,: An EOQ Model with substitutions between products. Journal of the Operations Research Society, 46(7), 887–891 (1995).
4. Sameer Prasad,: Classification of Inventory Models and Systems. International Journal of Production Economics, 34, 309–322 (1994).
5. Wedad Elmaghraby, Pinar Keskinocak,: Dynamic Pricing in the presence of Inventory Considerations: Research Overview, Current Practices and Future Directions. Management Science, 49(10), 128–1309 (2003).
6. Cengiz Haksever, John Moussourakis,: A model for optimizing multi product inventory systems with multiple constraints. International Journal of Production Economics, 97, 18–30 (2005).
7. S. Bose, A. Goswami, K.S. Chaudhuri,: An EOQ Model for Deteriorating Items with Linear Time Dependent Demand Rate and Shortages under Inflation and Time Discounting. Journal of the Operations Research Society, 46, 771–782(1995).
8. Hui Ming Wee,: Deteriorating Inventory Model with Quantity Discounts, Pricing and Partial Backordering. International Journal of Production Economics, 59, 511–518 (1999).
9. S.K. Goyal, B.S. Giri,: Recent Trend in Modeling of Deteriorating Inventory. European Journal of Operations Research, 134, 1–16 (2001).
10. Prakash L. Abad,: Optimal Price and Order size for a reseller under partial Backordering. Computers and Operations Research, 28, 53–65 (2001).
11. B.C. Giri, K.S. Chaudhuri,: Theory and Methodology: Deterministic Models of Perishable Inventory with stock dependent demand rate and non linear holding cost. European Journal of Operations Research, 105, 467–474 (1998).
12. Raida Abuizam,: Optimization of (s,S)Periodic Review Inventory Model with Uncertain Demand and Lead Time using Simulation. International Journal of Management and Information System, 15(1), 67–80 (2011).
13. Amir Ahmadi-Javid, Amir Hossein Seddighi,: A Location Routing Inventory Model for Designing Multisource Distribution Networks. Engineering Optimization, 44(6), 637–656 (2012).
14. Tom Jose V., Akhilesh Jaya kumar, Sijo M. T.,: Analysis of Inventory Control Techniques: A Comparative Study. International Journal of Scientific and Research, 3(3), 1–6 (2013).
15. Sara Shahpouri, ParvizFattahi, Ali Arkan, Kia Parsa,: Integrated Vendor Buyer Cooperative Inventory Model with controllable lead time, Ordering cost reduction and service level constraint. International Journal on Advanced Manufacturing Technology, 65, 657–666 (2013).
16. R. Venkateshwarlu, M.S. Reddy,: Time Dependent Quadratic Inventory Models under Inflation. Global Journal of Pure and Applied Mathematics, 10(1), 77–86 (2014).

DSL Approach for Development of Gaming Applications

Aadheeshwar Vijayakumar, D. Abhishek and K. Chandrasekaran

Abstract This research paper mainly concentrates on introducing DSL(Domain Specific language) approach in developing gaming applications. DSL approach hides the lower level implementation in C, C Sharp, C++, and JAVA and provides abstraction of higher level. The higher level of abstraction provided by the Domain Specific Language approach is error-free and easy to develop. The aim of this paper is to propose an approach to use GaML (Gamification Modelling Language, a form of DSL for gaming) for Unity based complex games efficiently in this paper. The paper doesn't focus on the How and Whys of the Gaming Modelling Language usage, but rather focuses on the run-time enforcement. At the end of the paper, survey has been made on total lines of code and time invested for coding using a case study. The case study proves that DSL approach of automated code generation is better than manual.

Keywords DSL · Gaml · Game development · Unit

1 Introduction

Computer game development has become a sophisticated subject, drawing on advanced knowledge in a number of areas of computer science, such as artificial intelligence and computer graphics. In this stream, people will learn about both the principles and practice of designing and developing modern computer games.

A. Vijayakumar (✉) · D. Abhishek · K. Chandrasekaran
Department of Computer Science and Engineering, National Institute of Technology,
Surathkal, Karnataka, India
e-mail: vkaadheeshwar95@gmail.com

D. Abhishek
e-mail: abhid95@gmail.com

K. Chandrasekaran
e-mail: kchnitk@ieee.org

© Springer India 2016
S.C. Satapathy et al. (eds.), *Information Systems Design and Intelligent Applications*, Advances in Intelligent Systems and Computing 433,
DOI 10.1007/978-81-322-2755-7_21

199

Anyone who have the brains, the talent, and the courage to build a game, a career in Computer Game Development [1] may be right for them. The glory, the money, the job security and the creative satisfaction can all be theirs, but only if people have what it takes to do the job. This paper proposes a component based strategy to model game development. The approach used is made available as a Domain-Specific Language (DSL) [2], which is given to the game designer in the form of models. It introduces GaML [3] for development of gaming application. The meta-model based language supports the designing of the development and deployment dependencies between different segments. The paper's primary focuses are as follows. Sub Division 2 illustrates the background knowledge of game engine Unity, and Gaming Modelling Language. Sub Division 3 illustrates the problem statement. Sub Division 4 presents the proposed model and technique for development of gaming applications. Sub Division 5 explains in detail the approaches along with case study. Sub Division 6 provides with the Evaluation of proposed approach through survey. Sub Division 7 concludes our work and also the future work needed.

2 Literature Review

2.1 Unity

Unity is a cross-platform game engine with a builtin IDE developed by Unity Technologies. It is used to develop video games for desktop platforms, web and web related plugins, consoles and cellular phones [4]. The Unity game engine is also often used for designing more complicated games. They make use of the properties of Unity such as physics, gravity, networking, cross compiling and other useful properties. The categories, under which these serious games generally appear, include:

- Layer 1: Product demonstration and configuration (e.g., Bike configurator, Gun Construction)
- Layer 2: Exercise simulations (e.g., US Navy Virtual Training, I.R.I.S.)
- Layer 3: Architectural visualization (e.g., Insight 3D, Aura Tower)
- Layer 4: Education and instruction (e.g., Micro Plant, Surgical Anatomy of the Liver)

2.2 Design Objectives of GaML

The paper defines four design objectives for the language in general. First, GaML [3] formalizes the gamification concepts stated in the requirement section to address

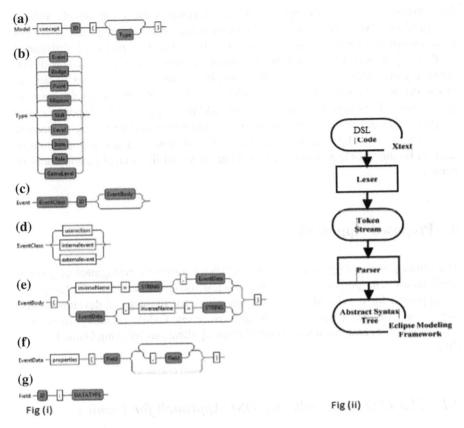

Fig (i)   Fig (ii)

Fig. 1 **i** GaML design. **ii** From code to model and vice versa

problem statement P1. A parser has to decide, whether an instance is well-formed according to the languages grammar or not. Second, a valid instance of GaML [4] should be automatically compilable to gamification runtimes, e.g., gamification platforms (P2). Third, GaML should be at least readable for domain experts with minor IT knowledge, e.g., consultants or designers. Fourth, GaML should be fully writable for IT experts and partially writable by domain experts which addresses problem statement P3. A brief picture is given in the Fig. 1i.

3 Problem Statement

The problem of development of Unity based gaming applications and its toughness has remained same over the years. The problem is in Unity, everything need to be developed from scratch. So the main focus is on proposing an approach so that anybody liking to develop a game can develop it using the models already generated by DSL experts. Another problem is that game development is always concentrated in gaming industries. The focus is also laid on making it distributed so that any person who is not much into programming but interested in developing a game with his own idea shall be able to do it. So the focus is laid on proposing ideas upon GaML also which is a form of DSL used for gaming applications. The challenge therefore is to making the gaming development easier by preventing developer from the toughness and errors and providing predefined templates through DSL. Another primary goal is to propose a desirable solution or strategy to this problem in order to facilitate easy and distributed development of games.

4 Proposed Approach

The primary goal is to provide a strong backbone for easy development of gaming applications. The proposed approaches suggest using Domain Specific Language [2] approach for developing gaming applications for those game developers who use Game Engines (like Unity). It also proposes GaML (Gaming Modelling Language) [3, 4] as an idea to develop gaming applications by using Game Engines (like Unity).

4.1 The Idea of Introducing DSL Approach for Gaming

The approach involves Domain specific language which makes the game designers to concentrate better on the gaming aspect. The approach involves several steps given below [2]. At first, the programmers specialized in DSL will produce class templates required for a specific type of game which they combine and say domain specific Language of a certain game. Game Developers do this in the form of GaML (Gamification Modelling language which is similar in concept to DSL in game development). Next, Game Designers make use of the language produced to model gaming applications. The major benefit of this approach is that the DSL provides power to game designers to model games by making use of higher level abstraction models for presenting gaming aspects. Compared to game development with a normal coding language such as C++, C Sharp or Java, coding details are not given to game Designers. The mixing of data abstracted models and the feature of code generation by itself gives the best design and reduces the time of development.

4.2 Code to Model Generation

Xtext is an Eclipse plug-in that supports domain specific language development. Due to this fact, many of Xtext functions are based on already established frameworks in Eclipse and on the features of Eclipse itself [5]. The final outcome of a DSL development in Xtext is again available as an Eclipse plug-in. Hence, it does not only provide a compiler and a code generator but a complete development environment. Furthermore, this can be adjusted to any specific requirements of a particular DSL. The advantage here is that many developers are familiar with such a common environment like Eclipse. A possible disadvantage, however, could be that this tool is directed to a target group of programmers due to its considerable level of complexity. Another special feature of Xtext is the use of a text editor as an input medium. Hence, the DSL code (GaML code) is entered by the end users as free text. They are thereby assisted by automatic code completion of the inputs, syntax highlighting and other features. A validation of their inputs is as well made using various criteria, which can be defined by the end-users themselves. However, due to the entering of the GaML code in the form of a free text, Xtext should at first convert the user's input into a semantic model. Afterwards, validations and transformations (e.g., target code generation) are carried out based on the semantic model [6]. This conversion is performed with the help of a lexer (lexical analyzer) and a parser. In the first instance and according to specific rules, the entered free text is decomposed by the lexer in individual related components; so-called tokens. Based on the lexers outcome, the parser interprets the users input and translates it into a semantic model (Abstract Syntax Tree) using the defined parser rules. For the projection of this model, the modeling constructs of the Eclipse Modeling Framework are used. Only when this transformation of the free text into a semantic model can be successfully finished, other transformations such as target code generation could be made.

4.3 Model to Code Transformation

The code generators were written using the Xtend language, which was specifically configured to enable programming based on text templates. The information provided by the semantic model is accessible within the Xtend program by using the Java classes, which represent the languages AST model. These classes are generated automatically via the Eclipse Modeling Framework [4]. Xtend is a statically typed programming language whose typing, however, is mostly done by type inference (i.e., automatic deduction of the type of expressions). Therefore, types are rarely explicitly declared. The Xtend language was specifically designed to improve some of the existing Java concepts. Upon execution, Xtend is not directly translated into Bytecode but Java code. This makes Xtend programs deployable on arbitrary platforms. Moreover, this has the advantage that the programmer can trace the Java programming constructs, which result from his code. Within Xtext, Xtend is used as a code generator for the implemented DSLs. The Xtend constructs that were especially intended to be used for code generation include dispatch functions, which are featured by Xtend, in addition to normal functions [7], whose use is done via a static binding through their parameter values. By using the keyword dispatch, various functions (having the same name) can be combined into a set of dispatch functions. Within these overloaded functions, it is decided through the type(s) of the input parameter(s), which function is called. The code snippet below shows two dispatch functions having the same identifier compile but different input parameters (due to demonstration purposes, the snippet shows a possible way to generate Java code from the GaML model [4]; though in this work JSON code is generated). One of the main tasks of the code generation is to put together long strings. In order to make this task as simple as possible and obtain a readable code as well, Xtend offers rich strings. To mark the use of such strings within the code, three single quotation marks are used at the opening and at the end (see code above). Within this template the objects attributes can be accessed. It is also possible to call other (self defined) functions and to use available features such as IF and FOR. To call such functions inside of the template, we need to put them between the signs ≪and≫; everything within this signs is thus executed as code.

Figure 2 explains, how proposed approach is better than older approach. In older approach the concept of using design principles was developed and the same person used to be a domain expert. He codes the requirement specific templates and uses it whenever needed, hence does not concentrate on the generalised version. This approach is error prone and time consuming as coding and designing is done by the same person, Fig. 2b describes this. In proposed approach the focus is laid on separating the designing of game from coding DSL templates for the game. Domain expert prepares the general purpose concept using DSL principles for gaming. These templates are error free so that any game developer can use it instead of coding from scratch. Game developer takes care of the design and development subcomponent of the game. It will be easier for him to develop games using concepts provided by Domain expert and faster than older method. Figure 2a describes this.

5 Implementation

The game development is a huge task for beginners as they are unaware of the higher levels of abstractions. The implementation is further explained using an instance of developing a 2D game in unity. The first phase of development involved

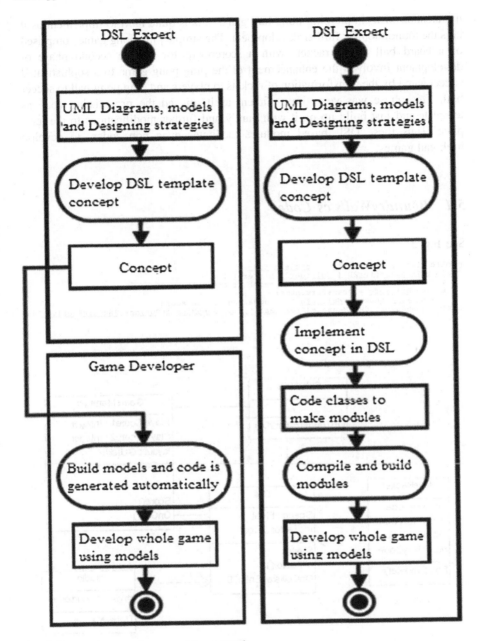

Fig. 2 Proposed approach versus older approach

the creation of a simple ping pong ball game, this is quite hard to implement as it lays the foundation for further development. The simple ping pong game comprised of a board ball and 2 rackets with a scoreboard for it. The second phase of development involved the enhancement of the ping pong game to a sophisticated soccer game by the transformation of rackets to players and ping pong ball to soccer ball. This phase was very time efficient as it utilised the pre-existing code as template and just build up on it with hardly and further modifications to the template. Thus these DSL based modules facilitates even non coders to develop high-end games.

5.1 BoundryWalls.cs Code

See Fig. 3.

```
public class BoundryWalls : MonoBehaviour {
  void OnTriggerEnter2D ( Collider2D hitter ) {
   if (hitter.name == "Ball"){ // to check if the ball collided
    string Wallinfo = transform.name;
    GameManager.Score ( Wallinfo );// updating of the score
  hitter.gameObject.SendMessage("resetBall",0.5f,SendMessageOptions.RequireReceiver );
   } }}
```

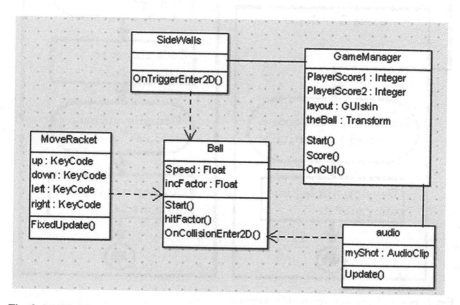

Fig. 3 Model driven approach for case study

5.2 GameManager.cs Code

```
public class GameManager : MonoBehaviour {
 public static int Points1 = 0, Points2 = 0;// to keep track of scores of 2 players
 public GUISkin layout;
 Transform theBall;
 void Start() { //initiating ball movement
  theBall = GameObject.FindGameObjectWithTag (" Ball ").transform;
 }
 public static void Score ( string wallID ) {
  if ( wallID == "rightWall" ) Points1++; // allot points based on side of collision
  else Points2++;
 }
 void OnGUI()
 {
  GUI.skin = layout;
  GUI.Label ( new Rect ( Screen.width / 2 - 150
                  - 12, 20, 100, 100 ), "" + Points1 );
  GUI.Label ( new Rect (Screen.width / 2 + 150
                  + 12, 20, 100, 100 ), "" + Points2 );
  if ( GUI.Button ( new Rect ( Screen.width/2
                  - 60, 35, 120, 53 ), "RESET" )) {
   Points1 = 0;Points2 = 0;
   theBall.gameObject.SendMessage (" resetBall ",
                  .5f, SendMessageOptions.RequireReceiver );
  }// End of the game (Winning pts)
 if ( Points1 == 15) {
  GUI.Label ( new Rect ( Screen.width / 2 - 150,
                  200, 2000, 1000), "Congrats Player1 " );
  theBall.gameObject.SendMessage (" hasWon ", null,
                  SendMessageOptions.RequireReceiver);
 }
 else if ( Points2 == 15) {
  GUI.Label ( new Rect ( Screen.width / 2 - 150,
                  200, 2000, 1000), "Congrats Player2 " );
  theBall.gameObject.SendMessage ( "hasWon",null
                  , SendMessageOptions.RequireReceiver );
 }
 }
}
```

5.3 MoveStriker.cs Code

```
#define tfm transform
#define tlt translate
public class MoveStriker : MonoBehaviour {//top and bottom keys ( manually set )
 public KeyCode top, bottom, left, right;
 void FixedUpdate ()
 { // Striker Movements
  if ( Input.GetKey( top ) ) { //upward
   tfm.tlt ( new Vector2 (0.0f, 0.25f) ); }
  if ( Input.GetKey( bottom ) ) { //downward
   tfm.tlt ( new Vector2 (0.0f, -0.25f) ); }
  if ( Input.GetKey( left ) ) { //leftward
   tfm.tlt ( new Vector2 (-0.25f, 0.0f) ); }
  if ( Input.GetKey( right ) ) { //rightward
   tfm.tlt ( new Vector2 (0.25f, 0.0f) ); }
 }
}
```

5.4   Ball.cs Code

```
#define vel velocity
public class Ball : MonoBehaviour {
 public float speed = 2.0f; // Initial speed
 public float incFactor = 0.2f; //amt to increase speed by each time
 void Start() {// to set ball with initial movement direction
  rigidbody2D.vel = V2.one.normalized * speed;
  speed += incFactor;
 }            // V2 is Vector2
 float collision_Component ( V2 bPos, V2 sPos,float StrikerHeight ) {// ascii art:
  // ⊔  1 <- top , 0 <- center ,-1 <- bottom of the Striker
  return( bPos.y - sPos.y )/ StrikerHeight;
 }
 void OnCollisionEnter2D ( Collision2D c ) {
  if ( c.gameObject.name == "StrikerLeft" ) {
   float y = collision_Component ( transform.position, c.transform.position,
 c.collider.bounds.size.y );// Calculate direction, make length=1
                    via .normalized
   V2 direc_ = new V2 ( 1, y ) .normalized;
   rigidbody2D.vel = direc_ * speed;
   speed += incFactor;
  }
  if ( c.gameObject.name == "StrikerRight" )
  {//if collision happens with the right striker// Calculate Collision Component
   float y = collision_Component ( transform.position,
   c.transform.position, c.collider.bounds.size.y );
                    via .normalized
   V2 direc_ = new V2 ( -1, y ).normalized;
   rigidbody2D.vel = direc_ * speed;
   speed += incFactor;
}}}
```

5.5   BackgroundMusic.cs Code

See Fig. 4.

```
public class BackgroundMusic : MonoBehaviour {
public BackgroundMusicClip myShot;
 void Update()
 {
  BackgroundMusic.clip = myShot;
  BackgroundMusic.Play ();
 }
}
```

This simple game was first phase of development and took much more time than expected, to make it bug free. The visual appearance is not that convincing and is certainly not worth the efforts. But it has all the modules like ball, racket, boundary sidewalls from which any kind of related 2d game like football, table tennis, hockey, cricket, tennis, soccer, badminton etc. can be developed by organising modules appropriately and giving desired background. The figure of ping pong ball game is shown and also soccer game which is built from the simple ping pong game in second phase in very less time uses modules built in first phase (Fig. 5).

This is the power of DSL. In this way it helps to build better, bug-free and complicated games using modules.

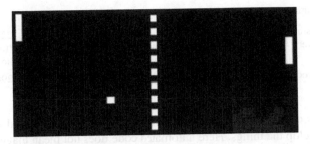

Fig. 4 Ping pong ball game developed using 18 h

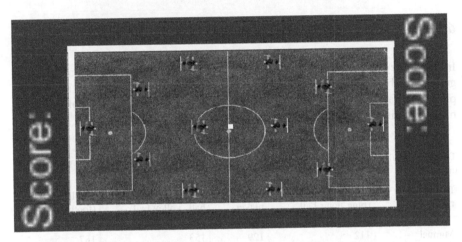

Fig. 5 Final KICKOFF game developed in an hour using models listed above

6 Evaluation of Proposed Approach (Survey)

The proposed DSL approach (developed in the form of GaML) can be evaluated based on following two factors:

- Amount of code
- Time required for development

The paper illustrates taking into consideration the above case study.

6.1 Amount of Code

Table 1 represents the amount of code generated automatically by DSL approach. This tells us how much of manual work can be avoided to develop a game. It can be observed that for every 4 player addition the number of line increases by 50. Hence

Table 1 Describing automated code generated for particular games

Games with distinct number of players	2 players (ping-pong ball, tennis)	8 players (polo)	12 players (basket ball, volley ball)	22 players (football, hockey)
Total lines of code	106	162	198	290

automated code generation is much better than manual coding. And even a non coder can develop the things. Here automated code does not mean that every player has same features, instead it means a slight modification in features of players.

6.2　Time Required for Development

In this survey comparison is made between the manual coding and automated code generation by DSL for various games involving teams of different number of players. It can be observed that automated code generation saves a lot of time than manual coding. Table 2 gives the picture of it (Fig. 6).

Table 2 Total time taken in minutes to develop a game for well prepared coder well versed in unity

Development type	2 players (ping-pong ball, tennis)	8 players (polo)	12 players (basket ball, volley ball)	22 players (football, hockey)
Manual	112	129	153	187
Automated (100+)	13	17	21	28

Fig. 6 Manual and automatic development

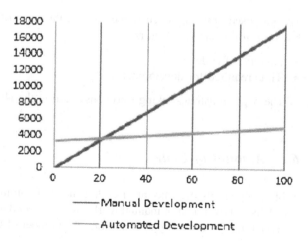

Manual Development

Automated Development

7 Summary and Conclusions

A case study is shown above on a game development team using Unity without using DSL and with using DSL. The time taken was 10 times better. And the GaML approach of development of game developed using DSL is much more efficient for a game designer and it gives higher level abstraction of the game so that developer can easily implement his ideas without having to care about underlying code. Further it is possible to distribute the work among DSL experts and Developers and develop the game without much bugs in it as the models provided by DSL expert to developer will be tested and used in many other gaming applications.

References

1. A.K. Baughman, W. Chuang, K.R. Dixon, Z. Benz, and J. Basilico. Deepqa jeopardy! gamification: A machine learning perspective. Computational Intelligence and AI in Games, IEEE Transactions on, 6(1):55–66, March 2014.
2. Sledziewski, Krzysztof, Behzad Bordbar, and Rachid Anane. "A DSL-based approach to software development and deployment on cloud." *Advanced Information Networking and Applications (AINA), 2010 24th IEEE International Conference on.* IEEE, 2010.
3. Herzig, Philipp, et al. "GaML-A modeling language for gamification."*Proceedings of the 2013 IEEE/ACM 6th International Conference on Utility and Cloud Computing.* IEEE Computer Society, 2013.
4. Matallaoui, Amir, Philipp Herzig, and Rudiger Zarnekow. "Model-Driven Serious Game Development Integration of the Gamification Modeling Language GaML with Unity." *System Sciences (HICSS), 2015 48th Hawaii International Conference on.* IEEE, 2015.
5. K. Bierre. Implementing a game design course as a multiplayer game. In Games Innovation Conference (IGIC), 2012 IEEE International, pages 1–4, Sept 2012.
6. Jennifer Thom, David Millen, and Joan DiMicco. Removing gamification from an enterprise sns. In Proceedings of the ACM 2012 Conference on Computer Supported Cooperative Work, CSCW '12, pages 1067–1070, New York, NY, USA, 2012. ACM.
7. Sebastian Deterding, Dan Dixon, Rilla Khaled, and Lennart Nacke. From game design elements to gamefulness: Defining "gamification". In Proceedings of the 15th International Academic MindTrek Conference: Envisioning Future Media Environments, MindTrek '11, pages 9–15, New York, NY, USA, 2011. ACM.

Digital Forensic Architecture for Cloud Computing Systems: Methods of Evidence Identification, Segregation, Collection and Partial Analysis

Digambar Povar and G. Geethakumari

Abstract Various advantages offered by cloud computing business model has made it one of the most significant of current computing trends like personal, mobile, ubiquitous, cluster, grid, and utility computing models. These advantages have created complex issues for forensic investigators and practitioners for conducting digital forensic investigation in cloud computing environment. In the past few years, many researchers have contributed in identifying the forensic challenges, designing forensic frameworks, data acquisition methods for cloud computing systems. However, to date, there is no unique universally accepted forensic process model for cloud computing environment to acquire and analyze data available therein. This paper contributes in three specific areas to expedite research in this emerging field. First is designing a digital forensic architecture for cloud computing systems; second is evidence source identification, segregation and acquisition; and finally methods for partial analysis of evidence within and outside of a virtual machine (VM).

Keywords Virtualization · Virtual machine · Cloud computing · Cybercrime · Digital forensics · Digital evidence · Cloudcrime · Cloud forensics

1 Introduction

Over the past few years, cloud computing is maturing and has revolutionized the methods by which digital information is stored, transmitted, and processed. Cloud computing is not just a hyped model but embraced by Information Technology

D. Povar (✉) · G. Geethakumari
Department of Computer Science and Information Systems, BITS Pilani, Hyderabad Campus, Jawaharnagar, Ranga Reddy District 50078, Telangana, India
e-mail: powar.d@hyderabad.bits-pilani.ac.in

G. Geethakumari
e-mail: geetha@hyderabad.bits-pilani.ac.in

© Springer India 2016
S.C. Satapathy et al. (eds.), *Information Systems Design and Intelligent Applications*, Advances in Intelligent Systems and Computing 433,
DOI 10.1007/978-81-322-2755-7_22

213

giants such as Apple, Amazon, Microsoft, Google, Oracle, IBM, HP, and others. Cloud computing has major concerns due to its architecture despite the technological innovations that have made it a feasible solution. It is difficult to fix the responsibility of a security breach in cloud due to the complex structure of the cloud services. In last decade, cloud computing security seems to be the most frequently surveyed topic among others by leading organizations such as IDC (International Data Corporation) and Gartner.

Recent attacks on cloud such as Sony Email hack, Apple iCloud hack, etc., proved the vulnerability in cloud platforms and require immediate attention for digital forensics in cloud computing environment. Cloud Security Alliance (CSA) conducted a survey related to the issues of forensic investigation in the cloud computing environments [1]. The survey document summarizes the international standards for cloud forensics, and integration of the requirements of cloud forensic into service level agreements (SLAs). In June 2014, NIST (National Institute of Standards and Technology) established a working group called NIST Cloud Computing Forensic Science Working Group (NCC FSWG) to research challenges in performing digital forensics in cloud computing platform. This group aims to provide standards and technology research in the field of cloud forensics that cannot be handled with current technology and methods [2]. The NIST document lists all the challenges along with preliminary analysis of each challenge by providing associated literature and relationship to the five essential characteristics of cloud. Our work focuses on some of the issues and/or challenges pointed out in the above two documents [1, 2].

The remaining part of the paper is organized as follows: In Sect. 1 we discuss an introduction cloud forensics. Section 2 provides the details of literature review. Section 3 emphasizes on the digital forensics architecture for cloud computing. Section 4 lists the methods of evidence source identification, segregation and acquisition. Section 5 describes the techniques for partial analysis of evidence related to a user account in private cloud. Finally we conclude the work and discuss future enhancements in Sect. 6.

1.1 Cloud Forensics

For traditional digital forensics, there are well-known commercial and open source tools available for performing forensic analysis [3–8]. These tools may help in performing forensics in virtual environment (virtual disk forensics) to some extent, but may fail to complete the forensic investigation process (i.e., from evidence identification to reporting) in the cloud; particularly cloud log analysis. Ruan et al., have conducted a survey on "Cloud Forensics and Critical Criteria for Cloud Forensic Capability" in 2011 to define cloud forensics, to identify cloud forensics challenges, to find research directions etc. as the major issues for the survey. The majority of the experts involved in the survey agreed on the definition "*Cloud forensics is a subset of network forensics*" [9]. Network forensics deals with forensic investigations of computer networks. Also, cloud forensics was defined by other researchers such as Shams et al. in 2013 as *"the application of computer forensic principles and procedures in a cloud computing environment"* [10]. The cloud deployment model under investigation (private, community, hybrid, or public) will define the way in which digital forensic investigation will be carried out. The work presented in this paper is restricted to IaaS model of private or public cloud infrastructure.

2 Literature Review

Dykstra et al. have used existing tools like Encase Enterprise, FTK, Fastdump, Memoryze, and FTK Imager to acquire forensic evidence from public cloud over the Internet. The aim of their research was to measure the effectiveness and accuracy of the traditional digital forensic tools on an entirely different environment like cloud. Their experiment showed that trust is required at many layers to acquire forensic evidence [11]. Also, they have implemented user-driven forensic capabilities using management plane of a private cloud platform called OpenStack [12]. The solution is capable of collecting virtual disks, guest firewall logs and API logs through the management plane of OpenStack [13]. Their emphasis was on data collection and segregation of log data in data centers using OpenStack as cloud platform. Hence, their solution is not independent of OpenStack platform and till date it has not been added to the public distribution (the latest stable version of OpenStack is *Kilo* released on 30th April 2015).

To our knowledge, there is no digital forensic solution (or toolkit) that can be used in the cloud platforms to collect the cloud data, to segregate the multi-tenant data, to perform the partial analysis on the collected data to minimize the overall processing time of cloud evidence. Inspired with the work of Dykstra and Sherman, we have contributed in designing the digital forensic architecture for cloud; implementing modules for data segregation and collection; implementing modules for partial analysis of evidence within (virtual hard disk, physical memory of a VM) and outside (cloud logs) of a virtual environment called cloud.

3 Digital Forensics Architecture for Cloud Computing

In this proposed research, we provide a digital forensic architecture for cloud computing platforms as shown in Fig. 1, which is based on the NIST Cloud Computing reference architecture [14] and Cloud Operating systems like OpenNebula, OpenStack, Eucalyptus, etc. Cloud Operating system mainly consists of services to manage and provide access to resources in the cloud. User access to cloud resources is restricted based on delivery models (IaaS, PaaS, or SaaS). Other than hardware and virtualization layer, user has access to all other layers in IaaS

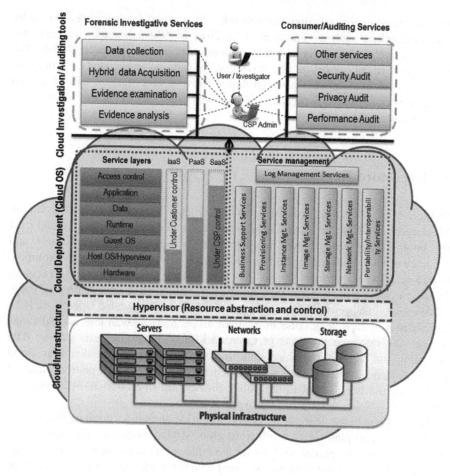

Fig. 1 Digital forensic architecture for cloud computing systems

model, whereas restricted access to PaaS and SaaS as depicted in the architecture. Therefore, our contribution is restricted to IaaS model.

Cloud provider may have external auditing services for auditing security, auditing privacy, and auditing performance. Our goal is to provide forensic investigative services for data collection, hybrid data acquisition, and partial evidence analysis. As shown in the figure, admin of CSP (Cloud Service Provider) can make use of Forensic Investigative Services directly whereas cloud user and/or investigator have to depend on the cloud admin. The suggested digital forensic architecture for cloud computing systems is generic and can be used by any cloud deployment model.

3.1 Cloud Deployment (Cloud OS)

For experimental purpose, we have set up an IaaS (Infrastructure-as-a-Service) cloud test bed using the two-node architecture concept of the OpenStack. The conceptual architecture uses two network switches, one for internal communication between servers and among virtual machines and another for external communication as shown in the Fig. 2. The controller node runs required services of OpenStack and compute node runs the virtual machines. Any number of compute nodes can be added to this test bed depending on the requirements to create number of virtual machines.

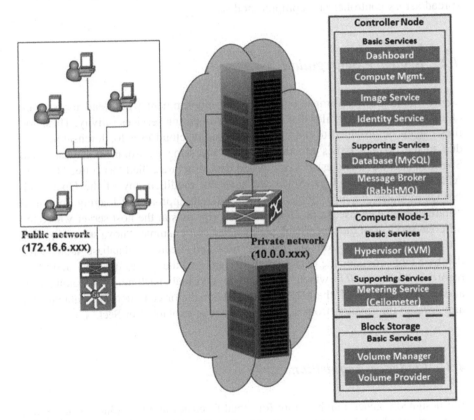

Fig. 2 Conceptual architecture of the private cloud IaaS

4 Digital Evidence Source Identification, Segregation and Acquisition

4.1 Identification of Evidence

The virtual machine is as good as a physical machine and creates lots of data in the cloud for its activity and management. The data created by a virtual machine includes virtual hard disk, physical memory of the VM, and logs. Virtual hard disk formats that different cloud provider may support include .qcou2, .vhd, .vdi, .vmdk, .img, etc. Every cloud provider may have their own mechanism for service logs (activity maintenance information) and hence there is no interoperability on log formats among cloud providers. The virtual hard disk file will be available in the compute node where the corresponding virtual machine runs. Cloud logs will be spread across controller and compute nodes.

4.2 Evidence Segregation

Cloud computing platform is a multi-tenant environment where end users share cloud resources and log files that store cloud computing services activity. These log files cannot be provided to the investigator and/or cloud user for forensic activity due to the privacy issues of other users in the same environment. Dykstra and Sherman [12] have suggested a tree based data structure called "hash tree" to store API logs and firewall logs. Since we have not modified any of the OpenStack service modules, we have implemented a different approach of logging known as "shared table" database. In this approach, a script runs at the host server where the different services of the OpenStack are installed (for examples "nova service"). This script mines the data from all the log files and creates a database table. This database table contains the data of multi-tenants and the key to uniquely identify a record is "*Instance ID*" which is unique to a virtual machine. Now, cloud user and/or investigator with the help of cloud administrator can query the database for any specific information from a remote system as explained in Sect. 4.3.

4.3 Evidence Acquisition

We designed a generic architecture for cloud forensics and the solutions are tested in the private cloud deployment using OpenStack (but may scale to any deployment model). The tools that are designed and developed for data collection and partial analysis will run on the investigator's workstation, whereas, the data segregation tool runs on the cloud hosting servers where the log files are stored. A generic view of the investigator's interaction with the private cloud platform is shown in Fig. 3.

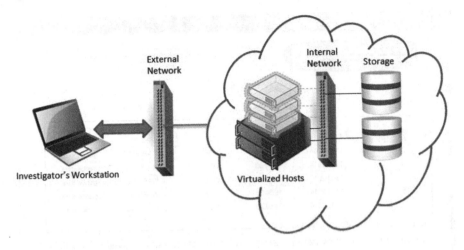

Fig. 3 Remote data acquisition in private cloud data center

Log data acquisition The segregated log data is collected using the investigator's workstation, i.e., a computer device where the acquisition and partial analysis tools are deployed. We have created a MySQL database with the name *logdb* and a table *servicelogs* under the database in the Controller node of OpenStack. The application screen shots for connecting to the database from investigator's machine and viewing the table content are shown in Figs. 4 and 5 respectively. The investigator can go through the table content and form a query based on ATTRIBUTE, CONDITION (==, ! =, <, <=, >, >=), and VALUE to filter the evidence required and download to the investigator's workstation if necessary as shown in Fig. 5.

Fig. 4 Connecting to cloud hosting server that stores the shared table database

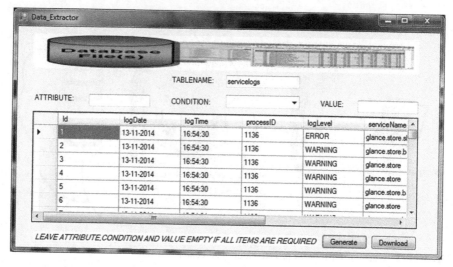

Fig. 5 Shared table with different attribute information

5 Partial Analysis (or Examination) of Evidence

The evidence examination and analysis approaches of traditional digital forensics cannot be directly applicable to cloud data due to virtualization and multi-tenancy. There is a requirement of "*digital forensic triage*" to enable cybercrime investigator to understand whether the case is worthy enough for investigation. Digital forensic triage is a technique used in the selective data acquisition and analysis to minimize the processing time of digital evidence. We now present the methods of partial analysis (also called evidence examination) required for virtual machine data.

5.1 Within the Virtual Machine

Using the examination phase at the scene of crime at different parts of evidence, we provide the investigator with enough knowledge base of the file system metadata, content of logs (for example content of registry files in Windows), and internals of physical memory. With this knowledge base, the investigator will have in-depth understanding of the case under investigation and may save a considerable amount of valuable time which can be efficiently utilized for further analysis.

Examination of file system metadata Once the forensic image of the virtual hard disk is obtained in the investigator's workstation, the examination of file system metadata or logs (for example registry file in Windows) will be started as shown in Fig. 6. Before using the system metadata extractor or OS log analyzer (for

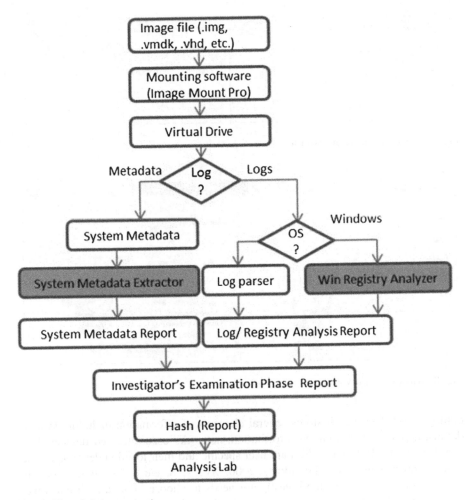

Fig. 6 Virtual disk examination process

example windows registry analyzer), the investigator has to mount the acquired virtual disk as a virtual drive.

After mounting, the virtual disk behaves like a drive where it is mounted. System metadata extractor as shown in Fig. 7 is used to list the metadata information of files and folders available in the different partitions of the virtual hard disk. For example, a machine where NTFS is used as file system, we have extracted metadata information of files/folders like MFT record No., active/deleted, file/folder, filename, file creation date, file accessed date, etc. This report may differ for various file systems (FAT32, EXT3, HFS, etc.).

Examination of registry files Windows operating system stores configuration data in the registry which is most important for digital forensics. The registry is a hierarchical database, which can be described as a central repository for

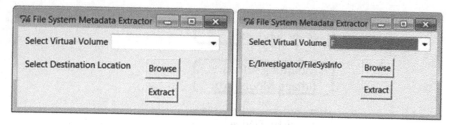

Fig. 7 File system metadata extractor

Fig. 8 Windows registry analyzer

configuration data (i.e., it stores several elements of information including system details, application installation information, networks used, attached devices, history list, etc. [15]). Registry files are user specific and their location depends on the type of operating system (Windows 2000, XP, 7, 8, etc.). To get the specific information from registry the investigator needs to choose, mounted virtual drive, Operating system, User, and the element of information to be retrieved as shown in Fig. 8. Based on the specific items selected, a sample report will be generated in the plain text format.

Examination of physical memory Physical memory (or RAM, also called as Volatile memory), contains a wealth of information about the running state of the system like running and hidden processes, malicious injected code, list of open connections, command history, passwords, clipboard content, etc. We have used volatility 2.1 [16] plugins to capture some the important information from physical memory of the virtual machine as shown in Fig. 9.

Apart from the selective memory analysis, we have implemented multiple keywords search using Boyer-Moore [17] pattern matching algorithm. The investigator can enter keywords using double quotes separated by comma as shown in the Fig. 10. For searching patterns, we have implemented regular expression search for *URL, Phone No., Email ID and IP address* as shown in the Fig. 11.

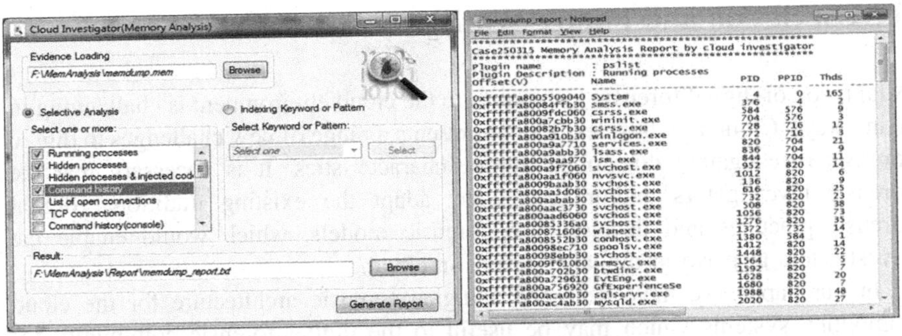

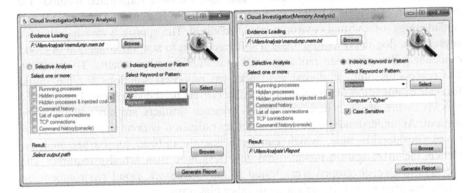

Fig. 9 Selective memory analysis

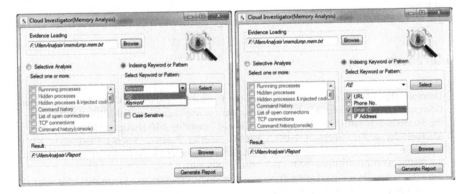

Fig. 10 Multiple keywords search (indexing)

Fig. 11 Multiple pattern search (indexing)

6 Conclusion and Future Work

Adaptation of digital forensic techniques to the cloud environment is challenging in many ways. Cloud as a business model presents a range of new challenges to digital forensic investigators due to its unique characteristics. It is necessary that the forensic investigators and/or researchers adapt the existing traditional digital forensic practices and develop new forensic models, which would enable the investigators to perform digital forensics in cloud.

In our paper, we have designed a digital forensic architecture for the cloud computing systems which may be useful to the digital forensic community for designing and developing new forensic tools in the area of cloud forensic; we have framed ways in which we can do digital evidence source identification, segregation and acquisition of evidentiary data. In addition, we have formulated methods for examination of evidence within (virtual hard disk, physical memory of a VM) and outside (logs) of a virtual environment called cloud. The approach we suggested for segregation (log data) will facilitate a software client to support collection of cloud evidentiary data (forensic artifacts) without disrupting other tenants. To minimize the processing time of digital evidence, we proposed solutions for the initial forensic examination of virtual machine data (virtual hard disk, physical memory of a VM) in the places where the digital evidence artifacts are most likely to be present. As understanding the case under investigation is done in a better way, it saves considerable time, which can be efficiently utilized for further analysis. Hence, the investigation process may take less time than actually required. The mechanisms we developed were tested in the OpenStack cloud environment. In future, we plan to test the solutions in other platforms.

Acknowledgments Our sincere thanks to Department of Computer Science and Information Systems of BITS Pilani, Hyderabad Campus, India, for providing us with the research environment.

References

1. Dominik Birk, Michael Panico: Mapping the Forensic Standard ISO/IEC 27037 to Cloud Computing. Available at (https://downloads.cloudsecurityalliance.org/initiatives/imf/Mapping-the-Forensic-Standard-ISO-IEC-27037-to-Cloud-Computing.pdf) [Accessed June 25th, 2015].
2. Michaela Iorga, Eric Simmon: NIST Cloud Computing Forensic Science Challenges. NIST Cloud Computing Forensic Science Working Group, June 2014. (available at: http://csrc.nist.gov/publications/drafts/nistir-8006/draft_nistir_8006.pdf.
3. Encase, https://www.guidancesoftware.com/products/Pages/encase-forensic/overview.aspx?cmpid=nav [Accessed June 25th, 2015].
4. Forensic Tool Kit (FTK), https://www.accessdata.com/solutions/digital-forensics/forensic-toolkit-ftk [Accessed June 25th, 2015].
5. X-Ways Forensics, http://www.x-ways.net [Accessed June 25th, 2015].
6. CyberCheck, http://www.cyberforensics.in/Products/Cybercheck.aspx [Accessed June 25th, 2015].

7. The Sleuth Kit (TSK), http://www.sleuthkit.org/sleuthkit [Accessed June 25th, 2015].
8. Digital Forensics Framework (DFF), http://www.digital-forensic.org/download [Accessed June 25th, 2015].
9. Ruan, K. et al.: Cloud forensics. Advances in Digital Forensics VII, pp.35–46, IFIP AICT 361, Springer (0211).
10. Shams Zawoad, Ragib Hasan: Cloud Forensics: A Meta-Study of Challenges, Approaches, and Open Problems. arXiv:1302.6312v1[cs.DC] (2013).
11. Dykstra, J. and Sherman, A.T.: Acquiring forensic evidence from infrastructure-as-a-service cloud computing: exploring and evaluating tools, trust, and techniques. Digital Investigation, Vol. 9, Supplement, pp.590–598, Elsevier (2012).
12. Dykstra, J. and Sherman, A.T.: Design and Implementation of FROST: Digital forensic tools for the OpenStack cloud computing platform. Digital Investigation, Vol. 10, Supplement, pp.87–95, Elsevier (2013).
13. https://www.openstack.org/software [Accessed June 25th, 2015].
14. Fang Liu et al.: NIST Cloud Computing reference architecture, Recommendations of the NIST, Special Publication 500–292 (2011).
15. Jerry Honeycutt: Microsoft windows registry guide (2005).
16. The Volatility Framework, https://code.google.com/p/volatility [Accessed June 25th, 2015].
17. Boyer, R.S., Moore, J.S.: A Fast String Searching Algorithm. Communications of the Association for Computing Machinery 20(10), pp. 762–772 (1977).

Brushing—An Algorithm for Data Deduplication

Prasun Dutta, Pratik Pattnaik and Rajesh Kumar Sahu

Abstract Deduplication is mainly used to solve the problem of space and is known as a space-efficient technique. A two step algorithm called 'brushing' has been proposed in this paper to solve individual file deduplication. The main aim of the algorithm is to overcome the space related problem, at the same time the algorithm also takes care of time complexity problem. The proposed algorithm has extremely low RAM overhead. The first phase of the algorithm checks the similar entities and removes them thus grouping only unique entities and in the second phase while the unique file is hashed, the unique entities are represented as index values thereby reducing the size of the file to a great extent. Test results shows that if a file contains 40–50 % duplicate data, then this technique reduces the size up to 2/3 of the file. This algorithm has a high deduplication throughput on the file system.

Keywords Deduplication · Hashing · Bloom filter · File system · Storage space

1 Introduction

Now-a-days the amount of data generated and used by different organizations is increasing very rapidly. So to store this huge volume of data a lot of space is required [1]. This exponential growth of data are creating problem for the data

P. Dutta (✉) · R.K. Sahu
Department of Computer Science & Engineering, National Institute of Science and Technology, Berhampur, India
e-mail: prasundutta_85@yahoo.com

R.K. Sahu
e-mail: rajeshsahu66@gmail.com

P. Pattnaik
Department of Information Technology, National Institute of Science and Technology, Berhampur, India
e-mail: pratikpriti2010@gmail.com

© Springer India 2016
S.C. Satapathy et al. (eds.), *Information Systems Design and Intelligent Applications*, Advances in Intelligent Systems and Computing 433,
DOI 10.1007/978-81-322-2755-7_23

centers. Another major problem associated with this huge amount of data is to move it from one place to another over the network. The problem drew attention of many researchers from the beginning of this century. Using data deduplication technique the cost associated with data storing and moving can be reduced significantly. So automatically deduplication technique has become a need to the information technology world. A lot of duplicate information is stored in the form of replicas to increase reliability of the system and to avoid data loss, which is a common scenario in modern enterprises. But the bitter truth is that the amount of storage cost increases as many replicas are made. Basically data backup in external systems are costly and power consuming. Recent survey shows that, although the storage cost has decreased but cost is the main reason to hinder the deployment of disk-based backup solutions. In backup systems a huge amount of duplicate or redundant data exists and which demands a lot of storage space. In data warehouses before storing data, data is cleaned and this is usually accomplished through an Extract-Transform-Load (ETL) process [2]. Data cleaning mainly deals with inconsistencies and errors. The problem of redundant information cannot be handled through this process.

There are numerous techniques to deal with data redundancy. Now-a-days organizations are mainly using data deduplication technique to deal with the problem. Data deduplication is also known as Intelligent Compression or Single-Instance Storage technique. The main idea behind data deduplication is to identify identical objects and eliminate its occurrences to reduce the overall size [1]. There are several ways to perform data deduplication namely, SQL based method [2], Chunk based method [3], and model using Genetic Algorithm. Although the main aim of deduplication is to solve space complexity problem, the time complexity of the algorithm also need to be taken care of. Initially researchers tried to solve the space complexity problem overlooking the time complexity problem but gradually the researchers tried to solve the space complexity problem taking care of the time complexity [1].

The rest of the paper is organized as follows; previous research works in this domain is described briefly in the second part of this paper. The part three of this paper shows merits and demerits of different techniques used to solve this problem. Part 4 and 5 describes the proposed algorithm and its implementation and result. Part 6 draws conclusion by highlighting future scope in this research domain.

2 Current State of Art

The literature review reveals that disk, flash, tape, cloud and indexing are the main five groups of application area where data deduplication technique has been applied by the researchers. Out of these five domains, researchers have mainly worked on the disks. In this paper researchers have mainly focuses on file system related data deduplication problem.

The researchers in paper [1] have described what de duplication is and they have specially focused on the problems during data back-up. Data backup is obvious to improve data availability and that leads to storing of redundant data. So data deduplication strategy need to be implemented in this scenario. Earlier this was done using SQL based sub query method, where using inner sub query and temporary tables distinct information was fetched to work on them. To remove same data stored in various representations, data cleaning process has been applied by many researchers [2]. In case of data warehouse, this data cleaning process is called ETL process. This ETL process is also called inline deduplication. The researchers in their work have highlighted redundant data storing in mail systems and have proposed solution explaining advantages of deduplication. The advantages include reduced storage cost as well as reduced power, space and cooling requirement of disk. One major advantage is reduced volume of data that is sent across a WAN [2]. The deduplication block-device (DBLK) and the application of bloom store and multilayer bloom filter (mbf) in dblk has been analyzed by the researchers in [3]. DBLK is a de duplication and reduction system with a block device interface and is used for detecting block wise de duplication. It has been observed that block wise deduplication creates a large amount of metadata which is managed by mbf. Using bloom filter the incoming data is verified about whether it is already existing or not. DBLK splits data into fixed sized blocks and computes collision free hash values, thus deduplicating and reducing data size. The researchers in paper [4] have primarily focused on data domain file system (DDFS), sparse indexing, and bloom filter to check the problem of bottleneck. Now-a-days for cloud environment deduplication runs on multi-core systems and also uses adaptive pipeline to deal with computational sub-tasks. In this case researchers have come up with two types of pipeline namely CHIE (Compression, Hash calculation, Duplication Identification, Encryption) and HICE (Hash calculation, Duplication Identification, Compression, Encryption). The researchers observed different types of pipeline bottleneck namely, hash bottleneck (hash calculation is the slowest stage in the pipeline), reduction bottleneck (reduction is the slowest stage in pipeline). In cloud environment the main aim of deduplication is to reduce the file size and thus to reduce bandwidth requirement for transmission [5]. The researchers have used binning and segmentation modules for this purpose. In [6] researchers have considered the effect of content deduplication in case of large host-side caches, where large amount of data and dynamic workloads are the main point of concern in case of virtual machine infrastructure.

3 Comparison Between Different Techniques

Researchers have come across different requirements of data deduplication and have tried to solve the problems of different scenario with variety of approaches. All approaches are not applicable to every scenario. The approaches have their own advantages and disadvantages. The overall scenario of different problem types and techniques to solve them along with their advantages and disadvantages are given in Table 1.

Table 1 Comparison between different deduplication techniques

Aims	Techniques	Advantages	Disadvantages
(1) To prevent duplication in databases	(a) Deduplication using co-related sub-query	Get rid of duplicates in a table	This technique can only be used if there is identity column in the target table
	(b) Deduplication using derived table	Keep only one instance of each row, regardless the value in the identity column	If more records are duplicated there should have more space in SQL server to store these data
	(c) Deduplication using temporary table	Distinct records can be separated from the table	Requires truncation related permissions
	(d) Deduplication by creating new table and renaming it	Distinct records will be generated	There should be enough space in data base in the default file group to hold all distinct records
	(e) Deduplication using merge statement	Common data warehousing scenarios can be handled easily by checking whether a row exists and then executing commands	The technique is only applicable in tabular concepts
	(f) Deduplication using hash	Makes de duplication process perform faster	Rare hash collision may occur where hash value is same for data sets that are not identical hence set elimination takes place
(2) To develop a technique to reduce latency of a deduplication system	(g) Venti	It is an archival system, works with a fixed block size	Will not work in other types except fixed block size
(3) To design a flash based KV store architecture called bloom store	(h) Foundation	It preserves snapshots of virtual machines and uses a bloom filter to detect redundancy	A false positive case can arise in a bloom filter and redundant may exist
(4) To develop a technique to reduce latency of a deduplication system	(i) TAPER	Find duplicates using content defined chunks	Content dependent process
	(j) Ocarina ECO	Uses a content aware method that works with pdf and zip files also. It extracts sub-files and removes redundancy	Content dependent process
	(k) MBF	Space efficient technology. Checks if an element is already a	It is possible to have false positive here, if it detects that the entry of data is an existing

(continued)

Table 1 (continued)

Aims	Techniques	Advantages	Disadvantages
		member of existing data set or not	member, it could be wrong
(5) Technique used to reduce/prevent de duplication in databases	(l) Chunk based method	It checks for duplicity node wise, so there is no chance of redundancy	For any slight change it checks again from the beginning for the new node
	(m) Normalization technique	It reduces database table size	It is only applicable for table format
	(n) Fingerprint summary	Keeps compact summary of each and every node	More space required to keep summary for each node
(6) To record duplicate detection	(o) Parsing technique	Makes it easy to correct and match data by allowing comparison of individual components	More comparisons as individual components need to be compared

4 Proposed Model

In this present research work the authors have considered deduplication process on file based systems only. Two main techniques used in file systems are hashing and bloom filter. Both hashing and bloom filter process have their own advantages and disadvantages. The authors in this work have considered different cases of these two processes. In this present work these two processes has been merged together in a single process where along with complete hashing process a part of the bloom filter has been used. The test result clearly shows the efficiency of the newly developed algorithm. This technique checks redundancy in an effective way. The proposed algorithm works with text files only. The algorithm accepts a folder containing text files. In the next step the algorithm checks for repetition of any word, number or symbol in these files individually. This process is done by bloom filter. Bloom filter produces a set of text files having no redundancy as its output. The file produced by this phase contains unique words, numbers and symbols. In this process the file size gets reduced to some extent but not in a large scale. The algorithm in its next phase tries to reduce the file size in large scale and for this the concept of hashing is used. In this phase the newly generated text files i.e. the output of the previous stage bloom filter are placed. The primary objective to this hashing step is to reduce the file size to a large extent, as well as it also checks for redundancy if available. In this stage the original values are replaced by unique hash values and the file size is reduced at a large scale. This hashing process is reversible i.e. we can get back the unique file from this hashed file.

The whole deduplication process is depicted in Fig. 1 through a flow chart diagram.

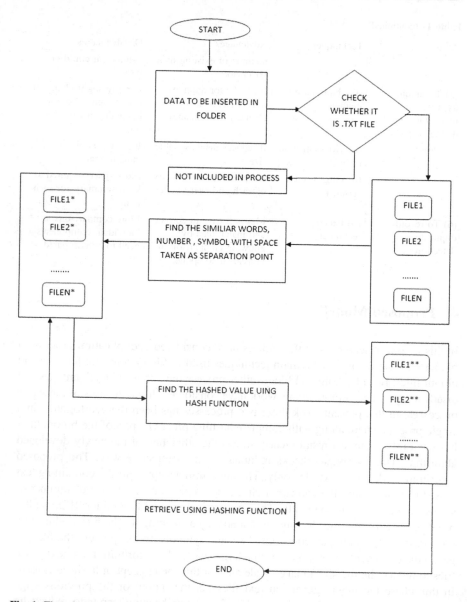

Fig. 1 Flowchart of the proposed deduplication algorithm

5 Implementation and Results

The aforementioned algorithm needs to be tested when it deals with a set of redundant data. For this purpose it has been implemented in 'C' language and has been tested with text files. There are mainly two steps in the algorithm, the first one is checking for redundancy and the second step is hashing. The first step mainly checks for repeated data and in the second step the algorithm replaces the unique data by corresponding hash values. The proposed algorithm can work for text files only. To prove the efficiency of the algorithm the result of the proposed algorithm has been compared with two well known techniques, namely Bloom Filter and Hashing. Basically the proposed algorithm combines the merits of these algorithms.

In the first step the algorithm checks whether the input file(s) is text file or not. If the algorithm is satisfied with the input then it checks for repetition of characters, number, and alphabet in a particular file. This step is carried out as a part of bloom filter algorithm. Different test cases shows that if the average input file size is 2 kb then in this step the size gets reduced to an average of 1.6 kb. This reduction in file size is due to removal of repetitions in a file. To reduce the file size further hashing process is used in the next step. Hashing generates the hash values as well as it checks for redundancy. Continuing with the previous example it is found that the average 1.6 kb files after hashing gets reduced to 1.1 kb file in an average. So overall the files with average size 2 kb is being reduced to 1.1 kb in an average. This is a large scale reduction indeed. The reduced file size will reduce the storage space requirement as well as it will help to transfer the file over the network if required. From the compressed file generated after the whole process the algorithm allows to get back the unique file which was generated after the first phase.

Table 2 summarizes the efficiency of the two step algorithm. The results shows the percentage reduction in size of the original text file containing duplicate data, when the bloom filter algorithm is applied alone or when the hashing process is applied alone and when the proposed algorithm combining two steps is applied. The result shows that the ratio of the original file size to the compressed file size when bloom filter algorithm is applied alone is 2.3:1 on an average. If the hashing process is applied alone then the ratio of the original file size to the compressed file size on an average is 1.3:1. In case of brushing algorithm the ratio is 3.8:1. The ratio proves the efficiency of the proposed algorithm.

Table 2 Experimental results

Number of files→	10	20	30	40	50
Original size (kb)	422	605	1034	1270	1587
Size after applying Bloom Filter algorithm alone (kb)	168	268	439	587	708
Percentage of compression over the original file size	60.18	55.7	57.54	53.77	55.4
Size after applying Hashing algorithm alone (kb)	316	473	779	1075	1270
Percentage of compression over the original file size	25.11	21.81	24.68	15.32	20.0
Size after applying Brushing algorithm (kb)	101	168	255	355	425
Percentage of compression over the original file size	76.06	72.23	75.33	72.04	73.21

6 Conclusion

Over the years researchers have fought with data deduplication in various ways. Each technique checks for redundancy in a different manner. Also it is observed that each technique is not applicable in all cases. After going through previous techniques the authors have proposed a new technique called 'Brushing' to deal with redundancy related to text files. The proposed algorithm is a two step process, in the first step it checks for duplicates and returns the files after removing them. In the second step the unique file is hashed and the file size gets reduced to a large extent. The test results prove about the efficiency of the algorithm. The algorithm is also able to retrieve the unique file from the final hash file. But in this technique the original unique file cannot be retrieved which was given as input to the system for deduplication. Because during the first phase the position and number of repetition times of (number, words, characters) of the original file were not maintained in the proposed algorithm to reduce the time requirement. Redundancy between two files has not been considered in the proposed algorithm because unique data of one file will be removed and cannot be recovered by the proposed algorithm. The proposed model looks to check redundancy only for text files as files of other formats like audio, video etc., needs different types of algorithm which do not matches with the proposed algorithm. In future researchers can try to solve the aforesaid problems.

References

1. Suprativ Saha, Avik Samanta, A brief review along with a New Proposed Approach of Data De duplication, ACER 2013, pp. 223–231 (2013).
2. Srivatsa Maddodi, GirijaV.Attigeri, Dr.Karunakar A.K, Data de duplication techniques and analysis, Third International Conference on Emerging Trends in Engineering and Technology, IEEE computer Society (2010).
3. Yoshihiro Tsuchiya, Takashi Watanabe, DBLK: De duplication for Primary Block Storage, IEEE (2011).
4. Jingwei Ma, Bin Zhao, Gang Wang, Xiaoguang Liu, Adaptive Pipeline for De duplication, IEEE 2012 (2012).
5. Amrita Upadhyay, Pratibha R Balihalli, ShashibhushanIvaturi, Shrisha Rao, De duplication and Compression Techniques in Cloud Design, IEEE (2012).
6. Jingxin Feng, Jiri Schindler, A De duplication Study for Host-side Caches in Virtualized Data Center Environments, IEEE (2013).

A New Approach for Integrating Social Data into Groups of Interest

Abdul Ahad, Suresh Babu Yalavarthi and Md. Ali Hussain

Abstract Our daily life is connected with various social network sites with large-scale public networks like Google+, WhatsApp, Twitter, or Facebook. For sharing and publishing, the people are increasingly connected to these services. Therefore, Social network sites have become a powerful tool of contents of interest, part of which may fall into the scope of interests of a given group. There is no solution has been proposed for a group of interest to tap into social data. Therefore, we have proposed an approach for integrating social data into groups of interests. This method makes it possible to aggregate social data of the group's members and extract from these data the information relevant to the group's topic of interests. Moreover, it follows a user-centered design allowing each member to personalize his/her sharing settings and interests within their respective groups. We describe in this paper the conceptual and technical components of the proposed approach.

Keywords Data mining · Social data · Social network sites · Groups of interest · Information sharing

A. Ahad (✉)
Department of CSE, Acharya Nagarjuna University, Guntur, Andhra Pradesh, India
e-mail: ahadbabu@gmail.com

S.B. Yalavarthi
Department of CSE, JKC College, Guntur, Andhra Pradesh, India
e-mail: yalavarthi_s@yahoo.com

Md. Ali Hussain
Department of ECM, KL University, Guntur, Andhra Pradesh, India
e-mail: alihussain.phd@gmail.com

© Springer India 2016
S.C. Satapathy et al. (eds.), *Information Systems Design and Intelligent Applications*, Advances in Intelligent Systems and Computing 433,
DOI 10.1007/978-81-322-2755-7_24

235

1 Introduction

Every day 3 billion kilobytes of data are produced and today high percentage of the data in the web server created within the last few years. Social web sites like Twitter, Facebook, and LinkedIn have become an important part in our daily life. For Communicating, sharing, and publishing millions of users are highly connected with each other to these websites. The very large amount of data (called social data) including shared information and conversation (e.g. messages, web contents) is increasingly created by users. The social network sites have been powerful sources of test and audio/video data of interest. Different people sharing the data by own interests. People join a group to gain knowledge which is built on every individual contribution. The contribution of more the members to the group, the more they can learn for themselves and each member is push the relevant information into the same or different group. Meanwhile, people to post any data directly on their social profiles, thus was letting it available on their own social networks. The social web sites and the groups of an authorized user are not identical and some groups can be formed within a social web site. Unfortunately, the information is missed by their group which can be shared by different users on different social networks. We propose a new approach for integrating social data into groups of interest. This method allows users to grouping their social data from different social networks like Twitter, LinkedIn and Facebook and to share some amount of the grouping data with their respective groups of interest. According to their own preferences, people are also able to personalize their sharing settings and interests within their respective groups [1–15].

2 Social Network Sites and Group of Interest

Social networking sites are open based web services whose main objective is to connect people. The social network sites allow individuals to generate a public profile and capitalization a list of users with whom they share a connection. In terms of number of active users, traffic volume, and number of generated contents there are a large number of social networking sites like Google+, Facebook, Twitter and LinkedIn available for users to choose. Gradually, Social networking sites have experienced and come out with new features to comply with users' upcoming demands such as creating tasks, sharing links, posting messages and sending instant or private data and so forth.

A group of interest may be a common project, a common goal, or a geographical location, or profession. A given community may furthermore contain many sub or nested communities. Communities of interest can be created and maintained on-line and/or off-line. Their forms also vary from small to very large, open communities on the Web such as the YouTube, Wikipedia and Flicker communities. In this paper, we are concerned by little sized communities of interest. The proposed approach should be considered as a bridge between social network sites and groups of interest with the objective of combining the strengths of both. It allows a given group of interest to serve the social data generated by its members to empower the internal information sharing process, in which every member can be an active contributor of his/her group while keeping using the social network sites normally. Especially, the interesting information published by the members on any social network site is automatically retrieved into the interested group without requiring any extra efforts.

3 Topologies

In this block, we present two data structures. The first structure allows representing the users and their social data aggregated from different social networking sites. The second structure allows representing the groups of interest, their interests and shared contents. First, in social networking sites, a user published the data or shared the data with others is known as User's Social data. The users comprise a wide range of information such as personal profile information, social connections, postings, interests, and so forth. For example, the user profile on Facebook is different than Twitter social cite. Facebook includes the basic information such as the name, birthday, photograph and relationship status; contact information such as mobile phone, landline phone, school mailbox, and address; and education and work information such as the names of schools attending/attended, and current employer. Twitter includes personal information such as interest, movies, books, favorite music and TV shows, and quotations. Therefore, to represent social data from different social network sites, it is necessary to define a common approach. We have built an adapted model based on activity stream which is a standardization effort for syndicating activities taken by users in different social network sites. By this model, people can have several social accounts (each social account contains a number of attributes identical to the Profile Information) from different social networks. It describes the four types of social activity such as: posts a post, receives a post, be friends with a social network member, and adds an interest. This is shown in Fig. 1.

Second, a group's shared information is made up of the information extracted from its members' social data which may contain sensible contents that the users do not want to retrieve. Structured data are numbers and words that can be easily categorized and analyzed. These data are generated by things like network sensors embedded in electronic devices, smart phones, and global positioning system (GPS) devices. It is important to give the users a control over what they are ready to share with the group instead of systematically sharing all of their aggregated social data. The growth of massive information is driven by the quick growing of advanced information and their changes in volumes and in nature (Fig. 2).

For example, let us consider the following Table 1; where the columns represent different social accounts (e.g. Twitter, Facebook, LinkedIn) and the rows represent different types of social data (e.g. Friends, Interests, Posts). As a member of the

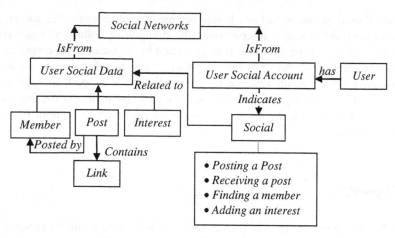

Fig. 1 Types of social activities

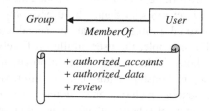

Fig. 2 Membership activities

Table 1 Authorized social data with profiles

		Authorized profile		
		Twitter	Facebook	LinkedIn
Authorized social data	Friends	No	No	No
	Interests	No	No	No
	Post	No	Yes	Yes

group, the user can freely choose which social accounts and which types of social data will be used to share with the group. The only rule is that the user has to open at least the Posts-type social data of one of his/her social accounts.

In the above table, the user decided to share his Facebook and LinkedIn accounts. Consequently, the Post-type social data from these two social accounts will by default be selected to match with the group's topics of interest in order to extract the relevant information. The final (3rd) attribute called "review" is optional and complementary to the two first ones.

The mining procedures need machine thorough computing units for information analysis and comparisons for giant data processing, as a result of quantity of knowledge is huge. So one laptop computer (PC) cannot handle a typical massive processing framework can rely upon cluster computers with a superior computing platform, with an information mining task being dead by running some parallel Computing tools, like Map Reduce or Enterprise management Language (ECL), on an oversized range of clusters. In data mining process, Semantic and Application knowledge refer to several aspect related to the rules, policies, and user information. The most important aspect in this approach contains information sharing and domain application knowledge.

To extract the interesting information from the user's social data and organize the shared contents within a given group of interest, we have applied a two-level model. The first level called topics correspond to the topics of interest of the group. The next level called selectors are technical specifications of the corresponding topics. The following diagram shows three types of Selector such as concept, keyword, and hash tag (Fig. 3).

The above characteristics make intense challenges for discover the useful knowledge from the Big Data. Exploring the Big Data in this scenario is help to draw a best possible illustration to uncover the actual sign of the object in actual way. Certainly, this job is not as simple as enquiring each member to designate his spirits about the object and then getting a skill to draw one single picture with a joint opinion and they may even have confidentiality concerns about the messages they measured in the information exchange procedure. In a centralized information system, the main target is on finding best feature values to represent every observation. This type of sample feature representation inherently treats each individual as an independent entity without considering of their social connections, which are the most important factors of the human society.

4 Proposed System

In the past, the term "Data Aggregation" has served as a catch all phrase for the massive amounts of knowledge on the market and picked up within the digital world. Today, massive information is being known as the rising power of the

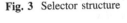

Fig. 3 Selector structure

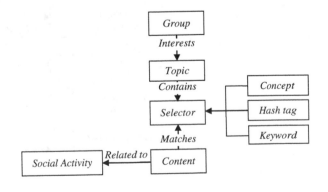

twenty first century and helps the maximum amount over nonsensicality, acting as an enormous high performance. We present two principal modules of a system that are able to support our approach. These are (see Fig. 4): (1) social data aggregation and (2) relevant information filtering. Both modules are detailed in the following subsections.

4.1 Aggregation

The user's gathering the social data from their subscribed social network sites in this social data aggregation module. We have created a number of programs, which are able to deal with the different application programming interfaces provided by the social network site providers known as "Aggregators". Each aggregator can request the corresponding APIs for the users' social data at any time by the user permission. This approach enables an extended access to most of the users' social data. Information sharing is a crucial goal for all systems relating multiple parties. The sensitive data is accessible by restricted cluster users and removes information fields such sensitive data cannot be pinpointed to a personal record.

4.2 Filtering

Filtering is a method by which we filtering the relevant information from the members' social data the information relevant to the groups' topics of interest. We have applied the information retrieval techniques and the process is composed of three main steps such as (a) social data indexing, (b) information searching, and (c) information indexing. The first step is to gather all the aggregated social data and

Fig. 4 Data aggregation and filtering

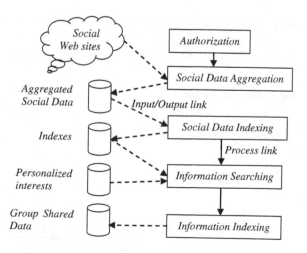

appending the social data containing external links with the content extracted from the referred web pages. The second method is searching the personalized interests by using suitable queries. Its main goal is to search them against the saved indexes in order to return matching contents. The queries are differently generated according to the types of selector such as Keyword-based selector, Hash tag-based selector, and Concept-based selector.

5 Future Work

This work focuses to manage a more detailed evaluation of the proposed system and studying the benefit and the feasibility of some advanced features beyond the information sharing enhancement purpose. At first, statistically prove the interest of the users toward our approach and started another test with a larger set of users which are 15 computer engineering students at the Lingayas University, Vijayawada. Also, we will send them after the observation period a questionnaire to fill out with their notes and opinions. The feedbacks will help us to evaluate our system and improve it as well. We have identified some possible advanced features such as New topic discovering, Members' interest profiling, Group-related decision making support, which are especially dedicated to the awareness of the group, which is the understanding of the activities of each member of the group.

The first step is to find the right member for a given task requiring specific knowledge and skills. The second step is to select the best candidates to form a good team. Finally putting all members of the group and their respective interests at the same level, we can obtain an updated overview of the group's available competency. This is shown in Fig. 5.

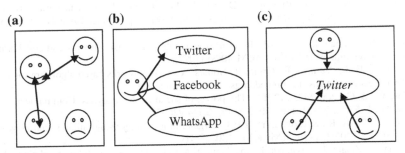

Fig. 5 a Member's relationship with other members. **b** Member's interested topics. **c** Topics and member's who are interested by it

6 Conclusion

While the term massive knowledge specifically associated with knowledge volumes, we introduced a user-centered approach for integrating social data into groups of interest. We have presented a social data model and a group model which are our first contribution. The first model allows representing the users and their social data aggregated from different social network sites. It is quite generic to handle the most important dimensions of social data available on social network sites, and is extensive to easily include the upcoming additional dimensions. The second model allows representing the groups of interest and their topics of interest and shared contents. It also contains features allowing each member to adapt his/her sharing settings and personalize his/her interests within a given group. Note that the topics of interests of the group are collectively added and specified by any member over the time thanks to a topic-selector structure.

References

1. Abbasi, M. M., and S. Kashiyarndi. "Clinical Decision Support Systems: A discussion on different methodologies used in Health Care." (2006).
2. Bhavani Thuraisingham, Data Mining for Security Applications, IEEE/IFIP international Conference on Embedded and Ubiquitous Computing, 2008.
3. Ozer, Patrick. "Data Mining Algorithms for Classification." (2008).
4. Abdel-Hafez, A., & Xu, Y. (2013). A survey of user modeling in social media websites. Computer and Information Science, 6(4), 59–71. http://dx.doi.org/10.5539/cis.v6n4p59.
5. Abel, F., Gao, Q., Houben, G., Tao, K. (2011). Analyzing temporal dynamics in twitter profiles for personalized recommendations in the social web, In Proceedings of the ACM WebSci'11, Koblenz, Germany (pp. 1–8).
6. Bonds-Raacke, J., & Raacke, J. (2010). MySpace and Facebook: Identifying dimensions of uses and gratifications for friend networking sites. Individual Differences Research, 8(1), 27–33.
7. Bojars, U., Passant, A., Breslin, J., Decker, S.: Social network and data portability using semantic web technologies. In: Proceddings of the Workshop on Social Aspects of the Web. (2008).
8. Kautz, H., Selman, B., Shah, M.: Referral web: combining social networks and collaborative ltering. Communications of the ACM 40(3) (1997).
9. Carmagnola, F., Vernero, F., Grillo, P.: Sonars: A social networks-based algorithm for social recommender systems. In: Proceddings of the 17th Internat.
10. Fan, W., & Gordon, M. D. (2014). The power of social media analytics. Communications of the ACM, 57(6), 74–81. http://dx.doi.org/10.1145/2602574.
11. Fire, M., Puzis, R., & Elovici, Y. (2013). Organization mining using online social networks. CoRR, abs/1303.3, 1–19. arXiv:arXiv:1303.3741v1.
12. Wang, Y., Zhang, J., Vassileva, J.: SocConnect: A user-centric approach for social networking sites integration. In: Proceedings of the International Conference on Intelligent User Interface (IUI) Workshop on User Data Interoperability in the Social Web. (2010).
13. Webster, A., Vassileva, J.: Personal relations in online communities. In: Adaptive Hypermedia and Adaptive Web-Based Systems, Dublin, Ireland, Springer LNCS (June 2006) 223–233.

14. Platt, J.C.: Fast training of support vector machines using sequential minimal optimization. In Schoelkopf, B., Burges, C., Smola, A., eds.: Advances in Kernel Methods: Support Vector Learning. MIT Press (1999).
15. Mitchell, T.M.: Machine Learning. McGraw-Hill (1997).

Estimation of Shape Parameter of the Radial Distribution of Cosmic Ray Shower Particles

Rajat K. Dey, Sandip Dam, Manabindu Das and Sabyasachi Roy

Abstract The problem of manipulating lateral distributions of secondary cosmic ray (CR) charged particles is the central in ground-based extensive air shower (EAS) experiments. The analytical approaches in obtaining the spectra of EAS electrons ($e^{\pm}$), muons ($\mu^{\pm}$), and hadrons ($h(\bar{h})$) in the cascade theory suffer from restricted applicability due to several assumptions or approximations adopted in the theory. Estimation of the shape parameter of the radial distribution of shower particles from simulated data can bypass all these bindings adopted in the theory and thereby improving the reliability of the method, even if it has normal dependencies upon hadronic interactions implemented in the simulation. We have various profile functions for the radial distribution of EAS particles in terms of observables called shape or slope or age. These parameters actually determine how number of shower particles or radial density changes with atmospheric depth or core distance. A more judicious estimation of such observables has been made by defining the local age or segmented slope parameter (LAP or SSP) in the work. Using simulated/experimental data, the radial dependence of LAP/SSP for $e^{\pm}$, $\mu^{\pm}$ and $h(\bar{h})$ particles is investigated aiming for the measurement of chemical composition of primary cosmic rays (PCRs).

Keywords Cosmic ray · EAS · Local age or slope parameter · Monte carlo simulation

R.K. Dey (✉) · S. Dam · M. Das · S. Roy
Department of Physics, North Bengal University, Darjeeling 734013, West Bengal, India
e-mail: rkdey2007phy@rediffmail.com

S. Dam
e-mail: sandip_dam@rediffmail.com

M. Das
e-mail: manavsanti@gmail.com

S. Roy
e-mail: methesabyasachi@gmail.com

© Springer India 2016
S.C. Satapathy et al. (eds.), *Information Systems Design and Intelligent Applications*, Advances in Intelligent Systems and Computing 433,
DOI 10.1007/978-81-322-2755-7_25

245

1 Introduction

Detailed studies of some important characteristics of PCRs especially its chemical composition around 3 PeV energy are very important for proper understanding of the origin of distinct spectral features exhibited by experimental data and to get an idea about the mechanism that pushes CRs to very high energies in the galaxy. The information about such high energy CRs at least on statistical basis can be achieved through the study of CR EASs, which contain different particles like $e^{\pm}$, $\mu^{\pm}$, $h(\bar{h})$ generated from nuclear interactions between PCR and atmosphere. The EAS experiments measure densities and arrival times of EAS particles in a particle detector array, mostly $(e^{\pm})$ and photons, at different lateral distances from which charged particles content, zenith and azimuthal angles of EAS events are obtained. Besides, some of the other EAS components such as $\mu^{\pm}$, $h(\bar{h})$ etc. are measured employing other detector types in conjunction with $e^{\pm}$. The measurements are usually interpreted in terms of primary components using MC simulations, which deal with the development of shower cascades generated from interaction mechanisms of primary with the atmosphere.

The theoretical description for the EAS initiated by primary $e^{\pm}$ and γ-photon was first given analytically in the cascade theory with the average longitudinal profile based on A and B approximations. From these considerations the radial distribution (RD) of EAS particles given by Nishimura and Kamata is well described through Nishimura, Kamata and Greisen (NKG) density function of EAS charged particles [1]. The universality property reveals that the shape/slope of EAS $e^{\pm}$ or γ-photons initiated by a nucleonic shower becomes almost identical with that estimated from average of many showers initiated by either $e^{\pm}$ or γ-ray. This brings an important concept on the origin of the universality but is not of course a simple numerical fluke.

One simple point to justify universality property taking showers induced by $e^{\pm}$, γ-ray and hadron, about 90 % of the total number of charged particles is $e^{\pm}$. In the hadron induced EAS a major part of the primary energy is transferred to γ-photons through π^0 and η-mesons production and decay in each hadronic interaction, resulting the generation of the electromagnetic part of the shower that dominates over other particles in the EAS reaching on the ground level. It is obvious that the universality for shower cascades of the same age has clearly limitations since it only works with dominant component such as $e^{\pm}$ but not all particles in the EAS. The showers induced by hadrons are different from those induced by electromagnetic components in respect of core structure and variety of EAS particles, and surely exhibits different density structures. However, in a MC calculation all universality features of the spectra cannot be easily deducible to that extent as achieved in analytic methods though a MC method avoids many limitations. The deviation of the analytic description from the MC solutions requires new results from high energy hadronic interactions as well and their implementation in the simulation [2]. The MC code CORSIKA with EGS4 option for electromagnetic interactions contributes results more closer to the experimental data.

The main parameters associated with an EAS at the given observation level are the electron size (N_e), muon size (N_μ) and hadron size (N_h), and the lateral (or transverse) shower ages ($s(e)$, $s(\mu)$ and $s(h)$) which describe the radial distribution shapes of $e^\pm$, $\mu^\pm$ and $h(\bar{h})$ in EAS. N_e and N_h are often related with the energy of the EAS initiating particle while N_μ is used for distinguishing primary particles. The present research work basically aims to explore for a relatively new and reliable estimate of shower age/slope parameters of lateral distributions of secondary particles to investigate their important characteristics and relative efficiencies to measure the chemical composition of primary CRs from the adopted simulation.

The foremost task in the work, however, is to estimate the lateral shower age (s) from simulated/observed particle densities. It is found from experiments that a single s obtained from shower reconstruction is inappropriate for the correct description of RD of EAS $e^\pm$ and more probably to $\mu^\pm$, and $h(\bar{h})$ particles as well as in all regions, which otherwise implies that the shower age depends upon radial distance. People then modified the NKG structure function to some extent but still can't describe RD of particles in all distances from the EAS core which otherwise means that s changes with radial distance r. One then needs the notion of LAP to describe RD more accurately. Instead a single s one should now calculate age/slope parameter within a small radial bin and eventually at a point [3]. Like NKG in case of $e^\pm$ other lateral density distribution functions are also in use for $\mu^\pm$ and $h(\bar{h})$ to estimate LAPs/SSPs and after studying their radial dependencies a single age/slope parameter has been assigned to each EAS event using RD profiles of $e^\pm$, $\mu^\pm$ and $h(\bar{h})$ components respectively.

The concept and analytical expressions for LAP and SSP of EAS are given in Sect. 2. In Sect. 3, we have given the method of our adopted simulation. The estimation of LAP and SSP parameters and hence results on the characteristics of the LAP and SSP will be given in Sect. 4. The final Sect. 5 gives some concluding remarks on the work.

2 Local Age or Segmented Slope Parameters

Using *Approximation B* related with energy loss mechanism of an EAS due to radiation process in the so called numerical method for a three dimensional EAS propagation by Nishimura and Kamata that gives radial density distribution of secondary particles/electrons advancing towards the ground in a homogeneous medium. The RD of EAS electrons or charged particles is well approximated by the famous radial dependent NKG function actually gave away first by Greisen [1], and which is

$$\rho(r) = C(s)N(r/r_0)^{s-2}(1+r/r_0)^{s-4.5},\tag{1}$$

where $C(s)$ accounts normalized factor.

This above important radial density function does not work if s experiences a variation with core distance, as verified in several experiments like Akeno, KASCADE and NBU. It otherwise suggests that s becomes a variable of core distance of an EAS. As a diagnosis to the problem an alternative approach was furnished bestowing the concept of LAP and SSP [3]. Assuming a pair of two adjacent positions, at r_1 and r_2, the LAP for NKG type density function $\rho(x)$ with $\left(x = \frac{r}{r_0}\right)$ in $[x_1, x_2]$:

$$s_{LAP}(1,2) = \frac{\ln(F_{12}X_{12}^2 Y_{12}^{4.5})}{\ln(X_{12}Y_{12})} \tag{2}$$

Here, $F_{12} = f(r_1)/f(r_2)$, $X_{12} = r_1/r_2$, and $Y_{12} = (x_1 + 1)/(x_2 + 1)$. When $r_1 \rightarrow r_2$, it actually yields the LAP, and is $s_{LAP}(x)$ (or $s_{LAP}(r)$) at a position:

$$s_{LAP}(x) = \frac{1}{2x+1}\left((x+1)\frac{\partial \ln f}{\partial \ln x} + 6.5x + 2\right) \tag{3}$$

The density function $\rho(r)$ now involves $s_{LAP}(r)$ instead of single s to fit $\rho_e(r)$ around an arbitrary point at a core distance r.

This representation of $s_{LAP}(r) \equiv s_{LAP}(1, 2)$ for mean distance $r = \frac{r_1 + r_2}{2}$ works nicely for observed data also (Just making $F_{12} = \rho(r_1)/\rho(r_2)$).

The radial variation of LAP using simulated density of shower particles agrees well with the variation obtained using experimental radial density distributions from the Akeno experiment. The present approach of estimating LAP/slope had been verified many occasions through several EAS experiments mentioned in [4].

Several air shower groups have proposed analytical expressions empirically to describe the radial distribution of muons from the EAS core, in analogy to the RD functions for electrons. These were basically constructed empirically or from the electromagnetic cascade theory with some approximations. In the intermediate energy range and radial distances like the present case the Hillas function (Eq. 4) used by Haverah park experiment has been chosen for the parametrization to the lateral distribution of muons for defining segmented slope $\beta_{local}(x)$ [5].

$$\rho_\mu(r) \propto (r/r_m)^{-\beta} \exp\left(\frac{-r}{r_m}\right) \tag{4}$$

For the reconstruction of experimental RD of muons, many EAS groups have used structure function proposed by Greisen with different Moliere radii and muon sizes in the formula (5) [6]. We have also verified the radial variation of the SSP or LAP using simulated and experimental data into it. Sometimes exact NKG-type structure function (1) only with N_μ instead of N_e has also been used.

$$f_\mu(r) = const.(r/r_G)^{-\beta}(1 + r/r_G)^{-2.5}, \tag{5}$$

with slope parameter $\beta = 0.75$ and Greisen radius $r_G = 320$ m.

For the reconstruction of an EAS with hadronic component the KASCADE experiment employed an NKG-type function as was used to describe the $e^\pm$ distribution and found a similar trend as the electron distribution with $r_0 = 10$ m [7]. For fixed r_0 the radial dependence of the age parameter can be computed using the expression (2) for the RD of hadrons. However, an exponential type function employed by Leeds group in their data analysis for the reconstruction of hadronic part has also been used here for studying the radial dependence of the $\beta_{local}(x)$. The characteristic function of the RD for the hadrons to obtain $\beta_{local}(x)$ brought from the Leeds group is the following:

$$\rho_\mu(r) = C \exp\left(\frac{-r}{r_m}\right)^\beta, \tag{6}$$

where C is a constant.

3 Simulation with Monte Carlo Code CORSIKA

We have used the MC simulation package CORSIKA of version 6.970 [8], by choosing the high energy interaction models QGSJet 01 ver. 1c [9] and EPOS 1.99 [10]. Hadronic interactions at low energy is performed by the UrQMD model [11]. Interactions among electromagnetic components are made by the simulation program EGS4 [12] that can be treated as an important model for the cascade development involving electrons and photons. The flat atmospheric model [13] which is valid within zenith angle 70°, is considered for the work. The simulated showers have been generated at the KASCADE and Akeno conditions. Nearly 7000 simulated showers have been produced for the study taking both Proton (p) and Iron (Fe) components of PCRs with energy bin 10^{14}–3×10^{15} eV coming only from directions falling in the range 0°–50°.

4 Results

4.1 Estimation of Local Age or Segmented Slope Parameter

The simulated $e^\pm$, $\mu^\pm$ and $h(\bar{h})$ density data have been analyzed to obtain $s_{local}(r)$ when NKG-type density profile functions are used. To compute SSPs ($\beta_{local}(r)$) corresponding to $\mu^\pm$ and $h(\bar{h})$ lateral distributions we have used density profile functions employed in Haverah park and Leeds experiments respectively. The Greisen function given in Eq. 5 is also used for the purpose.

4.1.1 Electron Component of EAS

The results found from the analysis show that LAP starts decreasing first up to a core distance of about 50 m, where LAP takes is lowest value, and again continues to rise up to 300 m and changes direction again. The simulated density corresponding to KASCADE location is used to study the radial variation of LAP as shown in Fig. 1 (Left) for p and Fe. Extracted observed density from published paper has also been used to estimate LAP in the figure. The error of the LAP in the concerned energy is different for different core distance ranges. It is about 0.05 for our considered radial distance range i.e. 5–300 m. Our simulation results corresponding to Akeno site along with observed results on LAP is given in the right figure Fig. 1 [14]. The lateral shower age estimated from traditional method using NKG function for p and Fe initiated showers is also included in the figure taking electron lateral density profile only.

4.1.2 Muon Component of EAS

The LAP has also been estimated with $\mu^{\pm}$ using NKG-type lateral distribution function and the radial variation of LAP as obeyed by $e^{\pm}$ still persists for muons as well. Here we have considered two Moliere radii (r_m) for studying radial variation of LAP for muons but obtained almost the same nature except with slightly different values. All these studies have been performed for p and Fe initiated MC data along with the KASCADE density data extracted from published work and, shown in Fig. 2. The same study has been repeated using Greisen proposed muon RD

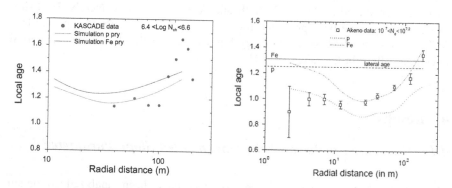

Fig. 1 *Left* Variation of LAP with core distance using MC and KASCADE density data. *Right* Same as *left figure* but for Akeno data. *Parallel lines* (*continuous* and *dotted lines*) to X-axis show single average ages for p and Fe with the combination QGSJet 1c and UrQMD models

Fig. 2 *Top* Variation of LAP with core distance estimated from MC p and Fe data for muons using NKG function. Comparison with KASCADE observed data is also included. *Bottom* The radial variation of SSP estimated from simulated p and Fe showers for muons using Greisen function. Comparison with KASCADE results obtained from observed lateral distribution data

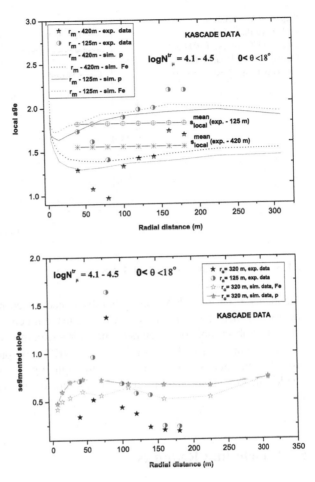

function for estimating SSP and radial variation of the parameter. The radial behaviour of the SSP is just opposite to the LAP, and is expected because if we compare Greisen density function with NKG then the SSP and LAP is related through the relation, $s(r) = 2 - \beta(r)$. Here, SSP has been estimated using simulated p and Fe data at $r_m = 320$ m only but the radial variation of SSP has made at $r_m = 320$ m and 125 m respectively using KASCADE data.

In a particular range of energy, zenith angle etc. EAS observables always come across fluctuations, and therefore it is more appropriate to an average LAP (s_{LAP}(average)), results from the average of all LAPs at various discrete points falling in the range 75–150 m. Such average ages are shown in the top of Fig. 2 by straight lines parallel to X-axis. Though radial variation has been studied approximately in the range 2–350 m for simulation but the observed variation is taken over a small radial interval of 40–185 m only due to some limitation.

Fig. 3 The variation of LAP
with core distance using
extracted KASCADE RD data
from published paper for
hadrons using NKG function

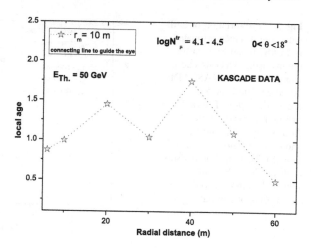

4.1.3 Hadron Component of EAS

Secondary $h(\bar{h})$ in an EAS are generally concentrated near the shower core. This
means that these particles of an EAS experience much fluctuations which certainly
will affect observable parameters of the hadronic component. In the present study of
the radial variation of LAP for $h(\bar{h})$ with NKG type profile function such inconsistent behaviour is exhibited clearly through Fig. 3. A more extensive study is
needed including simulated data in the future on the hadronic component of an
EAS.

5 Concluding Remarks

The LAP varies with distance from the EAS core with a decreasing trend showing a
lowest value nearly within the range 50–75 m for $e^{\pm}$. Beyond that it behaves in
opposite manner where instead of decreasing it further rises with core distance,
reaching a peak value nearly at ≈ 300 m, and then again changes its direction. In
case of $\mu^{\pm}$ such variation still persists but minimum of muon LAP is shifted to 75 m
range while maximum occurs at around 150 m. The nature of the SSP with radial
distance follows a reverse trend compared to LAP for $e^{\pm}$. The nature of the curves
remains unaltered even for different Greisen radii used in the formula for lateral
distribution.

The radial dependence of the LAP for $h(\bar{h})$ using KASCADE data seems very
unreliable. Further analysis with a large volume of simulated data is due in order to
predict a systematic behavior of the LAP. The LAP and SSP may offer good
solutions for more accurate and judicious estimation of the shape/slope parameters
in order to extract the nature of the PCRs. LAP is higher for Fe showers than p
initiated showers and might be useful for composition study of PCRs.

Acknowledgments Our sincere thanks to Dr. A. Bhadra of HECRRC, NBU for many useful suggestions. RKD thanks NBU for providing travel grant to attend the conference.

References

1. K. Greisen, Progress in Cosmic Ray Physics vol III (Amsterdam: North Holland) (1956).
2. P. Lipari, *Phys. Rev. D*, **79**, 063001 (2009).
3. J. N. Capdevielle and F. Cohen, *J. Phys. G: Nucl. Part. Phys.* **31** 507 (2006).
4. S. Tonwar, Proc. 17th Int. Cosmic Ray Conf. (Paris) vol 13 p 330 (1981).
5. A. M. Hillas et al, ICRC, Budapest 3 533 (1970).
6. K. Greisen, Annu. Rev. Nucl. Part. Sci. **10** 63 (1960).
7. T. Antoni et al Astro-ph/0505413 To appear in *Astropart. Phys.* (2005).
8. D. Heck J. Knapp J. N. Capdevielle G. Schatz and T. Thouw 1998 *The CORSIKA air shower simulation program, Forschungszentrum Karlsruhe Report FZK 6019* (Karlsruhe).
9. N. N. Kalmykov S. S. Ostapchenko and A. I. Pavlov *Nucl. Phys. B (Proc. Suppl.)* **52 17** (1997).
10. K. Werner et al *Phys. Rev. C* **74** (2006) 044902.
11. Bleicher M. et al, *J. Phys. G: Nucl. Part. Phys.* 25 (1999) 1859.
12. W. R. Nelson H. Hiramaya D. W. O. Rogers *Report SLAC* **265** (1985).
13. National Aeronautics and Space Administration (NASA) U.S. Standard Atmosphere Tech. Rep. NASA-TM-X-74335 (1976).
14. R. K. Dey and A. Bhadra, *Astropart. Phys.*, **44** 68–75 (2013).

Network Security in Big Data: Tools and Techniques

Pushpak Verma, Tej Bahadur Chandra and A.K. Dwivedi

Abstract With the time, Big Data became the core competitive factor for enterprises to develop and grow. Some enterprises such as, information industrial enterprises will put more focus on the technology or product innovation for solving the challenges of big data, i.e., capture, storage, analysis and application. Enterprises like, manufacturing, banking and other enterprises will also benefit from analysis and manage big data, and be provided more opportunities for management innovation, strategy innovation or marketing innovation. High performance network capacity provides the backbone for high end computing systems. These high end computing systems plays vital role in Big Data. Persistent and Sophisticated targeted network attacks have challenged today's enterprise security teams. By exploring each aspect of high performance network capacity, the major objective of this research contribution is to present fundamental theoretical aspects in analytical way with deep focus on possibilities, impediments and challenges for network security in Big Data.

Keywords Big data · Current networking scenario · Future networking scenario · Network security

P. Verma · T.B. Chandra (✉)
School of Information Technology, MATS University, Raipur 492001, CG, India
e-mail: tejbahadur1990@gmail.com

P. Verma
e-mail: verma.pushpak@gmail.com

A.K. Dwivedi
Government Vijay Bhushan Singh Deo Girls Degree College, Jashpur 496331, CG, India
e-mail: Anuj.ku.dwivedi@gmail.com

© Springer India 2016
S.C. Satapathy et al. (eds.), *Information Systems Design and Intelligent Applications*, Advances in Intelligent Systems and Computing 433,
DOI 10.1007/978-81-322-2755-7_26

1 Introduction

With the time, Big Data became the core competitive factor for enterprises to develop and grow. In the age of Big Data [1], data is generated from everywhere, some enterprises such as, information industrial enterprises will put more focus on the technology or product innovation for solving the challenges of big data, i.e., capture, storage, analysis and application. Enterprises like, manufacturing, banking and other enterprises will also benefit from analysis and manage big data, and be provided more opportunities for management innovation, strategy innovation or marketing innovation. High performance network capacity provides the backbone for high end computing systems. These high end computing systems plays vital role in Big Data. Persistent and Sophisticated targeted network attacks have challenged today's enterprise security teams.

Big Data analytics promises major benefits to the enterprises. Enterprises need to enable secure access to data for analytics, in order to extract maximum value from gathered information, but these initiatives can be a cause for big prospective risks. Handling massive amounts of data increases the risk with magnitude of prospective data breaches. Sensitive data are goldmines for criminals, data can be theft/exposed, it can violate compliance and data security regulations, aggregation of data across borders can break data residency laws. Thus secure solutions for sensitive data, yet enable analytics for meaningful insights, is necessary for any Big Data initiative [2]. Big data analytics will play a crucial role in future for detecting crime and security breaches [3].

2 Prior Research Works on Network Security for Big Data

Enterprises awash in flood of unstructured, semi structured and structured data, which introduced a multitude of security and privacy issues for organizations to contend with. Today's enterprise security teams focused and searching for the root causes of the attack often feels like looking for a needle in a haystack, but as per a white paper [4], getting valuable information in context of big data is more than "looking for the needles", security is a serious business and it is "eliminating the hay from the haystack". Security has traditionally been all about the defense. The term network security means providing security when data is on fly, i.e. over network.

Network traffic monitoring remains a decisive component of any enterprise's security strategy, but gaining context into the gigantic amounts of data collected from network, in a timely fashion, is still a hurdle for many enterprise security teams. Incident responders are eventually looking for possible ways to definitively identify threats for evaluating risk of infection and to take the necessary steps to remediate [5].

A new generation of methods and architectures designed specifically for big data technologies are needed that extract value from gigantic amounts of different data types through high-velocity capture, discovery and analysis. In its review, authors [6] illustrate efficient extraction of value from data and through a figure correlate three associated things: analytics, cloud-based distributed environment/deployment, and Networked Society, and these will be inextricably linked.

It is observed that data generated by the many devices having spatial and temporal characteristics, are part of the networked society. The emergence of complex networks and networks within networks are today's reality [7]. When network society, cloud computing and different phases associated with big data are correlated and viewed in a single sleeve, these two figures (Figs. 1 and 2) are originated in current and future context, because networking is currently in a transition phase, from layer-based approaches to layer-less approaches.

So from network security point of view focus should be assessed from current scenario to future requirements/aspects. Getting oneself abreast of current literature on Big Data and their idiosyncrasies with respect to security and privacy issues of/in Big Data is totally dependent on three Vs (variety, velocity and volume). Since a proliferation of data which is being generated by multitude of devices, users, and generated traffic, with incredible volume, velocity, and variety [8]. Authors of a research paper discussed characteristics, architecture and framework for Big Data [9]. As per authors of same research paper, a big data framework consist several layers, such as system layer, data collection layer, processing layer, modelling/statistical layer, service/query/access layer, visualization/presentation layer etc.

Authors of a paper [10] highlights that traditional security approaches are inadequate since they are tailored to secure small-scale static data. The three Vs of Big Data demands ultra-fast response times from security and privacy solutions/products.

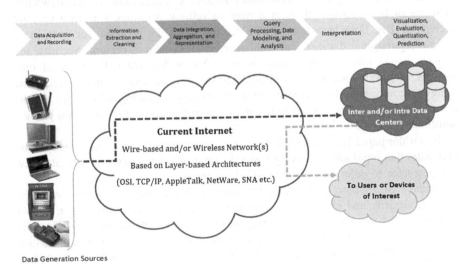

Data Generation Sources

Fig. 1 Big data aspects in current networking scenario

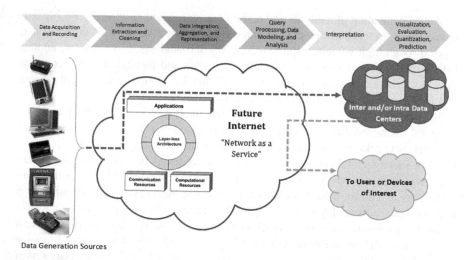

Data Generation Sources

Fig. 2 Big data aspects in future networking consequence

In same paper, authors highlights that these are the top 10 Big Data security and privacy challenges from Big Data point of view: needs secure computations in distributed programming frameworks, non-relational data stores demands best security practices, required security at data storage and transactions logs, validation/filtering is required at input end-points, real time security/compliance monitoring is required, scalable and composable privacy preserved data mining and analytics required, access control and secure communication must be cryptographically enforced, demands granular access control, required granular audits, and data provenance.

Due to dependency on Big Data and criticalness of data/information/knowledge in terms of human lives, there is a need to rethink particularly from a security viewpoint. Big Data breaches will be big too, with the potential for even more serious reputational damage and legal repercussions than at present. Security and privacy issues must be magnified by three Vs of Big Data. Diversity of data sources, streaming nature of data acquisition, distinct data formats, large scale heterogeneous networking environment, proprietary technologies/software are the big causes that why there is a need to think beyond traditional security and privacy solutions.

In a white paper [11], author focuses on intelligence-driven security for big data and states that rapid and massive growth information related to security creates new competencies to defend against the unknown threats.

Authors of another white paper [8], focuses that intelligence is necessary for tackling security and privacy issues related to big data. In the same paper authors suggested that these four steps are required for security intelligence: Data collection, Data integration, Data analytics and Intelligent threat and risk detection (which includes real-time threat detection/evaluation, pattern matching, statistical correlation, security analysis, and log data management, i.e. monitor and respond with the help of sophisticated correlation technologies).

3 Challenges for Network Security in Big Data

In new context of complex, social and nested networks, ubiquitous computing environment, privacy and personal security are increasingly at stake [7].

Traditional network perimeters are thawing as businesses reshape their organizations around emerging technologies such as clouds, social media, handheld computing devices, big data etc.—these sending risks escalating sky high and making it more difficult to defend against the increasing frequency and impact of attacks and ensure the safety of people tapping into data and applications.

Traditional security tools for example firewalls are good enough and mature but offer no protection from breaches which originate from within the firewall perimeter. Thus regular awareness and training for staffs are required on security and privacy issues with do's and don'ts.

While security solutions are emerging/enhanced prepared for the big data, but security teams may not. Data analysis is also an area where internal knowledge of the staff may be lacking.

As per the authors of a research paper [12], a serious attention is needed in term of security on ICT supply chain (all hardware and software involve in different phases of big data from production, storage and application) which is the carrier of big data.

As well as, from network point of view small and midsize businesses may not have adequate expertise and resources to concentrate on security issues related to networks. Therefore, they like to get it as a service from a third party which can be suspicious in many times.

4 Existing Tools and Techniques Best for Big Data Security Solutions

Highly distributed, redundant, and elastic data repositories makes big data architecture so critical. In current context and with future requirements, security challenges from big data point of view can be categorized at different level, such as: data level, authentication level, network level and rest of the generic issues. These are the some identified impediments discussed by an author [13] for network security in big data:

1. Vulnerability at the data collection end in term of different devices, insecure software, dishonest employees.

Information leak out, data corruption and D.o.S. (denial of service) are some identified treats at data collection end. Sensitive data can be accessed, can corrupt the data leading to incorrect results, and can perform D.o.S. leading to financial/creditable losses.

2. Input validation and data filtering problem during data collection process.

A clear hallucination/strategy is needed during input about trusted data/data sources, trust parameters, untrusted data/data sources as well as there should be clear mechanisms for filtering rogue or malicious data.

3. Data are partitioned replicated and distributed in thousands of nodes (mobility, outsourcing, virtualization, and cloud computing) for performance and business initiative reasons.

Heartbleed vulnerability [14] in the OpenSSL library was a biggest event/challenge for Security Industry in year 2014. For good security more attention is needed on vulnerabilities, exposure and de-identification issues as well.

As per the authors of a white paper [15], open source technologies, was not created with security in mind, they support few security features but this is inadequate in current context.

As per the authors of a white paper [4], big data can play a vital role in security management as well, security management foundational concepts involve three aspects:

- An agile "scale out" infrastructure must be able to retort and fit for scalable infrastructure and evolving security threats.
- Analytics and visualization tools to support security professionals. It includes from basic event identification with supporting details, trending of key metrics in addition to high-level visualization, reconstruction of suspicious files with tools to automate testing of these files, and full reconstruction of all log and network information about a session to determine precisely what happened.
- Threat intelligence to correlate causes/pattern/impact of treats visible inside organization with the currently available information about threats outside the organization.

With the help of a comprehensive enterprise information architecture strategy that incorporates both cybersecurity and big data, proper classification of risks/risk levels, enough and regular investment on emerging tools business houses/companies mitigate risks.

There are several steps that must be normally taken before deciding on a treat tackling tool for each use case. First, search for different currently popular solutions and do a survey regarding general pros and cons for each tool. Second, narrow the list down to few (two/three) candidate tools for each use case based on the fit between the tools' strengths and own specific requirements. Third, conduct a complete benchmarking and comparison test among the candidate tools using own data sets and use cases to decide which one fits on needs best. Usually, after these three steps, we found the best available tools to tackle the threats.

Big data will have great impact that will change most of the product categories in the field of computer security including solutions/network monitoring/authentication and authorization of users/identity management/fraud detection, and systems of governance, risk and compliance. Big data will change also the nature of the security controls as conventional firewalls, anti-malware and data loss prevention. Techniques such as attribute based encryption may be useful and necessary to protect sensitive data and apply access controls. In coming years, the tools of data analysis will evolve further to enable a number of advanced predictive capabilities and automated controls in real time.

Organizations should ensure that the continued investment in security products promote technologies that use approaches agile-based analysis, not static signature-based tools to threats or on the edge of the network.

Some commercial/proprietary products are available to tackle emerging threats associated with/for Big data, few of them are addressed as follows:

- IBM Threat Protection System [16], is a robust and comprehensive set of tools and best practices that are built on a framework that spans hardware, software and services to address intelligence, integration and expertise required for Big data security and privacy issues.
- HP ArcSight [8], another product that can strengthen security intelligence able to delivers the advanced correlation, application protection, and network defenses to protect today's hybrid IT infrastructure from sophisticated cyber threats.
- Another set of products (Identity-Based Encryption, Format-Preserving Encryption and many more) provided by Voltage Security Inc. [17], provides new powerful methods to protect data across its full lifecycle.
- RSA Security Management Portfolio [4], for Infrastructure, Analytics, and Intelligence can be another good option.
- Cisco's Threat Research, Analysis, and Communications (TRAC) tools [18] is also a good option in this category.

Except above mentioned tools a multitude of vendors play in this space with their respective tools.

An updated and good list of Network Monitoring platforms/tools [19] and vulnerability management tool [20] currently in use are available with open source and proprietary classification. These tools can be useful for some fruitful context.

5 Conclusion and Future Directions

Numerous literatures/white papers are available that focused/discussed on security and privacy issues associated with Big data. It is also observed that security professionals/companies apply most controls at the very edges of the network. Though, if attackers infiltrate security perimeter, they will have full and unrestricted access to sensitive big data. Some literatures/white papers/technical report suggest that placing controls as close as possible to the data store and the data itself, in order to create a more effective line of defense. Traditional security and privacy tools and techniques are even mature but unable to tackle new issues specifically associated with big data. An additional intelligence regarding pattern identification, layer based security, event based security, identification based security etc. is required parallel with traditional security approaches. As with the time detecting and preventing advanced persistent threats may be concretely answered by using Big Data style analysis. Tools and techniques will continuously demand enhancement with time as well. This research contribution present an analytical approach regarding possibilities, impediments and challenges for network security in Big Data and will be fruitful for individuals focusing on this emerging area of research.

References

1. S. Lohr, "The Age of Big Data", New York Times, Feb 11, 2012, http://www.nytimes.com/2012/02/12/sunday-review/big-datasimpact-in-the-world.html.
2. Voltage Security, "Big Data, Meet Enterprise Security", White paper http://www.voltage.com/solution/enterprise-security-for-big-data/.
3. Gartner-Research Firm, https://www.gartner.com.
4. "Gettting Real About Security Management And Big Data: A Roadmap for Big Data in Security Analytics", White Paper, RSAs and EMC Corporation, www.EMC.com/rsa.
5. Arbor Networks Blog on "Next Generation Incident Response, Security Analytics and the Role of Big Data", http://www.arbornetworks.com/corporate/blog/5126-next-generation-incident-response-security-analytics-and-the-role-of-big-data-webinar, Feb. 2014.
6. M. Matti and T. Kvernvik, "Applying Big-data technologies to Network Architecture", Ericsson Review, 2012.
7. G.F. Hurlburt, and J. Voas, "Big Data, Networked Worlds", Computer, pp. 84–87, April 2014.
8. "Big security for big data", Business White Paper, HP, December 2012.
9. F. Tekiner, and J.A. Keane, "Big Data Framework", In Proceedings of IEEE International Conference on Systems, Man, and Cybernetics, 2013.
10. A. Cardenas, Y. Chen, A. Fuchs, A. Lane, R. Lu, P. Manadhata, J. Molina, P. Murthy, A. Roy, and S. Sathyadevan, "Top Ten Big Data security and Privacy Challenges", White Paper, Cloud Security Alliance, November 2012. http://www.cloudsecurityalliance.org/.
11. S. Curry, E. Kirda, Sy, L. Stemberg, E. Schwartz, W. H. Stewart, and A. Yoran, "Big Data Fuels Intelligence-driven Security", RSA Security Brief, Jan. 2013.
12. T. Lu, X. Guo, B. Xu, L. Zhao, Y. Peng, H. Yang, "Next Big Thing in Big Data: the Security of the ICT Supply Chain", In Proceedings of the IEEE SocialCom/PASSAT/BigData/EconCom/BioMedCom, pp. 1066–1073, 2013.
13. J. Chauhan, "Penetration Testing, Web Application Security", Lecture Notes, Available at: http://www.ivizsecurity.com/blog/penetration-testing/top-5-big-data-vulnerability-classes/.
14. http://heartbleed.com/.
15. "The Big Data Security Gap: Protecting the Hadoop Cluster", A White Paper, Zettaset Company, California, USA, 2013.
16. http://www.ibm.com/software/security/.
17. http://www.voltage.com.
18. http://blogs.cisco.com/security/big-data-in-security-part-i-trac-tools/.
19. http://www.slac.stanford.edu/xorg/nmtf/nmtf-tools.html.
20. http://searchitchannel.techtarget.com/feature/Commercial-and-open-source-vulnerability-management-tools.

Design of Ripple Carry Adder Using 2-Dimensional 2-Dot 1-Electron Quantum-Dot Cellular Automata

Kakali Datta, Debarka Mukhopadhyay and Paramartha Dutta

Abstract Quantum-Dot Cellular Automata or QCA is an important name among the emerging technologies in the nanotechnology domain as it overcomes the serious technical limitations of CMOS. In this article, we have first designed a full adder using two-dimensional two-dot one-electron QCA cells. Then we have used this full adder to design a ripple carry adder. Finally, we have discussed the issues related to energy and power needed to drive the proposed architecture.

Keywords QCA · Majority voter · Full adder · Ripple carry adder · Coulomb's repulsion

1 Introduction

Lent and Tougaw [1] proposed a new technology. This technology is called quantum-dot cellular automata. It promises higher efficiency, higher computing speed, smaller size and longer lifetime. Moreover, the inherent drawbacks of CMOS technology such as limitation to the continued scaling, diminishing returns in switching performance and off-state leakage are overcome by QCA technology. QCA is a new method of information transfer, computation and transformation.

K. Datta (✉) · P. Dutta
Department of Computer & System Sciences, Visva-Bharati University,
Santiniketan 731235, West Bengal, India
e-mail: kakali.datta@visva-bharati.ac.in

P. Dutta
e-mail: paramartha.dutta@gmail.com

D. Mukhopadhyay
Department of Computer Science, Amity School of Engineering & Technology,
Amity University, Kolkata 700156, West Bengal, India
e-mail: debarka.mukopadhyay@gmail.com

© Springer India 2016
S.C. Satapathy et al. (eds.), *Information Systems Design and Intelligent Applications*, Advances in Intelligent Systems and Computing 433,
DOI 10.1007/978-81-322-2755-7_27

263

It provides a nano-scale solution. Thus it is now among the top six emerging technologies for future generation computers. In this paper we to use two-dimensional 2-dot 1-electron QCA cells. 4-dot 2-electron QCA cell has 2 electrons within 4 dots whereas 2-dot 1-electron QCA has only one electron within two dots. Moreover, it eliminates the four ambiguous configurations among the six possible configurations [2]. The complexity of wiring is reduced as Coulomb's principle is used to pass information in binary from one cell to another.

In the following Sect. 2, we discuss two-dimensional 2-dot 1-electron QCA cell structures, the mechanism of clocking and the basic gates. In Sect. 3.1, we propose a full adder design and verify the outputs. This full adder acts as the building block of the ripple carry adder. We have then designed a ripple carry adder in Sect. 3.2. In Sect. 4, we have used potential energy calculations to verify the outputs. We have analyzed our proposed design in Sect. 5. We have discussed the amount of energy and power required to operate the proposed architecture in Sect. 6 followed by conclusion in Sect. 7.

2 2-Dimensional 2-Dot 1-Electron QCA

A variant of 2-dimensional 4-dot 2-electron QCA is the 2-dimensional 2-dot 1-electron QCA. Here the cells are rectangular oriented, either horizontally or vertically, with two holes (or dots) at the two ends of the cell. One free electron may tunnel through between these two quantum dots. Figure 1a, b show the structures of the vertically oriented 2-dot 1-electron QCA cells with polarities 0 and 1 respectively whereas Fig. 1c, d the structures of the horizontally oriented 2-dot 1-electron QCA cells with polarities 0 and 1 respectively.

CMOS clocking synchronizes various operations. In 2-dot 1-electron QCA scenario clocking is somewhat different. It determines the direction in which the signal is flowing. It supplies energy to weak input signals. This helps the signal to propagate throughout the architecture [3, 4]. The different clock zones in a QCA architecture are represented by different colors. The color codes used here is shown in Fig. 2.

In [5], we get a lucid description of the clocking mechanism and the different building blocks namely the binary wire (Fig. 3a) the inverter using a cell of different orientation in between two cells of a binary wire (Fig. 3b), the inverters by turning

(a)   **(b)**   **(c)**   **(d)**

Fig. 1 2-dimensional 2-dot 1-electron QCA cells. **a** Vertically oriented cell with polarity '0'. **b** Vertically oriented cell with polarity '1'. **c** Horizontally oriented cell with polarity '0'. **d** Horizontally oriented cell with polarity '1'

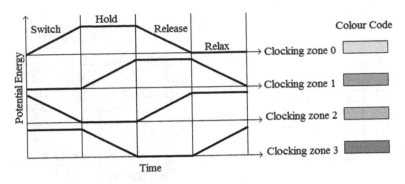

Fig. 2 The 2-dot 1-electron clocking

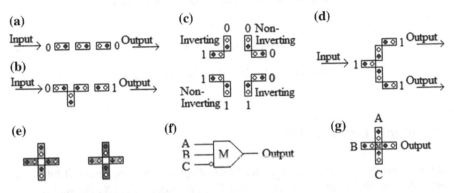

Fig. 3 The 2-dot 1-electron. **a** Binary wire. **b** Inverter with differently oriented cell. **c** Inverted/non-inverted turnings. **d** Fan-out. **e** Planar wire crossing. **f** Schematic diagram of majority voter. **g** Implementation of majority voter

at corners (Fig. 3c), a fan-out gate (Fig. 3d), crossing wires (Fig. 3e), the majority voter gate (Fig. 3f) and its implementation (Fig. 3g).

The output function for the majority voter gate is given by

$$\text{Output} = AB + A\overline{C} + B\overline{C} \qquad (1)$$

3 Proposed Designs

3.1 Full Adder Design as the Building Block

Two bits, A_i and B_i, are added along with the third bit of carry-in signal, C_{in_i} by a full adder. Figure 4a shows the full adder logic diagram and Fig. 4b shows its implementation. The output functions of the the full adder are given by the equations

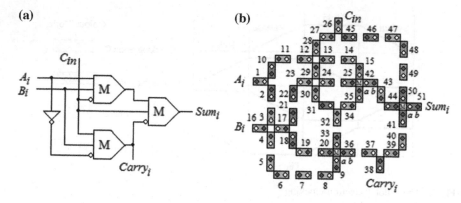

Fig. 4 **a** Full adder logic diagram. **b** 2-dot 1-electron QCA cell full adder

$$Sum_i = A_i \oplus B_i \oplus C_{in} \tag{2}$$

$$Carry_i = A_i \cdot B_i + A_i \cdot C_{in} + B_i \cdot C_{in} \tag{3}$$

3.2 Ripple Carry Adder

Figure 5 shows the ripple carry adder block diagram. We implement 4-bit ripple carry adder as shown in Fig. 6. A_i, B_i and C_{i-1} are added to get S_i, the sum and C_i, the carry $i = 0, 1, 2, 3$; S_i and C_i denote the resulting 4-bit binary sum and carry respectively. C_3 is equal to the final carry output, C_{out}.

4 Output Energy State Determination

In order to justify the outputs of 2-dot 1-electron QCA circuits, we take recourse to some standard mathematical functions as no simulator is available to us. To verify a circuit, we apply Coulomb's principle to calculate the potential energy [4]. Let q_1

Fig. 5 Ripple carry adder block diagram

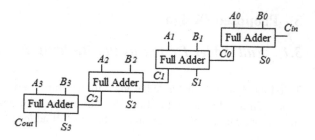

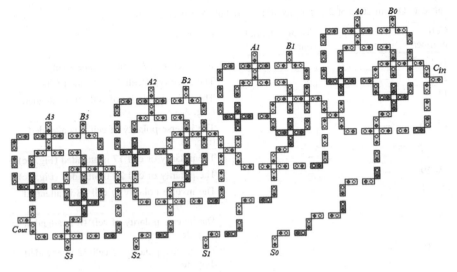

Fig. 6 2-dot 1-electron QCA ripple carry adder

and q_2 be two point charges separated by a distance r. The potential energy between them is

$$U = \frac{Kq_1q_2}{r} \tag{4}$$

K being the Boltzman constant. Therefore,

$$Kq_1q_2 = 9 \times 10^9 \times (1.6)^2 \times 10^{-38} \tag{5}$$

Then, an electron's potential energy, considering the effects of all the neighboring electrons is

$$U_T = \sum_{t=1}^{n} U_t \tag{6}$$

Since an electron can move in between the dots, it tends to move to a position where the potential energy is minimum. This property is used to establish the result. We have calculated the potential energies for all possible positions of a electron within cells and then we have chosen the one with least potential energy.

In Fig. 4b we have numbered the QCA cells in order to obtain the respective potential energies as shown in Table 1. Then we present the result obtained in Table 1 to obtain Table 2 showing the potential energies for the cells in architecture shown in Fig. 6.

Table 1 Output state of 2-dot 1-electron QCA Full Adder architecture

Cell number	Electron position	Total potential energy	Comments
1, 17, 24	–	–	Input cell C_{in}, B_i and A_i respectively
2–10	–	–	The polarity of cell 1 is attended (Fig. 3d)
11	–	–	The inverse polarity of cell 10 is attended (Fig. 3c)
12–16	–	–	The inverse polarity of cell 1 is attended (Fig. 3)
18–23	–	–	The polarity of cell 17 is attended (Fig. 3c)
25–29	–	–	The polarity of cell 24 is attended (Fig. 3c)
30–31	–	–	The inverse polarity of cell 28 is attended (Fig. 3b)
32	–	–	The inverse polarity of cell 18 is attended (Fig. 3c)
33–35	–	–	The inverse polarity of cell 32 is attended (Fig. 3c)
36	a	3.329×10^{-20} J	Electrons will latch at b as potential energy is less
	b	0.537×10^{-20} J	at this position
37–38	–	–	The polarity of cell 36 is attended (Fig. 3a)
39–42	–	–	The inverse polarity of cell 15 is attended (Fig. 3b)
43	a	-6.905×10^{-20} J	Electrons will latch at a as potential energy is less
	b	-0.294×10^{-20} J	at this position
44–45	–	–	The polarity of cell 43 is attended (Fig. 3)
46	–	–	The inverse polarity of cell 44 is attended (Fig. 3b)
47–49	–	–	The inverse polarity of cell 10 is attended (Fig. 3c)
50	a	5.891×10^{-20} J	Electrons will latch at b as potential energy is less
	b	0.537×10^{-20} J	at this position

5 Analysis of the Proposed Designs

High degree of area utilization and stability are the two main constraints for analyzing an QCA architecture [6].

All the input signal must enter the gate simultaneously with equal strength; majority voter gate output is obtained at the same or next clock phase; these conditions ensure that a design is stable. Figures 4 and 6 satisfy the above conditions to ensure that the proposed designs are stable.

Table 2 Output state of 2-dot 1-electron QCA full adder architecture

Cell number	Comments
1	Input cell C_{in}
2, 4, 6, 8	Input cells A_0, A_1, A_2, A_3 respectively
3, 5, 7, 9	Input cells B_0, B_1, B_2, B_3 respectively
10–22	Attains the polarity of the sum of A_0, B_0 and C_{in} (Fig. 4)
23–24	Attains the polarity of the carry of A_0, B_0 and C_{in} (Fig. 4)
25–33	Attains the polarity of the sum of A_1, B_1 and C_0 (Fig. 4)
34–35	Attains the polarity of the carry of A_1, B_1 and C_0 (Fig. 4)
36–40	Attains the polarity of the sum of A_2, B_2 and C_1 (Fig. 4)
41–42	Attains the polarity of the carry of A_2, B_2 and C_1 (Fig. 4)
43	Attains the polarity of the sum of A_3, B_3 and C_2 (Fig. 4)
44	Attains the polarity of the carry of A_3, B_3 and C_2 (Fig. 4)

Let a 2-dot 1-electron QCA cell be of length p nm and breadth q nm. Figure 6 shows that it requires 224 cells. The effective area covered is $224pq$ nm^2. The design covers $(30p + 29q) \times (22p + 20q)$ nm^2 area. Thus the ratio of utilization area is $566pq : (30p + 29q) \times (22p + 20q)$. We can summarize that the number of cells required is 224, area covered is 14,560 nm^2, the number of majority voter gates required is 12, the number of clock phases needed is 18.

6 Energy and Power Dissipation in the Proposed Design

From [7], we get the expressions for the different parameters, viz. E_m, the minimum energy to be supplied to the architecture containing N cells; E_{clk}, the energy to be supplied by the clock to the architecture; E_{diss}, energy dissipation from the architecture with N cells; v_2, frequency of dissipation energy; τ_2, time to dissipate into

Table 3 Different parameter values of our findings for 2-dot 1-electron QCA ripple carry adder

Parameters	Value
$E_m = E_{clk} = \frac{n^2 \pi^2 \hbar^2 N}{ma^2}$	1.598×10^{-17} J
$E_{diss} = \frac{\pi^2 \hbar^2}{ma^2}(n^2 - 1)N$	1.582×10^{-17} J
$v_1 = \frac{\pi \hbar}{ma^2}(n^2 - n_2^2)N$	4.630×10^{16} Hz
$v_2 = \frac{\pi \hbar}{ma^2}(n^2 - 1)N$	1.446×10^{15} Hz
$(v_1 - v_2) = \frac{\pi \hbar}{ma^2}(n_2^2 - 1)N$	4.486×10^{16} Hz
$\tau_1 = \frac{1}{v_1} = \frac{ma^2}{\pi \hbar(n^2 - n_2^2)}N$	2.159×10^{-17} s
$\tau_2 = \frac{1}{v_2} = \frac{ma^2}{\pi \hbar(n^2 - 1)}N$	6.911×10^{-16} s
$\tau = \tau_1 + \tau_2$	7.127×10^{-16} s
$t_p = \tau + (k - 1)\tau_2 N$	2.632×10^{-12} s

the environment to come to the relaxed state; v_1, incident energy frequency; τ_1, time required to reach the quantum level n from quantum level n_2 (here we have taken $n = 2$ and $n_2 = 2$); τ is the time required by the cells in one clock zone to switch from one to the next polarization; t_p, propagation time through the architecture; v_2–v_1, the differential frequency. We have calculated the same them in Table 3 for our proposed design of ripple carry adder. Here n represents the Quantum number, $\hbar$ is the reduced Plank's constant, m is the mass of an electron, a^2 the area of a cell, N cells are used to build the architecture which goes through k number of clock phases.

7 Conclusion

We have presented a design of a ripple carry adder with 2-dot 1-electron QCA cells and also analyzed the circuit. Due to non-availability of a simulator, we have used potential energy determination method based on Coulomb's repulsion principle to establish the outputs. We have also calculated the amount of energy and power required by the architecture.

References

1. C.S. Lent, P.D. Tougaw: A Device Architecture for Computing with Quantum Dots, Proceedings of the IEEE, 85, 4, 541–557 (1997).
2. Loyd R Hook IV, Samuel C. Lee: Design and Simulation of 2-D 2-Dot Quantum-Dot Cellular Automata Logic, IEEE Transactions on Nanotechnology, 10, 5, 996–1003 (2011).
3. Debarka Mukhopadhyay, Sourav Dinda, Paramartha Dutta: Designing and Implementation of Quantum Cellular Automata 2:1 Multiplexer Circuit, International Journal of Computer Applications, 25, 1, 21–24, (2011).
4. Paramartha Dutta, Debarka Mukhopadhyay: Quantum Cellular Automata based Novel unit Reversible Multiplexer, Advanced Scientific Publishers, Advanced Science Letters, 5, 163–168 (2012).
5. K. Datta, D. Mukhopadhyay, P. Dutta: Design of n-to-2n Decoder using 2-Dimensional 2-Dot 1-Electron Quantum Cellular Automata, National Conference on Computing, Communication and Information Processing. Excellent Publishing House, 2015, pp. 7791 (2015).
6. M. Ghosh, D. Mukhopadhyay, P. Dutta: A 2 dot 1 electron quantum cellular automata based parallel memory, Information Systems Design and Intelligent Applications, ser. Advances in Intelligent Systems and Computing. Springer India, 2015, vol. 339, pp. 627–636 (2015).
7. D. Mukhopadhyay, P. Dutta: A Study on Energy Optimized 4 Dot 2 Electron two dimensional Quantum Dot Cellular Automata Logical Reversible Flipflops, Microelectronics Journal, vol. 46, pp. 519–530 (2015).

Advanced Congestion Control Techniques for MANET

M.D. Sirajuddin, Ch. Rupa and A. Prasad

Abstract Mobile Ad hoc Network (MANET) is a wireless infrastructure less network in which nodes communicates with each other by establishing temporary network. One of the most important issues in the MANET is the congestion. It leads to the network performance degradation. In order to transmit real time data reliably in MANET, TCP can be used. But the congestion control techniques used by the TCP is inadequate for such networks because of node mobility and dynamic topology. Also the routing protocols designed for MANET do not handle the congestion efficiently. In this paper we proposed a new TCP congestion control scheme called TCP-R for detecting the congestion and proposed ADV-CC (Adhoc Distance Vector with Congestion Control) as a new dynamic routing algorithm to control congestion in MANET. ADV-CC improves the network performance than AODV due to its congestion status attribute as an additional feature. The main strength of this paper is performance results and analysis of TCP-R and ADV-CC.

Keywords Congestion · TCP · RTT · AODV · ADV-CC and NS2

M.D. Sirajuddin (✉)
Department of Computer Science and Engineering, JNTU Kakinada, Kakinada, Andhra Pradesh, India
e-mail: siraj.cs@gmail.com

Ch. Rupa
Department of Computer Science and Engineering, V.R. Siddhartha Engineering College, Vijayawada, Andhra Pradesh, India
e-mail: rupamtech@gmail.com

A. Prasad
Department of Computer Science, Vikrama Simhapuri University, Nellore, Andhra Pradesh, India
e-mail: prasadjkc@yahoo.co.in

© Springer India 2016
S.C. Satapathy et al. (eds.), *Information Systems Design and Intelligent Applications*, Advances in Intelligent Systems and Computing 433,
DOI 10.1007/978-81-322-2755-7_28

271

1   Introduction

Many applications have been widely used for transmitting real time data through wireless networks. In general wireless networks are of two types: one is infrastructure based networks and second one is infrastructure less networks. MANET belongs to the category of infrastructure less network.

MANET is a wireless dynamic network in which communication between nodes takes place without any fixed infrastructure. This network has dynamic topology [1]. Routing protocols and congestion control techniques designed for wired networks cannot be applied on MANET. In order to support real time data transmission with minimum delay we need to focus on congestion and routing in MANET [2, 3]. For reliable data transmission TCP is the best choice. But the problem is that, the congestion control scheme used by TCP cannot be applied on to the MANET. TCP cannot able to differentiate between packet losses due to congestion from the losses that causes due to link failure. Whenever packet loss occurs it considers it as due to congestion and decreases its congestion window [4]. This causes performance degradation of the MANET. There are various issues which require more research. Some of these issues include security, topology control, QOS, routing, power management, congestion control etc.

In this paper we considered congestion as the major issue related to MANET. Congestion occurs in a network if the number of packets sent to the network is greater than the capacity of the network. Congestion leads to the packet loss [5]. Although some other factors are also there which causes packet loss, such as mobility, link failures, interference, etc. If no appropriate congestion control is performed, it can lead to a network collapse and no data is transferred. In order to solve this issue many congestion control algorithms have been proposed for MANET. Existing AODV routing protocol does not consider the congestion for making routing decisions.

In this paper we considered the effect of congestion on TCP and as well as on AODV routing protocol. We proposed RTT based congestion control algorithm for TCP to handle congestion effectively and also we proposed ADV-CC algorithm to handle congestion. We analyzed the performance of our two algorithms in ns2.

The rest of paper is structured as follows. Section 2 describes the effect of TCP congestion control scheme on MANET. Section 3 represents AODV protocol and problems in AODV. Section 4 consists of proposed solution. Section 5 represents the simulation setup and results. Section 6 describes the conclusion.

2   Effect of TCP Congestion Control Scheme on MANET

TCP is most widely used protocol in the internet especially designed for the wired networks. In wireless network the performance of TCP is more important, because in such networks the possibility of error rate is very high. When packet loss occurs, the sender TCP decreases the congestion window directly without investigating the reason for the packet loss [4]. If a packet loss occurs due to the link errors, reducing the congestion window leads to the reduction in data transmission rate. This feature affects the throughput of the network. Due to this problem in TCP the existing AODV protocol do not perform well, if it is executed by considering the older version of TCP such as TCP-Reno, TCP-Vegas etc. [6, 7]. Due to this reasons the performance of MANET will get reduced. In order to improve the performance of MANET we need to design an intelligent congestion control scheme for TCP which will identify the reason for packet loss and then decide whether to decrease the congestion window or not. We found that only less number of authors have concentrated on the interaction between TCP and ad hoc routing protocols. In this paper we concentrated on the affect of TCP on AODV routing protocol.

3 AODV Routing Protocol

A Routing Protocol is required to transmit the packet from source to the destination. In MANET routing protocols are broadly divided into three main categories i.e. proactive, reactive and hybrid routing protocols. In this paper we considered one of the reactive protocols i.e. AODV [8].

Ad Hoc on Demand Distance Vector routing protocol [8, 9] is a reactive routing protocol which creates a route whenever it is required by the source node. To find a route to the destination, the source node broadcasts Route Request packets (RREQ). When RREQ reaches the destination a Route Reply (RREP) is sent back through the same path the Route Request was transmitted. A node receiving the RREQ packet sends route reply RREP packet only if it knows the destination address or if it is the intended destination. In above two cases, it unicasts the RREP back to the source. If not, it again broadcast the RREQ packet. A node can able to detect duplicate RREQ packets based on sequence number field in RREQ packet. Every node maintains a routing table which is updated periodically. Figure 1 shows the route establishment in AODV.

Fig. 1 Route discovery in AODV

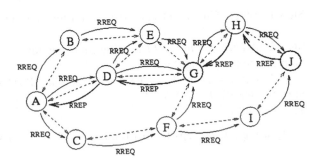

3.1 Issues in AODV

Existing AODV routing protocol do not handle the congestion effectively. When a new route is established, it will remain until the mobility or failure results in a disconnection. When congestion occurs during the packet transmission, the existing routing protocols do not seem to handle it effectively. Unlike well-established networks such as the Internet, a dynamic network like a MANET is expensive in terms of time and overhead, and it is mandatory to handle congestion effectively. We have identified the following limitations of AODV.

- Do not handle congestion efficiently.
- More routing overhead.
- Consumes more bandwidth for sending number of control packets to find new routes.
- Maintains only single path from source to destination.

All these limitations make AODV unsuitable for MANET. We need to enhance its performance by considering above all factors.

4 Proposed Solution

4.1 RTT Based TCP Congestion Control Technique (TCP-R)

The best technique to solve TCP congestion control problem is to use RTT and throughput [2–4]. In proposed scheme, for every packet TCP records the RTT and it also estimates the throughput. By using these two parameters the TCP can determine the status of the network and can able to detect the real cause of packet loss. The RTT between consecutive packets are compared. Increasing RTT and decreasing throughput signals the occurrence of congestion. The algorithm steps of TCP-R shown below.

Algorithm for TCP-R

Step 1 When Timeout Occurs, the reason for time out is identified by using the step 2.
Step 2 if($Throughput * RTT >$ congestion_window)
Step 3 then no Congestion
Step 4 Retransmit the lost packet with the same RTO value.
Step 5 else $RTO = 2 * RTO$
Step 6 End-if

By using above steps the performance of TCP in MANET can be improved.

4.2   Adhoc Distance Vector with Congestion Control (ADV-CC)

Existing AODV protocol do not handle congestion effectively. Many researchers have modified AODV routing protocol to handle congestion, but they considered older version of TCP such as TCP Reno, Vegas etc. [2, 3, 10]. Our proposed ADV-CC routing protocol enhances the existing AODV protocol by using congestion control technique. In ADV-CC protocol each node calculates the queue length to determine the status of the congestion [2, 6]. This congestion status is sent by each node to its neighbor periodically. Whenever the queue length reaches the threshold value it signals congestion. Whenever congestion occurs a warning is sent to the source [1]. When source receives a warning it reduces its data rate or selects non congested route for data transmission. In order to minimize the number of RREQ packets we included an extra field i.e. *CongStat* in a Hello packet. This extra field *CongStat* contains the congestion status of a node. Including an extra field in Hello packet saves extra control packets to be transmitted to indicate the congestion. The congestion status can be estimated based on queue length [2, 10]. The following equations are used to calculate the average queue length.

$$Minthreshold = 25\% \ of \ Total \ Buffer\_Size \tag{1}$$

$$Maxthreshold = 3 * Minthreshold \tag{2}$$

$$Avgquenew = (1 - Wq) * Avgqueold + Instant\_queue * Wq \tag{3}$$

where *Wq* is the queue weight, is a constant (*Wq* = 0.002 from RED, Floyd, 1997), and Instant_queueis instantaneous queue size.

$$QueueStatus = Instant\_queue - Avgquenew \tag{4}$$

If *QueueStatus* < *Minthreshold* indicates no congestion.
If *QueueStatus* > *Minthreshold*and Instant_queue < Maxthresholdindicates likely to be congested.
If *Instant_queue* > *Maxthreshold*, indicates congestion.
Based on above calculations *CongStat* field contains 0, 1 or 2. Where 0 indicates no congestion, 1 indicates likely to be congested and 2 indicates congestion.

As we know that whenever a node enters into a network it broadcasts Hello Packets to obtain its neighbors information. When a new node sends a hello packet it specifies its congestion status. When a neighbor node receives a Hello Packets it responds with its address and its congestion status. Once network is established Hello packets are transmitted periodically to obtain information about existence of neighbor nodes and their congestion status.

In our proposed routing protocol each node maintains information about it neighbor nodes in its routing table along with the congestion status. Consider the example below. In Fig. 2 node S is a source and node D is a destination. In

Fig. 2 Normal data
transmission from S to D

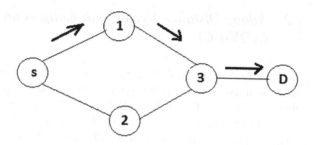

proposed AODV source node S maintains its neighbor nodes information along with its congestion status. Consider the primary route for data transmission from node s follow path S → 1 → 3 → D shown in Fig. 2. In this protocol source node maintains multiple paths for destination D.

Consider that node 1 is congested. Whenever it is congested it sends congestion status to node S. When a node S receives a congestion signal from node 1, then it selects next neighbor for data transmission i.e. node 2 shown in Fig. 3.

Figure 4 shows data transmission from node S to node D via path S → 2 → 3 → D.

Our proposed AODV routing protocol avoids initiation of route discovery process when node 1 gets congested. Instead of initiating route discovery process node S selects alternate path for transmitting data to node D. The multipath routing reduces the overhead of route discovery process of AODV [2]. Our new approach

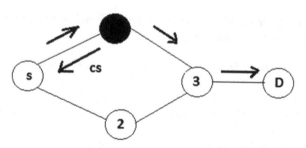

Fig. 3 Node 1 is congested and congestion warning from node 1 to node S

Fig. 4 Alternate path from
node S to node D

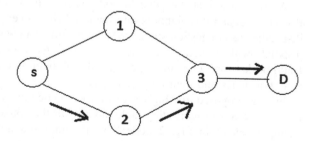

Fig. 5 Architecture of
ADV-CC

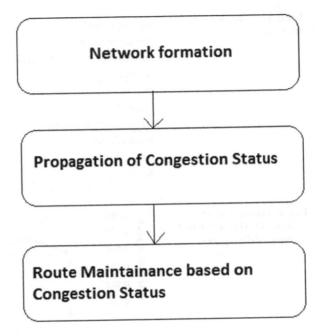

in this paper enhances the AODV by maintaining alternate paths that can be obtained during route discovery. This novel ADV-CC protocol enhances the performance by reducing the number of route discovery process. Figure 5 shows the basic steps involved in our proposed routing protocol.

5 Simulation Setup

To simulate our RTT based TCP congestion control algorithm over ADV-CC we used Network Simulator 2 [11]. We have considered the following simulation parameters shown in Table 1.

We have considered the following metrics to evaluate the performance of our proposed algorithms:

- **Routing Overhead** This metric describes the number of routing packets transmitted for route discovery and route maintenance.
- **Packet Delivery Ratio** The ratio between the amount of incoming data packets and actually received data packets.

Table 1 Simulation
parameters

Simulator used	NS2.34
Number of nodes	50
Dimensions of simulated area	800 × 600 m
Routing protocol	ADV-CC
Simulation time	50 s
Channel type	Channel/wireless channel
Radio-propagation model	Propagation/two ray round
Interface queue type	Queue/drop tail
Antenna	Antenna/Omni antenna
Source type	TCP-R

Fig. 6 Comparison of
AODV and ADV-CC based
on packet delivery ratio

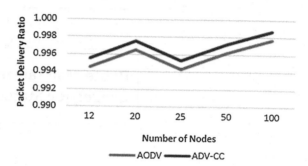

Fig. 7 Comparison of
AODV and ADV-CC based
on routing overhead

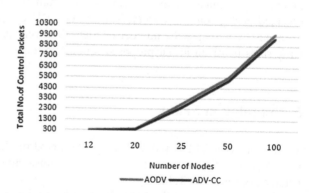

Figure 6 shows a graph between AODV and ADV-CC. We executed AODV by considering TCP Reno and ADV-CC by considering TCP-R. The graph shows that our proposed ADV-CC and TCP-R gives better Packet Delivery Ratio than the existing AODV.

Figure 7 shows the comparison between AODV and ADV-CC based on the routing overhead. It shows that our ADV-CC has less routing overhead than AODV.

6 Conclusion

Congestion is the major problem in MANET which leads to packet loss and degradation of network performance. Since AODV has no congestion control mechanism, we have modified existing AODV to handle congestion and also we have implemented TCP-R to improve the performance of the MANET. In this paper we executed ADV-CC by considering TCP-R. Our simulation results shows that ADV-CC gives better packet delivery ratio and requires less routing overhead than the existing AODV routing algorithm.

References

1. Phate N, Saxena M et al, Minimizing congestion and improved Qos of AODV using clustering in Mobile Ad hoc Network, ICRAIE, 2014 IEEE international conference, PP 1–5.
2. Kumar H et al, TCP congestion control with delay minimization in MANET, Information Communication and Embedded Systems, IEEE international conference 2014 PP 1–6.
3. Sahu C.P, Yadav M.P et al, Optimistic Congestion control to improve the performance of mobile ad hoc networks, IACC-2013, IEEE 3rd international conference, PP 394–398.
4. Sirajuddin, Ch. Rupa & A. Prasad, Techniques for enhancing the performance of TCP in wireless networks, Artificial Intelligence and Evolutionary Algorithms in Engineering Systems and Computing Vol 324, 2015, PP 159–167.
5. Bansal B et al, Improved routing protocol for MANET, Advanced Computing and Communication Technologies (ACCT), IEEE international conference 2015, PP 340–346.
6. Gupta H et al, Survey of routing based congestion control techniques under MANET, ICE-CCN 2013, IEEE international conference 2013, PP 241–244.
7. S. Sheeja et al, Cross Layer based congestion control scheme for Mobile Ad hoc Networks, IJCA journal Vol.67 No.9, 2013.
8. Bandana Bhatia et al, AODV based Congestion Control Protocols: Review, IJCSIT, Vol 5(3), 2014.
9. K G et al, A novel approach to find the alternate path for route maintenance in AODV, IJCSET 2013 Vol. 4.
10. T. Senthilkumar et al, Dynamic congestion detection and control routing in ad hoc network journal of King Saud University-Computer and Information Sciences (2013) 25, 25–34.
11. NS2 http://www.isi.edu/nsnam.

A Method of Baseline Correction for Offline Handwritten Telugu Text Lines

Ch. N. Manisha, E. Sreenivasa Reddy and Y.K. Sundara Krishna

Abstract In the field of offline handwritten character recognition, baseline correction is an important step at preprocessing stage. In this paper, we propose a baseline correction method for Offline handwritten Telugu text lines. This method deals with both baseline skew and fluctuated characters of offline handwritten Telugu text lines. This method is developed based on the main component of the character. This method has been tested on various handwritten Telugu text lines written by different people. Experimental results show that the effectiveness of the proposed method to correct the baselines of the offline handwritten Telugu text lines.

Keywords Telugu · Baseline · Handwritten · Offline · Skew · Fluctuated

1 Introduction

Based on the highest lifetime of the documents and easily transferable capability people are more interesting about digitizing physical documents. Digital document images will easily transferable and a single digital document image can be read by any number of persons at the same time, at different locations.

In the characters recognition process, pre-process the document images are very important. Because of improper pre-processed will increase the inaccurate recognition rate. One of the important steps in the preprocessing of handwritten based

Ch.N. Manisha (✉) · Y.K. Sundara Krishna
Krishna University, Machilipatnam, Andhra Pradesh, India
e-mail: ch.n.manisha@gmail.com

Y.K. Sundara Krishna
e-mail: yksk2010@gmail.com

E. Sreenivasa Reddy
Acharya Nagarjuna University, Guntur, Andhra Pradesh, India
e-mail: esreddy67@gmail.com

© Springer India 2016
S.C. Satapathy et al. (eds.), *Information Systems Design and Intelligent Applications*, Advances in Intelligent Systems and Computing 433,
DOI 10.1007/978-81-322-2755-7_29

281

document images is baseline correction. Correction of baseline is a crucial step at the stage of preprocessing for recognition of handwritten based document images. Since the correction of baseline will segment easily.

Many researchers were implemented various methods for correcting the entire document skew. But document skew correction is not enough. A specially hand-written characters preprocessing needs more attention. Because many handwritten based document images suffer with various baseline skews. Therefore the text baseline skew correction is necessary for handwritten based documents.

To correcting Offline handwritten text baseline is a very difficult task. Because offline handwritten documents does not have any dynamic information and the document images contains different handwritten styles.

In previous work we defined a hierarchical preprocessing model for handwritten based document images in [1]. It is necessary to implement a baseline correction method based on the languages. Since each language has its own structure of characters and the characters contains either a single or multiple components depends on the language.

Telugu language is one of the spoken languages in India. It is the language found in Andhra Pradesh and Telangana [2]. Telugu characters can be written from left to right. Due to the complex structure of Telugu characters, its baseline detection is a complicated task because it contains one to four unconnected components for each character. In this paper, we proposed a method to correcting the baseline for offline handwritten Telugu Text lines.

The rest of this paper is organized as follows: Reviews of Related work gives in Sect. 2. The proposed baseline correction method for Offline handwritten Telugu characters presents in Sect. 3. The experimental result gives in Sect. 4. The conclusion of this paper is draws in Sect. 5.

2 Related Work

Various methods were implemented on skew correction of the text baseline for different languages. Makkar et al. [3] Surveyed on various methods for skew estimation and correction for text. In [4] document skew corrected by Hough transform and corrected baseline skew. In [5] signature skew estimated by using projection profile analysis, center of Gravity and Radon Transform. Jipeng et al. [6] proposed a method based on the Hough transform to correcting Chinese characters skew. Gujarati language based printed and handwritten characters skew detected and corrected by Linear Regression in [7].

Cao et al. [8] introduced eigen-point concept for baseline skew correction. They consider eigen-point is the center at the bottom of the bounding box. Ramappa et al. [9] Kannada baseline skew corrected using the bounding box technique. Bahaghighat et al. [10] detected baseline by using a projected pattern of radon transform for printed and handwritten Persian or Arabic, English and Multilingual documents. In [11] baseline skew and fluctuations are corrected based on sliding window for estimation of vertical position of baseline. In [12] estimated primary baseline and local baseline for online Arabic script based languages. This method divided as diacritical marks segmentation, estimation of primary and local baselines.

Candidate baseline pixels are detected based on template matching algorithm in [13]. Baseline estimation and correction done in two stages. At the first stage writing path and the baseline of subwords were estimated and at the second stage baseline was corrected by template matching algorithm. Arabic script baseline skew corrected using horizontal projection histogram and directions features of Arabic sub words skeleton in [14]. Baseline handwritten word skew corrected on bank check dates images using weighted least squares and mathematical morphology in [15]. They reduce the use of empirical thresholds by applying the pseudo-convex hull.

3 Proposed Method

The Proposed method deals with both baseline skews and fluctuated characters of the offline handwritten Telugu Text Lines based on the main component of the character. In this method the main component we called as a blob. The proposed method has four steps. For the first we removed the noise from the document image with the existing methods developed by different researchers. In the second step we identified blobs. In the third step we corrected, the baseline skew and at the final step we corrected the baseline of the fluctuated characters. The Block diagram of the proposed method as shown in Fig. 1.

3.1 Preprocessing

Before correcting the baselines of offline handwritten Telugu text lines, it is necessary to apply other preprocessing steps to Offline handwritten Telugu document images.

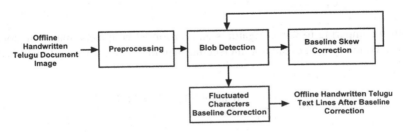

Fig. 1 Block diagram of proposed method

3.1.1 Noise Removal

Most of the document images suffer with noise because of degradation of document or malfunctions of scanner or camera. Using some filters like linear and nonlinear filters to reduce the common image noise from the document images. The common noise in handwritten based document images is underline noise. Bai et al. [16] Proposed a method to remove underlines in text document images.

3.1.2 Binarization

In this step image can be converted into a binary format. In this format zeros consider as black pixel and ones consider as white pixel. Otus threshold method [17] used for Binarization of the document image.

3.1.3 Document Skew Correction

A modified Hough transform used for detection and correction of document image skews in [18]. Najman [19] estimated document skew by using mathematical morphology.

3.2 Blobs Detection

Telugu characters contain one to four unconnected components. In this method each component identified with a bounding box [20]. In this method main character component considers as a blob. Before correcting baseline skews and correcting fluctuated characters, it is necessary to identify the blobs.

Let $C_{i,1}(x, y)$ is minimum 'x' and minimum 'y' values of the component 'i'. $C_{i,2}(x, y)$ is maximum 'x' and maximum 'y' values of the component 'i'. The components 'x' values intersect, then the components consider as the same character as shown in Eq. 1.

$$T_n = \begin{cases} C_j & \text{if } \{C_{i,1}(x), \ldots, C_{i,2}(x)\} \cap \{C_{j,1}(x), \ldots, C_{j,2}(x)\} \neq \phi \\ C_i & \text{else} \end{cases} \tag{1}$$

When comparing to vowel ligatures and consonant conjuncts, main component is taller and the 'y' values are not intersecting then the main component identified in Eq. 2.

$$B_i = \begin{cases} T_k & \text{if } \text{height}(T_j) < \text{height}(T_k) \\ T_j & \text{else} \end{cases} \tag{2}$$

In some cases consonant conjuncts and vowel ligatures taller than main component and 'y' values are intersecting then the tallest consonant conjunct starts above or equal from top one-fourth part of the main component. In case of vowel ligatures, the Main component starts from below the top one-fourth part of the vowel ligatures. In this case the blob identified by Eq. 3.

$$B_i = \begin{cases} T_j & \text{if } T_{k,1}(y) \leq T_{j,1}(y) + (\text{height}(T_j)/4)) \\ T_k & \text{else} \end{cases} \tag{3}$$

3.3 Baseline Skew Correction

In this step, select the components for the blob detection by Eq. 1. If the difference between $C_{i,2}(y)$ and $C_{i+1,1}(y)$ is more than the height of the largest component from $\{C_1, C_2, ..., C_i, C_{i+1}, ..., C_n\}$ then identify the blob among $\{C_1, C_2, ..., C_i\}$ with Eqs. 2 and 3. The baseline skew error shown in Fig. 2.

In this proposed method, consider two blobs B_1 and B_2 as top-left most blob and top-right most blob respectively. We detected the baseline skew angle based on the angle of the slope line in [21]. The baseline skew can be detected by 'α' in Eq. 4. The angle of the baseline skew can be identified by 'θ' in Eq. 5.

$$\alpha \leftarrow (B_{2,2}(y) - B_{1,2}(y))/(B_{2,2}(x) - B_{1,2}(x)) \tag{4}$$

$$\theta = \text{Tan}^{-1}(\alpha) \tag{5}$$

The negative angles can be converted into positive angles in [22]. The angle of baseline skew can be corrected by '$\Delta\theta$' in Eq. 6.

$$\Delta\theta = \begin{cases} \theta + 360 & \text{if } \theta < 0 \\ \theta & \text{else} \end{cases} \tag{6}$$

After baseline skew correction offline handwritten Telugu Text lines shown in Fig. 3. After baseline skew correction the text lines will easily segmented by horizontal projection profile [23].

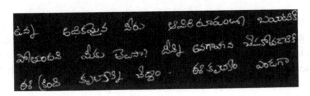

Fig. 2 Offline handwritten telugu text with baseline skew error

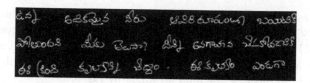

Fig. 3 After baseline skew correction offline handwritten telugu text lines

Fig. 4 Offline handwritten telugu text line contains fluctuated characters

Fig. 5 After fluctuated characters baseline correction in handwritten telugu text line

3.4 Fluctuated Characters Baseline Correction

This approach applied to single Telugu text lines. Based on the main component we corrected baseline of the fluctuated characters. Fluctuated character's text line shown in Fig. 4.

In this step intersection of all components consider for identification of the blobs. We used maximum 'y' value of the blob as a baseline position for text line. According to the maximum 'y' value of the blob, other blobs base line 'y' values were identified in Eq. 7. The baseline adjusted for all the blobs with Eqs. 8 and 9. After baseline correction of fluctuated characters in the text line shown in Fig. 5.

$$P_i = MAX\{B_{1,2}(y), B_{2,2}(y), \ldots, B_{blob\_count,2}(y)\} - B_{i,2}(y) \quad (7)$$

$$B_{i,1}(y) = B_{i,1}(y) + P_i \quad (8)$$

$$B_{i,2}(y) = B_{i,2}(y) + P_i \quad (9)$$

4 Experimental Results

We conduct experiments on 250 offline handwritten based Telugu text lines of document images for baseline skew correction and got a 94.4 % success rate and we conduct experiments on 1210 offline handwritten Telugu text lines for fluctuated characters baseline correction and got a 91.65 % success rate. Overall Success rate was 87.05 %. Analysis of experimental results is given in Table 1.

Table 1 Result analysis

Method	Total	Corrected	Success rate (%)
Baseline skew correction	250	236	94.40
Fluctuated characters baseline correction	1210	1109	91.65
Overall	1274	1109	87.05

Table 2 Comparative analysis

Authors	Method	Language
Cao et al. [8]	Eigen-point	English
Morillot et al. [11]	sliding window	English
Morita et al. [15]	The weighted least squares approach	English
Bahaghighat et al. [10]	Projected pattern analysis of Radon transform	Persian/Arabic, English
Abu-Ain1 et al. [14]	Horizontal projection histogram	Arabic
Majid Ziaratban et al. [13]	Cubic polynomial fitting algorithm	Farsi Arabic
Jipeng1 et al. [6]	Hough transform	Chinese
Shah et al. [7]	Linear regression	Gujarati
Ramappa et al. [9]	Bounding box technique, Hough transform, contour detection	Kannada
Proposed method	Bounding box, slope of line	Telugu

Existing baseline correction methods developed for different languages. The comparative analysis of different baseline correction methods are given in Table 2.

5 Conclusion

Correction of handwritten Telugu text base lines is a difficult task because of its complex structure. This paper contains an effective method for correcting the offline handwritten Telugu text baseline. We estimated and corrected the skew of the baseline and we corrected the baselines of the fluctuated characters. This method efficiently corrected the baseline skew and fluctuated character's baseline of offline handwritten Telugu text lines. The success rates for baseline skew correction and baseline correction of the fluctuated characters were 94.4 and 91.65 % respectively. Then the overall success rate was 87.05 %.

References

1. Ch. N. Manisha, E. Sreenivasa Reddy, Y. K. Sundara Krishna.: A Hierarchical Pre-processing Model for Offline Handwritten Document Images. International Journal of Research Studies in Computer Science and Engineering, 2(3), 41–45 (2015).
2. Telugu script, http://en.wikipedia.org/w/index.php?title=Telugu_script&oldid=655805709.
3. Naazia Makkar, Sukhjit Singh.: A Brief tour to various Skew Detection and Correction Techniques. International Journal for Science and Emerging Technologies with Latest Trends. 4(1), 54–58 (2012).
4. S. P. Sachin, Banumathi. K. L., Vanitha. R.: Database Development of historical documents: Skew detection and Correction. International Journal of Advanced Technology in Engineering and Science, 2(7), 1–10 (2014).
5. L. B. Mahanta, Alpana Deka.: Skew and Slant Angles of Handwritten Signature. International Journal of Innovative Research in Computer and Communication Engineering, 1(9), 2030–2034 (2013).
6. Tian Jipeng, G. Hemantha Kumar, H.K. Chethan.: Skew correction for Chinese character using Hough transform. International Journal of Advanced Computer Science and Applications, Special Issue on Image Processing and Analysis, 45–48 (2011).
7. Lipi Shah, Ripal Patel, Shreyal Patel, Jay Maniar.: Skew Detection and Correction for Gujarati Printed and Handwritten Character using Linear Regression. International Journal of Advanced Research in Computer Science and Software Engineering, 4(1), 642–648 (2014).
8. Yang Cao, Heng Li.: Skew Detection and Correction in Document Images Based on Straight-Line Fitting. Pattern Recognition Letters 24, no. 12, 1871–1879 (2003).
9. Mamatha Hosalli Ramappa, Srikantamurthy Krishnamurthy.: Skew Detection, Correction and Segmentation of Handwritten Kannada Document. International Journal of Advanced Science and Technology, Vol. 48, 71–88 (2012).
10. Mahdi Keshavarz Bahaghighat, Javad Mohammadi: Novel approach for baseline detection and Text line segmentation. International Journal of Computer Applications, 51(2), 9–16 (2012).
11. Olivier Morillot, Laurence Likforman-Sulem, Emmanuèle Grosicki.: New baseline correction algorithm for text-line recognition with bidirectional recurrent neural networks. Journal of Electronic Imaging, 22(2), 023028-1–023028-11 (2013).
12. Muhammad Imran Razzak, Muhammad Sher, S. A. Hussain.: Locally baseline detection for online Arabic script based languages character recognition. International Journal of the Physical Sciences, 5(7), 955–959 (2010).
13. Majid Ziaratban, Karim Faez.: A novel two-stage algorithm for baseline estimation and correction in Farsi and Arabic handwritten Text line. In: ICPR, 1–5 (2008).
14. Tarik Abu-Ain, Siti Norul Huda Sheikh Abdullah, Bilal Bataineh, Khairuddin Omar, Ashraf Abu-Ein.: A Novel Baseline Detection Method of Handwritten Arabic-Script Documents Based on Sub-Words. Soft Computing Applications and Intelligent Systems. 67–77 (2013).
15. Marisa E. Morita, Jacques Facon, Flávio Bortolozzi, Silvio J.A. Garnés, Robert Sabourin.: Mathematical Morphology and Weighted Least Squares to Correct Handwriting Baseline Skew. In: Fifth International Conference on Document Analysis and Recognition, 430–433 (1999).
16. Zhen-Long BAI and Qiang HUO: Underline detection and removal in a document image using multiple strategies., In: Proceedings of the 17th International Conference on Pattern Recognition, 578–581 (2004).
17. Otsu, N.: A Threshold Selection Method from Gray-Level Histograms. IEEE Transactions on Systems, Man, and Cybernetics, 9(1), 62–66 (1979).
18. Deepak Kumar, Dalwinder Singh.: Modified approach of hough transform for skew detection and correction in documented images. International Journal of Research in Computer Science, 2(3), 37–40 (2012).
19. Najman, Laurent A.: Using mathematical morphology for document skew estimation. In: Electronic Imaging 2004, International Society for Optics and Photonics, 182–191 (2003).

20. Chaudhuri, D., Samal, A.: A simple method for fitting of bounding rectangle to closed regions. Pattern recognition, 40(7), 1981–1989 (2007).
21. Slope of a Line (Coordinate Geometry), http://www.mathopenref.com/coordslope.html.
22. Lesson Trigonometric Functions of Angles Greater than 90 degrees, http://www.algebra.com/algebra/homework/playground/THEO-20091204.lesson.
23. Sangeetha Sasidharan, Anjitha Mary Paul.: Segmentation of Offline Malayalam Handwritten Character Recognition. International Journal of Advanced Research in Computer Science and Software Engineering, 3(11), 761–766 (2013).

A Technique for Prevention of Derailing and Collision of Trains in India

Shweta Shukla, Sneha Sonkar and Naveen Garg

Abstract Railways are widely used transportation in India because it is affordable to every class of the society. It serves as a great transport to cover large geographical distances over short period of time. It also contributes a large to the Indian economy. However, unexpected delays and accidents make it less reliable. In the era of innovation and technology, India must implement transport mechanism in railways to make it more reliable and attractive for investment, which incorporate such a system, in which the loco pilot can easily view the railway track and can also assess the presence of another train on the same track. This paper proposes a technique that will help the loco pilot to view the tracks and coordinates of the trains that could avoid derailing, delaying and collisions.

Keywords GPS · TCAS technology · Thermographic camera · Loco pilot

1 Introduction

In India, the rail transport system has undergone drastic change since its introduction in mid-19th century. In 1845, Sir Jamsetjee Jejeebhoy and Hon. Jaganath Shunkerseth formed the Indian Railway Association. The first rail track was laid down between Mumbai and Thane on 16th April 1853, first train journey held from Mumbai to thane in a 14 carriage long train. After independence, in 1947, India had a ruined railway network. Around 1952, railway lines were electrified to AC [1]. In 1985, steam locomotives were removed as it employs a large investment on coal

S. Shukla (✉) · S. Sonkar · N. Garg
Department of Information Technology, Amity University, Noida, India
e-mail: shwetashukla416@gmail.com

S. Sonkar
e-mail: snehasnkr@gmail.com

N. Garg
e-mail: ngarg1@amity.edu

S.C. Satapathy et al. (eds.), *Information Systems Design and Intelligent Applications*, Advances in Intelligent Systems and Computing 433,
DOI 10.1007/978-81-322-2755-7_30

and speed of the train was also not satisfactory, and then electric engines came into existence. Computerization of railway reservation began in 1987.

Till 1995, entire railway reservation became computerized. Metro trains then, came into picture. Kolkata became the first Indian city to have a metro train. In 2012, a tremendous feature was added into Indian railways, that is, TCAS Train Collision Avoidance System. The first train with TCAS technology was tested in Hyderabad. Currently, this feature is confined only to Shatabdi and Rajdhani [2].

Railway signaling system are designed to prevent any train from entering a stretch by another train track occupied by another train But, On February 2012, nearly 1,000 passengers of a Gwalior-bound narrow-gauge train escaped unhurt, when its engine derailed after hitting a tractor and its coaches got stuck over a canal bridge [2] (Fig. 1).

Hence manual signaling system between stations must be replaced with an automated signaling. Since, TCAS system is being used in India, but it has not reached to the root. Some of the accidents of such situation: On, 11 January 2012, five persons were killed and nine others, including a child, injured in a collision between the Brahmaputra Mail and a stationary goods train [3] (Fig. 2).

Fig. 1 Train derailing near a station [3]

Fig. 2 Train collisions [4]

Also, trains get delayed due to diverse weather conditions. These scenarios are becoming an everyday affair. Recently, On 24th may 2015 Muri express got derailed, near Vaishali, Ghaziabad [4].

2 Problems with Current System

The current railway transportation system possesses following issues:

2.1 **Derailing of trains** Since, loco pilot could not view tracks clearly, therefore there is always a chance of train got derailed which can cause huge destruction to lives.

2.2 **Collision of trains** Since, the loco pilot could not predict the presence of other train on the same track (no device is there that can help him to do so), therefore, train collision do take place [4].

2.3 **Delay in winters** Due to bad weather the loco pilot becomes unable to view tracks, for this reason to avoid any mishappening; he drives the train with a very slow speed. Every year in winters it is now a common scenario that the train got delayed by 20–24 h, which causes huge inconvenience to passengers.

2.4 **TCAS technology** With this technology, a huge amount is required to be invested, that is for running a train 1 km, about 1–2 lakhs rupees are required, that means cost of implementing it in trains would be very high [4]. This poses a great expenditure on Indian railways. Perhaps, this huge investment may be one reason that, inspite of the introduction of TCAS in India, in December, 2012, it has not been able to be implemented yet [5].

3 Proposed System

The problems with the current system, as discussed in Sect. 2, can be prevented using the system proposed in this paper.

With this system, train will be equipped with a thermographic camera (mounted on the engine), with the help of which the loco pilot can see the tracks and objects (like humans, animals etc.) in the surrounding even in diverse weather conditions of heavy rains and dense fog. A screen will be present on the dash board of the loco pilot showing resultant view of thermographic camera and coordinates of trains on the tracks as well that is, at a time the loco pilot can see the outside view and other trains nearby. Using that screen, the loco pilot can avoid the situations of derailing, delaying and collision of trains.

Figure 3 represents coordinates of the track and resultant view of thermographic camera–this screen will show coordinates of the track so that the train runs on the track only and not get derailed (Fig. 4).

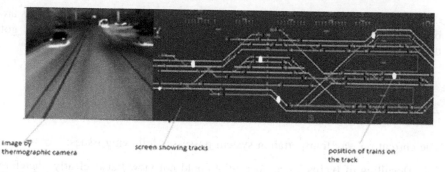

image by
thermographic camera   screen showing tracks   position of trains on
                                                                                  the track

Fig. 3 Screen showing view of thermographic camera and coordinates of trains on tracks [6, 8]

Fig. 4 GPS showing location of trains [9]

3.1 **GPS device** This device will track the location of the train with respect to the stations, it will be displayed on the screen.

3.2 **ICs on the engine and on the last coach** These will enable the loco pilot, to see the length of the train on the screen (i.e. position of engine and the last coach as well).

3.3 **Thermographic camera (FLIR HRC-series)** This camera will enable the loco pilot to see clearly see during diverse weather conditions like fog, rain etc. Also, it will enable loco pilot to view the tracks clearly till the range of 4–5 km throughout every season [6]. This camera cost round 65,000–75,000 rupees, which would be more cheaper than TCAS technology [7] (Fig. 5).

4 Issues Resolved by Proposed System

4.1 **No derailing of trains** The trains will not get derailed because the loco pilot can see the tracks clearly on the screen.

4.2 **No collision of trains** Two trains on the same track will not collide because, TCAS system will be present that will prevent trains from colliding.

Fig. 5 Thermographic camera [6]

4.3 **No delay in train timings during winters** Because of the Thermographic camera, the loco pilot can see clearly and the train will run with its usual speed.

4.4 **Less cost** Since cost incurred by thermographic camera is much less than TCAS technology, providing the better result also. Therefore, money will be saved using this system.

5 Conclusion and Future Work

This system once implemented will prevent from longer delays in train timings during winters as well as loss of human lives. There might be few hurdles in implementing this system that it is a driver-oriented system but, providing better results. The most important being the initial cost of implementing such a system and cost of maintaining and training for the same. However once the system is up and running, the cost will eventually decrease. However, its cost is cost is much less than the current system. In future, this system could be automated to make the system flaw-less.

References

1. Sandes, Lt Col E.W.C. (1935). The Military Engineer in India, Vol II. Chatham: The Institution of Royal Engineers.
2. http://indianrailway.org/ R.P. Saxena, Indian Railway History Timeline.
3. http://timesofindia.indiatimes.com/topic/Train-derailed.

4. http://timesofindia.indiatimes.com/india/Railways-to-install-indigenous-technology-to-prevent-collisions/articleshow/29832480.cms.
5. http://www.business-standard.com/article/current-affairs/railways-to-focus-on-train-collision-avoidance-system-114072600699_1.html.
6. http://en.wikipedia.org/wiki/Thermographic_camera.
7. http://www.flir.co.uk/cs/display/?id=42061.
8. https://en.wikipedia.org/wiki/Traffic_collision_avoidance_system.
9. http://www.indianews.com/2 wagons of goods train derail. Another goods train derails in Uttar Pradesh.

Temporal Scintimetric Characterization of Skeletal Hotspots in Bone Scan by Dr. V. Siva's Retention Ratio

V. Sivasubramaniyan and K. Venkataramaniah

Abstract Scintimetric Characterization of the distribution of the radiotracer Tc-99m MDP in the skeletal tissues such as Spines for the early detection of metastatic spread in the cases of Prostatic Carcinoma has been carried out. Temporal Scintimetry measures the counts in the region of interest in the scans done at two time intervals 4 and 24 h. A 4/24 h, Dr. V. Siva's retention ratio derived using Temporal Scintimetry in the characterization of skeletal hotspots in benign and malignant skeletal conditions is proposed and discussed.

Keywords Bone scans · Hotspots · Scintimetric characterization · Temporal Scintimetry · Dr. V. Siva's retention ratio

1 Introduction

Assessment of bone metastases by plain—X-ray is not reliable. The nuclear imaging with bone scintigraphy helps in the early detection of bony disease for many decades. 99mTechnetium—labeled diphosphonates used in the Bone Scan detects the pathological osteoblastic response occurring in the malignant cells. This technique is suitable for the whole body examination, at low cost, easy availability and high sensitivity. However, it lacks specificity [1].

The 99mTc-polyphosphate and its variants are adsorbed at crystal surfaces of the immature bone by ionic deposition. This process is facilitated by an increase in bone vascularity which enhances matrix localization. Thus the abnormal scintiscan

V. Sivasubramaniyan (✉)
Nuclear Medicine Department, Sri Sathya Sai Institute of Higher Medical Sciences, Prashantigram, Prasanthinilayam, Anantapur District 515134, Andhra Pradesh, India
e-mail: sivasubramaniyan.v@sssihms.org.in

K. Venkataramaniah
Physics Department, Sri Sathya Sai Institute of Higher Learning, Prasanthinilayam, Vidyagiri 515134, India

© Springer India 2016
S.C. Satapathy et al. (eds.), *Information Systems Design and Intelligent Applications*, Advances in Intelligent Systems and Computing 433, DOI 10.1007/978-81-322-2755-7_31

sites–focal hotspots represent areas of increased osteoblastic activity, to put it in a simple way. It has been shown that the bone Scintiscan is more sensitive than conventional X-rays in detecting focal disease of bone. The non-specificity of the results obtained is the main drawback of this method [2].

Why the metastasis from Prostatic Carcinoma mostly occur in bone has aroused significant interest in investigation. The first reason is the anatomical characteristics of the prostate gland. The venous drainage of the Prostate gland is occurs through a special collection of venous plexus termed as Baston's plexus. This is in direct continuity with the Lumbosacral plexus of veins. Because of this metastatic lesions from the advanced prostate cancer occur predominantly in the axial skeleton [3]. The metastatic involvement in the Skull, ribs and the appendicular bones could not be explained by this anatomical explanation.

To account for the extra-axial spread of metastatic lesions Paget proposed that the carcinoma prostate cells have mutated into the Osteotropic cell either by specific genetic phenotype formation or by the activation of specific cytokine and proteases. They have the capacity to be dormant till a favourable changes occur in the surroundings like a seed. Thus the 'seed and soil' theory gives some clue from another standpoint [4]. The increased levels of bone morphogenetic proteins (BMPs) and TGF-β in the prostate cancer cells had been established in bone metastasis [5–8]. These osteotropic metastatic seed cells attach to the bone endothelium more preferably than the endothelium of other tissues [9].

The Tc-99m MDP accumulates in the skeletal tissue because of the fact that the ionic radius of the radiopharmaceutical is similar to that of the Calcium Hydroxy appetite crystal. Hence it gets incorporated into the skeletal tissue. Thus even in the benign skeletal disorders like Paget's disease, Osteomyelitis and Post Traumatic sequences also focal skeletal hot spots do occur. Hence the real cause for the Skeletal Hotspots have to be ascertained only by invasive Biopsy and further imaging investigations as well.

Various Quantitative Parameters have been introduced to improve the Bone Scan Specificity. In addition the cross sectional imaging methods like BONE SPECT and fusion imaging methods like SPECT-CT and PET-CT have been established. In the present work we are proposing a new method of Scintimetric evaluation of bone scan for confirming or ruling out malignant or metastatic nature of the skeletal hotspots in a bone scan non-invasively.

The term SCINTIMETRY refers to the measurement of scintillations occurring in the region of interest in the scintigraphic images acquired in a study. Conventionally scintgraphic images are being interpreted by visual inspection comparing with the known normal and abnormal patterns with the help of Analog images and the photographic imprints of them. The advents of Digital imaging methods have enabled the quantification of the scintillations by counting the counts per pixels in the regions of interest. The Dicom compatible images have enabled their PACS transmission and post processing capabilities as well.

There are two types of Scintimetry. (1) Regional Scintimetry; Selecting the region of interest, drawing a region of interest–ROI over it and comparing it with either adjacent site or corresponding site on the contralateral side by drawing a similar ROI. This can also be termed as SPATIAL SCITIMETRY—Scintimetry related to two different sites. (2) Temporal Scintimetry: The margins of the focal hotspot in a bone scans taken at 4 and 24 h after the injection of the radiopharmaceutical are drawn with the drawing tool and maximum counts per pixel in those images are tabulated. The change in the focal hotspots in the bone scan with reference to time interval is ascertained by taking the ratio of the two values. This can be termed as Time BOUND SCINTIMETRY meaning Scintimetry of the single region of interest with respect to time.

There are several methods of quantitation for Scintimetric evaluations. The most widely used ones are (a) Local Uptake Ratio (b) Lesion bone ratios (c) bone soft tissue ratio (d) Percentage uptake measurements e. Standardized Uptake Value or SUV.

a. **Local uptake ratio** The simplest method of quantifying the uptake of the tracer in an area of interest is to express the count rate obtained in the area of the focal hot spot as a ratio with the count rate in an adjacent normal area of the same image [10]. Alternately a line profile curve evaluation at the lesion site can also be used [11]. The example of line profile evaluation of the epiphyseal regions of the two Knee joints is shown in Fig. 1.

b. **Lesion bone ratios L/B ratio** In this the term lesion refers to a focal area of abnormality and bone denotes a suitable area of normal bone. Condon et al. [12] and Vellenga et al. [13] have demonstrated the utility of this in detecting the visually imperceptible lesions and rib lesions.

c. **Bone/soft tissue ratio** Pfelfer et al. [14] have found these measurements useful in detecting patients with haematological malignancies. Constable and cranage [15] have used this ratio to confirm the presence of prostatic super scans.

d. **Percentage uptake measurements** This is calculated by referring the uptake in the area of interest to the administered dose [16] or to the activity of a known external standard by Meindok et al. [17].

Fig. 1 Line profile curve

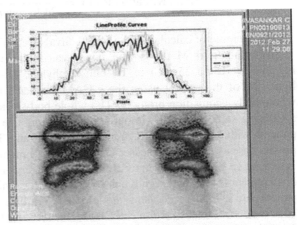

Fig.1 LINE PROFILE CURVE

All these methods will help in identifying the different count densities between lesion and the adjacent areas. They are not helpful in differentiating the benign and malignant lesions. Also the metabolic turn over in the hot spots cannot be inferred.

2 Proposed Method of Dr. Siva's Retention Ratio by Temporal Scintimetry

Israel et al. proposed a Temporal Scintimetry method [18]. They measured the ratio of Lesion count (L) and the Non Lesional area counts (NL) over the bone in the bone scan done at 4 h. The same ratio is repeated in the 24 h bone scan image also.

$$\text{TF ratio} = \frac{\text{L/N 24 h}}{\text{L/N 4 h}}$$

It has been shown that the measurements showed less or very minimal change in patients with degenerative disease and treated metastasis and very high increase in the metastatic lesions. The ratio resulted in decimal values only as the denominator is always high as per the Radioactive decay law.

In the present work we have proposed the 4/24 h Dr. V. Siva's Retention Ratio is to characterize the focal hot spots and to differentiate the metastatic lesions from the benign one the following procedure.

1. The maximum counts at the Focal hot spot or Lesion only is considered instead of the L/N ratio.
2. 4/24 h ratio is taken as against the 24/4 h ratio of Israel et al.

$$\text{Dr. V. Siva's Retention Ratio} = \frac{4 \text{ h } count \ at \ Focal \ Hot \ Spot}{24 \text{ h } count \ at \ Focal \ Hot \ Spot}$$

The example of calculating the Dr. V. Siva's Retention Ratio is shown in Fig. 2.

Materials and Methods The bone scan is done 4 h after the intravenous injection of 15 to 25 mCi of Tc-99m Methylene Di-Phosphonate with adequate hydration of the patient using the e-CAM Siemen's dual head gamma camera's whole body acquisition protocol. The whole body bone scan is repeated next day also using same protocol without fresh injection. The study group consists of 75 patients with biopsy proven Carcinoma Prostate and 32 patients with known and proven cases of Paget's disease, Osteomyelitis, Fracture, Avascular Necrosis and degenerative disorders. In the Carcinoma Prostate group in 53 patients metastatic involvement is seen and negative in 22. The 101 focal hotspots in various sites in 16 patients are characterized using the temporal Scintimetric method. Both the 4 and 24 h bone scan images are selected using the General Display protocol. Then with the help of the Region Ratio processing protocol the 4 and 24 h anterior and

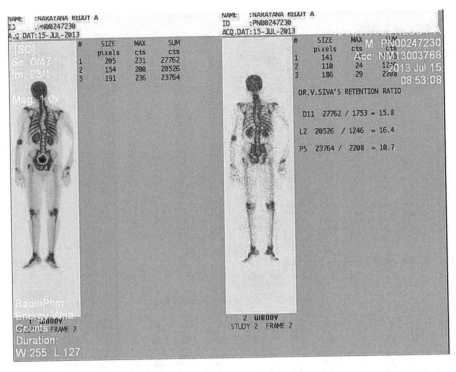

Fig. 2 Dr. V. Siva's retention ratio calculation procedure

posterior images are selected separately. Then maximum counts in the selected regions are tabulated. Then the 4/24 h Dr. V. Siva's retention ratio is derived by dividing the 4 h counts with 24 h counts along with the Israel's 24/4 h ratio as well. The same procedure is applied in the 92 lesions in the benign disease images also. The results are tabulated and scrutinized.

Results The mean value of 4/24 h Dr. V. Siva's retention ratio is found to be 11.5 ± 2.8 and that of 24/4 h Israel's ratio to be 0.08 ± 0.02 in the Carcinoma Prostate group. The mean value of 4/24 h Dr. V. Siva's retention ratio is found to be 4.8 ± 2.5. The whole body count ratios of 4/24 h and 24/4 h also showed correspondingly similar values. The Sites 4 h, 24 h counts and 4/24 h Dr. V. Siva's Retention Ratio in Malignant and Benign Hot spots are tabulated in Figs. 3 and 4.

Online Social Science Statistics calculator is used for statistical analysis of the values. The student T test for 2 independent values reveals that T value is 17.1. Then p value is <0.00001. The result is Significant at p value <0.05. The graphical estimation of dispersion of values shows significant difference between the benign and metastatic lesion as shown in Fig. 5.

In Fig. 5 the image I show the difference between Metastatic and benign lesions in dumbbell form. The line spread in II and the line spread with dispersions around it in III establishes the clear cut difference in the Retention Ratio between Metastatic and benign conditions. This proves that the temporal Scintimetric Characterization

METASTATIC HOT SPOTS DR.V.SIVA'S RETENTION RATIO DATA SHEET

SITE	4 hr	24 hr	4/24 hr	SITE	4 hr	24 hr	4/24 hr	SITE	4 hr	24 hr	4/24 hr	SITE	4 hr	24 hr	4/24 hr
Pubic sym	46947	4356	10.7	Rt foot	33202	3051	10.8	Rtrib	32810	2002	16.3	LTISC	29557	3080	9.5
D11	27762	1753	15.8	Lt.rib	141923	12909	10.9	Stem	21174	2044	10.3	PS	46947	4356	10.7
L2	20526	1246	16.4	D4	50374	3615	13.7	ttfem	16931	1383	10.6	D11	27762	1753	15.8
PS	23764	2208	10.7	Rt.Fem	16615	1128	14.7	rtrib	11538	1016	11.3	L2	20526	1246	16.4
rt.rib	21920	1629	13.4	rtrib	10087	902	11.1	ltfoot	7440	666	11	PS	23764	2208	10.7
L5	37105	2619	14.1	ltrib	9814	772	12.7	ltrib	16931	1076	15.7				
ltrib	14839	1124	13.2	scrot	6527	511	12.7	L3	21257	1004	21.1				
RT12	18141	1093	16.5	flgt	10427	707	14.7	L5	15304	1462	10.4		Mean		11.5
RT5	14826	1289	11.5	RLrib	11138	777	14.3	LPS	15688	1257	12.4		Std.Dev		2.8
RTRIB	6939	614	11.3	D12	9011	1259	7.1	RTFRONT	68905	6210	11				
L3	23488	2811	8.3	RTKNEE	5941	554	10.7	ltparei	17337	2095	8.2				
RSIJ	31138	3009	10.3	LTGT	7525	654	11.5	C7	40764	3922	10.3				
RTRIB	18853	1816	10.3	L3	89624	14065	6.3	RTCLAV	23189	1871	12.3				
L4	17642	1831	9.6	L4	97733	11546	8.4	Rtrib	17829	1959	9.1				
RTRIB	23790	1661	14.3	LTIC	156441	20594	7.6	Stern	90704	7813	11.6				
RTFRONT	145552	10598	13.7	RTIC	304334	29736	10.2	D12	81694	7033	11.6				
LTFOOT	63036	5315	11.8	L1	50118	5568	9	LTRIB	16182	1620	9.9				
LT PARI	15157	1307	11.5	L3	121252	15160	7.9	L2	44670	2542	17.5				
D5	21945	1561	13.9	LTSIJ	76088	7134	10.6	SAC	29937	3743	7.9				
RTFRONT	61375	4967	12.3	LTISC	56785	6137	9.2	RTIC	27544	3065	8.9				
LTFOOT	38417	1900	20.2	cal	17232	1286	14.2	LTPS	33425	4194	7.9				
RTRIB	20810	1990	10.4	ltshoul	13952	1187	11.7	LTFRON	19667	3291	5.9				
RT5RIB	37597	3765	9.9	ltrib	21298	2178	9.7	RT.FRON1	31028	2457	12.6				
rtrib	75984	7242	10.4	rtspg	61808	7489	8.2	IC	7189	958	7.5				
ltrib	61015	6486	9.4	ltocci	5376	396	13.5	RTOCCI	13041	957	14.5				
L4	25655	2538	10.1	ltshoul	8671	621	13.9	C3	22692	2441	9.2				
D1	56045	4049	13.8	ltrib	20672	2578	8	RTCLAV	12266	1350	9				
D7	82119	6260	13.1	lipr	9588	878	10.9	LTRIB	21919	2196	9.9				
RTSIJ	114114	9839	11.5	Stem	40265	3423	11.7	RTRIB	26034	2611	9.9				
RTISCH	49097	5684	10	rtknee	10183	739	13.7	L3	40277	4132	9.7				
RLrib	40830	2576	14	L5	29209	1902	15.3	SAC	35916	4299	8.3				
Stem	188301	13654	13.7	Lt rib	23286	2321	10	RIC	17560	2049	8.5				

Fig. 3 Malignant hot spots data sheet

BENIGN SKELETAL HOT SPOTS - DR.V.SIVA'S RETENTION RATIO DATA SHEET

SITE	4 HR	24 HR	4/24 HR	SITE	4 HR	24 HR	4/24 HR	SITE	4 HR	24 HR	4/24 HR	SITE	4 HR	24 HR	4/24 HR
RTPELVIS	75118	33473	2.2	RTFEM	33224	6333	5.2	RIC	19438	8695	2.2	RTHUM	33909	5223	6.5
RTHIP	63917	27070	2.3	rttibia	19005	9056	2	LTHIP	59797	31034	1.9		19005	9056	2
RTPELVIS	82117	69696	1.1	lttibia	25439	11719	2.1	RTFEM	45275	7092	6.3		109591	50724	2.1
LT SCALP	15118	6750	2.2	lttibia	25439	9633	2.6	LTHIP	69615	25866	2.7	LTCAL	243854	108618	2.2
RT HIP	76124	40931	1.8	rttibia	19005	6222	3	lttibia	109591	50724	2.1	LTHUM	124690	43796	2.8
LT SIJ	18094	7916	2.2	RTHUM	35683	4953	7.2	lttibia	95069	28498	2.6	LTCAL	277620	124442	2
LT HIP	34229	17238	1.9	LTHUM	23046	4176	5.5	L5	13361	5942	2.2	LTTF	171389	81407	2.1
LT FEM	40829	4684	8.7	RTFEM	106127	15325	6.9	D11	67768	12714	5.3	LT FEM	108611	24842	4.3
RT HUM	33909	5223	6.5	LTFEM	46119	10930	4.2	LTFEM	53967	6764	7.9	LTHIP	63999	9570	6.6
RT.SIj	79959	14036	5.6	LTHUM	23147	3648	6.3	LTFEM	167152	20289	8.2	LTHIP	35110	5144	6.8
LTILI	251561	36262	6.9	RTHUM	26244	3392	7.7	LTFEM	164381	250762	6.5	LTHIP	20650	3430	6
RTGT	24008	2115	11.3	LTFEM	41615	6800	6.1	rttibia	23273	3101	7.5	RT HIP	56508	7020	8
LTSIJ	27770	4989	5.5	RTFEM	62881	8761	7.1	rttibia	7656	1456	5.2	LTHIP	76695	11098	6.9
RTSI	281296	41146	6.8	RTFEM	200041	26113	7.6	lttibia	20318	3490	5.8	RTHIP	153044	19041	8
RGT	16867	2416	6.9	RTFEM	162078	27694	5.8	RTPELVIS	82177	69696	1.1	LTHIP	149205	17218	8.6
RTHIP	41435	9560	4.3	RTHIP	56077	18447	3	RTPELVIS	75118	33473	2.2	LTKNEE	57387	7285	7.8
LTHIP	23980	6121	3.9	LTHIP	61789	19557	3.1	RTHIP	63917	27070	2.3	LTKNEE	53289	5727	9.3
RTHIP	20685	5093	4	RTMAX	37279	11336	3.2	LTSCALP	15118	6750	2.2	LTKNEE	32629	3703	8.8
LTHIP	14614	2585	5.6	LTHUM	11768	1815	6.4	RTHIP	76124	40931	1.8	LTKNEE	383630	52433	7.3
RTFH	60344	8465	7.1	LTHUM	61627	8574	7.1	LSIJ	18094	7916	2.2	LTKNEE	1130826	143843	7.9
RTFH	36956	4567	8	LTHUM	104621	52271	2	LTHIP	34229	17208	1.9				
RTFH	163539	23860	6	LTHUM	89617	36802	2.4	LTSIJ	18291	9219	1.9		Mean		4.8
RTHIP	51688	19012	2.7	LTANKLE	64867	7641	8.5	LTISCH	39610	19302	2		Std.Dev		2.5
RTHIP	34336	12457	2.7	LTANLE	41806	4910	8.5	LTFEM	33800	9001	3.7				

Fig. 4 Benign hot spot data sheet

of the skeletal hot spots makes the differentiation between the Metastatic lesions and the benign ones easy and effective.

3 Discussion

Since the localization of the Tc-99m MDP radiopharmaceutical corresponds to the Osteoblastic activity mediated by Hydroxy Appetite crystals, the building blocks of the skeletal tissue, the retention ratio represents the metabolic turnover of the

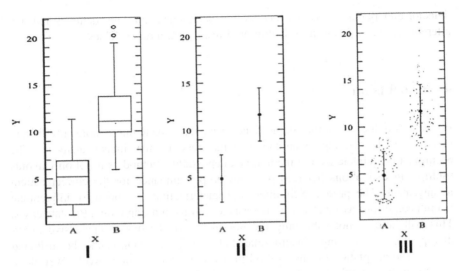

Fig. 5 The graphical representation of significant difference between two groups

skeletal tissue. Unlike in Israel's method in which the lesion counts and the Bone counts are taken into consideration, in this method the skeletal turnover in the hotspot only is considered.

In the Israel's ratio which is calculated by dividing the 24 h L/N ratio by the 4 h L/N ratio resulted in decimal values only as the denominator is always high as per the Radioactive decay law. Whereas in the Dr. V. Siva's retention ratio 4 h lesion count is divided by the 24 h lesion count which is always lower than the 4 h count gives a whole integer value which enables the Physicians to clearly understand the difference between various conditions. The lesser value of 4.8 ± 2.5 seen in the benign bone disorder is probably due to the normal or not much altered metabolic and cellular activity occurring in the skeletal matrix. The higher value of 11.5 ± 2.8 in the metastatic lesions of the Carcinoma of Prostate could be attributed to increased or accelerated skeletal metabolism due to cancerous proliferation and destruction. Thus inclusion of the 24 h bone scans acquisition and Scintimetric Characterization provides a one-step shop for differentiating between the Benign and Malignant Skeletal disorders. The simple reversal of the ratio from 24/4 to 4/24 h counts resulted in useful specific Numerical values to differentiate between various skeletal disorders. Hence they represent the skeletal metabolic turn over in the respective clinical conditions. Further analysis and scrutiny of all the hot spots in the remaining metastatic positive group in the light of this tool might be of interest.

The inherent limitations in this method are the marking of the focal hot spot must be identical or closely similar in the 4 and 24 h images. Improper and careless marking of them will affect the validity of the study. If an automatic tracing and measurement of the focal hotspots could be programmed and used operator dependent mistakes could be eliminated. The patient position, distance between the detector and patient body and acquisition parameters must be identical and any alteration in any one of these factors will result in improper values.

4 Conclusion

It can be concluded that the simple bone scan can be made into a more specific and useful tool by adding quantitative parameters in the image analysis. The Scintimetric Characterization method is a dependable method as it not only avoids multiple investigations but also provides proper guidance for the correct patient management. The present Scintimetric characterization is a boon to differentiate benign conditions associated with metastatic lesions in a Carcinoma Prostate cases. This will help in not subjecting the non-metastatic lesions to aggressive radiotherapy. Since it is a single institutional study much credentials cannot be attributed to this. The adaptation of this method in many other institutions world over alone can justify the significance of this novel method. Similarly the applicability of this method to PET scans has also to be contemplated.

References

1. Langsteger W, Haim S, Knauer M, Waldenberger P, Emmanuel K, Loidl W, Wolf I, Beheshti M. Imaging of bone metastases in prostate cancer: an update. Q J Nucl Med Mol Imaging. 2012 Oct; 56(5):447–58.
2. Brian C. Lentle, M.D., F.R.C.P. (C); Anthony S. Russell, M.B., F.R.C.P (C); John S. Percy, M.D., F.R.C.P. (Edin) (C); John R. Scott, M.Sc.; and Frank I. Jackson, M.B., C.R.C. P. (C) Bone Scintiscanning Updated. (Ann Intern Med. 1976; 84(3):297–303. doi:10.7326/0003-4819-84-3-297.
3. Batson OV: The function of the vertebral veins and their role in the spread of metastases. Clin Orthop Relat Res 312: 4–9, 1995.
4. Paget S: The distribution of secondary growths in cancer of the breast. Cancer Metastasis Rev 8: 98–101, 1989.
5. Masuda H, Fukabori Y, Nakano K, Takezawa Y, T CS and Yamanaka H: Increased expression of bone morphogenetic protein-7 in bone metastatic prostate cancer. Prostate 54: 268–274, 2003.
6. De Pinieux G, Flam T, Zerbib M, Taupin P, Bellahcene A, Waltregny D, Vieillefond A and Poupon MF: Bone sialoprotein, bone morphogenetic protein 6 and thymidine phosphorylase expression in localized human prostatic adenocarcinoma as predictors of clinical outcome: a clinicopathological and immunohistochemical study of 43 cases. J Urol 166: 1924–1930, 2001.
7. Shariat SF, Shalev M, Menesses-Diaz A, Kim IY, Kattan MW, Wheeler TM and Slawin KM: Preoperative plasma levels of transforming growth factor beta(1) (TGF-beta(1)) strongly predict progression in patients undergoing radical prostatectomy. J Clin Oncol 19: 2856–2864, 2001.

8. Klaus Strobel, Cyrill Burger, Burkhardt Seifert, Daniela B. Husarik, Jan D. Soykal, Thomas F. Hany Characterization of Focal Bone Lesions in the Axial Skeleton: Performance of Planar Bone Scintigraphy Compared with SPECT and SPECT Fused with CT AJR: 188, 467–474, May 2007.
9. Lehr JE and Pienta KJ: Preferential adhesion of prostate cancer cells to a human bone marrow endothelial cell line. J Natl Cancer Inst 90: 118–123, 1998.
10. Rosenthall L, Kaye M (1975) Technetium 99m Pyrophosphate kinetics and imaging in metabolic bone disease. J Nucl Med 16:33–39.
11. Lentle BC, Russell AS, Percy JS, Jackson FI(1977) The scintigraphic investigation of sacroiliac disease J Nucl Med 18:529–533.
12. Condon BR, Buchanan R, Garvie NW et al (1981) Assessment of secondary bone lesions following cancer of the breast or prostate using serial radionucleide imaging. Br. J Radiol 54: 18–23.
13. Vellenga CJLR, Pauwels EKJ, Bijvoet OLM (1984) Comparison between visual assessment and quantitative measurement of radioactivity on the bone scintigram in Paget's disease of bone. Eur J Nucl Med 9:533–537.
14. Pfeifer JP, Hill W, Bull U, Burkhardt R, Kirsch CM (1983) Improvement of bone scintigraphy by quantitative evaluation compared with X-ray studies and iliaccrest biopsy in malignant disease. Eur J Nucl Med 8: 342–345.
15. Constable AR, Cranage RW (1980) Recognition of superscan in prostatic bone scintigraphy. Br J Radiol 54: 122–125.
16. Hardy JG, Kulatilake AE, Wastie ML (1980) An index for monitoring bone metastases from carcinoma of the prostate. Br J Radiol 53: 869–873.
17. Mendiok H, Rapoport A, Oreopoulos DG, Rabinovich S, Meema HF, Meema S (1985) Quantitative radionucleide scanning in metabolic bone disease.Nucl Med Comm 6: 141–148.
18. Israel O, Front D, Frenkel A, Kleinhaus U(1985) 24 hour/ 4 hour ratio of technetium 99m methylene diphosphonate uptake in patients with bone metastases and degenerative bone changes. J Nucl Med 26: 237–240.

Executing HIVE Queries on 3-Node Cluster and AWS Cluster—Comparative Analysis

Shweta Malhotra, Mohammad Najmud Doja, Bashir Alam,
Mansaf Alam and Aishwarya Anand

Abstract Cloud Database Management System (CDBMS) is one of the potential services provided by various Cloud Service Providers. Cloud providers cope with different users, different data and processing or analysis of different data. Traditional Database Management Systems are insufficient to handle such variety of data, users and their requirements. Hence, at the conceptual layer of CDBMS, traditional SQL, Oracle and many more Database Languages are insufficient to provide proper services to their users. HIVE and Pig are the different types of languages which are suitable for the cloud environment which can handle such huge amount of data. In this paper, performance comparison of 3-Node cluster and Cloud Based Cluster provided by the Amazon Web Services is being done. We have compared the processing of structured data with the help of different queries provided by HIVE tool on 3-Node cluster and Amazon Web Service (AWS) cluster. It has been concluded that HIVE queries on AWS cluster gives better results as compared to 3-Node cluster.

Keywords Cloud database management system (CDBMS) · Amazon web service (AWS cluster) · 3 Node Hadoop cluster · HIVE

S. Malhotra (✉) · M.N. Doja · B. Alam · M. Alam
Jamia Millia Islamia University, New Delhi, India
e-mail: Shweta.mongia@yahoo.com

M.N. Doja
e-mail: mndoja@gmail.com

B. Alam
e-mail: babashiralam@gmail.com

M. Alam
e-mail: Mansaf_alam2002@yahoo.com

A. Anand
G D Goenka University, Sohna, Gurgaon, India
e-mail: Aishwarya.del.in@gmail.com

© Springer India 2016
S.C. Satapathy et al. (eds.), *Information Systems Design and Intelligent Applications*, Advances in Intelligent Systems and Computing 433,
DOI 10.1007/978-81-322-2755-7_32

307

1 Introduction

Database Management System [1, 2] is one of the pragmatic cloud service, cloud providers provides to their users. In providing this service cloud providers are facing many obstacles such as how to provide different Database services to a variety of large number of users and how to process and analyse huge amount of structured and unstructured data as simple RDBMS is inappropriate to solve the problem related to large amount of data. In the previous paper [1], we proposed 5 layered architecture of Cloud Database Management System. At the conceptual level of CDBMS languages like SQL, Oracle, MySQL etc. are insufficient to process or analyse variety of large amount of data. Cloud providers such as Amazon, Google, Yahoo, EMC, IBM, Microsoft etc. now with the help of Hadoop Platform analyse large amount of structured and unstructured data. Hadoop comprises of Map Reduce Programming model [3] which is used to handle large amount of different data sets. Queries on these datasets are performed with the help of HIVE-SQL like language that provides a SQL kind of interface on the top of Hadoop which is used to analyse such big amount of data. In this paper a performance comparison of simple 3-Node Hadoop cluster and Amazon AWS 3-Node cluster by issuing HIVE queries on structured data is being done. The whole paper is organized as follows. Sections 2–4 comprises of 3 Node cluster, Amazon AWS cluster, HIVE. Section 5 discusses about the results. Section 6 concludes findings of overall results.

2 3-Node Hadoop Cluster

Bigdata [4] is primarily unstructured data available in the form of log files includes aeroplane data, facebook data, electric poles data, stock market data, health care data, weather forecasting data etc. Earlier companies use oracle kind of RDBMS to process and analyse the data but Oracle was not able to process or scale that huge amount of data. Data coming to their server's grew from 10 s of GBs in 2006 to 1 TB per day in 2007 to 500 TB of data per day in 2013. Problems like huge number of users, terabytes of data generated per day, several queries of users per day and million photos shared per day were not handled by traditional RDBMS. Then companies started working with Hadoop.

Hadoop [5, 6] is a framework that provides solution to such Bigdata Problems as it provides both storage in the form of HDFS (Hadoop Distributed file System) and Processing in the form of YARN (Yet another Resource Allocator). It allows the distributed processing of such huge large amount of data sets across clusters of commodity computers using a simple programming model known as Map Reduce [3]. Cloud companies like Yahoo, Google, Facebook, Amazon, IBM, Microsoft etc. are continuously using Hadoop. Hadoop Distributed file System (HDFS) is based

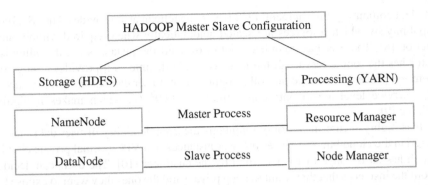

Fig. 1 Nodes in a cluster

on Master and Slave kind of configuration. Following is the description of Master and Slave nodes.

2.1 Nodes in Hadoop Cluster

- NameNode—A Master Process with Resource Manager contains metadata of the slave nodes and it also manages and maintains the blocks which are present in the Slave nodes.
- DataNode—A Slave Process with Node Manager are deployed on each machine which provides the actual storage and processing of data. They perform the task assigned to them by Resource Manager.

In a cluster there will be only one NameNode present and it can contain any number of DataNodes as shown in Fig. 1. In this paper we have implemented 3-Node Hadoop cluster with one NameNode and other two DataNodes.

Here user's job comes to the Resource Manager of the Master process for processing. Once the resource Manager gets the data then it forwards the request to Node Manager for Processing. Actual processing is being done on Node Manager. In Node Manager it creates Container which are the resources such as memory, processor etc. for processing the data. Map and Reduce Programs are performed on these containers for processing.

3 Amazon Cluster (AWS)

Amazon Elastic Compute Cloud (Amazon EC2) [7] provides computing capabilities in the cloud. It has a very simple service interface through which user can obtain and configure capacity efficiently which ensures that the capacity is used in a predictable manner. It provides a centralized control over all the resources and an efficient computing environment. EC2 is very beneficial as it provides Auto-Scaling capability by which resources are scaled automatically when required without any friction [8]. The user has complete control over all the instances i.e. the administrator has the root access to all the instances. It is flexible as it provides a range of memory, processor, storage and other options that a user can select from. Amazon EC2 service level agreements commitment is 99.95 %, which makes it highly reliable [9].

It is highly secure, it not only maintains the confidentiality of user data that is stored in the cloud but also places the user instances in a VPC (virtual private cloud) which has an IP range that is specified by the customer [10]. The users can decide about the instances that they want to keep private and the ones they want to expose to the world. There are security groups that allow users to specify and control the inbound and the outbound traffic to and from their instance. Amazon EC2 is not very expensive as the user pays on an hourly basis without any future commitment. It also provides Auto-Scaling, which lightens the customer from the burden of handling traffic spikes. Instances are available On-Demand and can also be reserved [10].

EMR is a framework that splits data into pieces and allows processing to occur and gathers results. Consider an e.g. where we have terabytes of data related to logs by a particular user. This data is split into many small pieces by EMR (Amazon Elastic Map Reduce), which is then processed in clusters. Finally, the results from all the cluster nodes are aggregated to get the final result as shown in Fig. 2.

Basically, the job flow under EMR consists of the following steps:

- **Cluster creation** Hadoop breaks a data set into multiple sets to process quickly on a single cluster node.
- **Processing** EMR executes a script such as java, HIVE and others to process the data in the data set.

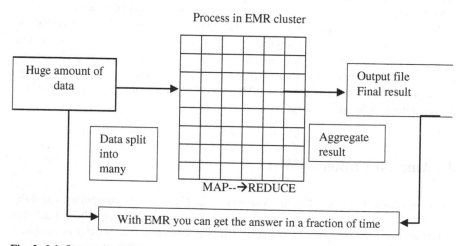

Fig. 2 Job flow under EMR

- **Chained step** Step is a unit work defined by EMR. EMR executes subsequent steps using data from previous step. Data is passes between steps using HDFS (Hadoop Distributed File System).
- **Cluster terminated** Output data is placed in Simple storage Service (S3) for subsequent use.

3.1 Nodes in EMR

Amazon EMR defines three roles for the servers in a cluster. They are

- **Master node** It consists of the master instances that manage the cluster. It distributes the work to different EC2 instances or the slave nodes.
- **Core nodes** They perform the work assigned to them by master node. Its stores the data in HDFS and runs the Task tracker daemon which performs the task assigned to the node.
- **Task nodes** These nodes can be added or subtracted to perform the work.

Core and Task instances read-write to S3. They take data from S3, process it and give the result back to S3 once the work of Map-Reduce is over [7].

4 HIVE

The whole journey of HIVE [11] started from Facebook. Facebook like social online sites wants to capture and analyse all the user activities and based on that they give the right recommendations to the users. Earlier companies use oracle kind of RDBMS to process and analyse the huge amount of data but Oracle was not able to process or scale the upcoming huge amount of data. Hence, Facebook developed the HIVE, which provides SQL type interface that can analyse terabytes of data which produced every day which runs on the top of Hadoop.

HIVE fills up the gap between the tools available in the market like Hadoop with Map Reduce and the kind of expertise users are having like user understand SQL kind of Languages very easily. With HIVE, there is no need to learn java and Hadoop API's. HIVE (SQL-style) query language provides a SQL kind of interface where users can write logics in SQL construct which will be converted into Map Reduce job with the help of Hadoop. Following are some features of HIVE

- Data warehousing package built on top of Hadoop i.e. HIVE is used to create tables, databases, views, partitioning, bucket in which user can restructure the dataset etc. It also provides schema flexibility where a user can alter table, move column etc.
- HIVE offers plugin custom code like JDBC/ODBC drivers.
- As it is similar to SQL. It is used for managing and querying structured data. Structured data like log processing, Data Mining Analysis, customer facing business intelligence queries.

5 Results and Discussions

In order to evaluate the performance of HIVE Queries with 90 MB of structured data on 3-Node Hadoop cluster and AWS cluster, We implemented 3-Node Hadoop cluster and AWS cluster with the same configuration. In Our Environment

- Machines are typically dual-processor ×86 processors running Ubuntu 14.04, with 4 GB of memory per machine.
- We have implemented both clusters with Hadoop 2.3.0 and HIVE 0.13.0 versions.
- A Cluster Consist of 3 Machines typically One NameNode (Master Machine) and the other two are DataNode (Slave Machine).

Table 1 is summarizing overall work that we did for comparing the two clusters Memory Processor Storage.

5.1 Queries and Result Discussion

1. Load and Describe
As shown in Fig. 3, when the query LOAD DATA is executed to load data from.csv file into the table on a 3-node Hadoop cluster it takes 9.654 s which is much more as compared to AWS cluster that takes 1.468 s. A query used for describing the table takes 0.199 s on a 3-node Hadoop cluster and 0.122 s on AWS cluster.

Table 1 Summary of time taken by 3-node cluster and AWS cluster for different HIVE queries

HIVE query performed	Time (3-node cluster) (s)	Time (AWS cluster) (s)	Result
Load data	9.654	1.468	Data is loaded from .csv file into the table
Describe txnrecords	0.199	0.122	Table description
Select count (*) from txnrecords	69.845	30.802	10,00,000 records were fetched
Select category, sum (amount) from txnrecords group by category	58.475	31.518	15 records were fetched according to the data
Select custno, sum (amount) as total from txnrecords group by custno order by total limit 10	103.454	54.825	10 records were fetched
Select count (distinct category) from txnrecords	55.128	27.681	15 records were fetched
Insert overwrite into local directory	44.457	23.558	Data is inserted into the local directory

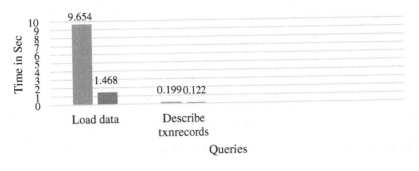

Fig. 3 Load data and describe—time in seconds

2. SELECT, Group By, Order By

A query used for selection operation also took more time when executed on a 3-node Hadoop cluster as compared to AWS cluster, like "Select count (*) from txnrecords". As shown in Fig. 4, when select queries such as aggregation, group by, order by were executed on AWS cluster, it took almost half the time for execution as it took on a 3-Node cluster.

3. Insert Overwrite into Local Directory

As shown in Fig. 5, when a query was executed to insert data into the local directory AWS cluster took 23.558 s which is very less as compared to the 3-Node cluster which took 44.457 s.

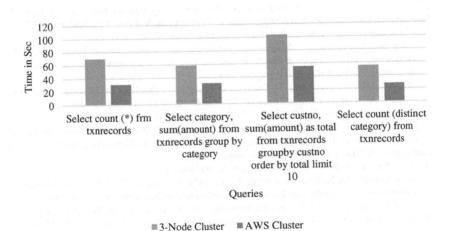

Fig. 4 SELECT, group by, order by—time in seconds

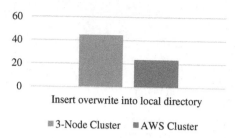

Fig. 5 Insert overwrite into local directory—time in seconds

6 Conclusion and Future Scope

In this paper, we have implemented 3-Node Hadoop cluster and AWS cluster and we have concluded that AWS cluster is comparatively easier to implement. 3-Node cluster is not only more complicated to implement but also it consumes more time to implement. It requires prior knowledge and understanding about complex Linux commands for implementation. We have evaluated the performance of both clusters by executing HIVE queries. For all Queries like Load Data, Describe, SELECT, Group By, Order By, AWS cluster is turned out to be more efficient with respect to time and resource allocation. In future, unstructured data analysis can be done with other languages like MapReduce and more.

References

1. M. Alam and K. Shakil.: Cloud Database Management System Architecture. In: UACEE International Journal of Computer Science and its Applications, Volume 3(1), 2013, pages 27–31.
2. AWS documentation; Auto scaling, http://aws.amazon.com/autoscaling.
3. J. Dean and S. Ghemawat.: Mapreduce: simplified data processing on large clusters. In OSDI'04. In: Proceedings of the 6th Symposium on Opearting Systems Design & Implementation (OSDI'04), 2004, pages 1–10.
4. L. Zhang, C. Wu, L. Zongpeng, C. Guo, C. Minghua and C.M. Lau. In: Moving Big Data to the Cloud: An Online Cost-Minimizing Approach. In: IEEE journal on selected areas in communications (2013), Vol 31, Issue 12, pages 2710–2721.
5. L. Huang, H. Shan, Chen and H. Ting-Ting.: Research on Hadoop Cloud Computing Model and its Applications. In: IEEE Third International Conference on Networking and Distributed Computing (ICNDC), 21–24 Oct. 2012, pages 59–63.
6. Apache: Apache Hadoop: http://hadoop.apache.org/docs/r2.7.1.
7. Amazon Elastic MapReduce, Developer Guide (API Version 2009-03-31), http://docs.aws. amazon.com/ElasticMapReduce/latest/DeveloperGuide/emr-how-does-emr-work.html.
8. AmazonEC2 Service Level Agreement, http://aws.amazon.com/ec2-sla/, Retrieved July 2012.
9. Amazon Virtual Private Cloud, Getting Started Guide, API Version 2013-10-15, http:// awsdocs.s3.amazonaws.com/VPC/latest/vpc-gsg.pdf.
10. Amazon EC2 Instance, http://aws.amazon.com/ec2/, Retrieved July 2012.
11. S. Mongia, M.N. Doja, B. Alam, and M. Alam.: 5 layered Architecture of Cloud Database Management System. In: AASRI Conference on parallel and Distributed Computing and Systems, Vol 5, Pages 194–199, 2013.

Performance of Speaker Independent Language Identification System Under Various Noise Environments

Phani Kumar Polasi and K. Sri Rama Krishna

Abstract Language Identification has gained significant importance in recent years, both in research and commercial market place, demanding an improvement in the ability of machines to distinguish languages. Although methods like Gaussian Mixture Models, Hidden Markov Models and Neural Networks are used for identifying languages the problem of language identification in noisy environments could not be addressed so far. This paper addresses the capability of an Automatic Language Identification (LID) system in clean and noisy environments. The language identification studies are performed using IITKGP-MLILSC (IIT Kharagpur-Multilingual Indian Language Speech Corpus) databases which consists of 27 languages.

Keywords Language identification · Indian languages · MFCC · GMM

1 Introduction

In natural language processing, LID is the process of classifying a language spoken by a person from a set of languages. Speech processing applications in man machine communications start to degrade when noise gets added to the original signal [1]. Henceforth, there exists a strong demand for escalating the capability of LID system in noisy environment.

Rest of the paper is organized as follows. Section 2 addresses prior work. Section 3 briefs the language database used for the study. Section 4 discusses the Feature extraction and model building. Section 5 gives the LID performance on the Indian Language database. Section 6 describes the performance of LID system in the presence of various noises with various SNR level. Section 7 concludes with a summary of the ideas proposed in this work.

P.K. Polasi (✉) · K. Sri Rama Krishna
ECE Department, V R Siddhartha Engineering College,
Vijayawada, Andhra Pradesh, India
e-mail: polasi.phani@vrsiddhartha.ac.in

© Springer India 2016
S.C. Satapathy et al. (eds.), *Information Systems Design and Intelligent Applications*, Advances in Intelligent Systems and Computing 433,
DOI 10.1007/978-81-322-2755-7_33

315

2 Prior Work

For a robust text-independent speaker identification the use of GMMs, primarily focusing on real time applications is analyzed in [2]. Zissman [3] and Foil [4] proposed Gaussian Mixture Classifier for language identification and extensive studies on the performance of LID systems using Hidden Markov Models (HMM) were made. GMM based text independent Identification and verification systems were evaluated and resulted in a low computational cost [2]. The performance of language identification of telephone speech messages is compared using [3]. Likelihood of recognizing languages with short duration and noisy speech segments was analyzed in [4]. Enhancement techniques for language identification systems were described in [5]. Hegde et al. [6] proposed modified group delay feature and achieved a performance of 3 % greater than the traditional MFCC. Spectral features such as MFCC, LPCC and PLP coefficients [7] were carried out in Language Identification. The mean, variance and weights of a GMM are estimated using ML estimation. Iterative EM algorithm was used for developing the maximum value [8]. Issues related to language and speaker identification of prosodic features extracted from the speech signal were addressed in [9]. Jothilakshmi et al. [10] explored the use of different types of acoustic features for Indian language identification using GMM, HMM and ANN. The performance of LID system using the MFCC featured extracted from block processing, pitch synchronous analysis and Glottal Closure Regions (GCR) was compared using different databases. The spectral and prosodic features from different levels for discriminating the Indian Languages was analyzed by [1].

3 Language Database

In this work, Indian LID system was analyzed using IITKGP-MLILSC which consists of 27 Indian regional languages. Sixteen languages were collected from news channels and TV talk shows, live shows and interviews are recorded. An average of 50 min of data is used for developing the language models. 60 test utterances, each of 5 s duration is used from each language to evaluate the language models.

4 Feature Extraction and Model Building

Language identification (LID) by a system generally involves three stages (i) Feature extraction, (ii) Modeling and (iii) Testing [11]. The feature extraction module is used to represent the precise information from the speech sample. In this work, vocal tract information is represented MFCC and is used to capture the language-specific information. MFCCs are derived from speech using the procedure in [11]. In this work, GMMs were designed for developing the language identification (LID) systems using spectral features. Gaussians mixtures are explored for modeling language specific features. For evaluating the model, two speakers from each language are considered for testing. In testing phase, the likeness scores between the unknown language and a stored model is calculated, and the model with maximum score is identified as the language.

5 LID Performance on Indian Language Database (IITKGP-MLILSC)

Performance of speaker independent LID system for a 32-Gaussian mixture model with different test durations is given in Table 1. Table 2 specifies the performance of speaker independent LID system for different Gaussian models. It is observed that a

Table 1 Performance of speaker independent LID system for a 32-Gaussian mixture model with different test durations

No. of mixture components	Average recognition performance % test duration (s)						
	5	10	15	20	25	30	60
Arunachali	71.6	62.5	60.7	65.0	62.5	58.3	66.7
Assamese	36.6	40.0	35.7	35.0	37.5	33.3	33.3
Bengali	45.0	50.0	50.0	50.0	50.0	50.0	50.0
Bhojpuri	70.0	77.5	82.1	85.0	81.3	83.3	100
Dogri	53.3	50.0	50.0	50.0	50.0	50.0	50.0
Gojri	56.7	50.0	50.0	50.0	50.0	50.0	50.0
Gujarati	76.7	85.0	92.9	100	100	100	100
Hindi	51.7	50.0	50.0	50.0	50.0	50.0	50.0
Kannada	68.3	62.5	67.9	65.0	56.3	50.0	50.0
Kashmiri	66.7	67.5	71.4	75.0	75.0	100	100
Konkani	88.3	90.0	96.4	95.0	100	83.3	100
Malayalam	40.0	62.5	53.6	70.0	68.8	91.7	100
Manipuri	96.7	97.5	96.4	95.0	100	100	100
Marathi	46.7	47.5	50.0	60.0	62.5	58.3	50.0
Mizo	78.3	77.5	75.0	80.0	75.0	66.7	66.7
Nagamese	100	100	100	100	100	100	100
Nepali	100	100	100	100	100	100	100
Oriya	96.7	97.5	100	100	100	100	100
Punjabi	90.0	95.0	100	95.0	100	100	100
Rajasthani	100	100	100	100	100	100	100
Sanskrit	100	100	100	100	100	100	100
Sindhi	45.0	50.0	57.1	55.0	56.3	50.0	50.0
Tamil	91.7	97.5	92.9	95.0	93.8	100	100
Telugu	100	100	100	100	100	100	100
Urdu	50.0	50.0	50.0	50.0	50.0	50.0	50.0
Average	73.2	74.9	75.9	77.2	77.3	78.1	79.0

Table 2 Performance of speaker independent LID system

No. of mixture components	Average recognition performance % test duration (s)						
	5	10	15	20	25	30	60
4	66.2	67.7	68.5	68.8	68.8	69.4	66.0
8	72.0	73.7	73.8	75.3	76.6	75.6	74.6
16	72.6	**75.1**	75.5	77.2	75.6	77.7	78.4
32	**73.2**	74.9	**75.9**	**77.2**	**77.3**	**78.1**	**79.0**
64	72.8	74.8	74.4	75.7	75.6	74.4	75.9
128	68.2	70.9	70.6	70.3	71.9	71.6	72.2

32-GMM provides 1 % higher identification compared to all other models for different test durations. Hence for the rest of the analysis, a 32 GMM is considered for LID system. From Table 1, languages namely, Chhattisgarhi, Konkani, Manipuri, Nagamese, Nepali, Oriya, Punjabi, Rajasthani, Sanskrit, Tamil and Telugu have a recognition performance greater than 80 %. Similarly there are 11 languages with their recognition performances ranging from 50 % to less than 80 %, and 5 languages with recognition performance less than 50 %. It is observed that, Assamese language has a recognition percentage of 35 % even for larger durations of test sample.

Moreover languages like Nagamese, Nepali, Rajasthani, Sanskrit and Telugu have a recognition percentage of 100 % for different test durations. In the case of Malayalam language, the recognition performance is found to be increasing as the duration of the test sample is varied in steps of 5 s. Other languages like Assameese, Bengali, Marathi and Sindhi have similar recognition performances at various test durations. From Table 1, the performance of the speaker independent LID system is compared for different mixture models with various test durations.

A total of 30 samples are considered for every language for a test duration of 5 s, it is found that the recognition performance of 32-Gaussian Mixture Model is 73 % which is higher by 1 % compared to all other models. Hence, the rest of the analysis is performed for test duration of 5 s for every sample.

6 Performance of LID System in the Presence of Different Noises

In most of the practical applications, the input speech is affected by environmental noise, causing a mismatch between the training (clean) and recognition (noisy) conditions. In our study, Buccaneer, Destroyer Engine, Factory, HF Channel and White noises are considered. These samples are collected from NOISEX 92 database. These noise samples are added to the clean data to generate the noisy speech. The performance of the LID system degrades completely with the noisy speech and hence cannot be used in real time applications like Automatic Speech Recognition (ASR) systems.

Table 3 LID performance in various noisy conditions

Noise type	Average recognition performance (%) SNR level				
	0 dB	5 dB	10 dB	15 dB	20 dB
Buccaneer	4.07	5.74	16.11	42.22	57.04
Destroyer engine	5.74	16.30	34.81	54.26	64.44
Factory	4.37	8.52	24.81	53.15	65.00
HF channel	7.07	17.78	38.52	49.44	65.56
White	4.63	9.44	12.78	19.26	35.56

Table 3 provides the LID systems' performance in different noisy environments analyzed at different SNR levels. It is obvious that, the performance of the system degrades when different types of noise at different SNRs are added individually to the clean speech samples during the testing phase. The performance on the language identification system is almost similar at a SNR level of 0 dB for any noise. At 5 dB, Buccaneer, Factory and White noises have similar performance. Column 1 specifies the different types of noise used during the course of study. Columns 2–6 specify different SNR levels considered during the course of present study. From the results of Table 3 it is evident that the LID systems' performance degrades by approximately 67 % at 0 dB SNR level when different types of noise are added to the test sample.

7 Summary and Conclusions

This paper addresses the LID system performance in noisy environments. Buccaneer, Destroyer Engine, Factory, HF Channel and White noises are considered in this study. From the experimental evaluation, an adverse recognition performance is observed when noisy data is added with the clean data. A degradation of 63 % is compared to clean LID performance. For this reason, it is proposed to apply enhancement techniques before testing phase.

Acknowledgments The authors are grateful to Dr K Sreenivasa Rao, Associate Professor and his team at School of Information Technology (SIT), IIT Kharagpur for providing IIT Kharagpur-Multilingual Indian Language Speech Corpus) databases which consists of 27 languages. We would also like to thank their anonymous suggestions and helpful discussions.

References

1. K. Sreenivasa Rao, Sudhamay Maity, and V. Ramu Reddy. "Pitch synchronous and glottal closure based speech analysis for language recognition."*International Journal of Speech Technology* 16.4 (2013): 413–430.
2. Reynolds, Douglas A., and Richard C. Rose. "Robust text-independent speaker identification using Gaussian mixture speaker models." *Speech and Audio Processing, IEEE Transactions on* 3.1 (1995): 72–83.

3. Zissman, Marc A. "Comparison of four approaches to automatic language identification of telephone speech." *IEEE Transactions on Speech and Audio Processing* 4.1 (1996): 31.
4. Foil, Jerry. "Language Identification Using Noisy Speech." *Acoustics, Speech, and Signal Processing, IEEE International Conference on ICASSP'86.*. Vol. 11. IEEE, 1986.
5. Goodman, Fred J., Alvin F. Martin, and R. Wohlford. "Improved automatic language identification in noisy speech." *Acoustics, Speech, and Signal Processing, 1989. ICASSP-89., 1989 International Conference on.* IEEE, 1989.
6. Hegde, Rajesh M., and Hema A. Murthy. "Automatic language identification and discrimination using the modified group delay feature." *Intelligent Sensing and Information Processing, 2005. Proceedings of 2005 International Conference on.* IEEE, 2005.
7. Ambikairajah, E., Li, H., Wang, L., Yin, B., & Sethu, V. (2011). Language identification: a tutorial. *Circuits and Systems Magazine, IEEE, 11(2)*, 82–108.
8. Dempster, Arthur P., Nan M. Laird, and Donald B. Rubin. "Maximum likelihood from incomplete data via the EM algorithm." *Journal of the Royal Statistical Society. Series B (Methodological)* (1977): 1–38.
9. Mary, Leena, and Bayya Yegnanarayana. "Extraction and representation of prosodic features for language and speaker recognition." *Speech communication* 50.10 (2008): 782–796.
10. Jothilakshmi, S., Vennila Ramalingam, and S. Palanivel. "A hierarchical language identification system for Indian languages." *Digital Signal Processing* 22.3 (2012): 544–553.
11. Maity, Sudhamay, Anil Kumar Vuppala, K. Sreenivasa Rao, and Dipanjan Nandi. "IITKGP-MLILSC speech database for language identification." *In Communications (NCC), 2012 National Conference on*, pp. 1–5. IEEE, 2012.

Multi-level Fusion of Palmprint and Dorsal Hand Vein

Gopal Chaudhary, Smriti Srivastava and Saurabh Bhardwaj

Abstract A novel multilevel level fusion of palmprint and dorsal hand vein is developed in this work. First feature level fusion is done on left and right hand palmprint to get feature fused vector (FFV). Next, the scores of FFV and veins are calculated and score level fusion is done in order to identify person. Hence both the feature level as well as score level fusion techniques have been used in a hybrid fashion. In the present work, feature fusion rules have been proposed to control the dimension of FFV. For palmprint, *IIT Delhi Palmprint Image Database version 1.0* is used which has been acquired using completely touchless imaging setup. In this feature level fusion of both left and right hand is used. For dorsal hand veins, *Bosphorus Hand Vein Database* is used because of the stability and uniqueness of hand vein patterns. The improvement of results verify the success of our approach of multilevel level fusion.

Keywords Fusion · Multimodal · Feature extraction · Identification

1 Introduction

Unimodal biometric system is based on a single trait and it suffers from various limitations such as spoof attacks and several other as stated in literature [1–4]. While a multimodal biometric system is created by fusing various unimodal systems to ensure the high performance of such biometric system as the evidences

G. Chaudhary (✉) · S. Srivastava
Netaji Subhas Institute of Technology, Delhi, India
e-mail: gopal.chaudhary88@gmail.com

S. Srivastava
e-mail: smriti.nsit@gmail.com

S. Bhardwaj
Thapar University, Patiala, India
e-mail: bsaurabh2078@gmail.com

© Springer India 2016
S.C. Satapathy et al. (eds.), *Information Systems Design and Intelligent Applications*, Advances in Intelligent Systems and Computing 433,
DOI 10.1007/978-81-322-2755-7_34

321

from different sources are combined together to avoid limitations of unimodal system [5]. These sources can be from different sensors based on a single biometric or different entities based on a single biometric, like palm feature vectors obtained from left and right hands or multiple biometric traits. The information retrieved from individual systems is combined using various schemes such as capturing same information using multiple sensors. For example, the audio samples of an individual are captured using a Hi-tech microphone and a Iphone. Also various traits of an individual are combined to yield a multimodal biometric system for establishing the identity of an individual. The cost of these systems is high since each modality requires separate sensor and data acquisition phase. For example, face and fingerprint of an individual are used for designing a multimodal system. Also, information from similar trait can be combined. For example, palm feature vectors obtained from left and right hands. These systems are cost-effective, because they require neither new sensors nor new algorithms for feature extraction. The single trait can be processed using multiple algorithms. If the biometric sample acquired is not of sufficiently good quality, then the samples from other sources can be banked upon to provide ample discriminatory information to ensure reliable decision-making. Noise in the sensed data from multiple traits has a lesser probability of affecting the performance of a biometric system. Multimodal biometric systems facilitate the choice of the modalities in a given situation.

The main issues and challenges in the design of multimodal biometric are: non availability of multimodal biometric database; the choice of modalities and the choice of fusion technology. Apart from these issues, the performance evaluation of biometric systems is essential in high security applications like defense, government sector, airports, and forensics and also in commercial applications like access control, mobile computing. Biometrics based authentication involves personal data, it may be possible that data collected may used for some unanticipated purpose. Privacy concerns are related to data collection, unauthorized use of recorded information and improper access to biometric records.

2 Palmprint Feature Extraction

Here both left and right palm images of *IIT Delhi Palmprint Image Database version 1.0* [6] are used in this work. For feature extraction of palmprint, all hand postures must be same to capture the same set of information from palms but in database there are variations in sample position. For this, coordinates of five fingertips, finger valleys and the centroid from each image are extracted and used to crop the region of interest (ROI) from an image of size 150 * 150. For enhancement of the palm pattern, an adaptive histogram equalization based on rayleigh distribution is employed on all the ROI. After pre-processing, the enhanced ROI images of size 150 * 150 are partitioned into non overlapping windows of size 15 * 15 each as shown in Fig. 1a. Thus a total of 100 windows are created from each image. Then the gaussian membership function [7, 8] is used for feature extraction.

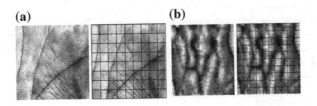

Fig. 1 ROI with their non-overlapping window partitions. **a** Palmprint images. **b** Dorsal hand vein image

The feature vector so obtained has a length of 100. The formula to extract GMF feature is $a_i = \frac{1}{K}\Sigma_{i=0}^{K} x_i u_i$ from ith window. Here, x_k is the pixel value at kth point of the window, $\bar{x}$ is mean pixel value and σ is the standard deviation of the window, $u_i = \frac{\exp - (x_k - \bar{x})^2}{2\sigma^2}$ is the membership function.

2.1 Dorsal Hand Vein Feature Extraction

Here *Bosphorus Hand Vein Database* [9] is used for dorsal hand vein feature extraction [10]. Same procedure of extraction of ROI of size 150 * 150, enhancement and preprocessings is followed here as explained in Sect. 2. Then the gaussian membership function based feature vector of length of 100 is extracted from each window of size 15 * 15 as shown in Fig. 1b.

3 Fusion Levels

Fusion of biometric modalities can be done at various levels. There are mainly four types of fusion levels; score Level; feature Level; sensor Level and decision Level.

4 Feature Level Fusion

At this level of fusion, combination of two feature sets must yield a new feature vector which would capture more individual relevant information and give high representation of a person. In feature level fusion, the evidences from different sources are combined together to avoid limitations of unimodal system. These sources can be from different sensors based on a single biometric or different entities based on a single biometric, like palm feature vectors obtained from left and right hands or multiple biometric traits. The information retrieved from individual systems is combined using various schemes such as capturing same information

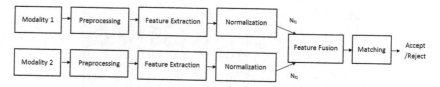

Fig. 2 Feature level fusion

GMF Feature vector of Left hand Palm P_{KL}				
$P_{KL}^{(1)}$	$P_{KL}^{(2)}$		$P_{KL}^{(i)}$	$P_{KL}^{(N)}$

Fused Feature vector				
ff(1)	ff(2)		ff(i)	ff(N)

GMF Feature vector of Right hand Palm P_{KR}				
$P_{KR}^{(1)}$	$P_{KR}^{(2)}$		$P_{KR}^{(i)}$	$P_{KR}^{(N)}$

Fig. 3 Feature fused vector (FFV)

using multiple sensors. For e.g., the audio samples of an individual are captured using a Hi-tech microphone and a Iphone. Feature vectors can be combined to form high-dimensional feature vectors when several feature vectors belong to different types [11] as shown in Fig. 2.

Suppose a given high-dimensional data set (raw ROIs) of individual biometric modality is denoted by $M = \{m_r\}_{r=1}^{q}$, where $m_r \in \Re^D$. After applying different image enhancement processing and feature extraction operations on each original ROI, we obtain N feature sets of each modality of P individuals. The feature vector so obtained here has a length of N = 100. After applying the gaussian membership function on Left and Right palmprint of a single individual, both palmprint are used as two different biometric modalities of $P = 100$ individuals. P_{KL}, P_{KR} are denoted as GMF feature vector of left hand and right hand of Kth individual respectively where $K = 1, \ldots, P$. Both P_{KL}, P_{KR} are fused together with the given feature level fusion rule in Eq. 1 denoted as Feature fused Vector (FFV) where $ff(i) \in \Re^D$ shown in Fig. 3.

4.1 Feature Level Fusion Rules

FFV is expressed as $FFV \leftrightarrow f(P_{KL}, P_{KR})$ where $f()$ is feature fusion rule which gives $ff(i)$ fused feature set where $ff(i) \in \Re^D$. Feature fusion rule must be correctly chosen that would better represent the person and dimensions of the fused vector must not increase much. So to control the dimensions of FFV, four feature fusion rule are proposed here which are analogous to score level fusion rules [12, 13].

Sum Rule

$$ff(i) = \left[[P_{KL}(i) + P_{RL}(i)]_{i=1}^{N}\right]_{K=1}^{P} \tag{1}$$

Product Rule

$$ff(i) = \left[[P_{KL}(i) * P_{RL}(i)]_{i=1}^{N}\right]_{K=1}^{P} \tag{2}$$

Modulus Rule

$$ff(i) = \left[[|P_{KL}(i), P_{RL}(i)|]_{i=1}^{N}\right]_{K=1}^{P} \tag{3}$$

Frank T-norm Rule

$$ff(i) = \left[\left[\log_b\left(\frac{1 + (b^{P_{KL}(i)} - 1)(b^{P_{RL}(i)} - 1)}{b - 1}\right)\right]_{i=1}^{N}\right]_{K=1}^{P} \tag{4}$$

5   Score Level Fusion of Two Biometrics

The block diagram depicting score level fusion is shown in Fig. 4. Previous studies shows that fusion at score level is most appropriate approach to multimodal biometrics and is most popular [14]. The matching scores (genuine and imposter) from the existing and proprietary unimodal systems can be easily utilized in a multimodal biometric system. The information (i.e. the match score) from prior unimodal evaluations of a biometric system can be used and this avoids live testing. The matching scores contain next level of rich information after the features of the input pattern. The scores generated by different matchers are easy to access and combine. This motivates combining information from individual biometric modalities using score level fusion.

Min-Max score normalization is done for making combination meaningful [15]. Let r_k denotes a set of matching scores where, $k = 1, 2, \ldots, n$ and $r_{k'} = \frac{r_k - \min}{\max - \min}$ which denotes normalized score.

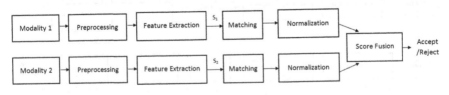

Fig. 4 Score level fusion

5.1 Score Fusion Rules

Various conventional and t-norm based fusion rules are given below. Let R_i be the matching score obtained from ith modality and R denotes the fused score or the combined score and N be the number of modalities.

1. Sum Rule: $R = R_1 + R_2 + \cdots + R_N = \sum_{i=1}^{N} R_i$
2. Product Rule: $R = R_1 * R_2 * \cdots * R_N = \prod_{i=1}^{N} R_i$
3. Hamacher t-norm: $R = \frac{R_1 R_2}{R_1 + R_2 - R_1 R_2}$
4. Frank t-norm: $R = \log_p \left(\frac{1 + (p^{R_1} - 1)(p^{R_2} - 1)}{p - 1} \right)$

6 Proposed Multi-level Fusion

The topology of proposed multi-level fusion approach is shown in Fig. 5. It combines the scores of feature fused vector (FFV) obtained by nearest neighbor algorithm and scores of dorsal hand vein. This includes advantages of both the feature level and score level fusions schemes. In this scheme, scores of feature fused vector (FFV) appeared from the feature fusion of left and right hand palmprints are combined with dorsal hand vein scores through the conventional score fusion rules as shown above.

7 Experiments and Results

To calculate the scores between the training and test sample, the K-nearest neighbor (KNN) classifier with Euclidean distance is trained with features obtained from each biometric modality with k-fold cross-validation. The score obtained by kNN classifier are used to verify the performance of the recognition system using the

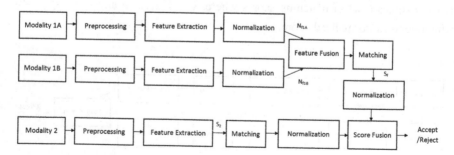

Fig. 5 Topology of proposed multi-level fusion

Table 1 False acceptance rate (FAR %) of individual modalities

Modality	False acceptance rate (FAR %)		Identification results
	0.1	1	
Right palm	88	92.5	93
Left palm	87	91	91.1
Dorsal hand vein	79	87	89

Receiver Operating Characteristic (ROC) curve between the Genuine acceptance rate (GAR) and false acceptance rate (FAR). Identification results of GMF based features of left and right hand palmprint for *IIT Delhi Palmprint Image Database version 1.0* database and GMF based features of Dorsal Hand Vein for *Bosphorus Hand Vein Database* is tabulated in Table 1.

ROC curves for GMF based features of left and right hand palmprint and their fused features for all mentioned rules are shown in Fig. 6. As it is seen in the plots the ROC curve of feature level fusion converges more rapidly as compared to the individual left hand and right hand palmprint showing the improvement in the performance of multimodal feature level fusion based biometric system. It is clear that with sum rule, at 0.1 FAR, GAR is 86.27 %, while for product rule, it is 90.82 %, for modulus rule, it is 89.83 % and using frank T-norm, the value comes out to be 91.76 %. Similarly for FAR = 1, GAR is 92.8 % for sum rule, while for product rule, it is 97.01 %, for modulus rule, it is 94.68 % and using frank T-norm, the value comes out to be 98.15 %. This shows that frank T-norm feature fusion rule outperform the other rules and converges to 100 % more rapidly. Identification results of feature level fusion of left hand and right hand palmprint for *IIT Delhi Palmprint Image Database version 1.0* database is tabulated in Table 2.

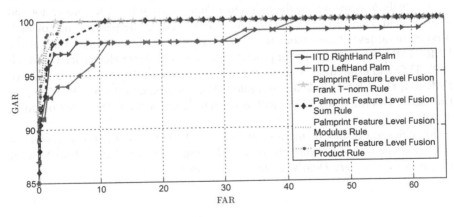

Fig. 6 ROC curves of feature level fusion of left hand and right hand palmprint

Table 2 Feature level fusion of left hand and right hand of IIT Delhi palmprint image database version 1.0

Rules	Sum rule		Product rule		Modulus rule		Frank T-norm	
False acceptance rate (FAR %)	0.1	1	0.1	1	0.1	1	0.1	1
Fused vector	86.27	92.8	90.82	97.01	89.83	94.68	91.76	98.15

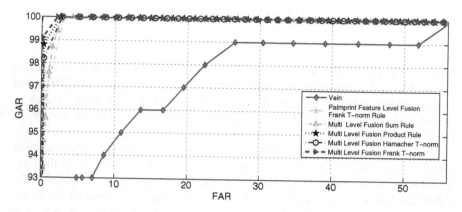

Fig. 7 ROC of multi-level fusion of Feature fused palm scores of *IIT Delhi palmprint image database version 1.0* and scores of dorsal hand vein of *Bosphorus hand vein database*

Receiver operating characteristic (ROC) curves for GMF based features of dorsal hand vein for *Bosphorus Hand Vein Database* is shown in Fig. 7. The ROC curves of proposed multi-level fusion are shown in Fig. 7. Results in Fig. 6 shows that Frank T-norm feature fusion rule based fused feature vector gives better performance to other feature fusion rules, that is why, Frank T-norm fused feature vector is normalized and then fused with GMF based features of dorsal hand vein by score level fusion. As it is seen in the ROC curve of multi-level fusion converges even more rapidly as compared to the individual Frank T-norm fused feature vector and Dorsal Hand Vein features showing the improvement in the performance of multimodal multi-level based biometric system. It is clear that with sum rule, at 0.1 FAR, GAR is 91.68 %, while for product rule, it is 98.2 %, for Hamacher T-norm, it is 97.33 % and using frank T-norm, the value comes out to be 99.3 %. Similarly for FAR = 1, GAR is 97.68 % for sum rule, while for product rule, it is 99.56 %, for Hamacher T-norm, it is 99.27 % and using frank T-norm, the value comes out to be 99.83.

Identification results of multi-level fusion of Feature fused palm scores of *IIT Delhi Palmprint Image Database version 1.0* and scores of Dorsal Hand Vein of *Bosphorus Hand Vein Database* are tabulated in Table 3.

Table 3 Multi-level fusion of feature fused palm scores of *IIT Delhi palmprint image database version* 1.0 and scores of dorsal hand vein of *Bosphorus hand vein database*

Rules	Sum rule		Product rule		Hamacher T-norm		Frank T-norm	
False acceptance rate (FAR %)	0.1	1	0.1	1	0.1	1	0.1	1
Fused vector	91.68	97.86	98.2	99.56	97.33	99.27	99.3	99.83

8 Conclusion

In multimodal biometric system, complementary information is fused to overcome the drawbacks of the unimodal biometric systems. The plots shown in the result section proves that the performance of multimodal biometric system is significantly improved as compared to the unimodal biometric systems. The performance of the biometric system gets further improved when feature level fusion of two biometric modalities i.e. left hand and right hand palmprint is performed. The performance of the biometric system gets enhanced when multi-level fusion of feature fused palm scores of *IIT Delhi Palmprint Image Database version 1.0* and scores of Dorsal Hand Vein of *Bosphorus Hand Vein Database* is performed. The fusion based on frank t-norm gives better results as compared to each unimodal biometric systems in both feature level fusion of left hand and right hand palmprint as well as multi-level fusion of feature fused palm scores and scores of Dorsal Hand Vein.

References

1. Finan, R., Sapeluk, A., Damper, R.I.: Impostor cohort selection for score normalisation in speaker verification. Pattern Recognition Letters 18(9), 881–888 (1997).
2. Wang, K.Q., Khisa, A.S., Wu, X.Q., Zhao, Q.s.: Finger vein recognition using lbp variance with global matching. In: Wavelet Analysis and Pattern Recognition (ICWAPR), 2012 International Conference on. pp. 196–201. IEEE (2012).
3. Rodrigues, R.N., Ling, L.L., Govindaraju, V.: Robustness of multimodal biometric fusion methods against spoof attacks. Journal of Visual Languages & Computing 20(3), 169–179 (2009).
4. Jain, A.K., Dass, S.C., Nandakumar, K.: Soft biometric traits for personal recognition systems. In: Biometric Authentication, pp. 731–738. Springer (2004).
5. Jain, A.K., Ross, A.: Multibiometric systems. Communications of the ACM 47(1), 34–40 (2004).
6. Kumar, A.: Incorporating cohort information for reliable palmprint authentication. In: Computer Vision, Graphics & Image Processing, 2008. ICVGIP'08. Sixth Indian Conference on. pp. 583–590. IEEE (2008).
7. Arora, P., Hanmandlu, M., Srivastava, S.: Gait based authentication using gait information image features. Pattern Recognition Letters (2015).
8. Arora, P., Srivastava, S.: Gait recognition using gait gaussian image. In: Signal Processing and Integrated Networks 2015 (SPIN 21015), 2nd International Conference on. pp. 915–918. IEEE (2015).

9. Yuksel, A., Akarun, L., Sankur, B.: Hand vein biometry based on geometry and appearance methods. IET computer vision 5(6), 398–406 (2011).
10. Lajevardi, S.M., Arakala, A., Davis, S., Horadam, K.J.: Hand vein authentication using biometric graph matching. IET Biometrics 3(4), 302–313 (2014).
11. Chin, Y., Ong, T., Teoh, A., Goh, K.: Integrated biometrics template protection technique based on fingerprint and palmprint feature-level fusion. Information Fusion 18, 161–174 (2014).
12. Hanmandlu, M., Grover, J., Gureja, A., Gupta, H.M.: Score level fusion of multimodal biometrics using triangular norms. Pattern Recognition Letters 32(14), 1843–1850 (2011).
13. Ross, A., Jain, A.: Information fusion in biometrics. Pattern recognition letters 24(13), 2115–2125 (2003).
14. Lip, C.C., Ramli, D.A.: Comparative study on feature, score and decision level fusion schemes for robust multibiometric systems. In: Frontiers in Computer Education, pp. 941–948. Springer (2012).
15. Jain, A., Nandakumar, K., Ross, A.: Score normalization in multimodal biometric systems. Pattern recognition 38(12), 2270–2285 (2005).

A Comparative Study on Multi-view Discriminant Analysis and Source Domain Dictionary Based Face Recognition

Steven Lawrence Fernandes and G. Josemin Bala

Abstract Human face images captured in real world scenarios using surveillance cameras won't always contain single view, instead they usually contain multi-view. Recognizing multi-view faces is still a challenging task. Multi-view Discriminant Analysis (MDA) and Source Domain Dictionary (SSD) are two techniques which we have developed and analyzed in this paper to recognize faces across multi-view. In MDA the faces collected from various views are reflected to a discriminant general space by making use of transforms of those views. SSD on the other hand is based on sparse representation, which efficiently makes the dictionary model of source data. It also signifies each class of data discriminatively. Both the developed techniques are validated on CMU-Multi PIE face database which contains 337 people recorded under 15 different view positions and 19 different conditions.

Keywords Face recognition · Multi-view discriminant analysis · Source domain dictionary

1 Introduction

Recognizing multi-view faces is widely studied by researchers over a decade [1–12] and it is still a challenging task. Researchers have used Viola-Jones method to detect multi-view of face but it cannot discriminate faces from non-faces and thus require many AdaBoost iterations to decrease the false positives [13]. To overcome this problem non-Haar-like features are added to Viola-Jones method. A method for automatically detecting multi-view face is presented in [14]. In this method for

S.L. Fernandes (✉) · G. Josemin Bala
Department of Electronics & Communication Engineering, Karunya University,
Coimbatore, India
e-mail: steva_fernandes@yahoo.com

G. Josemin Bala
e-mail: josemin@karunya.edu

© Springer India 2016
S.C. Satapathy et al. (eds.), *Information Systems Design and Intelligent Applications*, Advances in Intelligent Systems and Computing 433,
DOI 10.1007/978-81-322-2755-7_35

building the face detector and pose estimator modified learning methods are used. After the pre-processing stage coarse face detector is used which helps in cancelling the non-face regions. The obtained facial patches are classified into clusters. Another method for capturing side view of a person is presented in [15]. This method tries to recognize the face using the geometric features of the image. Curve Edge Maps (CEMs) are used in representing the face. CEMs are the set of poly-nomial curves containing curved region. Face CEMs determined by histograms of intensities is matched with face CEMs determined by histograms of relative posi-tions to perform the face recognition function. Both these methods [14, 15] are not found suitable to recognize faces under multi-view.

Among various proposed methods, the two most prominent methods to recog-nize faces in multi-view are Multi-view Discriminant Analysis (MDA) and Source Domain Dictionary (SSD). Section 2 presents Multi-view Discriminant Analysis. Section 3 presents Source Domain Dictionary. Section 4 presents Results and Discussions and Sect. 5 draws the Conclusion.

2 Multi-view Discriminant Analysis (MDA)

In Multi-view Discriminant Analysis (MDA) the samples collected from various views are reflected to a Discriminant general space by making use of the 'v' transforms of those views. One of the important task of MDA is to find out v linear transforms, say w_1, w_2, w_3, ..., w_v. Using these transforms, a sample part of 'v' views can be reflected into the discriminant general space.

Let's denote $X^{(j)} = \{X_{ijk} | i = 1, \ldots, c, k = 1, \ldots, n_{ij}\}$ as a small part from jth view (i.e. j = 1, 2, 3, ..., v), X_{ijk} = kth sample from jth view of ith class of dimension d_j. The part of samples obtained from v views are then reflected to the same discriminant general space using the view-specific linear transforms of each view. (i.e. using w_1, w_2, w_3, ..., w_v). The projection results are denoted as $y = \{Y_{ijk} = w_j^T | i = 1, \ldots, c; j = 1, \ldots, v; k = 1, \ldots, n_{ij}\}$. In the discriminant com-mon space, the between-class deviation should be kept high and within-class deviation should be kept low. Hence, the between-class deviation S_B^y from every view should be set to high and the within-class deviation S_W^y from every view should be set to low.

The above objective can be devised as a generalized Rayleigh's Quotient (w_1^*, w_2^*, ..., w_v^*) = arg $\max\limits_{w1,w2,\ldots,wv} \frac{Tr(S_B^y)}{Tr(S_W^y)}$. S_B^y represents the between-class scatter matrix whereas S_W^y represents the within class of scatter matrix. Both the scatter matrices of the samples in the general discriminant space are calculated as

$$S_W^y = \sum_{i=1}^{c} \sum_{j=1}^{v} \sum_{k=0}^{n_{ij}} (y_{ijk} - \mu_i)(y_{ijk} - \mu_i)^T \tag{1}$$

$$S_B^y = \sum_{i=1}^{c} n_i(\mu_i - \mu)(\mu_i - \mu)^T, \tag{2}$$

where $\mu_i = \frac{1}{n_i}\sum_{j=1}^{v}\sum_{k=1}^{n_{ij}} y_{ijk}$ gives the average value of all the sample of the class 'i' over all views in the common space, $\mu = \frac{1}{n}\sum_{i=1}^{c}\sum_{j=1}^{v}\sum_{k=1}^{n_{ij}} y_{ijk}$ is the mean of all samples over all views in the common space $n = \sum_{i=1}^{c} n_i$ symbolize the number of samples from all views and all classes. From (1) and (2), it is clearly found between-class deviations and within-class deviations are computed using the sample obtained from all the views. Along with the intra-view samples, inter-view samples are also collected. After obtaining the view-specific linear transforms, the sample obtained from different views can be compared after projecting them to discriminant common space respectively.

3 Source Domain Dictionary (SSD)

In Source Domain Dictionary (SSD) a set of source data $Y_S = [Y_1^s, Y_2^s, \ldots, Y_C^s] \in R^{d \times N_s}$, where Y_i^s = class 'i' samples, d is the dimension of the feature, N_s = total number of training samples is considered. Suppose coding coefficient matrix of Y_s over the dictionary $D \in R^{d \times m}$ is given as $X_s \in R^{m \times N_s}$, then it can be written as $Y_s = DX_s + E_s$, where E_s = reconstruction error, m = dictionary dimensions. In principle, it is possible to decompose $X_S = [X_1^s, X_2^s, \ldots, X_C^s]$, where $X_i^s \in R^{m \times N_i^s}$ is the sub-matrix that contain the coding coefficients related with samples $Y_i^s \in R^{d \times N_i^s}$ over the dictionary $D.N_i^s$ correspond to the sample number in Y_i^s.

In order to model the characteristics of the source data effectively, it is essential that the structured dictionary 'D' should exhibit a dominant reconstructive ability of the samples Y_s. In the Sparse Representation (SR) Framework, a sparse linear combination of the atoms in D is anticipated to signify any samples of Y_s, provided an over-complete dictionary D is offered and based on this, the penalty term for dictionary learning is produced by minimizing the reconstruction error E_s

$$\min_{D, X_s} E_s = \|Y_s - DX_s\|_F^2, \text{s.t.} \forall l, \|x_l^s\|_0 \leq T_0 \tag{3}$$

where X_l^s = column of X_s, T_0 = Sparsity Level.

From (3), it is noted that the maximum number of atoms in the dictionary D used for reconstructing the samples is limited by T_0. Upper limit of T_0 is the dictionary dimension, which is represented as m. Since the SR algorithm searches few of the most significant atoms in the dictionary D for sample reconstruction, the Sparsity level i.e. T_0 should be comparatively less than the value of m. The proper range of T_0 will be around half of the dictionary dimension. In this model, it is further necessary that each sub-dictionary D_i focus on modeling of the corresponding statistical property of subclass data, so that each class of data in the source view can be discriminately characterized by the structured dictionary D. First, each sub-dictionary D_i should be capable of representing the corresponding subclass samples Y_i^s. Decomposing the coding coefficients of Y_i^s over $D = [D_1, D_2, \ldots, D_C]$, as $X_i^s = \left[X_{i,1}^s; X_{i,2}^s, \ldots, ; X_{i,C}^s \right]$ where $X_{i,j}^s$ stand for the coding coefficient of Y_i^s equivalent to sub-dictionary D_j. The reconstruction item of the samples Y_i^s is computed as $\sum_{j=1}^C D_i X_{i,j}^s + E_i^s$. The first restriction is formed as $Y_i^s = D_i X_{i,j}^s + E_i^s$, which points out that the reconstruction items from other classes sub-dictionary D_j $(j \neq i)$ are avoided. Second, the reconstruction coefficients on D_i of samples from the class $j(j \neq i)$ should be nearly zero. The second constraint $r(D_i) = \sum_{j=1, j \neq i}^C \left\| D_i X_{j,i}^s \right\|_F^2$ is formed by presuming that $X_{i,j}^s$ as the coding co-efficient of Y_j^s. $r(D_i)$ should be less, which means that the association between D_i and $Y_j^s(j \neq i)$ is modified to be small. In this way, each Sub-dictionary D_i characterizes the corresponding subclass data Y_i^s discriminatively. Finally, the objective of the discriminatively training the structured dictionary is developed as

$$\min_{D, X_s} \|Y_s - DX_s\|_F^2 + \sum_{i=1}^C \left\{ \left\| Y_i^s - D_i X_{i,i}^s \right\|_F^2 + \alpha r(D_i) \right\} \quad \text{s.t.} \quad \forall l, \|x_l^s\|_0 \leq T_0, \quad \text{where}$$

α = positive scale parameter, α controls the dictionary D's discriminative power.

4 Results and Discussions

Face Recognition across multi-view is has various approaches [16–22]. Multi-view Discriminant Analysis and Source Domain Dictionary are two techniques which we have developed and analyzed in this paper to faces across varying poses. Both these techniques are validated using 337 subjects present under CMU-Multi PIE face database. These images are recorded on 15 different points of view and 19 different illumination conditions. Table 1 describes train and test face image used from CMU

Table 1 Train and test face image used from CMU multi-PIE face database

Total no. of train classes	337
No. of poses variations per train class	15
No. of illumination variations per train class	19
No. of images trained images per class	10
Total no. of trained images	$(337 \times 10) = 3370$
Test images per class	$(15 + 19) = 34$
Total no. of test images	$(337 \times 34) = 11{,}458$
Dimension	3072×2048

Table 2 Face recognition rate using MDA and SSD on CMU multi-PIE face database—first implementation using Intel 5th generation core i7-5557U processor NUC

Technique	Face recognition rate	Execution time
Multi-view discriminant analysis	10,770/11,458 = 93.99 %	06 h 37 min
Source domain dictionary	11,200/11,458 = 97.74 %	04 h 27 min

Table 3 Face recognition rate using MDA and SSD on CMU multi-PIE face database–second implementation using Intel 3rd generation core i5-4258U processor

Technique	Face recognition rate	Execution time
Multi-view discriminant analysis	10,774/11,458 = 94.03 %	07 h 13 min
Source domain dictionary	11,201/11,458 = 97.75 %	05 h 12 min

Table 4 Face recognition rate using MDA and SSD on CMU multi-PIE face database—third implementation using ODROID-XU4 powered by ARM technology

Technique	Face recognition rate	Execution time
Multi-view discriminant analysis	10,769/11,458 = 93.98 %	14 h 19 min
Source domain dictionary	11,200/11,458 = 97.74 %	11 h 52 min

Multi-PIE Face Database. Tables 2, 3 and 4 depicts the face recognition rate acquired on MDA and SSD during first, second and third implementation using MATLAB R2015b on Intel 5th generation Core i7-5557U processor NUC (8GB DDR3 RAM, 2TB 7200Rpm HDD), Intel 3rd generation Core i5-4258U processor (4GB DDR3 RAM, 1TB 7200Rpm HDD) and ODROID-XU4 powered by ARM big.LITTLE technology (2GB DDR3 RAM, Samsung Exynos 5422 Cortex—A15 2Ghz and Cortex—A7 Octal core CPUs).

Tables 2, 3 and 4 clearly indicates that the face recognition rate obtained on CMU-Multi PIE face database using SSD is 97.74 % (average of three implementations) is better than MDA on CMU-Multi PIE face database. Also the execution time taken by SSD is less when compared to MDA on CMU-Multi PIE face database.

5 Conclusion

Recognizing multi-view faces is a challenging task [16–22]. Multi-View Discriminant Analysis and Source Domain Dictionary are two techniques which we have developed and analyzed in this paper to faces across varying poses. In MDA the faces collected from various views are reflected to a discriminant general space by making use of transforms of those views. SSD on the other hand is based on sparse representation, which efficiently makes the dictionary model of source data. It also signifies each class of data discriminatively. Both the developed techniques are validated on CMU-Multi PIE face database which contains 337 people recorded under 15 different view positions and 19 different illumination conditions. From our analysis we have found that the face recognition rate obtained on CMU-Multi PIE face database using SSD is 97.74 % (average of three implementations) is better than MDA on CMU-Multi PIE face database. Also the execution time taken by SSD is less when compared to MDA on CMU-Multi PIE face database.

Acknowledgments The proposed work was made possible because of the grant provided by Vision Group Science and Technology (VGST), Department of Information Technology, Biotechnology and Science and Technology, Government of Karnataka, Grant No. VGST/SMYSR/GRD-402/2014-15 and the support provided by Department of Electronics & Communication Engineering, Karunya University, Coimbatore, Tamil Nadu, India and Sahyadri College of Engineering and Management, Mangalore, Karnataka, India.

References

1. Lee, H. S, Kim, D.: Generating frontal view face image for pose invariant face recognition. Pattern Recognition Letters, vol.27, no. 7, pp 747–754 (2006).
2. Steven L. Fernandes, G. Josemin Bala.: 3D and 4D Face Recognition: A Comprehensive Review. Recent Patents on Engineering. vol. 8, no. 2, pp. 112–119 (2014).
3. Gründig M. and Hellwich,O.: 3D Head Pose Estimation with Symmetry Based Illumination Model in Low Resolution Video. LNCS, pp. 45–53 (2004).
4. Steven L. Fernandes, G. Josemin Bala.: Development and Analysis of Various State of the Art Techniques for Face Recognition under varying Poses. Recent Patents on Engineering, vol. 8, no. 2, pp 143–146 (2014).
5. Balasubramanian, V.N., Krishna, S., Panchanathan, S.: Person-independent head pose estimation using biased manifold embedding. EURASIP J. Adv. Signal Process., pp. 1–15 (2008).
6. Steven L. Fernandes, G. Josemin Bala.: Recognizing Faces When Images Are Corrupted By Varying Degree of Noises and Blurring Effects. Advances in Intelligent Systems and Computing, vol. 337, no. 1, pp 101–108 (2015).
7. Steven L. Fernandes, G. Josemin Bala.: Low Power Affordable, Efficient Face Detection In The Presence Of Various Noises and Blurring Effects on A Single-Board Computer. Advances in Intelligent Systems and Computing, vol. 337, no. 1, pp. 119–127 (2015).
8. Steven L. Fernandes, G. Josemin Bala.: Recognizing facial images in the presence of various Noises and Blurring effects using Gabor Wavelets, DCT Neural Network, Hybrid Spatial Feature Interdependence Matrix. In: 2nd IEEE Int. Conf. on Devices, Circuits and Systems, Coimbatore (2014).
9. Steven L. Fernandes, G. Josemin Bala.: Recognizing Facial Images Using ICA, LPP, MACE Gabor Filters, Score Level Fusion Techniques. In: IEEE Int. Conf. Electronics and Communication Systems, Coimbatore (2014).
10. Steven L. Fernandes, G. Josemin Bala.: Robust Face Recognition in the presence of Noises and Blurring effects by fusing Appearance based techniques and Sparse Representation. In: IEEE Int. Conf. on Advanced Computing, Networking and Security, Surathkal (2013).
11. Steven L. Fernandes, G. Josemin Bala, et al.: A Comparative Study on Score Level Fusion Techniques and MACE Gabor Filters for Face Recognition in the presence of Noises and Blurring effects. In: IEEE Int. Conf. on Cloud & Ubiquitous Computing and Emerging Technologies, Pune (2013).

12. Ghiass, R.S., Sadati, N.: Multi-view face detection and recognition under variable lighting using fuzzy logic. In: Int. Conf. on Wavelet Analysis and Pattern Recognition, Hong Kong (2008).
13. Abiantun, R., Savvides, M.: Boosted multi image features for improved face detection. In: 37th IEEE on Applied Imagery Pattern Recognition Workshop, Washington DC (2008).
14. Ghiass, R.S., Fatemizadeh, E., Marvasti.: Designing an illumination effect cancelling filter in facial images for multi-view face detection and recognition in images with complex background. In: International Symposium on Telecommunications, Tehran (2008).
15. Quanbin LI., Xiaoming Wang., Jingao Liu.: A novel approach for multi-pose face detection by using skin color and FloatBoost. In: 8th World Congress on Intelligent Control and Automation, Jinan (2010).
16. Steven L. Fernandes, G. Josemin Bala.: Recognizing Faces Across Age Progressions and Under Occlusion. Recent Patents on Computer Science, UAE (2015).
17. Steven L. Fernandes, G. Josemin Bala.: A Novel Technique to Detect and Recognize Faces in Multi-view Videos. In: 19th International Conference on Circuits, Systems, Communications and Computer, pp 427–433, Greece, Europe (2015).
18. Steven L. Fernandes, G. Josemin Bala.: A Comparative Study to Recognize Surgically Altered Images. In: 19th International Conference on Circuits, Systems, Communications and Computer, pp 434–439, Greece, Europe (2015).
19. Deboeverie, F., Veelaert, P., Philips, W.: Best view selection with geometric feature based face recognition. In: 19th IEEE International Conference on Image Processing, Orlando, FL (2012).
20. Steven L. Fernandes, G. Josemin Bala.: Analysing State of the Art Techniques to Recognize Faces under Multimodal Biometrics. Advances in Intelligent Systems and Computing, vol. 381, no.3, pp 473–478, (2015).
21. Steven L. Fernandes, G. Josemin Bala.: Simulation Level Implementation of Face Recognition under Un-controlled Environment. Advances in Intelligent Systems and Computing, vol. 381, no. 3, pp 467–472, (2015).
22. Ying Ying., Han Wang., Jian Xu.:An automatic system for multi-view face detection and pose estimation. In: 11th International Conference on Control Automation Robotics and Vision, Singapore (2010).

Mining Maximal Efficient Closed Itemsets Without Any Redundancy

L. Greeshma and G. Pradeepini

Abstract Mining more relevant itemsets from various information repositories, which is an essential task in knowledge discovery from data that identifies itemsets with more interestingness measures (support and confidence). Due to the availability of data over Internet, it may retrieve huge number of itemsets to user, which may degrades the performance and increases time complexity. This paper proposed a framework called Analyzing All Maximal Efficient Itemsets to provide a condensed and lossless representation of data in form of rule association rules. We proposed two algorithms Apriori-MC (Apriori-Maximal Closed itemsets) and AAEMIs (Analyzing All Efficient Maximal Itemsets) by deleting non-closed itemsets. The proposed method AAEMIs regains complete relevant itemsets from a group of efficient Maximal Closed Itemsets (MCIs) without specifying user specified constraint and overcoming redundancy.

Keywords Frequent itemsets · Maximal closed efficient itemsets · Efficient mining · Association rule mining

1 Introduction

Due to wide availability of data over Internet is rapidly growing every year. Internet users are able to retrieve required information based on their specified intension. In general Information Retrieval System retrieves large of unlabeled data to the end users, which need to be preprocessed to a labeled data, which increases the time complexity. Knowledge discovery from data and data mining are multidisciplinary domain that mainly concentrates of achieving interesting rules from labeled data [1]. Through, the basic research oriented topic is miming frequent itemsets from various databases like datawarehouse, time series, sequential, relational, multimedia, object

L. Greeshma (✉) · G. Pradeepini
Department of CSE, K L University, Vaddeswaram, Andhra Pradesh, India
e-mail: greeshma243@gmail.com

G. Pradeepini
e-mail: pradeepini.gera@gmail.com

© Springer India 2016
S.C. Satapathy et al. (eds.), *Information Systems Design and Intelligent Applications*, Advances in Intelligent Systems and Computing 433,
DOI 10.1007/978-81-322-2755-7_36

339

relational, temporal, text databases and application domains like mobile environments, bioinformatics and web mining analysis.

Association rule is defined as $X \rightarrow Y$, which implies that if a particular transaction in a database contains itemset X then it will probably contain itemset Y [2]. For instance, if X purchases a computer then there is a chance that Y purchase software. The most significant association rules are evaluated by using interestingness measures like support, confidence, lift, and cosine. The association rules mining difficulty has become an important research area by concept of the Apriori algorithm i.e., mining frequent itemset with help of a candidate key generation. The mining relevant itemsets consist of two phases. In First phase, identifying frequent itemsets is essential for generating association rules [3]. In Second phase, validating these rules are important because most of obtained rules may have same antecedent part, which causes a redundancy. It also misleads the end user for taking right decision. One of problem to be overcome is to reduce complexity of resulted rule set. Normally, the number of rules obtained are high because of it many of data mining algorithm focuses on mining quantitative association rules on positive dependencies rather than negative dependencies but in certain scenarios we can correlate positive and negative dependencies [4]. Thus, time complexity for the algorithms increases which degrades the performance. Later on to represent frequent itemsets the concept of mining closed high utility itemset, compressed frequent pattern, crucial itemsets and frequent itemsets based on positive dependencies have been proposed in the literature [5, 6–9]. The problems associated with these association rules are those which are constructed from set of items by scanning database, but these rules still contain redundancy and also size of itemsets are shortened.

We aim to condense the size of an association rules without specifying user specified constraint and set of highly relevant rules are to be obtained without a lossless information which helps the analyst to take decision quickly. An interesting research query arises "As it is not conceivable to consider a closed and lossless information of an efficient closed itemsets motivated by the above-mentioned problems in Maximal efficient Closed Itemsets (MCIs) mining". Answering to this question is not an easy task. There are certain problems like previous algorithms may not be more efficient for extracting relevant data, merging frequent itemsets to form a MCIs may produce lossy representation of data which may not be useful for end users. We showed that proposed two algorithms named Apriori-MC (Apriori-based algorithm for mining Maximal efficient Closed itemset) and AAEMIs (Analyzing All Efficient Maximal Itemsets). Related work is discussed in Sect. 2. Section 3 outlines necessary prerequisites for the representation of MCIs and proposed methods. Section 4 describes about experimental results. Conclusion is given in Sect. 5.

2 Related Work

There are several existing algorithms to identify frequent itemsets by scanning database for each and every transaction and most prominent among them is Apriori Algorithm. In order to mine MCIs, Pasquier et al. [10] first proposed a hierarchical searching methodology with the help of Apriori properties over the subset pattern known as CLOSE. Pei et al. [3] proposed CLOSET using the dense data structure known as Frequent Pattern-Tree that is done by mining maximal itemsets without generating candidate keys and inheriting the divide and conquer approach from Frequent Pattern-Growth algorithm [11]. Grahne and Zhu [12] proposed next version of CLOSET+ termed as FP-CLOSE reduces iterative traversal over FP-Tree. By mining frequent itemset using vertical data format has been proposed by Zaki and Hsiao [5]. These rules can be evaluated by its coverage and accuracy. Moreover, the rule set obtained may not preserve the same conclusion ability as the entire rule set, and as a result the conclusion is not derivable. Pasquier et al. [10] produced rules having marginal antecedent and highest consequent part considered as a lossless non-redundant data. Rules obtained may not retain the same conclusion, as they do not sustain any added information to the user. Cheng et al. [13] provides the non-redundant rules by pruning them using the concept of user specified threshold value. However, their method does not moderate the number of association rules for having minimal set of items. Among all the origins proposed in the literature, produces association rules with minimum confidence. Major drawback is, it is not necessary to produce rule sets, which must be less than user specified threshold value. All these related works discussed here are meant for fetching highly relevant rules without any redundancy and lossless information to the end users. However, this paper focuses on retrieving Maximal efficient Closed Itemsets that can be adapted for multiple domains.

3 Prerequisites and Proposed Algorithms

In this section we provide fundamental concepts and definitions, which are required for describing proposed algorithms.

3.1 Maximal Efficient Itemsets Mining

Definition 1 (*Total efficiency of an itemsets in a database D*) The total efficiency of an itemset with transaction T_{ID} is represented as te (I_i, T_{ID}) and formulated as

$$te(I_i, T_{ID}) = wt(\{I_i\}, D) \times q(\{I_i\}, T_{ID}). \tag{1}$$

Let A be the total efficiency in transaction T_{ID} is defined as te $(\{A\}, T_{ID})$ and formulated as summation of total efficiency of each item in set and calculated as

$$\Sigma_{I_i \epsilon A} \, te(\{I_i\}, T_{ID}). \tag{2}$$

Definition 2 (*Maximal Efficient Itemsets*) Let A be Maximal efficient itemset is described as iff e(A) $\geq$ min_efficiency.

Table 1 Transactional
database

TID	Transaction	Support value
T1	$I_6 I_1 I_2 I_3$	5
T2	$I_3 I_1 I_2 I_3 I_3$	8
T3	$I_1 I_4 I_2 I_4$	8
T4	$I_3 I_3 I_5$	5
T5	$I_1 I_4 I_2 I_4 I_4$	11

Table 2 Profit associated
with the itemsets

Items	I_1	I_2	I_3	I_4	I_5	I_6
Weight	1	1	2	3	1	1

Definition 3 (*Complete set of MEIs in the database*) The set of itemsets denoted as
S and a function is represented as

$$f_M(S) = \{A | A \in S, e(A) \geq \text{min\_efficiency}\}. \tag{3}$$

The Maximal Efficient complete itemsets M in D (M $\subseteq$ D) and formulated as
$f_M(D)$.

Example 1 (Maximal Efficient Itemsets) Let Table 1 represent a relational database
containing five transactions. Each tuple contains group of items and have a transactional value. Let Table 2 represents Weight associated with items in transactional
database.

From Tables 1 and 2, the total efficiency of the item $\{I_4\}$ in the transaction T3 is
te $(\{I_4\}, T_3) = $ wt $(\{I_4\}, D) \times q$ $(\{I_4\}, T3) = 3 \times 2 \Rightarrow 6$. The efficiency of $\{I_2, I_4\}$ in
T3 is e $(\{I_2, I_4\}, T3) = e$ $(\{I_2\}, T3) + e$ $(\{I_4\}, T3) = (1 \times 1) + (3 \times 2) = 7$. Similarly
the efficiency of $\{I_2, I_4\}$ in T5 is e $(\{I_2, I_4\}, T5) = $ te $(\{I_2\}, T5) + e$ $(\{I_4\},$
T5) $= (1 \times 1) + (3 \times 3) = 10$. The total efficiency of $\{I_2, I_4\}$ in Supermarket
transactional database is te $(\{I_2, I_4\}) = e$ $(\{I_2, I_4\}, T3) + $ is e $(\{I_2, I_4\},$
T5) $= 7 + 10 = 17$. If min_efficiency threshold value is 10 then the itemset $\{I_2, I_4\}$ is
maximal whose value is greater than min_efficiency.

3.2 Mining Closed Itemsets

We discuss properties and terminology related to mine closed itemsets.

Definition 4 (*Closure of an itemset*) Let T_{ID} represents transaction of ith items and
S be the set of Association Rules with maximum confidence. Associative closure
$T_{ID}^+ = \cap S \in $ Supp_count(i) T_{ID}, where Supp_count(i) denotes support count of an
item in a database such that i $\in$ B and $T_{ID}^+ \subseteq A$ where A $\rightarrow$ B $\in$ S.

Definition 5 (*Maximal efficient Closed Itemsets*) Let A be Maximal efficient Closed Itemset is as follows:

$$MC = \{A | A \in S, A = T_{ID}^+(A), e(A) \geq min\_efficiency\}. \tag{4}$$

Non-closed efficient itemset A is defined as iff $A \in M \cap A \not\subseteq T_{ID}^+$. For example, If min_efficiency threshold value is 10 then set of closed MEIs in Table 1 is $MC = \{\{I_3\}, \{I_1, I_2, I_3\}, \{I_1, I_2, I_4\}\}$.

Property $\forall A$, *where A is non-closed maximal efficient itemset, their exit $B \in MC$ such that $B = (T_{ID}^+(A)) \cap (e(A) < e(B))$.*

Proof $\forall A \in S, \exists B \in T_{ID}^+$ such that supp_count(A) = supp_count(B). Since $A \in M$ and $A \not\subseteq T_{ID}^+$, $e(A) \geq$ min_efficiency and A is subset of B. Therefore, for all $A \in S$ supp_count(A) = sup_count(T_{ID}^+) $\Longleftrightarrow g(A) = g(T_{ID}^+(A))$ and total efficiency of an itemset in a database.

3.3 Algorithms for Mining Maximal Efficient Closed Itemsets

We proposed two significant algorithms Apriori-MC (An Apriori-based algorithm for mining Maximal efficient Closed itemsets) and AAEMIs (Analyzing All Efficient Maximal Itemsets) for mining. However the two algorithms are based on total efficient itemsets properties that vary from [5, 6–9].

The Apriori-MC Algorithm

This is standard algorithm, which requires two Phase I and Phase II. In First phase, scans database to find k-Itemset, which are used to generate (k + 1) candidate key [1]. Each itemset in a database having a supp_count no less than abs_min_efficiency is included to the set of (k + 1) MEIs L_k + 1. In Second phase algorithm removes non-closed itemsets in L_k + 1 by the following process. By the definition of maximal efficient closed itemsets. B ⊆ A such that (B ∈ L_k) ∩ (supp_count (A) = supp_count (B)). If true, B is discarded from Lk because B is not a maximal closed itemset according to Definition 5. If false, B is retained and identified as "closed" because it is added as maximal efficient closed itemset (MCIs).

Efficient Retrieval of Maximal Itemsets

This section employs divisive hierarchical strategy (i.e. top-down approach) named as Analyzing All Efficient Maximal Itemsets (AAEMIs) for retrieving all MCIs and their total efficiency from set of complete MCIs. The algorithm takes input parameters as total efficient threshold min_efficiency, a set of Complete MCIs and Largest length of Itemsets in MC.

Analyzing All Efficient Maximal Itemsets that outputs complete MCIs. $M = U_{i=1}^{k}$ with respect to min_efficiency where M_i indicates set of MEIs of length i.

First, the set M_{LL} is assigned to MC_{LL}, where the symbolization MC_i denotes the set of i-itemsets in MC. For each iteration, initially i is assigned to LL-1 during lines 2 in algorithm and this process is continued till the end of line 20 in the algorithm. After initialization for each iteration $e_i \in MC_i$ if total efficiency of ith itemset A < min_efficiency then it discards the ith itemset from MCi which designates the subset of ith itemset in the form of an association rules as $B \geq A - \{e_i\}$. Otherwise attach the total efficiency of ith itemset to MC_i that results (i + 1)-tem subsets. If $B \in MC_i$ or $B \notin MC_i$ and supp_count (A) > supp_count (B) then B is concatenated with $MC_i - 1$ and set count to the support count of A (Property 1), i.e., supp_count (A) = supp_count (B). Therefore repeat this process until complete set of MCIs is retrieved (Fig. 1).

ALGORITHM: AAEMIs

Input: Largest length of itemset in MC; min_efficiency
MC={ MC_1,MC_2,............MC_L}.
Output: M: Complete set of MCIs

1. $M_{LL} := MC_{LL}$
2. for (i=LL-1; i > 0; i--) do
3. {
4. for (for each ith- itemset A={$e_1, e_2, e_3, \ldots \ldots e_i$) $\in MC_i$ do
5. {
6. if(te (A) < min_efficiency) then remove A from MC_i
7. else
8. add A to its te (A) to M
9. for each item $e_i \in A$ do
10. {
11. B:= A-{e_i}
12. te (B) = te (A) – V (A, e_i)
13. if (te(B) ≥ min_efficiency) then
14. {
15. if B $\in MC_{i-1}$ & supp_count(A) > supp_count (B) then
16. supp_count(A) = supp_count (B)
17. }
18. else{
19. $MC_{i-1} = MC_{i-1}$ U B
20. supp_count (A) = supp_count (B)
21. }}}}

Fig. 1 AAEMIs algorithm

4 Experimental Results

Experiments are done on supermarket dataset to evaluate the performance of algorithms. The parameters required are defined in Table 4. Table 3 represent the number of patterns retrieved from supermarket dataset (Fig. 2).

Table 3 Parameters

Parameter	Value
T: # Transactions	100 K
L: Average transaction length	6
I: # Unique items	500
S: Average size of MCIs	5
Q: Maximum #items purchased	3

Table 4 Number of patterns retrieved

Minimum efficiency	# Candidates for Apriori-MC
0.8	18
0.6	35
0.4	341
0.2	16,194

Fig. 2 Execution time on supermarket dataset

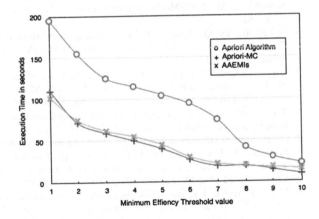

5 Conclusion

The challenges of ARM are to decrease the computational cost and to decrease the amount of labeled data as well as the retrieving relevant association rules. To overcome these difficulties it is divided into two phases. First, generates maximum number of frequent itemsets. Latter mining association rule is an important phase. Lot of work has been proposed to improve the efficiency. We proposed two algorithms for the retrieval of Maximal Efficient Itemsets named as Apriori-MC (Apriori-Maximal Closed itemset) and AAEMIs (Analyzing All Efficient Maximal Itemsets) by discarding non-closed itemsets. Apriori-MC accomplishes incremental search where ith-itemset helps us to find out (i + 1) itemsets to identify i-frequent itemsets by deleting irrelevant itemsets. In order to increase efficiency of retrieving resultant data from a closed maximal itemsets, which is represented in the form of association rules, proposed an algorithm named as AAEMIs produces condensed and lossless information without any redundancy.

References

1. T Joachims "Optimizing Search Engines Using Click through Data," Proc. Eighth ACM SIGKDD Int'1 Conf. Knowledge Discovery and Data Mining (KDD' 02), pp. 133–142, 2002.
2. Agrawal R, Imielin ski T Swami A (1993) Mining association rules between sets of items in large databases. In: Proceedings of the 1993 ACM SIGMOD international conference on management of data (SIGMOD' 93), pp. 207–216.
3. Pei J, Han J, Mao R (2000) CLOSET: An efficient algorithm for mining frequent closed itemsets. In ACM SIGMOD workshop on research issues in, data mining and knowledge pp 21–30.
4. Srikant R, Vu Q, Agrawal R(1997) Mining association rules with item constraints. In: Proceedings of the 3rd international conference on knowledge discovery in databases and data mining, pp 67–73.
5. Zaki MJ (April 2002) Hsiao CJ (2002) CHARM: An efficient algorithm for closed itemsets mining. In: Grossman R, Han J, Kumar V, Mannila H, Motwani R(eds) Proceedings of second SIAM international conference on data mining, Arlington, VA, pp 457–463.
6. T Hamrouni, key roles of closed sets and minimal generators in concise representation of frequent patterns Intell Data Anal., vol. pp. 581– 2012.
7. C Lucchese S. Orlando and R Perego "Fast and memory efficient mining of frequent closed itemsets," IEEE Trans. Knowl. Data Eng., vol., pp. 21–36 Jan. 2006.
8. Y Liu, W Liao and A. Choudhary, "A fast high utility itemsets mining algorithm," in Proc. Utility –Based Data Mining Workshop, 2005, pp. 90–99.
9. M J Zaki and C J Hsiao, "Efficient algorithm for mining closed itemsets and their lattice structure," in IEEE Trans. Knowl. Data Eng., vol., pp. 462–478, Apr.2005.
10. Pasquier N, Taouil R, Bastide Y, Stumme G, Lakhal L (2005) Generating a condensed representation for association rules. J Intell Inf Syst 24(1): 29–60.
11. Han J, Pei J, Yin Y(2000) Mining frequent patterns without candidate generation. In: Proceeding of the 2000 ACM SIGMOD international conference on management of data (SIGMOD' 00) pp 1–12.

12. Grahne G, Zhu J (2005) Fast algorithm for frequent itemset mining using FP-trees. IEEE Trans Knowl Data Eng 17(10) (1347–1362).
13. Cheng J, Ke Y, Ng W(2008) Effective elimination of redundant association rules. DataInKnowlDisc 16(2): 221–249.

An Interactive Freehand ROI Tool for Thyroid Uptake Studies Using Gamma Camera

Palla Sri Harsha, A. Shiva, Kumar T. Rajamani, Siva Subramanyam
and Siva Sankar Sai

Abstract Thyroid Uptake is a procedure that requires an injection of radiotracer/radio-isotope into the patient's blood stream. After injecting 2 millicuries of Technetium-99m pertechnetate radio-isotope, thyroid images are acquired. This uptake requires a special purpose camera called Gamma Camera. Thyroid uptake study provides both the functional, structural information. It is used for the diagnosis of various thyroid disorders. Thyroid uptake is calculated depending on the counts. Counts are nothing but the total number of intensity values present in the selected region of interest. LEAP (Low Energy All Purpose) collimator is used in the Gamma Camera which can handle only photons of lower energies. Technetium-99m pertechnetate is used having an emission energy of 140 keV. Thyroid uptake scan study using Gamma Camera has to be calibrated at each organization. In our super specialty hospital, it has been standardized that the uptake value greater than 2.5 % is considered as Hyperthyroidism, and value between 0.5 and 2.5 % is considered as Normal and the value less than 0.5 % is considered as Hypothyroidism. An Interactive freehand ROI tool was developed in Matlab R2013a as an alternative to the software existing in the Department of Nuclear Medicine, SSSIHMS. This ROI throws light on understanding the image data and calculating the Glomerular Filtration Rate. The GFR is calculated using GATES formula. The tracer uptake is obtained from both left and right thyroid

P.S. Harsha (✉) · A. Shiva · S.S. Sai
Sri Sathya Sai Institute of Higher Learning, Puttaparthi, India
e-mail: harshasri439@gmail.com

A. Shiva
e-mail: shiva.amruthavakkula@gmail.com

S.S. Sai
e-mail: sivasankarasai@sssihl.edu.in

K.T. Rajamani
Robert Bosch, Puttaparthi, India
e-mail: kumartr@gmail.com

S. Subramanyam
Sri Sathya Sai Institute of Higher Medical Sciences, Puttaparthi, India
e-mail: sivasubramaniyan.v@sssihms.org.in

© Springer India 2016
S.C. Satapathy et al. (eds.), *Information Systems Design and Intelligent Applications*, Advances in Intelligent Systems and Computing 433,
DOI 10.1007/978-81-322-2755-7_37

349

lobes by manually drawing a ROI separately. Developed tool was tested on 30 real time thyroid cases with expected thyroid disorders. The uptake value obtained from the developed tool are compared with the values of the existing software in SSSIHMS.

Keywords Interactive freehand · Matlab · Hyperthyroidism · Radiotracer · Syringe · Anticubital

1 Introduction

Nuclear Medicine imaging modalities such Single Photon Emission Computed Tomography are non-invasive. Functional imaging technologies had been producing good results for identification and diagnosis of aliments related to thyroid, renal, cardiac, bone study etc. [1, 2]. Thyroid study is done by using a Gamma Camera that could capture gamma radiation and convert into equivalent electronic signal for acquiring the images.

The thyroid scan and uptake provide structural and functional information of the thyroid. It is a gland located in the neck region which controls the body metabolism essentially responsible for converting food into energy. Thyroid uptake scan study is useful for the diagnosis and treatment of Hyperthyroidism. Technetium-99m pertechnetate is used for the thyroid uptake study. Apart from technetium, Iodine-131 with an oral administration of 25 microcuries and Iodine-123 (sodium Iodide) is also given orally in administered quantity of 200–400 millicuries [3].

2 Thyroid Uptake Procedure

2.1 Acquisition and Imaging

Thyroid uptake scan study was done after injecting technetium-99m pertechnetate of 2 millicuries through an intravenous injection. A Dual Head Angle Gamma Camera is used for the acquisition of thyroid images with a parallel hole LEAP (Low Energy All Purpose) collimator (anterior view) [4]. After injection, the patient is positioned under the Gamma Camera detector in supine with patients head tilted back, to expose the thyroid region to the detector of gamma camera. The following acquisition parameters were used: anterior view, 128 × 128 matrix and a zoom factor of 1.78.

Each patient data consists of four 2 dimensional 16 bit static images by name Full Syringe, Thyroid scan, Empty Syringe and Anticubital as shown in figure. The delayed static images are acquired 2–3 h after post injection.

The procedure is as follows:

1. Full Syringe: Syringe with radiotracer is imaged before injection for 1 min.
2. Thyroid is imaged for 10 min.
3. Empty Syringe: Post injection, Some amount of tracer is left after injection, which is imaged for 1 min.
4. Anticubital: Region where the tracer is injected in the body is imaged for 1 min.

2.2 Uptake and Calculations

The study of thyroid uptake is useful for the diagnosis and treatment of Hyperthyroidism. The thyroid counts were determined by manually drawing region of interest (ROI) around the left and right thyroid lobes separately. The background counts are determined by selecting the regions on either shoulders for subtraction using freehand ROI tool in Matlab R2013a. The counts of the Full Syringe, Thyroid, Empty Syringe, and Anticubital are obtained by manually drawing the ROI in the Fig. 1a–d.

Counts are obtained by summing up all the pixels from the selected region of interest of the two dimensional images. Thyroid Uptake was calculated by the following equation (Fig. 2).

$$Uptake(\%) = \frac{Thyroid\ cpm - Thyroid\ background\ cpm}{Standard\ cpm} \times 100 \qquad (1)$$

$$Standard\ cpm = Full\ Syringe - Empty\ Syringe - Anticubital \qquad (2)$$

cpm refers to Counts per minute. Since the thyroid image is acquired for a period of 10 min. In order to get the counts per minute, the thyroid counts obtained are divided by ten.

The medical expert performs the segmentation of the thyroid manually by drawing the ROI (Region of Interest) [5] separately on left and right thyroid lobes and the software in return gives the Uptake and counts along with the volume of

(a)   (b)   (c)   (d)

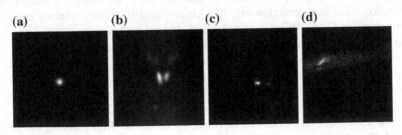

Fig. 1 a Full syringe. **b** Thyroid scan. **c** Empty syringe. **d** Anticubital

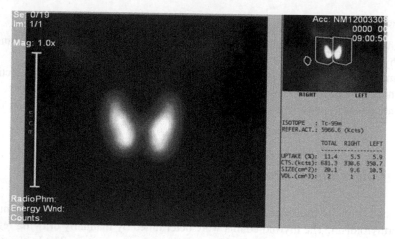

Fig. 2 User interface of thyroid uptake software used at SSSIHMS

thyroid, the standardization used in the SSSIHMS. The results obtained from the developed freehand ROI tool is compared with the results of the existing software present in the hospital.

2.3 Proposed Interactive Freehand ROI Tool

The Siemens Dual Head Gamma Camera has a LEAP collimator which can handle only photons of lower energies. Tech-99m which has an emission energy around 140 keV is used as radiotracer. In order to use Iodine-131 as a radio isotope with emission energy around 364 keV requires a Higher energy order with High resolution. A software tool similar to the existing tool present in the SSSIHMS was developed which would enable the radiologist to use the developed tool as a backup software for the Thyroid Uptake studies using Matlab R2013a [6]. Matlab is a proprietary software and a fourth generation programming language and an interactive environment developed by Math Works used by millions of scientists and engineers worldwide. It includes toolboxes like communications, control systems and signal and image processing. Matlab is used for Image Enhancement, Registration, and Segmentation etc. Matlab is compatible with various operating systems such as MAC, Windows, and Linux. Figure 3 shows the various windows of the developed inter active free hand ROI tool. Figure 3a shows Welcome window of the developed tool, Fig. 3b shows the Full Syringe window, Fig. 3c shows the Left Thyroid window, Fig. 3d shows the Right Thyroid window, Fig. 3e shows the Empty Syringe Window, Fig. 3f shows the Anticubital window. Figure 4 shows the windows of freehand ROI markings on the 16 bit DICOM images. Figure 4a shows the ROI marking of the Full Syringe, 4b shows the ROI marking of the Left Thyroid, Fig. 4c shows the ROI marking of the Right Thyroid, Fig. 4d shows the

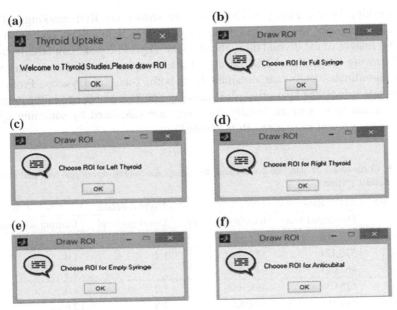

Fig. 3 Developed tool windows. **a** Welcome window. **b** Full syringe window. **c** Left thyroid window. **d** Right thyroid window. **e** Empty syringe window. **f** Anticubital window

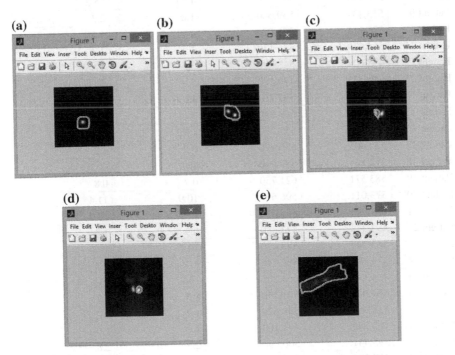

Fig. 4 Developed tool ROI windows. **a** Full syringe ROI window. **b** Left thyroid ROI window. **c** Right thyroid ROI window. **d** Empty syringe ROI window. **e** Anticubital ROI window

ROI marking of the Empty Syringe, Fig. 4e shows the ROI markings of the Anticubital. Counts are obtained from the above ROI markings on the 16 bit DICOM images of the thyroid through which the Thyroid Uptake can be calculated. After drawing the freehand ROI on the interested region the minimum and maximum co-ordinate values are obtained from both x-axis and y-axis. From this co-ordinate values, the respective pixel intensity values are obtained in the range of the minimum to maximum. Finally the counts are calculated by summing up all pixel values present in selected ROI (Table 1).

Table 1 Comparison of standard counts and thyroid uptake values: developed tool and existing thyroid uptake software

Patient ID	Std. counts		Thyroid uptake	
	Developed tool	Existing software	Developed tool	Existing software
Patient 1	567,468	602,960	16.2	16.3
Patient 2	520,124	579,860	1.1	1.4
Patient 3	462,047	540,100	2	2.3
Patient 4	435,058	491,100	1.2	1.6
Patient 5	440,181	500,900	2.8	2.6
Patient 6	495,582	580,040	20.9	20.9
Patient 7	423,978	461,480	8.2	8.6
Patient 8	505,193	534,320	14.1	15.2
Patient 9	513,437	559,660	1.8	2.3
Patient 10	536,111	575,020	4	4.5
Patient 11	554,674	535,170	1	1.3
Patient 12	495,604	548,690	3.5	4
Patient 13	440,373	470,520	5.8	7
Patient 14	518,586	563,780	12.7	13
Patient 15	543,167	589,240	3.1	4.1
Patient 16	534,070	557,270	13.9	13.6
Patient 17	535,605	586,910	0.8	1.2
Patient 18	552,864	580,040	19.1	20.9
Patient 19	583,571	621,750	0.7	0.8
Patient 20	573,011	596,660	10.9	11.4
Patient 21	649,794	681,670	13	12.9
Patient 22	497,373	573,880	0.9	1.8
Patient 23	563,240	588,470	10.2	10.6
Patient 24	379,190	562,540	8.4	8.7
Patient 25	433,144	464,200	13.8	13.5
Patient 26	489,021	501,250	8.5	8.7
Patient 27	518,456	548,620	1.7	2
Patient 28	462,280	491,100	1.1	1.6
Patient 29	395,441	500,990	2.5	2.8
Patient 30	427,294	535,170	0.8	1

3   Results

3.1   Discussion and Conclusion

An Inter active freehand ROI tool was developed for the Thyroid Uptake Studies using Gamma Camera in the Department of Nuclear Medicine at Sri Sathya Sai Institute of Higher Medical Sciences, Prasanthigram. The developed tool was tested on the 30 thyroid patients in real time with definite or suspected disease. The results of the tested 30 patients are tabulated in the above tabular form. The counts obtained from the developed tool are close to the values obtained the existing software but are not same. This is because the counts values differ from software to software and also depends on the precise drawing of ROI on the interested region in the image. Statistical analysis (P value) was also done for both developed tool and existing software and found to be 0.805 which is greater than the significant value 0.005 i.e. P value is greater than 0.005. This difference is considered to be not statistically significant.

References

1. David S. Cooper, Gerard M. Doherty, Revised American Thyroid Association Management Guidelines for patients with Thyroid Nodules and Differentiated Thyroid cancer vol. 19, no. 11, 2009.
2. Magdy M. Khalil: Basic Sciences of Nuclear Medicine. Springer verlag Heidelberg 2011.
3. Maria Lyra, John Striligas, Maria Gavirilelli, Christos Chatzijiannis Thyroid Volume determination by single photon tomography and 3D processing for activity dose estimation, IEEE International Workshop on Imaging Systems and Techniques, Chania, Greece, September 2008.
4. Jia-Yann Huang, Kun-Ju Lin, Yung-Sheng Chen Fully automated computer-aided volume estimation for thyroid planar scintigraphy, Computers in Biology and Medicine 43 (2013) 1341–1352.
5. P Van Isselt J.W, De Klerk J.M, van Rijk P.P, van Gils A.P, Polman L.J, Kamphuis C, Meijer R, Beekman F.J, Comparison of methods for thyroid volume estimation in patients with Graves disease, European Journal. Nucl. Med. Mol. Imaging 30 (4) 2003 525–531.
6. Website: www.mathworks.com.

Literature Survey on Intrusion Detection Systems in MANETs

Pooja Kundu, Neeti Kashyap and Neha Yadav

Abstract Mobile ad hoc networks are wireless networks consisting of mobile nodes with no boundary. Nodes are free to move and the network is dynamic. Unique features of these networks serve as benefits as well as drawbacks and give chances to attackers. Intrusion occurs when a malicious node tries to enter the network and misuses the resources. Several attacks and intrusion detection techniques are discussed in this paper.

Keywords Ad hoc networks · Intrusion detection · Manets · Security · Wireless

1 Introduction

Wireless networks provide scalability and reduced costs. Mobile ad hoc network (MANET) is one of the most significant applications of wireless networks. They have different and unique characteristics such as dynamic topology, self-configuration and maintenance, fast and easy to deploy, cheap and absence of centralized authority. The density of nodes keeps changing. Communication method is flexible because a MANET has self-organizing property. Nodes are equipped with transmitter as well as receiver. No expensive base-station is required. Such networks find their application in military, rescue operations, vehicular computing and offsite-meetings of a company. Memory, bandwidth and battery-power of nodes are limited.

There is a risk of different types of attacks. They can be categorized into active and passive. Active attacks are those which modify or drop the data packets and cause distortion in the traffic. Network is monitored by a malicious node in passive attacks. One or more layers of routing protocol are attacked. Network-layer attacks are Sybil, black hole and gray hole. Transport layer is attacked in session hijacking

P. Kundu (✉) · N. Kashyap · N. Yadav
Kolkata, India
e-mail: poojakundu9999@yahoo.com

© Springer India 2016
S.C. Satapathy et al. (eds.), *Information Systems Design and Intelligent Applications*, Advances in Intelligent Systems and Computing 433,
DOI 10.1007/978-81-322-2755-7_38

357

Table 1 Category of attacks

Active	Compromised routing logic	Black hole, gray hole, sybil, sleep deprivation
	Network traffic distorted	Packet dropping, packet replication, DoS
Passive	Traffic analysis, location disclosure, eavesdropping	

and repudiation attack, worms, mobile virus and Sybil attack application layer. Table 1 presents the different type of attack.

The paper is organized as follows. Section 2 is comprised of different architectures and techniques for intrusion detection. Section 3 presents the research done in this field and summarizes the work in a table. Finally Sect. 4 concludes this study.

2 Intrusion Detection Systems

With the evolution of wireless technology security has always been a major concern. Although there are many intrusions prevention techniques like authentication and encryption but they do not prove to be sufficient to guarantee security. Intrusion detection systems (ID) provide another level of security. They detect malicious nodes and try to respond accordingly. IDS used for traditional networks cannot be used for MANET because of the difference in architecture. An activity is defined to be abnormal when it attempts to compromise the integrity, availability and confidentiality. If an attack is detected IDS take an action to inform other nodes and try to recover. IDS can be classified according to audit data collected, technique used, and architecture followed. Audit data is collected by each node as there is no centralized management. According to the data collected by nodes IDS is classified into host-based or network-based. Host-based IDS depends upon the operating system to audit data. Network-based IDS relies on the analyses of network traffic and data packets. Figure 1 presents the classification of IDS [1].

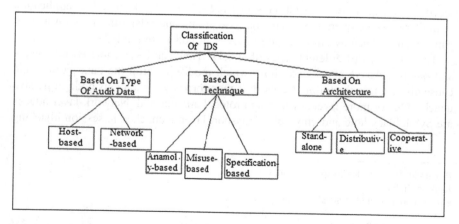

Fig. 1 Classification of IDS

2.1 Challenges Faced by IDS in MANET

Characteristics of MANET not only make it a unique application but also make it difficult for IDS to adhere to these features. IDS used by traditional networks are not suitable for a MANET. Following are some of the challenges faced by IDS when designed for a MANET.

Lack of central audit points: In traditional networks, routers, gateways and switches act as audit points. MANETs lack central management and central audit points.

Dynamic topology: The mobility of nodes in a MANET causes dynamic topology. The relationship of a node with its neighbors keeps changing. The IDS is required to be flexible to easily adapt changes in topology.

Resource constraint: Resources like battery power, bandwidth, and memory are limited and they cause problems like dead nodes, selfishness of nodes, and limited transmission power.

Assumption of routing algorithm: Most of the routing protocols used in a MANET are based upon global trustworthiness. It is assumed that all the nodes are cooperative [2]. Malicious nodes take advantage of this trust. It is easier for an attacker to intrude the network. Some nodes of the network act selfishly and do not cooperate with others.

From the above mentioned characteristics it is concluded that IDS must be distributive, scalable, flexible and cooperative.

2.2 Architecture Used for IDS

The architecture plays an important part in describing how efficiently the IDS can be applied to the network.

Stand-alone IDS: Local intrusion detections are performed at each node of the network. None of the nodes send alert information regarding the attacks detected by them. The information at each node may not be enough to provide the security. Thus it is not preferred in most of the applications.

Distributive and Cooperative IDS: Each node of the network participates in the intrusion. The distributive nature of MANET demands IDS to be distributed. Local as well as global intrusion detection is performed. The IDS at each node detects the intrusion locally. When the collected evidence by a node is strong enough then the node takes responsive action regarding the attack. All nodes in a network cooperate to take action towards the attack.

Hierarchical IDS: Every node is not equal and has different responsibility. Network is divided into clusters and some nodes are selected as cluster-heads. Cluster-heads are responsible for performing different roles for their respective clusters. Every node detects local intrusion. It is the duty of cluster-heads to respond according to the attacks detected in a cluster. A hierarchy of steps is followed to combat the attack detected. The disadvantage of implementing this architecture is that the classification of nodes in a cluster and election of leaders for the clusters result in an overhead as well as difficult to decide [3].

2.3 Technique for Intrusion Detection

Anomaly-based ID: A normal behavior is established by considering some parameters of the network. Auditing of data is done to find out the normal behavior. Normal profiles so decided are matched to detect any abnormal behavior. Suspicious activity may or may not be an attack. This decision is also taken by ID. These normal profiles must be updated with time. No prior knowledge of attacks is required. It is capable of detecting novel attacks. The drawback of this method is that normal profiles are not easy to be decided.

Misuse-based ID: Misuse-based intrusion detection called signature-based detection or knowledge-based detection. Known attack profiles are efficiently detected by this method. Signatures of known attacks are matched with the detected malicious activity in order to detect attacks. There is one drawback of this technique that novel attacks cannot be detected.

Specification-based ID: This technique uses both the above mentioned techniques to detect attacks in a network. Firstly normal and abnormal behaviors are differentiated from each other. These profiles are then used while monitoring the network. Since normal profiles are built manually, it is a time-consuming task and may contain errors.

3 Review of Intrusion Detection Algorithms

In 2000, Marti et al. proposed two techniques watchdog and pathrater. In watchdog technique the node promiscuously snoops and overhears the data packet sent by neighbors. For example, in Fig. 2, node A sends the data packet to node B which in return relay the packet to node C. A buffer is maintained by node A and it overhears the packet sent by node B to node C. If the packet sent by node B does not match with the packet stored in the buffer of A or the node B failed to relay the packet, failure counter of B is increased. Those nodes whose failure counter is above the threshold value are excluded by pathrater from the routing path. The watchdog technique is not able to work in following 6 situations: receiver collisions, ambiguous collisions, false misbehavior, collusions, limited transmission power and partial dropping [4]. The technique used in watchdog is extended by Nasser et al. in 2007 and proposed an algorithm named ExWatchdog [5]. This algorithm partitions the malicious nodes and then proceeds to protect the network. Any node which falsely report other node is detected but correctness of report is cannot be confirmed if malicious node exist in all the routing paths.

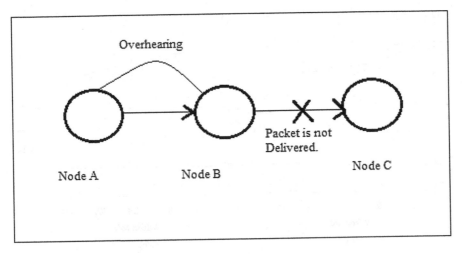

Fig. 2 Failure counter of node B is increased

In spite of detecting malicious nodes, TWOACK [6] detects misbehaving links. Although this algorithm is capable of detecting intrusion in case of limited trans-mission power and receiver collision but misbehaving nodes get more chances as they might be connected to different links. It fails to detect in presence of false misbehavior report and forged acknowledgement packets. AACK [7] intrusion detection is an enhancement to TWOACK. Adaptive ACKnowledgement (AACK) is used as a default mode to reduce the routing overhead of TWOACK scheme. Two modules are used namely enhanced TACK and AACK, and an algorithm is used to switch between these modules.

In 2013, EAACK [8] algorithm is proposed in which acknowledgement packets are digitally signed by using DSA algorithm to make them secure. It also solves the issue of false misbehavior. Packet delivery ratio (PDR) is the actual number of packets received by the respective destination to the number of packets sent by the source. Figure 3 shows the PDR vs. malicious nodes percentage in the network. In Fig. 3a ExWatchdog (EX) increases the packet delivery ratio as compared to Watchdog (W). In Fig. 3b EAACK outperforms TWOACK (TA) and AACK (AA).

Application layer attacks are resolved by the algorithm proposed by Chang et al. [9]. Mobile agents are used to augment the intrusion detection. They are dispatched on the demand of nodes of the network or periodically by MA server. MAs are responsible for detecting anomaly, adding normal profiles and signature of attacks, verifying the IDS agents on the nodes. Each MA performs only one task. They can be captured by malicious node or get lost. Wang et al. [10] presented a mechanism design-based secure leader election model which uses a distributed approach. While electing leaders for clusters some nodes may act selfishly to save their resources. To prevent this incentives are given in the form of reputation so that none of the nodes lie about its resources. Lifetime of the network is prolonged by this algorithm.

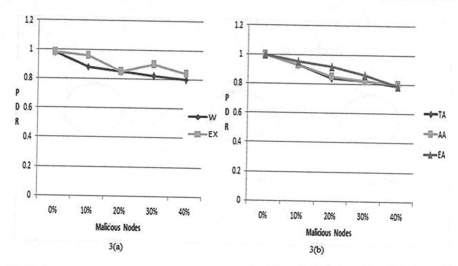

Fig. 3 Comparison of algorithms on the basis of PDR and malicious nodes

A novel scheme is proposed in [11] which combine support vector machines (SVM) and fuzzy integral. The fuzzy integral integrates multiple classifiers so that classification at low bit rate occurs efficiently. Two layer protection is provided by the scheme proposed in [12]. Prevention-based and detection-based approaches are combined and work optimally because both the schemes share information with each other. Multimodal biometrics is used for continuous authentication.

Nodes are divided into non overlapping zones in ZBIDS [13]. It is a 2-level scheme. Every node has IDS which work collaboratively and generate alerts. Gateway nodes are responsible for collaboration of the alerts inside a zone and generating an alarm in case of attack. Many false alerts are avoided by them. Trade-off between security and power are considered in power-aware intrusion detection [14]. Genetic programming (GP) is used for evolving programs in order to detect known attacks, while considering varying mobility of nodes.

Some algorithms are proposed for a specific type of attacks. Malicious nodes keep changing their identity in Sybil attack so that they do not get detected. Lightweight Sybil attack detection [15] scheme is proposed which do not use any extra hardware or third party. All the data packets are dropped in gray hole attack. It is a type of Denial of Service (DoS) attack. To detect such malicious nodes a scheme "A novel gray hole detection" [16] is proposed in 2007. Each node is expected to create a proof that it has received the message. If destination informs the source that few packets are not received then the checkup algorithm is invoked to detect malicious node. Table 2 summarizes the review of various intrusion detection algorithms.

Table 2 Review of IDS

Algorithm name	Attack resolved	Method used	Contribution	Advantage	Limitation
Watchdog and Pathrater [4]	Routing attacks	Promiscuously hearing other nodes	Nodes detect malicious behavior of neighbors	Throughput increased, robust	Fails to detect misbehavior in few cases
ExWatc hdog [5]	Routing attacks	Partition the malicious nodes	Detected falsely reporting nodes	Decreased Overhead	Report may be incorrect
TWOA CK [6]	Packet dropping attacks	Detect misbehaving link	Solved the problem of transmission power, receiver collision	PDR improved	Overhead, false misbehavior attack
AACK [7]	Network layer attacks	Switch between TWOACK and AACK	Solved the problem of transmission power, receiver collision	Overhead reduced, detection efficiency increased	Gray hole attack undetected
EAACK [3]	Packet dropping attacks	ACK packets are digitally signed	transmission power, false misbehavior, receiver collision	PDR improved	Overhead sometimes
MA-based application-layer IDS [9]	Application layer attacks	Detect malicious node, defend network	Both anomaly and misuse detection	Augments nodes ID'S capability	MAs can be lost
SVMFN [11]	Network attacks	Multi class SVM and fuzzy integral	Accuracy of ID and generalization performance improved	Identify network attacks at changeable bit rate	Overhead
Optima 1 combined ID and Authentication [12]	Various attacks	Problem is formulated as POMDP	Continuous authentication and ID in an optimal way	Multi-layer protection	Solving POMDP intractable
ZBIDS [13]	Routing disruption attack	Markov chain based local anomaly	Cooperation based approach	Local ID as well as global ID	Overhead

(continued)

Table 2 (continued)

Algorithm name	Attack resolved	Method used	Contribution	Advantage	Limitation
Power aware ID [14]	Flooding and route disruption attack	Genetic programming algorithm is used	Trade-off can be made on	Optimization based approach	Trade-offs affects die ID
Sybil attack detection [15]	Sybil attack	Received signal strength (RSS) algorithm is used	Standalone scheme or an add-on to existing schemes	Lightweight	Variable transmit power un considered
Grav Ho-le Attack Detection [16]	Gray hole attack	Aggregate signature is used	Packet delivery rate improved	Low bandwidth overhead	Each malicious node is not detected

4 Conclusion

MANETs are applicable in several fields and require providing secure communication in a cost-effective manner. In this survey, many proposed intrusion detection systems are reviewed. Their methods, contributions, advantages and limitations are discussed in brief. Efficient IDS must consider the following factors architecture, techniques for detecting intrusions and sources of audit data. Categories of attacks are also discussed. The presented algorithms in this research show that there are one or more limitations. Thus, a lot of research work is still needed to provide a safe environment. In future we would like to research a more robust and efficient IDS with lesser overhead.

References

1. Anantvalee, Tiranuch, and Jie Wu. "A survey on intrusion detection in mobile ad hoc networks." Wireless Network Security. Springer US, 2007. 159–180.
2. L. Buttyan and J. P. Hubaux, Security and Cooperation in Wireless Networks. Cambridge, U.K.: Cambridge Univ. Press, Aug. 2007.
3. Zhang, Yongguang, Wenke Lee, and Yi-An Huang. "Intrusion detection techniques for mobile wireless networks." Wireless Networks 9.5 (2003): 545–556.
4. S. Marti, T. J. Giuli, K. Lai, and M. Baker, "Mitigating Routing Misbehavior in Mobile Ad hoc Networks", in Proc. 6th ACM International Conference on Mobile Computing and Networking,, Boston, USA, August 2000, pp. 255–265.
5. Nasser, Nidal, and Yunfeng Chen. "Enhanced intrusion detection system for discovering malicious nodes in mobile ad hoc networks." Communications, 2007. ICC'07. IEEE International Conference on. IEEE, 2007.
6. K. Balakrishnan, J. Deng, and P.K. Varshney, -TWOACK: Preventing Selfishness in Mobile Ad Hoc Networks,‖ Proc. IEEE Wireless Comm. and Networking Conf. (WCNC '05), Mar. 2005.
7. Al-Roubaiey, Anas, et al. "AACK: adaptive acknowledgment intrusion detection for MANET with node detection enhancement." Advanced Information Networking and Applications (AINA), 2010 24th IEEE International Conference on. IEEE, 2010.
8. Shakshuki, Elhadi M., Nan Kang, and Tarek R. Sheltami. "EAACK—a secure intrusion-detection system for MANETs." Industrial Electronics, IEEE Transactions on 60.3 (2013): 1089–1098.
9. Chang, Katharine, and Kang G. Shin. "Application-Layer Intrusion Detection in MANETs." System Sciences (HICSS), 2010 43rd Hawaii International Conference on. IEEE, 2010.
10. Mohammed, Noman, et al. "Mechanism design-based secure leader election model for intrusion detection in MANET." Dependable and Secure Computing, IEEE Transactions on 8.1 (2011): 89–103.
11. Li, Huike, and Daquan Gu. "A novel intrusion detection scheme using support vector machine fuzzy network for mobile ad hoc networks." Web Mining and Web-based Application, 2009. WMWA'09. Second Pacific-Asia Conference on. IEEE, 2009.
12. Liu, Jie, et al. "Optimal combined intrusion detection and biometric-based continuous authentication in high security mobile ad hoc networks." Wireless Communications, IEEE Transactions on 8.2 (2009): 806–815.

13. Sun, B., Wu, K., and Pooch, U.W., (2006). Zone-Based Intrusion Detection for Mobile Ad Hoc Networks, International Journal of Ad Hoc & Sensor Wireless Networks, 3, 2.

14. Şen, Sevil, John A. Clark, and Juan E. Tapiador. "Power-aware intrusion detection in mobile ad hoc networks." Ad hoc networks. Springer Berlin Heidelberg, 2010. 224–239.

15. Abbas, Sonya, et al. "Lightweight Sybil attack detection in MANETs." Systems Journal, IEEE 7.2 (2013): 236–248.

16. Xiaopeng, Gao, and Chen Wei. "A novel gray hole attack detection scheme for mobile ad-hoc networks." Network and Parallel Computing Workshops, 2007. NPC workshops. IFIP International Conference on. IEEE, 2007.

Apply of Sum of Difference Method to Predict Placement of Students' Using Educational Data Mining

L. Ramanathan, Angelina Geetha, M. Khalid and P. Swarnalatha

Abstract The purpose of higher education organizations is to offer superior education to its students. The proficiency to forecast student's achievement is valuable in affiliated ways associated with organization education system. Students' scores which they got in exam, can be used to invent training set for dominate learning algorithms. With the academia attributes of students such as internal marks, lab marks, age etc. it can be easily predict their performance. After getting predicted results, improvement in the performance of the student to engage with desirable assistance to the students has to be processed. Educational Data Mining (EDM) offers such information to educational organization from educational data. EDM provides various methods for prediction of student's performance, which improve the future results of students. In this paper, by using their attributes such as academic records, age, and achievement etc., EDM is used for predicting the performance about placement of final year students. As a result, higher education organizations will offer superior education to its students.

Keywords Data mining · Educational data mining · Sum of difference · Prediction

L. Ramanathan (✉) · A. Geetha · M. Khalid · P. Swarnalatha
School of Computing Science and Engineering, VIT University, Vellore 632 014, Tamilnadu, India
e-mail: lramanathan@vit.ac.in

A. Geetha
e-mail: angelina@bsauniv.ac.in

P. Swarnalatha
e-mail: pswarnalatha@vit.ac.in

© Springer India 2016
S.C. Satapathy et al. (eds.), *Information Systems Design and Intelligent Applications*, Advances in Intelligent Systems and Computing 433,
DOI 10.1007/978-81-322-2755-7_39

367

1 Introduction

The basic idea behind data mining is obtaining the knowledge from the immense set of data, which is useful and favorable. There are various method used for knowledge discovery from huge data such as classification, clustering, prediction, association rule etc. In data mining, classification is the simple employ data mining technique that utilizes a group of pre-sorted example to create a model which can classify the immense set of data. Using decision tree and neural network, can be classified the huge amount of data. Clustering is the other technique in data mining, where used to find similar type of object. It is easily determine the small and deep patterns. It is not so cheap. So in preprocessing step clustering can be used for attribute selection. KNN algorithm is used for clustering. For prediction can be used regression technique. It defines a relationship between single or more than one dependent variables and independent variables. For prediction and classification same model can be used like decision tree or neural network. EDM is an application of data mining, which offer necessary information to educational organization, which hides in educational data of students. From the past and operational data inherit in the database of educational organization, the data can be gathered. EDM introduces a various fields such as prediction, clustering, distillation of human judgment, discovery with model, relationship mining. Using techniques such as Decision tress, MultiLayer Perceptron, Neural networks, Bayesian network, support vector regression and NaiveBayes simple algorithm can be describing many types of knowledge like association rules, classification and clustering. The main objective of this paper is to predict the placement of student by using similarity measure with mathematical method which is called sum of difference (SOD). SOD is used to analyze the performance of students using their attributes such as academic records and to find a method more accurate result for prediction.

2 Literature Survey

In data mining, to predict the performance of student there are various data mining tool, where we have to set the default parameter of various algorithms to gain the highest accuracy, which is hard for a non-technical person. In recent year, some works have done on automatic parameter tuning. They have taken 14 different educational dataset and focused how to increase precision the result by using parameter tuning of J48 algorithm, and compared also accuracy when using three basic characteristics (sum of examples, sum of features and sum of modules) numerical attributes over categorical attributes [1]. For performance prediction, temporal data of student is important for performance prediction. The performance of student is improved after studying relevant skill, which is not count in recent work. Pardos have done work on data leakage by using two prediction algorithms:

linear regression and random forest. By including K-means clustering technique he claimed to improve prediction accuracy [2].

In other research work, individual information of a student's performance is used to forecast rather than historical data of other student. Obviously it is cleared that learning level of each level are different, so it is not fit to judge performance of a student based on others. In proposed model he used two skills: how adept a student is in executing the task and what is the difficulty level of the task [3]. Additionally Sen [4] is used CRSIP-DM to predict test score of student with four prediction model, and found that C5 algorithm is best among rest of algorithm. By using various attributes of student, he analyzed the sensitivity of attributes, which are much important for prediction. Baker [5] introduced a model which is used to predict the preparation for future learning of student. Students have ability to learn new thing with the help of his existing knowledge, and they have skill to learn quickly new skill. He used knowledge engineering methods with data mining.

This model needs small data of student. Heffeman [6] compared between single algorithm and ensemble method with RMSE and correlation value with respect accuracy. Each algorithm generates better result with different aspect. So he mixed more than single algorithms and predicts post score of student to achieve better accuracy. The idea behind the model is called tabling, in which checked the student's response to solve same pattern of question, which he already faced. Vera [7] worked on imbalanced data, in which the sum of examples in '1' module is greatly superior to the sum of examples in additional module or further modules, which decreases the accuracy of result. He used various classification algorithm with supervised data filter technique SMOTE with best attributes among all attributes without affecting reliability. Akcapinar [8] used 10-cross fold validation method with two different random forest regression model to find high accuracy. Schoor [9] explored social activities of a group rather than single person activity. This model provides to group a collaborative learning task and analyzed the group performance and relationship between members. Wang [10] used the knowledge tracing model to find the accuracy of student first response answer and to improve the accuracy of prediction. Student first response may be affected by his skill or by guess about success. On the other hand it may be affected by forget or slip about unsuccessful. Gowda [11] proposed a new model to predict the post score of student with two factors: to detect the time by time learning and to improve the student's slipping and guessing. In this work slipping is the best model because improving the slip model improves the performance of student than other model.

3 Methodology

To predict the student's placement status is not so easy task, because the placement depends on various factors. Sometime a student who have good academic record and still not placed, on the other hand a student who has not good in his academic life he got placed. Sometime it depends on luck and time and the frequency of companies which comes in college campus. It is hard to analysis using above factor to predict the student's placement. We have dataset of students' academic record; we can analysis the pattern from given dataset, which affect the placement status. Second thing is to collect the large dataset which is also a difficult task. If we have large real dataset, we can analysis more preciously result then the result would be more accurate. Third thing is to collect a lot of information about student, which is called attributes, as we are not sure which attributes effect the output, so if we have a lot of attributes we can find precious result, which will be more accurate our model for prediction. Similarity measure is used to find the pattern in the given object. There are various mathematical methods, which is used to find the pattern from the given data set. SOD is a one of them, which is used to find the similarity from given dataset.

The theory of rough set [11] model can be applied to discuss with definite or nominal data prediction that consists vagueness, applied to get fuzzy-rules, purpose with uncertainty and vagueness.

The superiority of hybrid algorithm (RIFCM) with other clustering methods for better performance have been introduced [12, 13] and tested through some experimental study.

The fuzzy rough set deals with membership method which efficiently handles the overlap divisions of rough sets that discuss with definite or nominal prediction. This involves vagueness and partial in class definition [7]. The combination of probable and possible membership methods of FCM (Fuzzy C-Means) may yield an apt able data model.

3.1 Attribute List

Gender	Male, Female
Category	General, OBC, SC/ST, other
Academic gap	0–10
Grade in 10th exam (G10E)	0–100
Grade in 12th exam (G12E)	0–100
Grade in B.Tech exam (GBTECHE)	Poor, average, good
Extra technical course (ECC)	Yes, no
Grade in M.Tech exam (GMTECHE)	0–100
Placed	Yes, no

3.2   Data Selection

The data have been generated from a Google page by filling pupils(s). The preliminary data consists of the performance resume, collected from a totality of 50 pupils(s) with 11 given characteristics that comprise of gender, category, academic gap, grade in 10th exam (G10E), grade in 12th exam, no. of arrear (NOAF), grade in BTech exam (GBTECHE), English communication skill, extra technical course (ECC), grade in M.Tech exam (GMTECHE), placed. The statistics consists of different kinds of standards either string or numeric standard. The statistics were then treated to create the procedures.

3.3   Data Pre-processing

After the preliminary test on the data, removal of gender, category, academic gap is made. This is in accordance with the number of missed values by considering '1' sample as a part of the data cleaning process.

3.4   Data Transformation

Conversion of data to a method is made to execute the process of data mining very easily once cleaning procedure gets over. Also conversion of large values to small values has been made due to difficult in handling any large data values. The final data can then be given for mining which undergo pre-process and transformation process.

3.5   Architecture

In the architecture we prepare input dataset for statistical analysis. Based on similarity measure, we find a pattern in given attribute, which is used for prediction for placement. There are various mathematical methods to find pattern for prediction. In this work, we have chosen SOD method to find pattern for prediction of student's placement. Based on mathematical method we implemented it in C# language, which predict the student placement. We used most important attribute which useful for

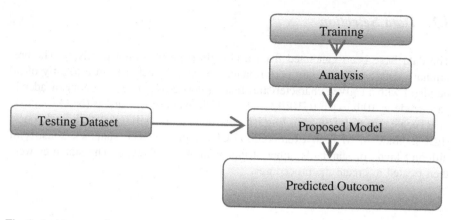

Fig. 1 Architecture of proposed model

prediction and set the priority of each attribute. For each attribute we chosen a reference point in the range of 0.0–1.0 based on its priority. Now we have chosen other reference point for each attribute in the range of value of given attribute and subtract it with each value respectively. Then we have added of each subtracted value. Based on SOD value we find a cut-off value where placement value altar (Fig. 1).

3.6 Proposed Algorithm

See Figs. 2, 3 and Table 1.

Algorithm [P, I, j, Z, T, SOD]
  1. Analysis the most important parameter P from the input dataset.
  $P = \{P1, P2.......Pi\}$ where i = number of most important parameter.
  2. Choose appropriate reference value Z for each attribute Pi based on priority for normalization.
        Where $Z = \{Z1, Z2........Zi\}$ where i=number of most important parameter.
  3. Multiply by reference number Zi to each parameter value Pij.
        Where J = number of record.
  4. Choose another reference value T in the range of parameter value for each attribute Pi based on priority.
        Where $T = \{T1, T2........ Ti\}$ where i=number of most important parameter.
  5. Subtract by reference number Ti to each parameter value Pij.
        Where J = number of record.
  6. Calculate SODi , Sum of differences of Pi.
  7. Based on SODi value analysis the result and set the value for prediction.
  8. End.

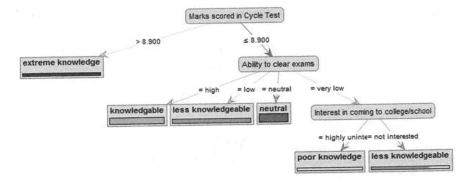

Fig. 2 Decision trees based on good knowledge

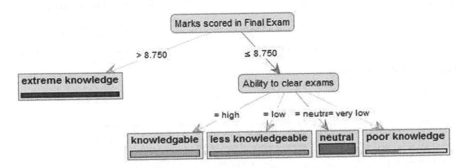

Fig. 3 Random forest based on good knowledge

Table 1 Comparison of decision tree and random forest algorithms

Algorithm : Decision tree based on Good Knowledge as Label	Algorithm: Random Forest based on Good Knowledge as Label:

The sample labels to be treated as base node where assign base node is assigned to BN.
1. Use BN to do
2. Characteristics C + Cut off Value T
   2.1 Then divide the sample nodes to BN to 2 subsets as SSL and SSR
   2.2. This is to increase the label accuracy in the subsets
3. Allocate (C,T)to BN
4. If SSL and SSR too small to be divided
   4.1 Add child lead nodes LL and LR to BN
   4.2. Label the leave nodes with the frequent label in SSL and SSR one to one
5.else
   5.1. Add child notes BNL and BNR to B N
   5.2. Allocate SSL and SSR to them one to one.
   5.3. Do this method for BN=BNL and BN=BNR
	Same procedcure for random forest also except the characteristics to be considered as given below:
	
	"Characteristics C
	(between a RANDOM subset) + Cut off Value T"
	
	Random Subset characteristics:
	
	Draw random iteratively for every node such that at each dimensional instances, typical subset = round of the square root of dimensional with round value of log . This increases diversity and minimizes calculation head of the algorithm

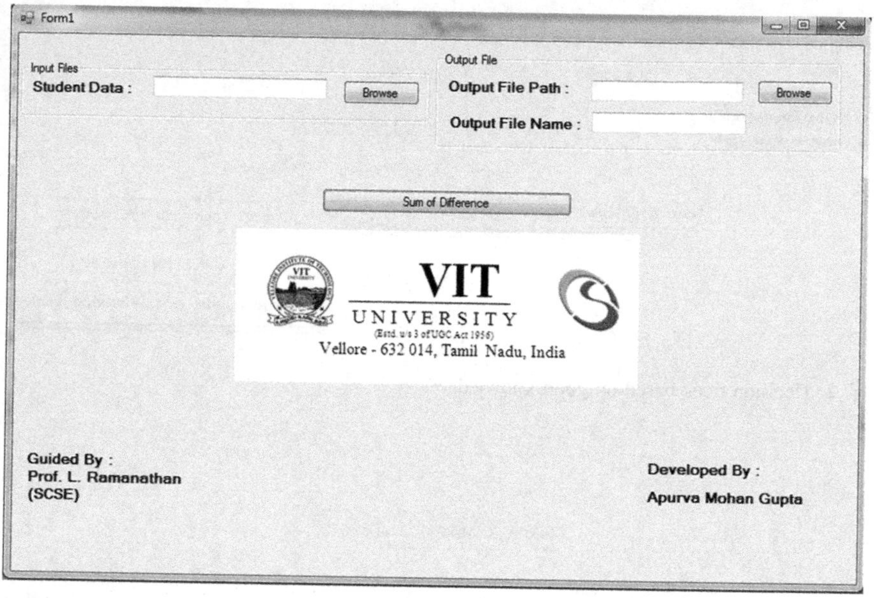

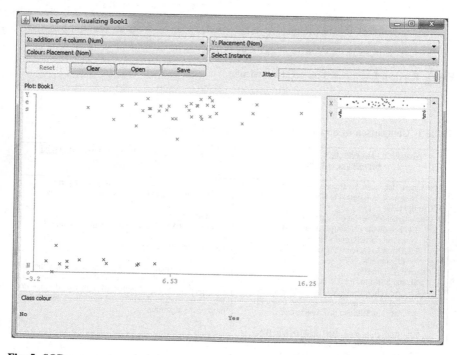

Fig. 4 User interface screen

Fig. 5 SOD-score versus placement

4 Implementation

Implementation is done using C# a programming language as front end. The C# is used to create a user interactive page and MS Visual Studio 2012 as the run time environment. First of all, run the application. A user interface screen will be generated as given figure below. In left top corner, a browse button is used for selecting the input file. If there is no file selected, a warning message will be displayed that select an input file. In right top corner, a browse button is used for set the path for output file. Below of it, there is a field for set the name for output field. After selecting input file, chose the path for the output file where user want to save the output file. User can give the name for the output file. If user does not provide name and set the path for the output file, output file will be save in C drive by default. After selecting the input file and set the path for the output file, a notepad file is generated which contains the predicted result.

5 Result

In above figure, X-axis shows SOD value and Y-axis shows the placements value (yes, no).in graph. Red cross indicates yes value and blue cross shows no value of student's placement. It is obvious from the figure, when the SOD value increase till 4; it indicates no value of placement. When the value goes high than 4, it indicates yes value of placement. So we can say that if the SOD of a student is below 4, he will not place, if no then he will get placed (Figs. 4 and 5).

The paper deals with three algorithms for comparison of prediction that lead to factual figures. The algorithms used are decision tree and random forest based on knowledge base. As a whole, the proposed algorithm gives efficient results on comparison with other algorithms.

6 Conclusion

In this paper, SOD method has been applied which yield the objective and retrieved the outlines from the given dataset. The paper presented a proposed model using SOD method to ascertain maximum necessary attributes between items in a given dataset. SOD found a pattern from the given dataset with statistical analysis. The paper simplified the value of attributes, because it is hard to deal with large value, and then normalize to find a pattern.

The proposed model efficiently predicts the placement of student. This is obvious that placement is not so easy to predict because it depends on many attributes, even we have considered four attributes. The paper selected the best combination of attributes. This combination works well for given dataset. From the given dataset, the results given are quite enough to predict the placement status of student. Finally, our proposed model with SOD is found to be more admirable in terms of efficiency. However, future research may focus more attributes with different dataset and other clustering algorithms.

Acknowledgments The author thankfully acknowledges the Apurva Mohan Gupta and others for their support for the completion of the manuscript.

References

1. Molina, M. M., Luna, J. M., Romero, C., & Ventura, S.: Meta-learning approach for automatic parameter tuning: a case of study with educational datasets. In Proceedings of the 5th international conference on educational data mining. pp. 180–183, (2012).
2. Pardos, Z. A., Wang, Q. Y., & Trivedi, S.: The real world significance of performance prediction. (2012).
3. Thai-Nghe, N., Horváth, T., & Schmidt-Thieme, L: Factorization models for forecasting student performance. In Proceedings of the 4th international conference on educational data mining. pp. 11–20, (2011).
4. B. Sen, E. Ucar, D. delen: Predicting and analyzing secondary education placement-test scores. (2012).
5. Baker, R. S. J. D., Gowda, S. M., & Corbett, A. T.: Automatically detecting a student's preparation for future learning: Help use is key. In Proceedings of the 4th international conference on educational data mining. pp. 179–188, (2011).
6. Pardos, Z. A., Gowda, S. M., Baker, R. S. J. D., & Heffernan, N. T.: Ensembling predictions of student post-test scores for an intelligent tutoring system.. In Proceedings of the 3rd international conference on educational data mining pp. 299–300, (2011).
7. Marquez-Vera, C., Romero, C., & Ventura, S.: Predicting school failure using data mining. In Proceedings of the 4th international conference on educational data mining. pp. 271–275, (2011).
8. J. Akcapinar, G., Cosgun, E., & Altun, A.: Prediction of perceived disorientation in online learning environment with random forest regression. In Proceedings of the 4th international conference on educational data mining, pp. 259–263, (2011).
9. Schoor, C., & Bannert, M.: Exploring regulatory processes during a computer-supported collaborative learning task using process mining. Computers in Human Behaviour, 28(4), pp. 1321–1331, (2012).
10. Wang, Y., & Heffernan, N. T.: Leveraging first response time into the knowledge tracing model. In Proceedings of the 5th international conference on educational data mining. pp. 176–179, (2012).
11. Gowda, S. M., Rowe, J. P., Baker, R. S.J. D., Chi, M., & Koedinger, K. R.:Improving models of slipping, guessing, and moment-by-moment learning with estimates of skill difficulty. In Proceedings of the 4th international conference on educational data mining. pp. 199–208, (2011).

12. Swarnalatha P., Tripathy B.K.: A Comparative Study of RIFCM with Other Related Algorithms from Their Suitability in Analysis of Satellite Images Using Other Supporting Techniques. Kybernetes, Emerald Publications.vol. 43, No 1, pp. 53–81, (2014).
13. Tripathy B.K., Rohan Bhargava, Anurag Tripathy, Rajkamal Dhull, Ekta Verma, Swarnalatha P.: Rough Intuitionistic Fuzzy C-Means Algorithm and a Comparative Analysis. In: Proceedings of ACM Compute-2013. pp. 21–22, (2013).

Robust Color Image Multi-thresholding Using Between-Class Variance and Cuckoo Search Algorithm

Venkatesan Rajinikanth, N. Sri Madhava Raja
and Suresh Chandra Satapathy

Abstract Multi-level image thresholding is a well known pre-processing procedure, commonly used in variety of image related domains. Segmentation process classifies the pixels of the image into various group based on the threshold level and intensity value. In this paper, colour image segmentation is proposed using Cuckoo Search (CS) algorithm. The performance of the proposed technique is validated with the Bacterial Forage Optimization (BFO) and Particle Swarm Optimization (PSO). The qualitative and quantitative investigation is carried out using the parameters, such as CPU time, between-class variance value and image quality measures, such as Mean Structural Similarity Index Matrix (MSSIM), Normalized Absolute Error (NAE), Structural Content (SC) and PSNR. The robustness of the implemented segmentation procedure is also verified using the image dataset smeared with the Gaussian Noise (GN) and Speckle Noise (SN). The study shows that, CS algorithm based multi-level segmentation offers better result compared with BFO and PSO.

Keywords Color image segmentation · Otsu's function · Cuckoo search · Gaussian noise · Speckle noise

V. Rajinikanth (✉) · N. Sri Madhava Raja
Department of Electronics and Instrumentation Engineering,
St. Joseph's College of Engineering, Chennai 600119, Tamilnadu, India
e-mail: rajinikanthv@stjosephs.ac.in

S.C. Satapathy
Department of Computer Science and Engineering, Anil Neerukonda Institute
of Technology and Sciences, Visakhapatnam 531162, Andhra Pradesh, India

© Springer India 2016
S.C. Satapathy et al. (eds.), *Information Systems Design and Intelligent
Applications*, Advances in Intelligent Systems and Computing 433,
DOI 10.1007/978-81-322-2755-7_40

379

1 Introduction

Image segmentation is a preliminary image processing procedure, largely employed to extract essential information from gray scale and color images in various fields [1–4]. Multi-level thresholding is one of the image segmentation techniques, extensively considered to split an image into multiple regions or objects in order to discover and interpret any meaningful information within the image.

In the multi-level thresholding process, a threshold value (T) is chosen using a favorite signal processing scheme, which separates the image into various clusters. For RGB images, finding the best possible threshold (T), which separates the image into foreground and background, remains a really significant step in image segmentation. Comprehensive evaluations on existing thresholding procedures can be found in the literature [5, 6].

In this paper, Otsu's function based global thresholding scheme is considered for the multi-level segmentation of 512 × 512 sized RGB image dataset. The combination of Otsu's between-class variance and heuristic methods, such as Particle Swarm Optimization (PSO) [7, 8], Bacterial Foraging Optimization (BFO) [9], Firefly Algorithm [10, 11], and Cuckoo Search (CS) [12, 13] have been presented in the literature. In this work, CS algorithm based RGB image segmentation is attempted and compared with the PSO and BFO using familiar image quality measures, such as Mean Structural Similarity Index Matrix (MSSIM), Peak Signal to Noise Ratio (PSNR), Normalized Absolute Error (NAE) and Structural Content (SC) [14, 15].

The robustness of the proposed multi-level segmentation is assessed on the RGB image dataset with an introduced Gaussian Noise (GN) and Speckle Noise (SN) for $T = \{2, 3, 4, 5\}$. Results of this study confirm that, multi-level segmentation based on CS algorithm offers better result compared with the alternatives.

2 Multi-level Thresholding

Otsu's between-class variance based image thresholding was primarily proposed back in 1979 [16]. This procedure offers the optimal threshold of a given image by maximizing the Otsu's function. This method already demonstrated its effectiveness on grey scale [7–10] and color images [11].

In this paper, Otsu's between-class variance is considered for color image segmentation by the assistance from its RGB histogram. In RGB space, each color pixel of the image is a mixture of Red, Green, and Blue (RGB) and for that same image, the data space size is $[0, L - 1]^3$ $(R = [0, L - 1], G = [0, L - 1],$ and $B = [0, L - 1])$. A detailed description of RGB image segmentation using Otsu can be found in [2, 11].

The multi-level segmentation proposed in this work is to search for an optimal value of t_j^C, which maximizes the objective function (J_{max}) of each image component $C = \{R, G, B\}$ defined as;

$$\varphi^C = \max_{1 < t_i^C <,...,L-1} \sigma_B^{C^2}(t_j^C) \qquad (1)$$

Due to its complexity, solving the RGB segmentation problem may require a larger computation time compared with the gray scale images. Hence, heuristic algorithm assisted segmentation procedures are used as alternatives for the existing traditional analytical procedures [11].

3 Overview of Cuckoo Search Algorithm

Cuckoo Search (CS) algorithm was originally proposed in 2009, by mimicking the breeding tricks of parasitic cuckoos [17, 18]. Due to its competence, CS was adopted by the researchers to solve the multi-level segmentation problems for gray scale images using the Otsu [12], Tsallis [13] and Kapur [19] function.

The mathematical expression of the CS considered in this study is given below:

$$X_i^{(t+1)} = X_i^{(t)} + \alpha \oplus Levy(\lambda) \qquad (2)$$

where $X_i^{(t)}$ is the initial position, $X_i^{(t+1)}$ is the updated position, α is chosen as 1.2 and $\oplus$ is the symbol for entry wise multiplication.

In this work, Levy Flight (LF) based search methodology is considered to update the position of the agents. Detailed description about LF can be accessible from [17, 18]. It is a random walk in which the search steps can be drawn using the following Levy distribution [12]:

$$Levy \sim u = t^{-\lambda} \quad for \quad (1 < \lambda \leq 3) \qquad (3)$$

In this work, the CS based optimization search is adopted to find the optimal thresholds for 512×512 sized RGB image dataset and its performance was compared with the PSO [8] and enhanced BFO [8] algorithms existing in the literature.

4 Result and Discussions

This section presents the experimental results obtained for Original Image (OI) and image with noise.

In the proposed method, the following algorithm parameters are assigned during the heuristic search: Number of agents (N) = 15; dimension of search = T; iteration number (t) = 500; stopping criteria = J_{max}. In order to get the best possible threshold value, the experiment is repeated 30 times for every image with a preferred threshold (T) and the mean value among the trials is considered as the optimal value. The multi-thresholding procedure is implemented on RGB image dataset available at [20].

In this study, 512 × 512 sized RGB images, such as Aerial, Bridge, Cactus, and Geckos are chosen for the study. In order to investigate the robustness of the segmentation scheme, these images are stained with Gaussian Noise (white noise of mean 0 and variance 0.01) and Speckle Noise (regularly scattered random noise with mean 0 and variance 0.05) [21, 22]. The performance measures, such as MSSIM [15], NAE, SC, and PSNR [14] are chosen to confirm the superiority of the segmentation process.

The image dataset and the corresponding histograms are presented in Table 1. In these histograms, the x-axis represents the RGB level and the y-axis denotes the

Table 1 Image dataset and the corresponding RGB histograms

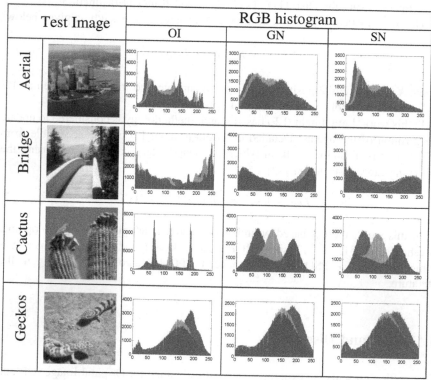

pixel level. Due to the noise, the pixel level of the image is significantly changed and its histogram is completely dissimilar compared with the original histogram of the image. This table also shows that, the impact of GN is more on the histogram compared with SN.

Multi-level segmentation is initially performed on the Aerial image using the CS algorithm for $T = \{2, 3, 4, 5\}$ and the results are presented in Tables 2 and 3. Table 2 shows thresholded images and its histograms for $T = \{2, 3, 4, 5\}$ for OI, GN and SN. Later, the segmentation procedure is carried out using PSO and BFO algorithms for all images with a chosen threshold value.

From Table 3, it can be noted that, CS algorithm offers better MSSIM compared with the PSO and BFO. The computation time (CPU time in secs) of the algorithms are computed using Matlab's tic-toc function. The average CPU time (mean $\pm$ standard deviation) obtained with various algorithms for the image dataset is as follows; $OI_{CS} = 41.28 \pm 0.06$; $GN_{CS} = 55.64 \pm 0.02$; $SN_{CS} = 54.05 \pm 0.04$;

Table 2 Segmented aerial image and optimal thresholds for T = 2–5

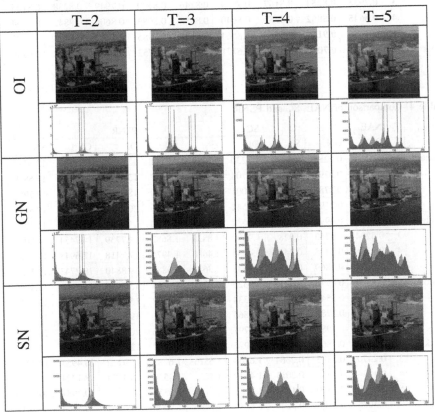

Table 3 Comparison of MSSIM value for CS, PSO and BFO

Image	T	MSSIM								
		CS			PSO			BFO		
		OI	GN	SN	OI	GN	SN	OI	GN	SN
Aerial	2	**0.3875**	**0.4208**	**0.3750**	0.3749	0.3995	0.3816	0.3845	0.4187	0.3738
	3	**0.5095**	**0.6190**	**0.5565**	0.4996	0.6016	0.5338	0.4996	0.6128	0.5526
	4	**0.6604**	**0.7818**	**0.6743**	0.6472	0.7822	0.6712	0.6574	0.7805	0.6711
	5	**0.7767**	**0.8651**	**0.7721**	0.7694	0.8557	0.7704	0.7711	0.8628	0.7694
Bridge	2	**0.4109**	**0.3452**	**0.3076**	0.4086	0.3431	0.3048	0.4085	0.3450	0.3038
	3	**0.6020**	**0.5816**	**0.5561**	0.5993	0.5784	0.5547	0.6007	0.5784	0.5527
	4	**0.6945**	**0.7177**	**0.7056**	0.6904	0.7150	0.6991	0.6883	0.7143	0.7039
	5	**0.7784**	**0.8193**	**0.7865**	0.7726	0.8117	0.7859	0.7715	0.8159	0.7857
Cactus	2	**0.5388**	**0.4933**	**0.4937**	0.5346	0.4895	0.4904	0.5362	0.4896	0.4912
	3	**0.6750**	**0.7399**	**0.6967**	0.6733	0.7364	0.6928	0.6738	0.7374	0.6948
	4	**0.8663**	**0.7773**	**0.8154**	0.8649	0.7751	0.8122	0.8658	0.7758	0.8128
	5	**0.8904**	**0.9074**	**0.8786**	0.8884	0.9012	0.8775	0.8895	0.9006	0.8792
Geckos	2	**0.5874**	**0.6361**	**0.6472**	0.5868	0.6349	0.6468	0.5864	0.6358	0.6375
	3	**0.7953**	**0.8254**	**0.8167**	0.7939	0.8246	0.8159	0.7949	0.8238	0.8125
	4	**0.8645**	**0.8862**	**0.8751**	0.8640	0.8858	0.8683	0.8606	0.8842	0.8694
	5	**0.8696**	**0.9169**	**0.9038**	0.8664	0.9106	0.8985	0.8677	0.9174	0.8997
Avg		**0.6817**	**0.7083**	**0.6788**	0.6771	0.7028	0.6752	0.6785	0.7058	0.6756

Table 4 Comparison of NAE, SC and PSNR

	T	NAE			SC			PSNR		
		OI	GN	SN	OI	GN	SN	OI	GN	SN
Aerial	2	0.5164	0.5284	0.5447	2.7230	2.9429	2.8511	13.2264	12.9622	12.8215
	3	0.4083	0.3791	0.4040	1.9708	1.9367	1.9623	15.5972	15.9728	15.5859
	4	0.2947	0.2788	0.3142	1.6001	1.6237	1.6310	18.2926	18.6653	17.7961
	5	0.2226	0.2195	0.2479	1.4542	1.4620	1.4746	20.5815	20.7871	19.8640
Bridge	2	0.4982	0.4998	0.4973	2.8637	2.9561	2.9650	11.1224	11.0061	11.1821
	3	0.3294	0.3253	0.3289	1.9000	1.8820	1.8682	14.7739	14.7454	14.8069
	4	0.2346	0.2355	0.2468	1.4634	1.4688	1.4975	17.7418	17.6948	17.5528
	5	0.1837	0.1869	0.1978	1.3538	1.3627	1.3888	19.8840	19.7743	19.5337
Cactus	2	0.4663	0.4974	0.5217	2.8323	2.8786	2.9745	13.3915	12.7762	12.4749
	3	0.3878	0.3543	0.3663	2.1334	2.0065	2.0274	15.1963	15.7854	15.5915
	4	0.2308	0.2783	0.2681	1.5865	1.7020	1.6211	19.1970	17.9014	18.1166
	5	0.2104	0.1988	0.2076	1.4797	1.4480	1.4526	20.4207	20.6823	20.3033
Geckos	2	0.3724	0.3863	0.4135	2.1925	2.1563	2.2322	12.3545	11.8476	11.3528
	3	0.2470	0.2585	0.2664	1.6208	1.6516	1.6783	15.9112	15.4953	15.3207
	4	0.1834	0.1911	0.2011	1.4044	1.4255	1.4538	18.4774	18.1724	17.8579
	5	0.1529	0.1510	0.1584	1.3132	1.3134	1.3336	20.0463	20.2981	20.0170
Avg		0.3086	0.3106	0.3240	1.8682	1.8886	1.9008	16.6384	16.5354	16.2611

$OI_{PSO} = 65.75 \pm 0.04$; $GN_{PSO} = 58.11 \pm 0.03$; $SN_{PSO} = 61.31 \pm 0.05$; $OI_{BFO} = 65.28 \pm 0.05$; $GN_{BFO} = 63.93 \pm 0.04$; $SN_{BFO} = 61.25 \pm 0.02$. Hence, this study also confirms that, due to the LF strategy, CS offers smaller CPU time compared to PSO and BFO algorithms.

Table 4, presents the image quality measures, such as NAE, SC and PSNR for OI, GN and SN attained with CS algorithm. From Tables 3 and 4, it can be observed that, the average (Avg) values of the image quality measures are approximately similar for the OI, GN and SN. This confirms that, the proposed image segmentation procedure is robust and offers approximately similar result for the clear and noisy images.

5 Conclusions

In this paper, a multi-level thresholding is presented for RGB image dataset using CS and Otsu's function. This work finds the optimal threshold for a chosen image with a chosen T value. The proposed segmentation technique is compared with other heuristic algorithms, such as PSO and BFO and the performance is using image quality measures, such as CPU time, MSSIM, NAE, SC and PSNR. The robustness of the segmentation scheme is also verified by considering GN and SN corrupted image dataset. The experimental results confirm that the CS assisted segmentation procedure offers better results on the original and noise corrupted image dataset.

References

1. Larson, E. C., Chandler, D. M.: Most apparent distortion: Full-reference image quality assessment and the role of strategy, Journal of Electronic Imaging, 19 (1), Article ID 011006 (2010).
2. Ghamisi, P., Couceiro, M. S., Martins, F. M. L., and Benediktsson, J. A.: Multilevel image segmentation based on fractional-order Darwinian particle swarm optimization, IEEE Transactions on Geoscience and Remote sensing, 52(5), pp. 2382–2394, (2014).
3. Kalyani Manda, Satapathy, S. C., Rao, K. R.: Artificial bee colony based image clustering, In proceedings of the International Conference on Information Systems Design and Intelligent Applications 2012 (INDIA 2012), Advances in Intelligent and Soft Computing, 132, pp. 29–37, (2012).
4. Manickavasagam, K., Sutha, S., Kamalanand, K.: Development of Systems for Classification of Different Plasmodium Species in Thin Blood Smear Microscopic Images, Journal of Advanced Microscopy Research, 9, (2), pp. 86–92, (2014).
5. Sezgin, M., Sankar, B.: Survey over Image Thresholding Techniques and Quantitative Performance Evaluation, Journal of Electronic Imaging, 13(1), pp. 146– 165, (2004).
6. Tuba, M.: Multilevel image thresholding by nature-inspired algorithms: A short review, Computer Science Journal of Moldova, 22(3), pp. 318–338, (2014).
7. Akay, B.: A study on particle swarm optimization and artificial bee colony algorithms for multilevel thresholding, Applied Soft Computing, 13 (6), pp. 3066–3091, (2013).

8. Rajinikanth, V., Sri Madhava Raja, N., Latha, K.: Optimal Multilevel Image Thresholding: An Analysis with PSO and BFO Algorithms, Aust. J. Basic & Appl. Sci., 8(9), pp. 443–454, (2014).
9. Sathya, P. D., Kayalvizhi, R.: Modified bacterial foraging algorithm based multilevel thresholding for image segmentation, Engineering Applications of Artificial Intelligence, 24, pp. 595–615, (2011).
10. Raja, N. S. M., Rajinikanth,V., Latha, K.: Otsu Based Optimal Multilevel Image Thresholding Using Firefly Algorithm, Modelling and Simulation in Engineering, vol. 2014, Article ID 794574, p. 17, (2014).
11. Rajinikanth, V., Couceiro, M. S.: RGB Histogram Based Color Image Segmentation Using Firefly Algorithm, Procedia Computer Science, 46, pp. 1449–1457, (2015). doi:10.1016/j. procs.2015.02.064.
12. Abhinaya, B., Raja, N. S. M.: Solving Multi-level Image Thresholding Problem—An Analysis with Cuckoo Search Algorithm, Information Systems Design and Intelligent Applications, Advances in Intelligent Systems and Computing, 339, pp. 177–186, (2015).
13. Agrawal, S., Panda, R., Bhuyan, S., Panigrahi, B. K.: Tsallis entropy based optimal multilevel thresholding using cuckoo search algorithm, Swarm and Evolutionary Computation, 11, pp. 16–30, (2013).
14. Grgic, S., Grgic, M., Mrak. M.: Reliability of objective picture quality measures, Journal of Electrical Engineering, 55(1–2), pp. 3–10, (2004).
15. Wang, Z., Bovik, A. C., Sheikh, H. R., Simoncelli, E.P.: Image Quality Assessment: From Error VisibilitytoStructural Similarity, IEEE Transactions on Image Processing, 13(4), pp. 600– 612, (2004).
16. Otsu, N.: A Threshold selection method from Gray-Level Histograms, IEEE T. on Systems, Man and Cybernetics, 9(1), pp. 62–66, (1979).
17. Yang, X. S., Deb, S.: Cuckoo search via Lévy flights. In: Proceedings of World Congress on Nature and Biologically Inspired Computing (NaBIC 2009), pp. 210–214,. IEEE Publications, USA (2009).
18. Yang, X. S: Nature-Inspired Metaheuristic Algorithms, Luniver Press, Frome, UK, 2008.
19. Brajevic, I., Tuba, M., Bacanin, N.: Multilevel image thresholding selection based on the Cuckoo search algorithm. In: Proceedings of the 5th International Conference on Visualization, Imaging and Simulation (VIS'12), pp. 217–222, Sliema, Malta (2012).
20. http://vision.okstate.edu/?loc=csiq.
21. Oliva, D., Cuevas, E., Pajares, G., Zaldivar, D., Perez-Cisneros, M.: Multilevel Thresholding Segmentation Based on Harmony Search Optimization, Journal of Applied Mathematics, vol. 2013, Article ID 575414, p. 24, (2013).
22. Oliva, D., Cuevas, E., Pajares, G., Zaldivar, D. andOsuna, V.: A Multilevel Thresholding algorithm using electromagnetism optimization, Neurocomputing, 139, pp. 357–381, (2014).

Social Media: A Review

Gauri Jain and Manisha Sharma

Abstract This paper is an insight about the online social media which is the most common form of media now a days and is been used most widely. It throws light on types of social media, various types of users and its major functional blocks.

Keywords Social media · Social media users · Functional blocks

1 Introduction

During the last decade, rapid developments have taken place in the content of world-wide web and the usage of internet. With the increase in internet usage, data transfer speeds and advancement of mobile devices, instant messaging, ecommerce, internet banking and social media have become part of the daily life of the urban population all over the world.

The objective of this paper is to provide an insight about the various types of online social media, its users and the functional blocks.

During the last 5 years social media has evolved a great deal (from a monologue driven blogs) and has become far more interactive, dynamic and different than the traditional media like TV, movies and newspapers etc. In traditional media, "experts" or "authorities" decide which information should be produced and how it should be distributed and to whom. The information creation and delivery is one way, from centralized producers to masses. Majority of users in this set-up do not take part in content creation or delivery process.

In social media, the main differentiator is that the users can be a consumer and as well as a producer. At any instant, millions of users actively participate on various

G. Jain (✉) · M. Sharma
Banasthali Vidhyapith, Banasthali, India
e-mail: jain.gauri@gmail.com

M. Sharma
e-mail: manishasharma8@gmail.com

© Springer India 2016
S.C. Satapathy et al. (eds.), *Information Systems Design and Intelligent Applications*, Advances in Intelligent Systems and Computing 433,
DOI 10.1007/978-81-322-2755-7_41

social media channels and they share content on the topics of their interest. Ability of sharing content and network in real-time has been boosted the usage of social media and helped it succeed. Below shows some interesting statistics about the usage of social media [1]:

- 40 million photos are uploaded to Instagram per day (Instagram)
- 2013: 24 % of online teens use Twitter, up from 16 % in 2011 (Pew Internet Research)
- 1 in 10 young people rejected for a job because of their social profile (on device)
- 751 million monthly active Facebook mobile products users (Facebook).

2 Social Media

Oxford dictionary defines social media as "Websites and applications that enable users to create and share content or to participate in social networking". Technically speaking, Social Media is a group of Internet based applications that builds on the ideological and technological foundations of web 2.0, and that allows the creation and exchange of user generated Content [2]. In simple words, social media website allow user to register and create his/her own profile, adds friends to their circle and interact with them on a regular basis. Users thus in a way generate their social circle in a form of a graph where nodes are the different users and arcs is the relationship between the users.

3 Types of Social Media

At present, social network sites like facebook, Linkedin are more popular current forms of social media. Figure below gives an indicative comparison of popularity of various social media channels [3] (Fig. 1).

However, in addition to the above channels, there are many types of online platforms within the vast domain of social media. The section below gives an overview of the social media platforms.

3.1 Collaborative Projects

Collaborative Projects bring together different types of users having similar interests and they work together as a team. The main aim of these types of projects is to work efficiently as a team. Collaborative projects differ mainly in two ways one in which individual user edits the already posted text according to his knowledge like wikis. Second type of collection is one in which group of users rate the collection of links

Fig. 1 Social media site usage

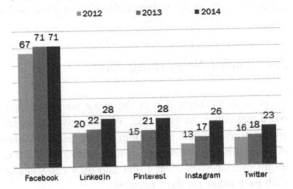

Social media sites, 2012-2014

% of online adults who use the following social media websites, by year

2012   2013   2014

Pew Research Center's Internet Project Surveys, 2012-2014. 2014 data collected September 11-14 & September 18-21, 2014. N=1,597 internet users ages 18+.

PEW RESEARCH CENTER

or media contents like social bookmarking sites. One of the most popular collaborative projects is online encyclopedia Wikipedia which is currently available in more than 230 languages, reaching out to millions of people online. Similarly, some of the most popular bookmarking sites are Delicious, Pinterest etc.

3.2 Blogs

Blogs, which represent the earliest form of Social Media, are special types of websites that usually display date-stamped entries in reverse chronological order [4]. Basically, controlled by an individual, blogs can be commented on by other users. It started with individuals sharing his/her personal feelings and now has taken a form where many companies are even using to get feedback on their products, product pre-launch feedback, employee satisfaction survey; common people are using it for social protests etc. Most popular media used in blogs is text.

3.3 Content Communities

Content Communities form the platform for users to come together and share different type media content ranging from text, images, videos etc. The content on these communities attract the users because of the ease of access of these communities across millions of users. Some very popular content communities are YouTube, Slideshare etc. These communities are platform for many companies to share information since it is access by majority number of users. Many users use it as a stage for marketing, obtaining popularity. For e.g. many musicians are releasing their music videos online using some of these communities.

3.4 Social Networking Sites

As soon as we come across the name social networking sites Facebook is the first word that comes to our mind. We can say that social networking sites are the sites where users connect with other online users by inviting or accepting their invitations. In this way a network of user is formed. Each user has its own network which might overlap with other users and they can share their ideas, pictures, videos etc with each other on their network.

3.5 Virtual Game World

Virtual worlds are platforms that replicate a three dimensional environment in which users can appear in the form of personalized avatars and interact with each other as they would in real life [2]. These social media simulate the environment and behave according to the user who is in the virtual world. Here users from all over the world can participate in the virtual game simultaneously and they form its integral part over a period of time. There are very strict rules users have to abide to be a part of virtual game world. One such virtual game world is Sonys playstation which is very popular amongst younger generation. One main disadvantage of this is indulging too much in these activities may affect your original personality.

3.6 Virtual Social Worlds

These types of virtual world are similar to virtual game world but differ in the sense that there are no strict rules according to which users are required to behave. Each user can choose their personality according to their own personal choice.

They appear in virtual 3D environment where only some basic physical restrictions like gravity apply. The most prominent example of virtual social worlds is the Second Life application founded and managed by the San Francisco-based company Linden Research Inc. They have virtual money, virtual saleable items, houses etc. They are platform for marketing for many companies, research for many organizations.

4 Types of Social Media Users

4.1 The Quiet Follower

This type of users silently observes your activities but does not interact with you. They are interested only in certain type of material. For e.g. a Facebook user who has liked your post but has not commented on your posts. You have to make an effort in order to get response from them. Effort may be in the form of some interesting posts or status which might provoke him to respond. These types of users helps to increase your visibility since your posts are shown in their news feeds.

4.2 The Casual Liker

This type of user is little more active then the quiet follower. He likes or shares your posts occasionally so we should start encouraging these types of users so that their interest in your activities and posts increases. This will in turn enhance your visibility and therefore good for a company who is marketing about its product through this media.

4.3 The Deal Seeker

These types of social media users are on lookout for online deals, discount coupons, contests, promotional activities etc. A whole lot of users come under this category. They follow the users who provide them with good online deals. As a result they share your posts and tweets in large number and in a way more and more people are able to view your deals and products.

4.4 The Unhappy Customer

Social media is a strong platform for unhappy customers to vent out their feelings. These types of users may write a complaint about your company or product on social networking sites. It is very easy to harm the reputation of the company in front of loads of users on the network, so it is very essential that complaints be addressed in proper manner and as quickly as possible.

On the other hand, sometimes company is not ready to listen to customers complaint directly, so when a user complaints via this platform they may come across other users having similar complaints. They might together put forwards their complaints more strongly then as an individual. This way it might be useful for customers.

4.5 The Negative Detractor/Ranter

These types of social media users uselessly posts complaints or text that is not in any way related to the companys business just to harm companys reputation or for their own satisfaction. While tackling these types of users the companys representative needs to be cautious enough so as to not harm companys reputation. Best way is to ignore these people or to give diplomatic replies.

4.6 The Cheer Leader

These are ideally the best users. They are users with a positive mind and perspective. They would not post any offensive material and would encourage all your posts by sharing, liking or commenting on them. They are best people to get you some publicity among their network.

5 Functional Blocks

5.1 Identity

The identity block refers to the amount or extent to which users put forward their information on social media sites. This information may be personal details like name, age, date of birth, gender, profession, location and also other personal information that can be implied from details like the hobbies, the group that he/she has joined. For example, on Facebook users have the ability to like and dislike post, such as actor/actresses or hobbies, from which other people might have some idea about the user interests and personality.

Sometime users use fake identity for their social presence, which might be because they want to hide their real self. It is very important to carefully choose the information one wants to share and the one he didnt want, since it might lead to compromising ones privacy (Fig. 2).

Fig. 2 Social media site usage

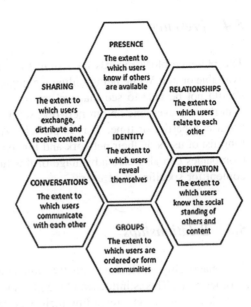

5.2 Conversation

In this functional block the social media user chooses the setting according to which he/she wants to converse with other users on his network. Different users have different motive of using the media. Some might used it as leisure activity for passing time by making new friends, other might communicate seeing it as a business opportunity or socially responsible people might view it as a platform for spreading some important message to the masses. Companies might saw the social media as a way to communicate with individual customer. This might also help in the marketing of their products. Based on this users might choose to share their post to a special group of people by default unless until they specify or company might choose to share their posts with their top 20 clients only or otherwise.

5.3 Sharing

The objective of sharing block is to exchange or broadcast the content over the network by a user. It might range from sharing your views, ideas, music or some web page that he liked and want others to see it. Sharing helps us to know about the likes and dislikes of other users. It is very important in a way that it helps strengthens the bond between the various users of the social media which might not be possible when interacting face to face. Sharing might be seen as a way of publicity for companies who are using online marketing strategies.

5.4   Presence

The presence block regards the extent to which users can know if other users are available or not. While being online we might want that other users on our network should not be able to see that we are online. This might be we dont want to be disturbed by everyone on our network, we just want to converse with some specific users at a particular time. This is particularly the case with the users who have a long list of users on their network and for some of them he is a quiet observer only. Companies can also take advantage of this functional block by displaying specific content for specific set of users.

5.5   Relationships

The relationship block represents the amount that users can relate to and are related to other users. This is influenced by and comes as a result of the information that is provided by other users regarding their personal information, likes and dislikes, and allows them to potentially meet people that they would essentially like to meet. The relationship block is made up of different aspects of social media sites where relationships are essential, such as skype, that allows users to communicate with and talk to people that they already know, or sites like Twitter, where often users do not know each other, and relationships hardly matter.

5.6   Reputation

The reputation is the functional block that represents the user identity in the eyes of other users. Reputation of a particular user is built on the basis of their personal information, the content of their posts, the time difference between consecutive posts and how often they share the content of other user. The companies may like that their content be shared by top clients of their companies so that they get more and more exposure from the trusted clients. The users with bad reputation may further harm companys reputation. The posts with better reputation may get more shares and more likes than otherwise.

5.7   Groups

The group functional block is the formation of communities by like minded users or the users having similar interest. These groups may share the information pertaining about a specific topic of their interests. They may join together to raise their concern about a particular social problem. The group may have a moderator who control and monitor group activities but this may not be true with each group. Un-moderated groups may also occur. Users can form group amongst their network also. They might want to differentiate their friends from their family members. Company can form a group of employees discussing about their upcoming product and pre-launch product reviews are good feedback for the product.

6 Conclusion

Social media has become an important part of the life of urban population. While social media is about creating and sharing information, they are also being used as a means to keep up with friends and relatives, especially those who are living far away or across the time zones.

Further, majority of users focus on sharing personal information, the large user base is enabling a social impact through real time sharing of social, natural and political events, news, and marketing information. Some of these have created a revolutionary impact on politics and business of the world, which sometimes is positive and at others negative.

Today, if an urban adult is not using social media, then S/he is missing out on critical personal and geo-political updates. However, while social media has beneficial aspects it can also have negative outcome if used by socially irresponsible people or groups as details of personal lives and personalities are accessible to a larger group of people. Hence, social media needs to be analyzed and one should choose the type of social media carefully and carefully follow the rules of privacy and social etiquette to reap the benefits.

References

1. Why is Social Media So Addictive, http://www.cmswire.com/cms/customer-experience/why-social-media-is-so-addictive-and-why-marketers-should-care-022276.php.
2. Kaplan and Haenlein, 2010, A. Kaplan, M. Heinlein, Users of the world, unite! The challenges and opportunities of social media, Business Horizons, 53 (1) (2010), pp. 5968.
3. Social Media Update 2014, http://www.pewinternet.org/2015/01/09/social-media-update-2014/.
4. OECD (2007), Participative Web and User-Created Content: Web 2.0, Wikis and Social Networking. OECD Publishing, Paris.

Bayesian-Fuzzy GIS Overlay to Construe Congestion Dynamics

Alok Bhushan Mukherjee, Akhouri Pramod Krishna
and Nilanchal Patel

Abstract Complex systems such as transportation are influenced by several factors which are subjugated by uncertainty and therefore, it has non-linear characteristics. Consequently, transition in congestion degree from one class to another class can be abrupt and unpredictable. If the possibility of transition in congestion degree from one class to another can be determined, then adequate measures can be taken to make transportation system more robust and sustainable. The present paper demonstrates the efficacy of Bayesian-Fuzzy GIS Overlay to construe congestion dynamics on a test data set. The test data set consists of congestion indicators such as Average Speed (AS) and Congestion Index Value (CIV) of different routes for the test study area. The results succeeded in representing the probable transition in congestion degree from one class to another with respective Bayesian probabilities for different classes.

Keywords Complex system · Uncertainty · Bayesian-Fuzzy GIS overlay · Congestion index value · Transportation · Congestion dynamics

1 Introduction

Transportation is a complex system which is dependent on several factors ranging from cognitive behavior, environment, and land use pattern to factors residing in local domain. Largely aforementioned factors' behavior is uncertain in nature. Consequently solutions to transportation problems need to address the elements of uncertainty involved in the system [11]. However the term "uncertainty" is not easy

A.B. Mukherjee (✉) · A.P. Krishna · N. Patel
Department of Remote Sensing, Birla Institute of Technology, Mesra, Ranchi, India
e-mail: alokbhushan@bitmesra.ac.in

A.P. Krishna
e-mail: apkrishna@bitmesra.ac.in

N. Patel
e-mail: npatel@bitmesra.ac.in

© Springer India 2016
S.C. Satapathy et al. (eds.), *Information Systems Design and Intelligent Applications*, Advances in Intelligent Systems and Computing 433,
DOI 10.1007/978-81-322-2755-7_42

397

to define. There are different factors which may occur in future, and can significantly affect the operation of a complex system. Therefore the behavior of a complex system cannot be forecasted on the basis of present framework and hence makes it unpredictable in nature [1]. Research investigations pertaining to traffic congestion ranges from traffic flow analysis to location theory. Studies on traffic flow theory are basically falls in the realm of prediction modelling [7]. Urban traffic performance's improvement is a major challenge to the transport authorities across the world. Furthermore, congestion is a consequence of several factors and its characteristics is dependent on spatial-temporal factors. Therefore the nature of congestion can vary significantly with spatial-temporal variation. Different congestion indicators such as average speed, travel time, congestion index, time moving and proportion stopped time were explained by the authors in their study. Furthermore, utility of Global Positioning System (GPS) in the measurement of congestion was described in the investigation [3]. Investigation to understand small world characteristics of spatial networks using spatial autocorrelation techniques was performed by [17]. It was observed from the investigation that small world characteristics are dependent on network structure and its dynamics [17]. The probable reasons responsible for lethargic transport infrastructure in developing countries were analyzed by [4]. Further, outlined the fact that congestion across the major cities prevails and consequently, environmental degradation is alarming in developing countries [4]. The study by [2] outlined the limitations of models which were designed to express the spatial-temporal phenomena. For example, models were effective in representing the events when homogeneity exists in the data. However, it fails to represent scenario when there is heterogeneity in the relationship among the variables. Therefore, the authors proposed a model that can perform dynamic assessment of the heterogeneity in traffic data employing proposed STARIMA model and dynamic spatial weight matrix. Furthermore, investigations performed by [5, 6, 9, 13, 15, 16] focused on different dimensions of traffic flow analysis and travel behavior. However, none of the aforementioned investigations used the potential of Geographical Information System (GIS) in analyzing the transportation system despite the fact that transport is primarily a spatially dominant phenomenon. The aforementioned limitation was overcome by the studies of [8, 10, 12, 18] as they have employed the potential GIS technology in their investigations. However none of the above mentioned investigations attempted to assess the probable transition in congestion degree from one class to another. The abrupt transition from one class to another class can disturb the stability of the whole system. Consequently, it can create chaotic situation all across the transportation network from the perspective of smooth traffic flow.

Since there are several factors that can influence the characteristics of congestion and therefore, sudden change in the behavior of congestion cannot be ruled out. The present research investigation demonstrated the potential of Bayesian-Fuzzy GIS Overlay in modelling the probable transition of congestion from one state to another. The efficacy of the proposed technique was validated using a traffic data set consisting of factors such as Average Speed (AS) and Congestion Index Value (CIV).

2 Methodology

The research methodology of the present research investigation consists of three different levels. First, it determines probability of congestion for different classes i.e. High Congestion, High-Moderate Congestion, Moderate Congestion and Low Congestion for two different variables i.e. Congestion Index Value and Average Speed using Bayesian probability. Then peer layers of the two variables i.e. Congestion Index Value and Average Speed were integrated using GIS Weighted Sum Overlay to identify routes vulnerable to High Congestion, High-Moderate Congestion, Moderate Congestion and Low Congestion respectively. Finally, results obtained from the integration of peer layers of the variables i.e. Congestion Index Value and Average Speed were compared with fuzzy continuous membership values of the variables i.e. Congestion Index Value and Average Speed for different aforementioned classes to determine the possibility of transition of congestion degree from one class to another. The research methodology flowchart is shown in the Fig. 1.

2.1 Quantification of Probability of Different Congestion Classes Using Bayesian Probability [14]

Bayesian inferences enable us to use the past information (prior probabilities) in the current model and determine posterior probabilities. Moreover hypothesis is designed on the basis of data. Therefore prediction of some event under uncertain conditions become possible since it considers historical information. The strength of Bayesian probability lies in its capability of deriving information from available data based on condition rather than combinatorial pattern of factors. Therefore it succeeds representing the behavior of events in an uncertain environment.

The Bayes theorem is represented as follows:

$$P\left(\frac{Ak}{B}\right) = \frac{P(Ak \cap B)}{P(A1 \cap B) + P(A2 \cap B) + \cdots + P(Ak \cap B)} \tag{1}$$

where

$P(A_k \cap B) = P(A_k) * P\left(\frac{B}{Ak}\right)$

$P(A_k)$ Probability of kth event

B   Conditional statement imposed on the events

$P\left(\frac{B}{Ak}\right)$ Conditional Probability

Application of Bayesian theorem in the present investigation constitutes of following steps:

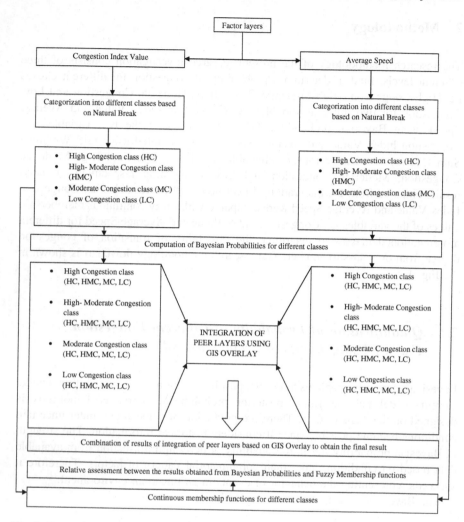

Fig. 1 Research methodology flowchart

- First, significant factors that can be instrumental in understanding the dynamics of congestion were identified and selected for further investigation. Selected factors were Congestion Index Value (CIV) and Average Speed (AS) respectively.
- The aforementioned selected factors were categorized into different equal interval Congestion classes' i.e. High Congestion, High-Moderate Congestion, Moderate Congestion and Low Congestion. The equal gap of Class Interval (CI) of the above mentioned classes was determined with the following equation:

$$EG = \frac{Max.(X) - Min.(X)}{\text{Total Number of Classes}} \qquad (2)$$

The range of different congestion classes are as follows:

$$\text{Range of Low Congestion Class (LC)} = \{l, l_1\} \qquad (3)$$

$$\text{Range of Moderate Congestion Class (MC)} = \{l_1, l_2\} \qquad (4)$$

$$\text{Range of High-Moderate Congestion Class (HMC)} = \{l_2, l_3\} \qquad (5)$$

$$\text{Range of High Congestion Class (HC)} = \{l_3, u\} \qquad (6)$$

where
X: It is the value of either CIV or AS
l: Min.(X)
EG: Equal gap value of class intervals
l_1: $l + EG$
l_2: $l_1 + EG$
l_3: $l_2 + EG$
u: Max.(X)

- Having categorized the variables i.e. CIV and AS into different classes' i.e. High Congestion, High-Moderate Congestion, Moderate Congestion and Low Congestion, the Prior Probabilities of different classes for each of the variables i.e. CIV and AS were determined. The prior probability of different classes' (P (C)) was determined with the following equations:

$$P(HC) = \frac{n_1}{N} \qquad (7)$$

$$P(HMC) = \frac{n_2}{N} \qquad (8)$$

$$P(MC) = \frac{n_3}{N} \qquad (9)$$

$$P(LC) = \frac{n_4}{N} \qquad (10)$$

where n_1, n_2, n_3, and n_4, are number of routes in High Congestion Class, High-Moderate Congestion Class, Moderate Congestion Class, Low Congestion class respectively and P (HC), P (HMC), P (MC), and P (LC) are the prior probabilities for High Congestion Class, High-Moderate Congestion Class, Moderate Congestion Class, Low Congestion class respectively, and N represents total number of routes considered for the study.

- Having determined the prior probabilities of different congestion classes', the Conditional Probabilities or Belief Measures of different classes' for each of the variables were determined using the following functions:

For Congestion Index Value (CIV) variable: The Conditional Probabilities of different classes' were determined:

High Congestion Class:

$$P\left(\frac{CIV}{HC}\right) = \frac{l_3}{u} \tag{11}$$

$$P\left(\frac{CIV}{HMC}\right) = \frac{l_2}{u} \tag{12}$$

$$P\left(\frac{CIV}{MC}\right) = \frac{l_1}{u} \tag{13}$$

$$P\left(\frac{CIV}{LC}\right) = \frac{l}{u} \tag{14}$$

High-Moderate Congestion Class:

$$P\left(\frac{CIV}{HC}\right) = \frac{l_3}{u} \tag{15}$$

$$P\left(\frac{CIV}{HMC}\right) = \frac{l_2}{l_3} \tag{16}$$

$$P\left(\frac{CIV}{MC}\right) = \frac{l_1}{l_3} \tag{17}$$

$$P\left(\frac{CIV}{LC}\right) = \frac{l}{l_3} \tag{18}$$

Moderate Congestion Class:

$$P\left(\frac{CIV}{HC}\right) = \frac{l_3}{u} \tag{19}$$

$$P\left(\frac{CIV}{HMC}\right) = \frac{l_2}{u} \tag{20}$$

$$P\left(\frac{CIV}{MC}\right) = \frac{l_1}{l_2} \tag{21}$$

$$P\left(\frac{CIV}{LC}\right) = \frac{1}{l_2} \tag{22}$$

Low Congestion Class:

$$P\left(\frac{CIV}{HC}\right) = \frac{u - l_3}{u} \tag{23}$$

$$P\left(\frac{CIV}{HMC}\right) = \frac{u - l_2}{u} \tag{24}$$

$$P\left(\frac{CIV}{MC}\right) = \frac{u - l_1}{u} \tag{25}$$

$$P\left(\frac{CIV}{LC}\right) = \frac{u - 1}{u} \tag{26}$$

where

$P\left(\frac{CIV}{HC}\right)$ degree of CIV when there is high congestion

$P\left(\frac{CIV}{HMC}\right)$ degree of CIV when there is high-moderate congestion

$P\left(\frac{CIV}{MC}\right)$ degree of CIV when there is moderate congestion

$P\left(\frac{CIV}{LC}\right)$ degree of CIV when there is low congestion

For Average Speed (AS) variable: In similar manner as determined for the variable Congestion Index Value (CIV), the conditional probabilities for the variable Average Speed (AS) was also computed for different classes':

High Congestion Class:

$$P\left(\frac{AS}{HC}\right) = \frac{1}{l_1} \tag{27}$$

$$P\left(\frac{AS}{HMC}\right) = \frac{1}{l_2} \tag{28}$$

$$P\left(\frac{AS}{MC}\right) = \frac{1}{l_3} \tag{29}$$

$$P\left(\frac{AS}{LC}\right) = \frac{1}{u} \tag{30}$$

High-Moderate Congestion Class:

$$P\left(\frac{AS}{HC}\right) = \frac{1}{l_1} \tag{31}$$

$$P\left(\frac{AS}{HMC}\right) = \frac{l_1}{l_2} \tag{32}$$

$$P\left(\frac{AS}{MC}\right) = \frac{l_1}{l_3} \tag{33}$$

$$P\left(\frac{AS}{LC}\right) = \frac{l_1}{u} \tag{34}$$

Moderate Congestion Class:

$$P\left(\frac{AS}{HC}\right) = \frac{1}{l_1} \tag{35}$$

$$P\left(\frac{AS}{HMC}\right) = \frac{1}{l_2} \tag{36}$$

$$P\left(\frac{AS}{MC}\right) = \frac{l_2}{l_3} \tag{37}$$

$$P\left(\frac{AS}{LC}\right) = \frac{l_2}{u} \tag{38}$$

Low Congestion Class:

$$P\left(\frac{AS}{HC}\right) = \frac{l_1 - 1}{u} \tag{39}$$

$$P\left(\frac{AS}{HMC}\right) = \frac{l_2 - 1}{u} \tag{40}$$

$$P\left(\frac{AS}{MC}\right) = \frac{l_3 - 1}{u} \tag{41}$$

$$P\left(\frac{AS}{LC}\right) = \frac{u - 1}{u} \tag{42}$$

where

$P\left(\dfrac{AS}{HC}\right)$ degree of AS when there is high congestion

$P\left(\dfrac{AS}{HMC}\right)$ degree of AS when there is high-moderate congestion

$P\left(\dfrac{AS}{MC}\right)$ degree of AS when there is moderate congestion

$P\left(\dfrac{AS}{LC}\right)$ degree of AS when there is low congestion

- Finally the values of Prior Probabilities and Conditional Probabilities of different classes' of the variables i.e. CIV and AS were substituted in the Eq. 1 to compute the Bayesian probabilities of different congestion classes' for each of the variables i.e. CIV and AS.

2.2 Integration of Peer Layers of the Significant Factors

The Peer layers of the CIV and AS variables were combined using GIS Weighted Sum Overlay to determine route's probability with different classes'. For example, the layer of High Congestion Class of CIV was integrated with the High Congestion Class of AS to determine variation in the degree of High Congestion Class for different routes of the road network. In similar manner, the other peer layers were also integrated.

Multi-criteria analysis using GIS Weighted Sum Overlay:

The Multi-criteria analysis using GIS establish a relationship between spatial data and result. The relationship between the spatial data and the result is defined on the basis of decision rules. The Weighted Sum Overlay is one of the multi-criteria methods that can be employed in the realm of spatial decision making. However there are few things that need to be considered while spatial decision making using GIS. For example, data acquisition capability of GIS and furthermore its efficiency in storing and manipulation of data must be considered. The GIS Weighted Sum Overlay constitutes of steps such as definition of objectives, formulation of alternatives, assignment of decision variables to different alternatives, analysis of alternatives, prioritization of alternatives, assessment of sensitivity of result to different alternatives, and finally making decision on the basis of results.

2.3 Assessment of the Possibility of Transition in Congestion Degree of Different Routes

As discussed in the Sect. 2.2, routes with probabilities for different classes' i.e. High Congestion Classes, High-Moderate Congestion Class, Moderate Congestion Class and Low Congestion Class were determined using GIS Weighted Sum Overlay. Now the routes with probability for each class was compared with their membership values of aforementioned congestion classes for the both variables i.e. CIV and AS.

The membership functions were designed for different classes' for the both variables:

For Congestion Index Value (CIV):

High Congestion Class: Range $= \{l_3, u\} = >$ mean $(m) = \frac{l_3 + u}{2}$

$$f(y) = \begin{cases} y \geq m, 1 \\ y < m, \frac{y}{m} \end{cases} \tag{43}$$

High-Moderate Congestion Class: Range $= \{l_2, l_3\} = >$ mean $(m_1) = \frac{l_2 + l_3}{2}$

$$f(y) = \begin{cases} y \leq m_1, \frac{y}{m_1} \\ m_1 < y < m, \frac{(m-y)}{m} \\ y \geq m, 0 \end{cases} \tag{44}$$

Moderate Congestion Class: Range $= \{l_1, l_2\} = >$ mean $(m_2) = (l_1 + l_2)/2$

$$f(y) = \begin{cases} y \leq m_2, \frac{y}{m_2} \\ m_2 < y < m, \frac{(m-y)}{m} \\ y \geq m, 0 \end{cases} \tag{45}$$

Low Congestion Class: Range $= \{l, l_1\} = >$ mean $(m_2) = (l + l_1)/2$

$$f(y) = \begin{cases} y \leq m_3, 1 \\ m_3 < y < m, \frac{(m-y)}{m} \\ y \geq m, 0 \end{cases} \tag{46}$$

For Average Speed (AS):
High Congestion Class: Range $= (l, l_1) = >$ mean $(m_4) = (l + l_1)/2$

$$f(z) = \begin{cases} z \leq m_4, 1 \\ m_4 < z < u, \frac{(u-z)}{u} \end{cases} \tag{47}$$

High-Moderate Congestion Class: Range $= \{l_1, l_2\} = >$ mean $(m_5) = (l_1 + l_2)/2$

$$f(z) = \begin{cases} z \leq m_4, 0 \\ m_4 < z < m_5, \frac{z}{m_5} \\ z > m_5, \frac{(u-z)}{u} \end{cases} \qquad (48)$$

Moderate Congestion Class: Range $= \{l_2, l_3\} = >$ mean $(m_6) = (l_2 + l_3)/2$

$$f(z) = \begin{cases} z \leq m_4, 0 \\ m_4 < z < m_6, \frac{z}{m_6} \\ z > m_6, \frac{(u-z)}{u} \end{cases} \qquad (49)$$

Low Congestion Class: Range $= \{l_3, u\} = >$ mean $(m_7) = (l_3 + u)/2$

$$f(z) = \begin{cases} z \leq m_4, 0 \\ m_4 < z < m_7, \frac{z}{m_7} \\ z > m_7, 1 \end{cases} \qquad (50)$$

3 Results and Discussion

The present research investigation is divided into three main segments i.e. computation of Bayesian probabilities for different classes for the factors, 'AS' and 'CIV' representing status of congestion of respective classes, application of GIS Weighted Sum Overlay using Bayesian Probabilities Values to identify routes which are more likely congested and finally, the results of GIS Weighted Sum Overlay were compared with the continuous fuzzy membership function's values to assess the likelihood of transition in congestion degree from one class to another under uncertain environment.

The Bayesian probabilities of different congestion classes for the factors AS and CIV is presented in the Tables 1 and 2 respectively. Table 3 shows the categorization of routes into different classes obtained from the GIS Weighted Sum Overlay operation. Finally, Table 3 contains the geometric mean values of the continuous fuzzy membership functions for the factors AS and CIV. The probable transition in congestion status from one class to another can be observed from the Tables 1 and 2 respectively. For example, in the High Congestion class, the probability of high congestion for the routes 3, 4, 5, 7, 8, 19, and 21 is 0.44 and its transition probability to other classes i.e. High-Moderate, Moderate or Low Congestion classes is 0.44, 0.07 and 0.05 respectively (Table 1). Interestingly, for the Low Congestion class, the probability of low congestion (0.33) is lower than the probability of High-Moderate Congestion (0.41) that means there is high likelihood that routes in this class which are low congested in general scenario can be transited

Table 1 Bayesian probabilities of different classes for the variable average speed (AS)

Classes	Route ID	H	HM	M	L
High congestion class (H)	3	0.44	0.44	0.07	0.05
	4				
	5				
	7				
	8				
	19				
	21				
High moderate congestion class (HM)	1	0.24	0.60	0.1	0.05
	2				
	6				
	9				
	10				
	15				
	16				
	20				
	22				
	23				
	24				
Moderate congestion class (M)	11	0.36	0.13	0.30	0.20
	12				
	13				
Low congestion class (L)	14	0.02	0.41	0.24	0.33
	17				
	18				

Table 2 Bayesian probabilities of different classes for the variable congestion index value (CIV)

CLASSES	Route ID	H	HM	M	L
High congestion class (H)	4	0.43	0.55	0.19	0.02
	5				
	19				
High moderate congestion class (HM)	7	0.35	0.42	0.18	0.05
	9				
	20				
	21				
Moderate congestion class (M)	1	0.30	0.21	0.45	0.04
	2				
	10				
	15				
	16				
	22				
	23				
	24				

(continued)

Table 2 (continued)

CLASSES	Route ID	H	HM	M	L
Low congestion class (L)	3	0.01	0.04	0.31	0.63
	6				
	8				
	11				
	12				
	13				
	14				
	17				
	18				

Table 3 Results of GIS weighted sum overlay (routes in different classes)

Classes	Route ID
High congestion class	4
	5
	19
High moderate congestion class	1
	2
	7
	8
	10
	15
	21
	20
	22
	23
	24
Moderate congestion class	11
	12
	3
	8
	6
	13
Low congestion class	14
	17
	18

to High-Moderate Congestion class (Table 1). Table 3 consist of routes categorized into different classes based on GIS Weighted Sum Overlay using Bayesian Probabilities Values and Table 3 shows the continuous membership values for each class. Route 1 is in the High-Moderate Congestion class (Table 3) with possibility values 0.39, 0.45, 0.82, and 0.64 for High Congestion class, High-Moderate Congestion class, Moderate Congestion class and Low Congestion class respectively. That means, there is high likelihood that route 1 which is specifically in the High-Moderate Congestion class can move into the Moderate Congestion class. In similar manner, the probable transition of congestion degree from one class to another class can be determined for different routes based on Bayesian-Fuzzy GIS Overlay.

4   Conclusions

The objective of the present research investigation is to quantify the dynamics of transition in congestion degree from one state to another. The Bayesian-Fuzzy GIS Overlay technique was proposed to assess the dynamics of congestion. Application of Bayesian approach helped in representing the unintuitionistic scenario of congestion. While on the other hand, GIS Overlay represented the combinatorial analysis of factors pertaining to traffic flow using Bayesian probabilities. Finally, fuzzy theory helped to identify the possible transition of congestion for one class to another for specific routes as Bayesian probabilities represent probable transitions between congestion states for different classes.

The proposed technique i.e. Bayesian-Fuzzy GIS Overlay to construe congestion dynamics succeeded in providing a range of possibilities of transition in congestion degree from one state to another state for different routes precisely. The inferred information from the results of Bayesian-Fuzzy GIS Overlay can be used for transport management.

Acknowledgements The authors are grateful to the Vice Chancellor, Birla Institute of Technology, Mesra, and Ranchi for providing research facilities to perform the investigation in Geographic Information System (GIS) & Digital Image Processing (DIP) labs of the Department of Remote Sensing, BIT Mesra, Ranchi.

References

1. Batty, M., Morphet, R., Masucci, P., Stanilov, K.: Entropy, complexity, and spatial information. In: J. Geogr. Syst. 16., 363–385, doi:10.1007/s10109-014-0202-2, (2012).
2. Cheng, T., Wang, J., Haworth, J., Heydecker, B., Chow, A.: A dynamic spatial weight matrix and localized space–time autoregressive integrated moving average for network modeling. In: Geogr. Anal. 46, 75–97, (2014).
3. D'Este, G.M., Zito, R., Taylor, M.A.P.: Using GPS to Measure Traffic System Performance. In: Comput-Aided Civ. Inf. 14, 255–265, (1999).
4. Gwilliam, K..: Urban transport in developing countries.: In: Transport. Rev. Vol. 23, No. 2., 197–216, (2003).

5. Hamilton, A., Waterson, B., Cherrett, T., Robinson, A., Snell, I.: The evolution of urban traffic control: changing policy and technology. In: Transport. Plan. Techn. 36:1., 24–43, doi:10.1080/03081060.2012.745318, (2012).
6. Jost, D., Nagel, k.: Traffic jam dynamics in traffic flow models. In: 3rd Swiss Transport Research Conference, Monte Verita/Ascona. (2003).
7. Lago, A.: Spatial models of morning commute consistent with realistic traffic behavior. PhD Dissertation, Institute of Transportation Studies, University of California at Berkeley. (2003).
8. Lopes, S.B., Brondino, N.C.M., Silva, A.N.R.D.: GIS-Based analytical tools for transport planning: spatial regression models for transportation demand forecast. In: ISPRS Int. J. Geogr. Inf. 3, 565–583; doi:10.3390/ijgi3020565, (2014).
9. Lovelace, R., Ballas, D., Watson, M.: A spatial micro simulation approach for the analysis of commuter patterns: from individual to regional levels. In: J. Transport. G. 34, 282–296, (2014).
10. Miller, H.J., Wu, Y-H., Hung, M-C.: GIS-Based dynamic traffic congestion modeling to support Time-critical logistics. In: IEEE. Published in the Proceedings of the Hawai'i International Conference on System Science, Maui, Hawaii. (1999).
11. Ottomanelli, M., Wong, C.K..: Modelling uncertainty in traffic and transportation systems. In: Transportmetrica. 7:1, 1–3, doi:10.1080/18128600903244636, (2009).
12. Prathap, R.C., Rao, A.M., Durai, B.K., Lakshmi, S.: GIS application in traffic congestion management. Proceeding of International. Conference. on Recent Trends in Transportation, Environmental and Civil Engineering. (2011).
13. Salonen, M., Toivonen, T.: Modelling travel time in urban networks: comparable measures for private car and public transport. In: J. Transport. G. 31, 143–153, (2013).
14. Stat Trek, http://stattrek.com/probability/bayes-theorem.aspx.
15. Watanbe, M.: Traffic dynamics and congested phases derived from an extended optimal-velocity model. In: Eng. 6, 462–471, (2014).
16. Xin, W., Levinson, D.: Stochastic congestion and pricing model with endogenous departure time selection and heterogeneous travelers. 86th Annual Meeting of the Transportation Research Board in Washington, DC, 21–25. (2007).
17. Xu, Z., Sui, D.Z.: Small-world characteristics on transportation networks: a perspective from network autocorrelation. In J. Geogr. Syst. 9:, 189–205, (2007).
18. You, J., Kim, T.J.: An integrated urban systems model with GIS. In: J. Geogr. Syst. 1:305, 321, (1999).

A Hybrid Clustering Technique to Improve Big Data Accessibility Based on Machine Learning Approaches

E. Omid Mahdi Ebadati and Mohammad Mortazavi Tabrizi

Abstract Big data is called to a large or complex data from traditional ones, which is unstructured in many case. Accessing to a specific value in a huge data that is not sorted or organized can be time consuming and require a high processing. With growing of data, clustering can be a most important unsupervised approach that finds a structure for data. In this paper, we demonstrate two approaches to cluster data with high accuracy, and then we sort data by implementing merge sort algorithm finally, we use binary search to find a data value point in a specific range of data. This research presents a high value efficiency combo method in big data by using genetic and k-means. After clustering with k-means total sum of the Euclidean distances is $3.37233e+09$ for 4 clusters, and after genetic algorithm this number reduce to 0.0300344 in the best fit. In the second and third stage we show that after this implementation, we can access to a particular data much faster and accurate than other older methods.

Keywords Hybrid clustering · Data mining · k-means algorithm · Genetic algorithm · Machine learning

E.O.M. Ebadati (✉)
Department of Mathematics and Computer Science, Kharazmi University,
#242, Somayeh Street, Between Qarani & Vila, Tehran, Iran
e-mail: omidit@gmail.com

M.M. Tabrizi
Department of Knowledge Engineering and Decision Science, Kharazmi University,
#242, Somayeh Street, Between Qarani & Vila, Tehran, Iran
e-mail: mortazavie@outlook.com

© Springer India 2016
S.C. Satapathy et al. (eds.), *Information Systems Design and Intelligent Applications*, Advances in Intelligent Systems and Computing 433,
DOI 10.1007/978-81-322-2755-7_43

413

1 Introduction

Information technology could ascertain for human beings, many approaches to ease the life for them. Modern services such as: Internet, e-finance, web data and etc. every information service that can work better, need a data. These services need data and also generate data. For many problems like data collection, analysis, storage, and the application of data that can't be solved by traditional information processing technologies should find other solutions.

Everyday information systems produce a large volume of data. The New York Stock Exchange produces 1 terabyte (TB) of trading data every day; Facebook produces more than 10 TB of data every day; Twitter generates more than 7 TB of data every day. In the future, the data volume of the whole world will be doubled every 18 months. The number will reach 35 (zettabytes) ZB in 2020. These information force data storages entering to "Zetta" volume. There are many definitions of big data, but a definition is: data that exceeds the processing capacity of conventional database systems [1]. There are management and analytics challenge in traditional data from Volume, Velocity, Variety, and Value that named "4 V" [2]. Big data is unstructured and structured data, and if we can manage it, it also can help to improve the company's operations to make faster and more intelligent decisions.

2 Literature Review

Now a day the importance of big data and computing on big data is inevitable. It uses in majority scales in the world. The leader companies like Microsoft, Apple, Facebook, Google and etc. are offering various applications that can help governments, organizations and companies to store their data in the cloud and reduce their current maintenances. For process big data, cloud computing can help us with high flexibility [3], but still there is a gap After massive data generated daily by people and their management and computing. For example, Facebook stores terabyte data of users like images, videos, audios, comments. First of all, many of these data are unstructured and we should create a structure for these big data, it could be more useful with the use of "Map Reduce Framework" to analysis big data [4]. Young has proposed an approach and research for using big data to HIV Prevention, he worked on the data, which prepared from mobile, social media and mobile applications [5].

Clustering is a method for grouping related data that are utilized by many applications. For data clustering, there exists many different approaches that are used in various situations and depend on the type of data it uses C-means [6], k-means [7, 8], genetic algorithm [9–11], binary [12], hierarchy [13]. However, the combination methods also can be more useful and create a better method with fewer incompetency. In this regards, a hybrid of genetic algorithm and k-means were proposed by Rahman and Islam [14]. Villalba and its colleagues [15] proposed clustering on smart phone image in forensic use. They offer a solution for recognizing fraud and type of the mobile that take photos with clustering method. Also, there is much research about pattern recognition that related to clustering [16].

Genetic Algorithm (GA) is a meta-heuristic approach, which may combine with other methods and create a powerful solution for problems. In many techniques like k-means, we should define the number of clusters but GA is more useful as the number of clusters dependent on data. In this paper, we develop a GA as we need genes to start and these genes choose randomly from an initial population, then for all chromosomes, fitness function calculates and in every iteration the best chromosome chooses for producing a better population. In this method at each step it close to correct centering of clusters [14].

One of the common approaches for clustering is k-means that it can cluster data into a K number of groups and every data in its own cluster has a minimum distance, but has a maximum distance with a data in another cluster, the goal is to minimize Sum of Squared Error (SSE) for each cluster and usually Euclidian distance use in this approach [7]. In another paper apply this method to classify a dataset of 10 years data for hydro-power and power generation from river [17]. Pavithra and Aradhya show a method that detects multilingual text in image and video [18], and there is another application of clustering for text, based on k-means [19].

After clustering a big data set, search in massive data can be difficult and need so much time for the processing. Many approaches are proposed in different researches and every method has their own advantages. One of the general attitudes is Linear Search (LS). This search is arraying data structure, cell by cell. The complexity of this search algorithm is $O(n)$, and this method is a specific case of Brute-Force Search (BFS). Other approach in searching is Binary Search (BS) that needs a sorted array and for the first step it compares input value with the middle cell of array, if input value and value in middle cell of array was equal, it returns index and algorithm end, but if the input value was greater from middle cell value, it waiver from left side and again compare the input value with the middle cell of right side of array, and so on until there is no cell to half, the complexity of this method is $O(\log(n))$ and this approach is derived from divide and conquer algorithm [20]. In another technique after initial centroid, use a Step Size of Movement (SSM) and in each step of all data, calculate the sum of data and SSM and if don't occur any improvement, make half SSM and again summation data with SSM for all data until finding the best centroid [21]. To continue, we use k-means at first, and with GA, we try to reach high accuracy, and with the merge-sort and BS we search clusters.

3 Proposed Work

One of the problems with big data is clustering and finding a specific data value in datasets with terabyte size. Over time, every database size is increasing and accessing of spatial data becomes harder. Certainly, access to a data in chaos dataset is very hard. In this paper, we are proposing two methods to approach to a particular data. First, we should grouping data and after data is clustered, we can search required data in a specific group. With this method, we don't need to search all of the datasets.

To apply this idea, we use k-means for clustering at the first step. It is an NP-hard problem, which algorithm can cluster the input data $x = \{x_1, x_2, \ldots, x_k\}$ and it tries to minimize SSE in each step. At first we should initial the number of clusters and every k cluster are exclusive $S = \{S_1, S_2, S_3, \ldots, S_k\}$, $\bigcup_{k=1}^{K} S_k = S, S_i \cap S_j = \emptyset$, where $1 \leq i \neq j \leq k$ by minimizing the SSE:

$$SSE = \sum_{k=1}^{K} \sum_{x_i \in S_k} \|x_i - C_k\|_2^2 \tag{1}$$

where $\|.\|_2$ is a Euclidean norm and C_k is the center of each cluster. The Euclidean norm usually used to calculate distance between two points. The Euclidean distance between two points $x = \{x_1, x_2, \ldots, x_n\}$ and $y = \{y_1, y_2, \ldots, y_n\}$ calculate as follows:

$$d = \sqrt[2]{\sum_{i=1}^{n} (x_i - y_i)^2} \tag{2}$$

At first, every center can have random value and every input data belongs to each center. For the next steps for each cluster calculate the mean of belonging points and update value of centers and again each input data belongs to the centers. These operations repeat for n step that we can determine the value of n or until we have no change in SSE.

The output of this algorithm is centers of each cluster $C = \{c_1, c_2, \ldots, c_k\}$. Given an initial set of k-means $m_1, m_2, \ldots, m_i$. This method proceeds following two steps:

1. Assignment step: in this step assigns each input data to the nearest cluster

$$S_i^{(t)} = \left\{ x_p : \left\| x_p - m_i^{(t)} \right\|^2 \leq \left\| x_p - m_j^{(t)} \right\|^2 \forall j, 1 \leq j \leq k \right\} \tag{3}$$

2. Update step: calculate the means of each point in every cluster and update the value of centers

$$m_i^{(t+1)} = \frac{1}{\left| S_i^{(t)} \right|} \sum_{x_j \in S_i^{(t)}} x_j \tag{4}$$

We use this method for beginning and also for finding the initial value of the center.

Another step is using GA to have accurate centers. After we find centers of clusters in k-means method, use these points for initial population with genetic

Fig. 1 Approach methods

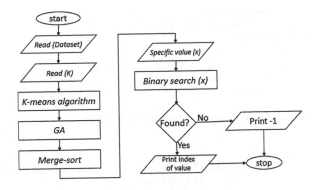

method. GA is a heuristic method that mimics the process of natural selection and it starts with some data as population of candidate solutions. At fitness, it tries to improve the answers. Every data in population called chromosomes or genotype. The evolution usually starts from a population of randomly generated individuals, but in this paper, we used the output value from k-means. This operation is an iterative process. Every fitness of genotype in each generation will be calculated and best answers moved to next generations (Fig. 1).

This method searches the problem space. In this approach, we use the Euclidian distance for fitness function, and the goal is minimizing the distance between centers in each cluster, see in formula (2).

Commonly, this method ending with different conditions, when the maximum number of generations has been produced, or a satisfactory fitness level has been reached for the population. The next step is searching in these clusters to find the different point a particular value (data) that in this method, we used BS. We should sort each cluster, after that, BS returns the index of a value in an array.

The array is indexed where $K(0) \leq K(1) \leq \cdots \leq K(n-1)$ (Fig. 2).

For sorting these arrays, there are many methods and with attention to type of data, we can use a suitable approach. For these data, we recommend merge sort. This method has complexity $O(n\log(n))$ in worst, average and best case (Figs. 3 and 4).

Then, after sorting the array, for finding a specific value in the array we use the BS (Fig. 5).

We use the BS algorithm in each cluster that we want to obtain the index of valuable data. If data found it return the index of data, else, it return -1. In best case performance is $O(1)$ and in worst and average case performance is $O(\log(n))$.

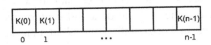

Fig. 2 Presentation of binary search array

```
mergesort(L = a1,a2,...,an: list of real numbers)
if (n == 1) then return L
else
       m = ⌊n/2⌋
       L1 = a1,a2,...,am
       L2 = am+1,am+2,...,an
return merge(mergesort(L1),mergesort(L2))
```

Fig. 3 Pseudo-code of merge sort algorithm

```
merge(L1,L2: sorted lists of real numbers)
     O = emptylist
     while (L1 or L2 is not empty)
          if (L1 is empty)
               move smallest(L2) onto the end of O
          else if (L2 is empty)
               move smallest(L1) onto the end of O
          else if (smallest(L1) < smallest(L2))
               move smallest(L1) onto the end of O
          else move smallest(L2) onto the end of O
     return O
```

Fig. 4 Pseudo-code of the implemented merge algorithm

Fig. 5 Pseudo-code of implemented binary search algorithm

```
while (low <= high)
     mid = [(low + high)/2]
     if (var < x(mid))
          high=mid-1
     elseif (var > x(mid))
     low = mid+1
     else
          index=mid
          return index
     end
end
index= -1
return index
```

4 Analysis and Design of the Proposed Model

In this paper, we use "perfume_data" as a dataset, which has a massive data. Table 1 represents a short view of this dataset [22].

In the first step, we use k-means to cluster data for $k = 4$ and now we have 4 centers. In order to have a view of the data and centers, we draw 14 columns in vertical and 14 columns in horizontal of the chart that shows shown in Fig. 6 (Table 2).

Table 1 Short view of dataset

Ajayeb	64,558	64,556	64,543	64,543	64,541	64,543	64,543	64,541
Ajmal	60,502	60,489	61,485	60,487	61,485	61,513	60,515	60,500
Amreaj	57,040	57,040	57,040	58,041	58,041	58,041	58,041	57,042
Aood	71,083	72,087	71,091	71,095	71,099	72,103	71,099	72,099
Asgar_ali	68,209	68,209	68,216	68,216	68,223	68,223	68,223	68,223
Bukhoor	71,046	71,046	71,046	71,046	71,046	71,046	71,046	71,046
Buberrry	61,096	61,096	60,093	60,092	60,093	60,093	61,096	61,096
Dhealaod	68,132	69,137	69,137	68,137	68,137	69,142	69,142	68,137
Junaid	71,590	71,575	71,574	71,560	71,560	71,559	72,573	71,559

Fig. 6 *k*-means clustering result

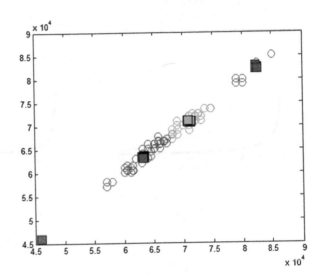

Table 2 Centers after implementation of *k*-means

Col 1	Col 2	Col 3	Col 4	Col 5	Col 6	Col 7
63158.22	63156.44	63266.11	63377.33	63264.33	63377.11	63266.56
46,014	46,014	46,014	46,014	46,014	46,015	46,015
70955.71	71243.14	71101.86	70814.71	70959.14	71243.43	71244.86
82149.67	82145.67	82,146	82142.33	82,479	82142.33	82,157

A sample of clustering with *k*-means centers is as follows:

The total sum of the Euclidean distances in *k*-means is 3.37233*e*+09 and we can use this number as accuracy of centers, and we should try to minimize this number to reach the best result. So, we use these centers for start GA and we practice these data as population of candidate solutions. In this paper, we proposed 4000 generations for genetic method and after 5 runs we have (Fig. 7).

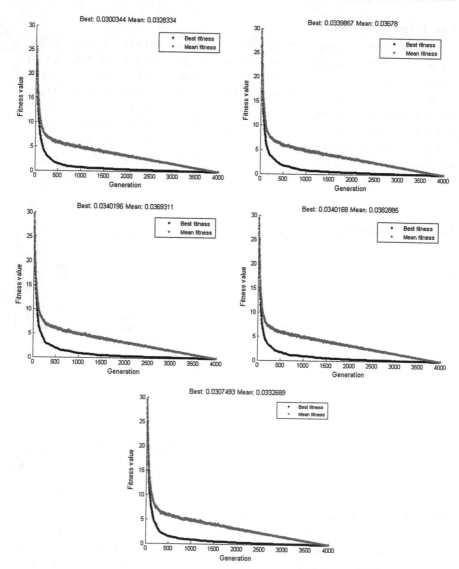

Fig. 7 GA clustering results

The results in Table 3.

So, in Table 3, we can see the change of accuracy from 3.37233e+09 to 0.0300344 in best fit and 0.0328334 in mean for the first run with variance 3.1891e-06 and 4.6957e-06. This change shows that accuracy is increased and now we have to make centers accurate. After genetic clustering, centers in short view are shown in Table 4.

Table 3 The result of accuracy in five runs

	1	2	3	4	5	Mean	Variance
Best	0.0300344	0.0339867	0.0340196	0.0340168	0.0307493	0.0325614	3.1891e-06
Mean	0.0328334	0.03678	0.0369311	0.0382885	0.0332689	0.0356204	4.6957e-06

Table 4 GA clustering centers

Col 1	Col 2	Col 3	Col 4	Col 5	Col 6	Col 7
63157.52	63155.74	63265.41	63376.63	63263.63	63376.41	63265.85
46009.4	46009.4	46009.4	46009.4	46009.4	46010.4	46010.4
70954.7	71242.12	71100.84	70813.7	70958.13	71242.41	71243.84
82146.92	82142.93	82143.26	82139.59	82476.26	82139.59	82154.26

After clustering, we have the minimum and maximum of each cluster and we can search on a specific range of huge data instead search all data. Actually, in each searching, with respect to the number value that we want to search, operation of sorting and searching, just done in one cluster, therefore only $\frac{1}{k}$ of data search is done and this method is much faster than to search all data value points. After clustering with attention to the value of the input data and the range of clusters, we can easily find the number of clusters (k) and only sort and search that cluster.

We implement our proposed model on a system with 8 GB RAM and Intel Core i5 M460 2.53 GHz CPU, the time of sorting and searching of a specific data value is like as follows:

In Table 5, we choose some data from clusters and calculate the sort and search time in 10 runs, and we select minimum number of times in each 10 runs. In a different system, above these values may have different results for timely consumption. As we saw in our proposed method, we can access to specific data in very short time with high accuracy. If data exist in the cluster it returns row number and index, else it returns −1.

Table 5 Results of implementation the approach method

Value	Cluster	Row	Index	Time (s)	Is data exist in cluster?
60,093	1	4	17	0.000117	Yes
59,140	1	–	−1	0.000132	No
46,014	2	1	3	0.000098	Yes
46,013	2	–	−1	0.000101	No
68,137	3	4	7	0.000131	Yes
69,500	3	–	−1	0.000145	No
82,440	4	1	27	0.000100	Yes
82,999	4	–	−1	0.000135	No

5   Conclusion and Future Work

With attention to the increasing volume of data, the big data presents are more important role day to day. One of the most important operations on a data is classification and clustering. This paper proposed a method to cluster data with a high accuracy, sorting and searching in a part of huge data instead of the whole dataset. This paper has explored, with combining several approaches, which data clustering is the most accurate ones. At first we use k-means for initial clustering centers and then use GA to improve accuracy of centers, after that, with respect to needed data value, sort and search the specific cluster will take its place and finally get the best accuracy of 0.0300344. For the future of this work, the proposed method can combine to another algorithm like memetic or a hybrid technique of that.

References

1. Tian, W.D. and Y.D. Zhao, Optimized Cloud Resource Management and Scheduling: Theories and Practices. 2014: Morgan Kaufmann.
2. Gupta, R., H. Gupta, and M. Mohania, Cloud Computing and Big Data Analytics: What Is New from Databases Perspective?, in Big Data Analytics. 2012, Springer. p. 42–61.
3. Hashem, I.A.T., et al., The rise of "big data" on cloud computing: Review and open research issues. Information Systems, 2015. 47: p. 98–115.
4. Fadiya, S.O., S. Saydam, and V.V. Zira, Advancing big data for humanitarian needs. Procedia Engineering, 2014. 78: p. 88–95.
5. Young, S.D., A "big data" approach to HIV epidemiology and prevention. Preventive medicine, 2015. 70: p. 17–18.
6. Liu, Z.-g., et al., Credal c-means clustering method based on belief functions. Knowledge-Based Systems, 2015. 74: p. 119–132.
7. Jain, A.K., Data clustering: 50 years beyond K-means. Pattern Recognition Letters, 2010. 31 (8): p. 651–666.
8. Ebadati E, O.M. and S. Babaie, Implementation of Two Stages k-Means Algorithm to Apply a Payment System Provider Framework in Banking Systems, in Artificial Intelligence Perspectives and Applications, R. Silhavy, et al., Editors. 2015, Springer International Publishing. p. 203–213.
9. Liu, Y., X. Wu, and Y. Shen, Automatic clustering using genetic algorithms. Applied Mathematics and Computation, 2011. 218(4): p. 1267–1279.
10. Razavi, S., et al., An Efficient Grouping Genetic Algorithm for Data Clustering and Big Data Analysis, in Computational Intelligence for Big Data Analysis, Springer International Publishing. 2015, p. 119–142.
11. Ebadati E., O.M., et al., Impact of genetic algorithm for meta-heuristic methods to solve multi depot vehicle routing problems with time windows. Ciencia e Tecnica, A Science and Technology, 2014. 29(7): p. 9.
12. Barthélemy, J.-P. and F. Brucker, Binary clustering. Discrete Applied Mathematics, 2008. 156 (8): p. 1237–1250.
13. Alzate, C. and J.A. Suykens, Hierarchical kernel spectral clustering. Neural Networks, 2012. 35: p. 21–30.

14. Rahman, M.A. and M.Z. Islam, A hybrid clustering technique combining a novel genetic algorithm with K-Means. Knowledge-Based Systems, 2014. 71: p. 345–365.
15. Villalba, L.J.G., A.L.S. Orozco, and J.R. Corripio, Smartphone image clustering. Expert Systems with Applications, 2015. 42(4): p. 1927–1940.
16. Yu, J., et al., Image clustering based on sparse patch alignment framework. Pattern Recognition, 2014.
17. Adhau, S., R. Moharil, and P. Adhau, K-Means clustering technique applied to availability of micro hydro power. Sustainable Energy Technologies and Assessments, 2014. 8: p. 191–201.
18. Pavithra, M. and V.M. Aradhya, A comprehensive of transforms, Gabor filter and k-means clustering for text detection in images and video. Applied Computing and Informatics, 2014.
19. Yao, M., D. Pi, and X. Cong, Chinese text clustering algorithm based k-means. Physics Procedia, 2012. 33: p. 301–307.
20. Lipschutz, S., Data Structures With C (Sie) (Sos). Vol. 4.19–4.27. McGraw-Hill Education (India) Pvt Limited.
21. Hatamlou, A., In search of optimal centroids on data clustering using a binary search algorithm. Pattern Recognition Letters, 2012. 33(13): p. 1756–1760.
22. UCI Machine Learning Repository: Perfume Data Data Set. 2002–2003 [cited 2015; Available from: https://archive.ics.uci.edu/ml/datasets/Perfume+Data.

Efficient Iceberg Query Evaluation in Distributed Databases by Developing Deferred Strategies

Vuppu Shankar and C.V. Guru Rao

Abstract With the rapid increase of the distributed databases the fast retrieval of the data from these databases are playing an important role. The bitmap index technique is well known data structure to provide fast search from large collections. The iceberg queries are frequently used where small output is required from large inputs. Recently, the bitmap indices with compression technique (WAH) are being utilized for efficient iceberg query evaluation. However, the results of these techniques are available only with standalone database. In this paper, we present an effective iceberg query evaluation by developing deferred strategy in distributed databases. This reduced the AND operations performed excessively in existing algorithm. The proposed algorithm executes in both the ways of data shipping and query shipping. The experimental results are verified and compared with existing algorithm

Keywords Iceberg query · Distributed databases · Bitwise logical operations · Data shipping · Query shipping · Threshold (T)

1 Introduction

The iceberg queries are basically a kind of retrieval techniques, which are first introduced by scientist Fang et al. [1]. These are frequently used when small output is required from large collections of databases. The syntax of an IB query on a relation R (C1, C2... Cn) is given Fig. 1.

V. Shankar (✉)
Department of Computer Science and Engineering,
Kakatiya Institute of Technology & Science, Warangal 506015, Telangana, India
e-mail: vuppu.shankar@gmail.com

C.V. Guru Rao
Department of Computer Science and Engineering, S. R. Engineering College,
Warangal 506371, Telangana, India

425

Fig. 1 Template of writing
an IBQ

select Ci, Cj, ..., Cm, AGG(*)
from TABLE R
group by Ci, Cj..., Cm
having AGG (*) > = T;

The above syntactic query fires on a relation R, and then column values that are above user desired threshold are selected as output of iceberg query. Many researchers had developed a variety of techniques for execution of iceberg queries. These are simple counter, sorting, sampling, coarse counting, Hybrid and tuple scan based methods etc. However, the researches reveal that the above methods had consumed large execution time and disk space due to scanning of database at least once.

In order to reduce the scanning time of database tuples and disk space a new kind of database index called bitmap was proposed by Spiegler and Mayan [2]. These bitmaps were used for solving iceberg queries efficiently [3–6]. A naïve bitmap index technique was used bit indices to answer IBQ. In this technique, more bitwise-AND operations were conducted while answering IBQ. Due to this large number of operations, the execution time and disk space was also large. After that, a static bitmap pruning technique was developed which prunes the bitmap vectors whose 1's count are less than specified threshold (T). However, this was not completely utilized the anti-monotone property of iceberg queries. Next, technique i.e., Dynamic index based pruning was developed to prune the bitmap vectors whose 1's count less than user specified threshold. This pruning step was applied before and after bitwise-AND operations performed. But, this technique was unable to reduce bitwise-AND operations completely whose resultant will be an empty vector. After that, an IBM Scientist whose name is He et al. [3] had developed a vector alignment algorithm which solves the problem of conducting the bitwise-AND operation between bitmaps whose resulting vector is empty.

In order to avoid the null AND operations between pair of bit vectors, these two vectors are examined for its alignment. The two vectors are said to be aligned whose first 1 bit positions (FBP) are same. While finding the alignment between two vectors under consideration, large numbers of push and pop operations were done. However, some of them were identified as conducted in excess. Therefore, the compressed bitmap index technique was not so effective and the iceberg query evaluation was challenging due to two major parameters as listed below:

1. The speed of the evaluation
2. The space required for the evaluation.

In addition, iceberg query evaluation in a distributed database environment makes it most challenging as the literature reveals only results on standalone databases. Therefore, the existing compressed bitmap index technique [3] is first proposed to be implemented in distributed databases. Consequentially, the proposed algorithm of compressed bitmap index technique [3] is modified and being proposed as "Modified Compressed Bitmap Index Technique" for efficient iceberg query evaluation in distributed databases. The modified compressed bitmap index algorithm is responsible to produce the output from large collection of input data available in distributed databases by either shipping the data to the central server or shipping the iceberg query to every client. At first, we present the proposal of modified compressed bitmap index algorithm using data shipping.

Section 2 presents a survey on the related work carried out on the topic under investigation. The proposal made on the problem of iceberg query evaluation in distributed databases is explained at Sect. 3. The Sect. 4 describes the implementation details of proposed solution. The experiments that are conducted on the implementations are presented at Sect. 5. The Sect. 6 brings out the results obtained when experimented on the implementation and analyzed. The research work carried out is concluded in the Sect. 7 followed by references.

2 Related Work

In this section, we are here to present a review of related work of IBQ evaluation in two sub sections. In first sub section i.e. 2.1, we, mainly focused on the IBQ evaluation without using bitmap index technique. The second sub section i.e. 2.2, which focuses the evaluation techniques of IBQ using bitmap index technique.

2.1 Iceberg Query Evaluation Without Using Bitmap Index Technique

The iceberg queries were first proposed by Fang et al. [1], in his first proposal i.e., The simple counter method, the distinct column values were selected by firing the IBQ against to the database table and then count the number of values. If this cunt value was above threshold (T) selected as IBQ result, else count was ignored. This technique was efficient when database was small. However, it took large execution time and consumed more disk space to evaluate an IBQ. The next technique was developed and referred as sort technique. The concept of sorting was to sort the records before use them for its evaluation. Then sorted records were sent to IBQ evaluation as input. The effectiveness of sorting the records reduced the IBQ evaluation time and disk space that was consumed. In the next technique, a few records were considered for IBQ evaluation as sample. Hence, it was referred as sampling technique. This improved an execution time and disk space over previous techniques. But, this was provided not accurate results. However, this technique was compromised with the approximate results. After that, they were also developed the partitioning technique by Bae et al. [5, 7]. The partitioning of the database into many numbers of partitions and each partition was examined for valid IBQ result. The IBQ result from each partition was added to final IBQ result and compared with threshold. If this was above threshold sent to output. Else result were ignored. This was good and effective for small databases. As database size was increased the number of partitions or partition size was also increased.

All the above techniques reviewed in sub section A, were consumed large execution time and as well as disk space for its evaluation. Therefore, another technique that was used for IBQ evaluation known as index technique that was generally provide a fast search of records from large collection of database records. The various bitmap indexing techniques were reviewed in the next sub section. i.e., B.

2.2 Iceberg Query Evaluation Using Bitmap Index Technique

In this sub section, we present the review of various bitmap index techniques that were used for IBQ evaluation. First, a naïve technique was used for an evaluation of IBQ. In this technique, the bit vectors were extracted from the different column values as input and sent to IBQ evaluation. A bitwise-AND operation was performed between them. The number of 1 bits were counted. If this number was above threshold were declared as IBQ result and sent to output. The remaining bit vectors were also examined in the same way until all the bit vectors were completed. This technique was inefficient, since it required the performing of huge number of AND operations while answering the IBQ. The next technique developed was referred as static bitmap pruning technique. In this technique, a pruning of bitmaps were done before performing the any AND operation between bit vectors. With this, the number of bitmaps were reduced that were not eligible for AND operation. Hence, the net effectiveness of IBQ evaluation was increased. However, this was not completely removes the ineligible bit vectors whose contribution was not leading to IBQ result. There was another technique referred as dynamic index pruning technique. In this, the prunings of bitmaps were done twice. i.e. before and after performing AND operation. The next technique i.e. alignment algorithm was developed by IBM scientist He et al. [8], He was used the bit vectors in an efficient way to answer IBQ, and they were avoided from the conduction of AND operation. This was by finding the first 1 bit position of each vector and sent to alignment algorithm that was used to test whether inputted vectors were aligned or not. If they aligned, sent them to perform AND operation. Else, they were ignored. But, in this technique, the observation was done towards the checking of alignment process and was found as, this technique took place a large numbers of push and pop operations were to be performed. Thereby the IBQ evaluation was slow down. The next section presents a proposal to effectively prune the bitmap vectors by differing push-pop operations in the above stated process.

3 Proposed Research Work

The proposed research work is done in two subsections. In the first subsection i.e., 2.1, the block diagram of the proposed research work model is described. In the second sub section i.e., 2.2 an algorithm for deferred strategy to avoid excessive push-pop operations is proposed. **Problem statement**: In finding the alignment between two vectors under consideration a large numbers of push and pop operations were found to be conducted. But, some of them were identified as conducted in excess. Due to large number of these ineffective push-pop operations the bitmap vectors are entered unnecessarily into PQs for many times. Subsequently, the IBQ evaluation time will be large. Therefore, the ineffective push-pop operations are deferred before entering into its PQs. This is described in two sub sections.

Block Diagram of Proposed Model: Data Shipping

In the Fig. 2 block diagram, the term DB indicates database, PQ is priority queue. A, B are aggregate attributes. A1, A2, A3, ... An, and B1, B2, B3,... Bm are bitmap vectors, WAH is Word Aligned Hybrid technique, AND denotes bitwise-AND operation, XOR denotes Bitwise-XOR operation, "r" is resultant vector of Bitwise-AND operation, IBQ is iceberg query and T is threshold. In the above Fig. 2, the attributes A, B are passed to WAH technique in order to generate bitmap vectors and compress them. The compressed vectors are placed in corresponding PQs according to their FBP. FindNextAlignedVector module is used to determine the alignment between two bitmap vectors. If they are aligned, send them to IBQ Evaluation module of computation block, else sent them to sub module "deferred push-pop". This module saves push and pop operations that are excessively performed by deferring them. Otherwise, pop vector and push it into its PQ. After that, these aligned vectors are popped out from both PQs and passed to IBQ Evaluation module as input.

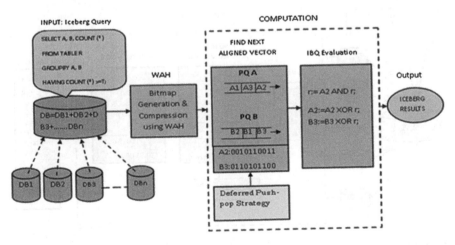

Fig. 2 Block diagram of deferred push-pop while data shipping

B. Algorithm

FindNextAlignedVectorsByDeferredPushPop
Input:Priority queues, Bitmap vectors, Threshold (T)
Output: Aligned vectors.
Step 1: begin
Step 2: generate bitmap vectors for every aggregate attribute
Step 3: push these bitmaps into corresponding priority queues prioritized by their first 1 bit position (fbp)
Step 4: do (
Step 5: If (fbp(A2)<fbp(B3))
Step 6: (
Step 7: find nextfbp(A2)
Step 8: if(fbp(A2)==fbp(B3)) return A2 , B3
Step 9: pop and push A2 into PQA
Step 10:)
Step 11: else
Step 12: (
Step 13: find nextfbp(B3)
Step14: if(fbp(A2)==fbp(B3)) return A2, B3
Step 15: Pop and push B3 into PQB
Step 16:) while fbp(A2≠fbp(B3)
Step 17:if fbp(A2)==fbp(B3);
return A2, B3
Step 18: end

Block Diagram of Proposed Model: Query Shipping

In Fig. 3, the term DB indicates database, PQ denotes priority queue. A, B are aggregate attributes, A1, B1 are bitmap vectors, WAH is Word Aligned Hybrid technique. AND represents Bitwise-AND operation, XOR represents Bitwise-XOR operation, "r" is resultant vector of bitwise-AND operation, IBQ is iceberg query and T is threshold. The above block diagram is proposed to be operated under query shipping. In above block diagram, the iceberg query comes to the database (DB) and is passed to other databases DB1, DB2, DB3 and DB4. At each of these databases the

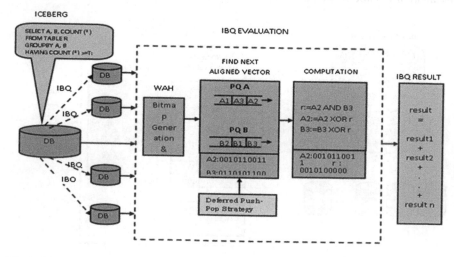

Fig. 3 Block diagram of deferred push-pop while query shipping

IBQ is processed same as in Fig. 2 in order to defer excessive push and pop operations. Then results at these databases are now combined to yield iceberg results.

4 Implementation

The above said proposals are implemented in the following four modules. These modules are developed using JAVA as frontend and the data is backend in MS-Acess, SQL-Server and Oracle.

1. Generate Bitmap
2. First1bit Position
3. FindNextAlignedVectorby deferring push-pop operations
4. Iceberg query evaluation

   1. **GenerateBitmap**: This module accepts the records of the database table containing columns mentioned in IB query as input. A bitmap vector is generated for each distinct value of an aggregate attribute, as marking 1 if it present in that record, otherwise 0 is marked. These bitmaps are generated using WAH technique [].
   2. **Fisrt1BitPosition**: This module accepts bitmap vector as input and returns the position of the first 1-bit.
   3. **FindNextAlignedVectorbydeferringpush-pop**: This module is used to find next aligned vectors from two PQs. The two bitmap vectors are said to be aligned, if the first 1 bit positions are same. The PQs are adjusted until the corresponding top most bitmap vectors are aligned by deferring unnecessary pop and push operations. The First1BitPosition module is called to find the first 1 bit position of a vector when the vectors are not aligned. This module returns null vectors if it fails in finding next aligned vectors.
   4. **IcebergQueryEvaluation**: This is the main module. It pushes the bitmap vectors, which are generated using GenerateBitmap module into PQs based on its first 1 bit position, which is returned from First1BitPosition module. The FindNextAlignedVector module is recursively called until either of the PQs is empty. In every next aligned vectors the bitwise-AND operation is conducted to get the intermediate result.

5 Experimentation

The above implemented modules are experimented in both data shipping and as well as query shipping. The above modules are experimented on the same setup used by modified compressed bitmap index algorithm and are explained from Sl no 4.

These 1 lakh records are fetched from 4 computers by loading the respective Java Data Base Connectivity (JDBC) drivers and establishing JDBC connections into client site. The main module called IcebergQueryEvaluation module first it invokes GenerateBitmap module. This module accepts an aggregate attributes and its distinct values of a record as input for bitmap generation. It considered of having m distinct values and n number of tuples of that attribute. Therefore, we need to construct an m × n matrix of bitmap vectors. If the distinct value of an aggregate attribute is present in the row of a matrix then bitmap vector is marked as a 1, otherwise 0. The bitmap vector matrices are occupying huge space. Hence, to get the memory efficiency these bitmaps are given for WAH technique [9] to compress. In this technique, the bitmap vectors are divided into number of words. Each word comprises of 32 bits. After that, these compressed words are given as input to next module i.e., First 1 Bit Position. This takes each compressed word and decomposes them into sequence of 1's and 0 bits. Then it is searched for position of first 1 bit. If the first bit of a word is 1, then it returns that position. If it is 0 bit, goes to next position of the word. The process repeats until the last word of compressed bitmap. After that, these bitmaps are placed according to their first 1 bit position in priority queues. The vectors in the PQs are input to next module i.e., Find Next Aligned Vector by Deferring Push-Pop. The two top positioned bitmaps from both priority queues are drawn to find the alignment of those vectors. Now, the first 1 bit positions of those vectors are compared. If they are same, then vector pair is said to be aligned and return them for further action. Otherwise, it defers the push and pop operations by selecting the vector which has smaller first 1 bit position and finds its next first 1 bit position. Once again the new first 1 bit position is compared with first 1 bit position of opposite vector. If they are same, return them for further action. Else pop that vector from its priority queue and push it into corresponding priority queue. The same process is repeated for either of the priority queues gets empty. Next, these aligned vectors are pushed to the Iceberg Query Evaluation module for an operation. In this module, the bitwise-AND operation is conducted on these aligned vectors. If two bits of the same position are 1, then resultant bit is 1, otherwise 0. After AND operation, the number of 1 bits in resultant vector is counted. This count value is compared with user given threshold. If it is above that, declared that vector pair is an iceberg result and send to output. Thereafter, a bitwise-XOR operation is also performed to test whether this vector pair is useful for next iceberg result or not. For conducting a bitwise-XOR operation, the two vectors are considered as input, one vector, which is previously used in bitwise-AND operation and another vector is resultant vector of bitwise-AND operation. In this XOR operation, the bit positions of two bit vectors are compared. If they are opposite bits then resultant bit is 1, otherwise 0. The number of 1 bits in resultant vector of XOR operation is above threshold are forwarded to re-enter into PQs. The same process repeats until either of the PQ becomes empty. These experiments are repeated for different thresholds as well as number of tuples in each time.

6 Results

The results of above algorithm are compared with modified compressed bitmap index algorithm. The observations from that comparison are, the deferred push-pop algorithm is showing better performance than modified algorithm in terms of its execution times.

Next, we present the analysis on execution times conserved by each proposed algorithm by computing gain parameter between them. For that, we summary the results in their execution times in deferred push-pop, as against the modified compressed bitmap index algorithm are noted for every experiment that was performed in data shipping. The gains in percentage of the summary results are presented at Table 1.

$$\% \text{ gain in Exe. time} = ((\text{Exe. time})\text{existing} - (\text{Exe. time})\text{proposed}/(\text{Exe. time})\text{existing}) \times 100$$

$$(1)$$

The r denotes correlation and is considered as given in Eq. (1) and reproduced as hereunder r = $\sum XY/SQRT$ ($\sum X2*\sum Y2$). The $\sum X$ and $\sum Y$ are sum of thresholds utilized in experiments and sum of gain in execution time of combined push-pop and bitwise-XOR operations to find the correlation and resulted in r_1 = 0.5946 = 59.46 %.

Now it can be understood that from the above correlation coefficients (r) the proposed deferred push-pop, algorithm consumes 40.54 %, of execution time, required in modified compressed bitmap index algorithm.

The execution times gained in deferred push-pop, against the modified compressed bitmap index algorithm are noted for every experiment that was performed in query shipping. The summary results of execution times of above algorithms and are presents at table 2.

Table 1 Percentage gain

Threshold (x)	gEPP (%) (y_1)	X = x − x' x' = 550	$Y_1 = y_1 − y'$ y' = 12.96
100	42.85	−450	+24.89
200	10.90	−350	−2.06
300	11.5	−250	−1.46
400	15.22	−150	+2.26
500	10.45	−50	−2.51
600	6.746	+50	−6.214
700	5.033	+150	−7.927
800	8.26	+250	−4.7
900	7.23	+350	−5.73
1000	11.45	+450	−1.51
Σx: 5500		Σy1:129.639	

Table 2 Summary of results in execution time while query shipping

Threshold (x)	gEPP(%) (y₁)	X = x − x' x' = 550	Y₁ = y₁ − y' y' = 10.28
100	52.99	−450	+42.71
200	10.44	−350	+0.12
300	10.0	−250	−0.28
400	11.5	−150	+1.22
500	6.66	−50	−3.62
600	3.84	+50	−6.44
700	0.629	+150	−9.65
800	0.725	+250	−9.55
900	0.9	+350	−9.38
1000	5.12	+450	−5.16
Σx: 5500	Σy1:102.8		

The first column is threshold value specified for each experiment and it is denoted by (x). The second, column represent gEPP, indicating percentage gain in execution time in experiments conducted on algorithm of deferred push-pop, The last two columns represents variables X and Y, considered for statistical computations required in determining correlation between thresholds and combined deferred push-pop operations obtaining is given the formula for correlation at Eq. (1) is considered. The percentage gain is computed by the following equation.

The r denotes correlation and reproduced as hereunder r = ΣXY/SQRT (ΣX2*ΣY2)). The ΣX and ΣY are sum of thresholds utilized in experiments and sum of gain in execution time of combined are submitted to find the correlation and resulted in r1 = 0.6828 = 68.28 %.

Now it can be understood from the above correlation coefficients (r), the proposed deferred push-pop algorithm consumes 31.72 % execution time required in modified compressed bitmap index algorithm.

7 Conclusion and Future Scope

Hereunder, we conclude the work which is done for efficient iceberg query evaluation in distributed databases while data ships and query shipping. The considerable execution time and disk space is reduced by reduction of the number of push-pop operations performed in existing technique. This is by deferring excessive push-pop operations while finding alignment between two vectors.

The above deferred algorithms for iceberg query evaluation can be extended from two aggregate attributes to multiple aggregate attributes. These are also used for other than count function such As SUM, AVG, RANK and DENSE RANK etc. These algorithms are extended for parallel processing of iceberg query when query is shipped to multiple databases.

References

1. M. Fang, N. Shivakumar, H. Garcia-Molina, R. Motwani and J.D. Ullman. "Computing Iceberg Queries Efficiently". In VLDB, pages 299–310, 1998.
2. Spiegler I; Maayan R "Storage and retrieval considerations of binary databases". Information processing and management: an international journal 21(3): pages 233–254, 1985.
3. D.E. Knuth, "The Art of Computer Programming: A Foundation for computer mathematics" Addison-Wesley Professional, second edition, ISBN NO: 0-201-89684-2, January 10, 1973.
4. G. Antoshenkov, "Byte-aligned Bitmap Compression", Proceedings of the Conference on Data Compression, IEEE Computer Society, Washington, DC, USA, Mar 28–30,1995, pp. 476.
5. Jinuk Bae, Sukho Lee, "Partitioning Algorithms for the Computation of Average Iceberg Queries", Springer-Verlag, ISBN:3-540-67980-4, 2000, pp: 276–286.
6. K. P. Leela, P. M. Tolani, and J. R. Haritsa. "On Incorporating Iceberg Queries in Query Processors", in DASFAA, 2004, pages 431–442.
7. J. Baeand, S. Lee, "Partitioning Algorithms for the Computation of Average Iceberg Queries", in DaWaK, 2000.
8. Bin He, Hui-I Hsiao, Ziyang Liu, Yu Huang and Yi Chen, "Efficient Iceberg Query Evaluation Using Compressed Bitmap Index", IEEE Transactions On Knowledge and Data Engineering, vol 24, issue 9, sept 2011, pp. 1570–1589.
9. P.E. O'Neil and G. Graefe. "Multi-Table Joins Through Bitmapped Join Indices". SIGMOD Record, 24(3):8–11, 1995.

Unique Identifier System Using Aneka Platform

Karishma Varshney, Rahul Johari and R.L. Ujjwal

Abstract This paper is regarding the development of a console application named 'Unique Identifier System (UIS)' used to create a unique identification number (UID) for a person at the time of his/her birth. Instead of having various identity proofs like PAN card, Driving license, Aadhaar card, Voter ID card, Ration card etc., the person will be recognized by its UID all over the world. This UID will be stored on a cloud to be accessed all over the globe. For hosting the application on Cloud, we are using the Aneka tool, a Cloud Application Development Platform http://www.manjrasoft.com/manjrasoft_downloads.html which will make all the UIDs available globally.

Keywords UIS · UID · Aneka · Cloud · Identity · Cloud computing

1 Introduction

A unique identifier (UID) is a numeric or alphanumeric string that is associated with a single person across the world. It is a random number which is virtually impossible for a person to identify. The UID approach is designed on an on-line system, where centrally data is stored and authentication is done online. UID number is issued to an individual by capturing and storing some of his/her personal details that will uniquely identify the person all over the globe [1].

K. Varshney (✉) · R. Johari · R.L. Ujjwal
USICT, Guru Gobind Singh Indraprastha University, Sector 16-C, Dwarka,
New Delhi 110075, India
e-mail: karishma.varshney2990@gmail.com

R. Johari
e-mail: rahul@ipu.ac.in

R.L. Ujjwal
e-mail: ujjwal_rl@rediffmail.com

© Springer India 2016
S.C. Satapathy et al. (eds.), *Information Systems Design and Intelligent Applications*, Advances in Intelligent Systems and Computing 433,
DOI 10.1007/978-81-322-2755-7_45

The Unique Identifier System (UIS) is an application developed using C sharp programming language on Aneka Platform. UIS allow you to assign a unique identification number (UID) which will help to recognize the person globally instead of having various different identification proofs like PAN card, Driving license, Aadhaar card, Voter card, Ration card and many more.

2 How UIS Work

As soon as the child takes birth, his/her complete information will be disclosed by the parents to the hospital authority. Hospitals will then transfer the details to Municipal authority. The Municipal authority uses the UIS software for creating the UID for the child and a unique identity number will be assigned to him/her.

The details being asked by the UIS are as follows:

- Country code
- State code
- City code
- Block number
- House number
- Date of birth
- Gender

This will create a 15 digit code that will help to identify the person like its country he/she belong to, state in which he/she resides, city code etc. (see Fig. 1).

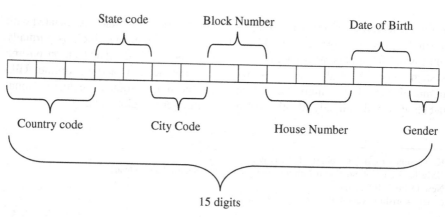

Fig. 1 Format of UID created

3 Motivation for Doing Work

Presently every person has more 5–6 identity proofs like PAN card, Ration card, Driving license, Aadhaar card, Voter card etc. It is difficult to maintain such a large database for a population of billion of people across the globe. It is wastage of memory to store details for a single person in different format. Secondly the authentication for each of the proof requires huge amount of computation again leading to wastage of resources.

So there is a need to create a single identity proof for person that will work all over the world. For this purpose UIS is being developed for creating a UID for each individual. Thus reducing the size of databases as well as instead of authenticating so many identity proofs, authentication of only one UID is required, thus saving the consumption of resources.

4 Methodology Adopted

4.1 Flow Chart

See Fig. 2.

4.2 Algorithm

```
1.  Let uid be a string variable.
2.  uid = Null
3.  Enter the country_code
4.  if country_code lies in 100 to 999
5.          uid = country_code
6.          goto line 8
7.  else goto line 3
8.  Enter the state_code
9.  if state_code lies in 10 to 99
10.         uid = uid + state_code
11.         Goto line 13
12. else goto line 8
13. Enter the city_code
14. if city_code lies in 10 to 99
15.         uid = uid + city_code
16.         goto line 18
17. else goto line 13
```

```
18. Enter the block_code
19. if block_code != 'Special Character'
20.          uid = uid + block_code
21.          Goto line 23
22. else goto line 18
23. Enter the house_code
24. if house_code lies in 1 to 999
25.          uid = uid + house_code
26.          goto line 28
27. else goto line 23
28. Enter the dob_code
29. if dob_code lies in 1 to 31
30.          uid = uid + dob_code
31.          goto line 33
32. else goto line 28
33. Enter the gender
34. If gen_code entered correctly
35.          uid=uid+gen_code
36.           goto line 38
37. else goto line 33
38. return uid
```

Unique identity number is generated using seven if-else blocks and stored in the variable '*uid*'. Lines 4–7, 9–12, 14–17, 19–22, 24–27, 29–32 and 34–37 are if-else blocks involved in the creation of uid. So running time of algorithm is constant, i.e., $O(1)$.

As the unique identification code generated is of 15 bytes and is stored in a string variable '*uid*' which is returned in line 38, the algorithm has constant space complexity, i.e., $O(1)$.

5 Related Work

5.1 Yoganandani et al. [2]

The paper talks about how Cloud Computing has been evolved through years. Starting from the definition of Cloud Computing, it describes the elements of Cloud computing like resource pooling, broad network access and rapid elasticity. Followed by this, the author describes the cloud services such as SaaS, PaaS, IaaS, cloud deployment models which includes public, private and hybrid cloud, cloud computing platforms and lastly the security issues and challenges in cloud computing.

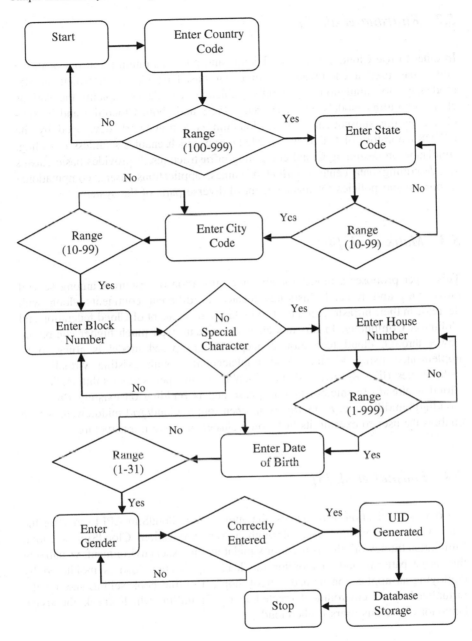

Fig. 2 UIS system flow

5.2 Khurana et al. [3]

In order to use Cloud resources efficiently and gain maximum profit out of it, the author has used a Cloud analyst simulator based on Cloudsim which enables modeling and simulations in cloud's ambience. The paper explains the various cloud computing models like IaaS, SaaS, PaaS and about CloudSim and its simulation in the subsequent topics. CloudSim is a framework developed by the GRIDS laboratory of University of Melbourne which enables seamless modeling, simulation on designing Cloud computing infrastructures. It provides basic classes for describing data centers, virtual machines, applications, users, computational resources, and policies for management of diverse parts of the system.

5.3 Malik et al. [4]

This paper proposes a model for effective utilization of resources among several universities and research institutes located in different continents along with decrease in their infrastructural cost. It includes the concept of Cloud federation and Volunteer Computing. In this model, institutes can avail much higher computing power through cloud federation concept. The proposed model is going to be implemented using Virtual Cloud implementation with existing virtualization technologies (like Xen). It will be available as an open source solution. Virtual Cloud is already in implementation phase and is building on existing *ProActive* cloud/grid middleware. *ProActive* is an open source cloud/grid middleware, which enables the user to execute its tasks on a cluster or cloud infrastructure.

5.4 Koushal et al. [5]

For optimal utilization of resources, the author used CloudSim API to monitor the load patterns of various users through a web application. CloudSim is Cloud Simulation framework that is used for simulating the exact environment, visualizing the usage patterns and to monitor the load. Application load is monitored by deploying an application named "ebookshop", a online book selling site, on the virtualized cloud environment created using CloudSim which check the usage behavior of various users in the cloud.

5.5 Arackal et al. [6]

This paper deals with Scientific Cloud Computing which means that the user has to focus on the application without being concerned about the infrastructure required and the installation process. Scientific Cloud is based on Infrastructure as a Service (IaaS) model which provides high performance computing virtual clusters as well as virtual machines on demand. For providing Scientific Cloud computing, the author presents a web based job submission mechanism named SciInterface. SciInterface provides a graphical user interface for the user to interact with the scientific cloud and submit their application for processing. Some of its features are user privileges mapping (privileges are assigned to user when he/she create account for submitting their job), job submission (submitting the job for processing), job monitoring (every job submitted is mapped with a job id through which status of the job can be checked) and output logs (activities performed for processing the job can be checked).

6 Simulation Performed

UIS is a console application implemented using Microsoft Visual Studio Professional 2013 [7] and the cloud environment is created using Aneka Platform, version 3.0, developed by Manjrasoft Pty Ltd. A table named UISdata with columns sno, uid and dob for storing the serial number, uid and date of birth respectively is created using Microsoft SQL Sever 2012 [8].

```
Enter Your Country Code:
145
You Entered:145

Enter Your State Code:
23
You Entered:23

Enter Your City Code:
78
You Entered:78

Enter Your Block Number:
A3
You Entered:A3

Enter Your House Number:
234
You Entered:234

Enter Your Date of Birth(only day):
16
You Entered:16

Enter gender:
M
You Entered:M


Your UNIQUE IDENTIFICATION NUMBER (UID):
1452378A3234160

Record inserted and saved
```

Fig. 3 Output screen

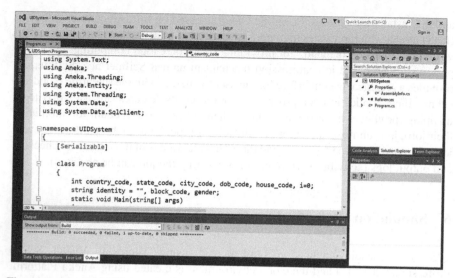

Fig. 4 Visual studio code snapshot

After entering the details like country code, state code, city code, block number, house number, date of birth and gender, the UIS will generate a 15 digit code i.e., the UID, unique for an individual (see Figs. 3 and 4).

7 Conclusion and Future Work

UIS is a console application for creating a unique identification number for an individual which will serve as an identity for the person all over the world. It's a 15 digit code in which each bit of the code depict unique information about the person like where the person belong to, from which state, its gender and many more.

The UIS is at the verge of development. A small application generated so far has been showcased in the paper but there is more to be done like handling of records of billion of population over the cloud, security issues need to be addressed, availability of resources for running the application and many more.

References

1. Aneka tool, a Cloud Application Development Platform, http://www.manjrasoft.com/manjrasoft_downloads.html.
2. PS Yoganandani, Rahul Johari, Kunal Krishna, Rahul Kumar, Sumit Maurya, C.: Clearing The Clouds On Cloud Computing: Survey Paper. In: International Journal of Recent Development in Engineering and Technology, pp. 117–121, (2014).

3. Sumit Khurana, Khyati Marwah, C.: Performance evaluation of virtual machine (vm) scheduling polices in cloud computing. In: 4th ICCCNT. (2013).
4. Sheheryar Malik, Fabrice Huet, Denis Caromel, C.: Cooperative cloud computing in Research and Academic Environment using Virtual Cloud. In: IEEE. (2012).
5. Sandya Koushal, Rahul Johri, C.: Cloud Simulation Environment and Application Load Monitoring. International Conference on Machine Intelligence Research and Advancement. In: IEEE. pp. 554–558, (2013).
6. Vineeth Simon Arackal, Aman Arora, Deepanshu Saxena, Arunachalam, B, Prahlada Rao, B B. C.: SciInterface: A web based job submission mechanism for scientific cloud computing. In: IEEE. (2013).
7. Microsoft Visual Studio Professional 2013, https://www.visualstudio.com/.
8. Microsoft SQL Server 2012, https://www.microsoft.com/en-in/download/.

A Novel Graphical Password Authentication Mechanism for Cloud Services

M. Kameswara Rao, T. Usha Switha and S. Naveen

Abstract Password provides high security and confidentiality for the data and also prevents unauthorized access. So, the most popular authentication method which is the alphanumeric passwords that provides security to users which are having strings of letters and digits. Due to various drawbacks in text based passwords, graphical password authentication is developed as an alternative. In graphical password authentication, password is provided based on the set of images. For users it is easy to remember images than text and also graphical passwords provide more security when compared to text based. There are two techniques in graphical passwords. They are Recognition based technique and Recall based technique. To provide more security to user a new idea has been proposed by combining Recognition based and Recall based techniques in this paper.

Keywords Graphical password · Authentication · Recognition based technique · Recall based technique

1 Introduction

Now-a-days, providing system security for the user has become more important. So password is provided for authentication. To provide security there are different authentication mechanisms and among them alphanumeric passwords which provides high security are also known as text based passwords. In text based system password contains a string of letters and digits. So, these text based passwords are strong enough and that password complexity is directly proportional to password

M. Kameswara Rao (✉) · T. Usha Switha · S. Naveen
Department of Electronics and Computer Engineering, K L University, Guntur,
Andhra Pradesh, India
e-mail: kamesh.manchiraju@kluniversity.in

T. Usha Switha
e-mail: usha.tadepalli11@gmail.com

© Springer India 2016
S.C. Satapathy et al. (eds.), *Information Systems Design and Intelligent Applications*, Advances in Intelligent Systems and Computing 433,
DOI 10.1007/978-81-322-2755-7_46

security. However, text based passwords are easily guessable, easy to guess through dictionary attacks, brute force attacks, key logger, social engineering etc. To provide more security an alternative solution for text based alphanumeric passwords A graphical password authentication has been developed. In this authentication system user have to select the password from a set of images in a specific order. The image that should be selected can be of any type such as image of nature, flower, place, person etc. Users can remember or recognize the images better than the text. But shoulder surfing is an eminent problem in graphical passwords. The meaning of shoulder surfing is looking over someone's shoulder to know passwords. To overcome this problem both the Recognition based and Recall based techniques are used. In Recognition based technique a set of images are presented to user for selection and at the time of authentication correct images should be clicked in sequential order. In Recall based technique user has to reproduce the same that are created or selected at the time of registration.

2 Related Works

Graphical password schemes are categorized into Recognition based and Recall based graphical password schemes. In Recognition based techniques, at the time of registration a user is given a set of images and he or she gets authenticated by recognizing and identifying the images. In Recall based techniques, a user is asked to reproduce that images in a sequential order in which he or she have created at the registration stage. Dhamija et al. [1] proposed a graphical password authentication scheme in which a user is asked to select a sequence of images from a set of random pictures. Later, for authentication user has to select the pre-selected images. Sobrado and Birget [2] proposed a graphical password technique which acts as a shoulder surfing resistant, in which system displays a specific number of pass objects (pre-selected by the users) from many objects that are given. In Man et al. [3] algorithms a number of images are selected by user as pass-objects. Each pass-object with a unique code has several variants. During authentication process several scenes are presented before user. Jansen et al. [4] proposed a mechanism based on graphical password for mobile devices. Takada and Koike [5] discussed a similar technique for mobile devices. Real user corporation developed a Pass-face algorithm [6] where user is asked to select four images from database as their password.

Basic types of Recall based techniques are reproduce a drawing and repeat a selection. Reproduce a drawing group includes DAS (Draw-a-Secret), Passdoodle method Sykuri method etc. Jermyn et al. [7] proposed Draw-a-secret method in which user has to draw their password which should be unique. Passdoodle method which was proposed by Goldberg et al. [8] consists of text based or hand written designs, usually drawn on touch sensitive screen with a stylus. Syukri et al. [9] proposed a system where user has to draw their signature using mouse for authentication. Blonder [10] designed a graphical password scheme in which user has to click on several location on an image by which password is created. Wiedenbeck et al. [11] proposed PassPoint Method which is the extension of Blonder's idea by eliminating the boundaries which are predefined and allows the arbitrary images for use. Passlogix Method [12] was also based on Blonder's idea. In this method user must click on different items in the image in correct sequential order for authentication. For each item invisible boundaries are defined so as to detect whether that particular image is clicked or not using mouse. Other related works are specified in [13–15].

3 Proposed Scheme

The proposed method is the combination of both recognition and recall based password techniques. The recall based is defined on the reproduction of picture selected at the time of registration. The various phases of the developed password are to correctly protect the user's information and restrict other than them to login to his account. The various phases of the password creation and execution are as follows:

Options Screen The main options are given on this screen so that entire handling will become easy for the user to reach any stage easily without difficulty. This is the first and most important part of the application which makes the user a user friendly application. All options are maintained at this screen.

Registration At the time of login the user must provide the unique ID number given at the time of registration. Then by following the recalling of the password sequence of the age and immediate selection of the age sequenced image should be done. That sequence of age and the image must match to the registered user unique ID password.

Login At the time of registration the user has to provide the details like user name, phone number, Gmail then he must select a sequence of age's group along with one photo at every sequence of age selection. Then by selecting the save button then the entire details of the user will be saved by giving a user id, which becomes the key attribute at the login time (see Fig. 1).

Recovery System By the help of Gmail account as a reference the recovery system is done. As these images are stored in the data base by the help of them, the names of the images are sent to the registered mail id and again the access can be provided. This phase plays an important role to recover the user's account.

Steps Involved

1. First the user must use the register option to get use of any particular site.
2. After selecting the registration phase then give the user details like user name, email, phone number etc.,
3. Then he must register the password both recall and recognition based mechanism.
4. The recall phase is for the age selection of particular age sequence.

Fig. 1 User login screen

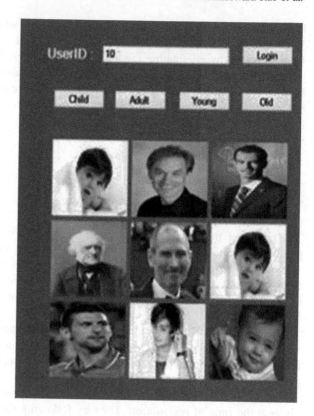

5. At every age selection the respective recognition based image must also be selected.
6. Then after completing the age sequence with respective to image of particular age sequence select the save button which completes the registration phase.
7. The immediate mail consisting of the sequence and the images as password is sent for recovery or protection purpose.
8. Then the user must go to login screen by selecting the login button.
9. Here he must reproduce the correct user id number which he had registered at the time of registration.
10. Then followed by the recalls age sequence and the respective recognition images in exact sequence for perfect login.
11. Any sequence change or image change may leads to restrict or prevent from the web site entry.

All the set of images are stared in the separate data bases of respective age groups of child, adult, young and old. Those data bases are consisting of images for recognition of nine images each age group. These images of 36 which are 9 of each age group must be shuffled to provide complexity and improve protection. The image selection for every recalled age group will proceed in four levels to complete

Fig. 2 Image database for child, young, adult, old categories

the entire correct password to be given. These selecting of 9 images from the database is done in a particular order as given below (see Fig. 2).

At first level the 9 images are selected as the one image of the 9 is from the 4 images of the registered time and the remaining images are taken two from each databases of the age sequence. Then the next second level of 9 images consists of second image of the second of the four registered images with the remaining two images of the databases of age sequences. This process continues for the four levels.

4 Security and Usability Study

A user study was conducted involving 24 graduate students to study usability, Security and login times for the proposed scheme after a learning session on the proposed scheme. The average login time for the proposed scheme is 32.3 s. The security of the proposed scheme depends on the change in the order of the image category selection among child, young, adult and old images. Increasing the levels in the selection of images will provide additional security.

5 Conclusions

Thus by combining both the recall and recognition based methods, more protection is given to the user account. So the login user's web services are protected by using this locking system. So by combining this two we will make the unknown user to strictly avoid any hacking techniques as it has both intermixed technology. So the user account will now get the maximum security from the hackers as their technology cannot handle when two techniques are intermixed.

References

1. Brown, A. S., Bracken, E., Zoccoli, S. and Douglas, K. (2004). Generating and remembering passwords. Applied Cognitive Psychology, 18, 641–651.
2. Paivio, A., Rogers, T. B., and Smythe, P. C. (1976) Why are pictures easier to recall then words? Psychonomic Science, 11(4), 137–138.
3. Alsulaiman, F. A. and Saddik, A. E., 2006, "A Novel 3D Graphical Password Schema", Proceedings of 2006 IEEE International Conference on Virtual Environments, Human-Computer Interfaces and Measurement Systems.
4. Suo Xiaoyuan Suo, Ying Zhu and G. Scott. Owen. "Graphical passwords: a survey," Proceedings of the 21st Annual Computer Security Applications. 2005, 463–472.
5. S. Gaw and E. W. Felten, "Password Management Strategies for Online Accounts," in ACM SOUPS 2006: Proc. 2nd Symp. on Usable Privacy and Security, pp. 44–55.
6. D. Florêncio and C. Herley, "A large-scale study of web password habits," in WWW'07: Proc. 16th International Conf. on the World Wide Web. ACM, 2007, pp. 657–666.
7. D. Weinshall, "Cognitive Authentication Schemes Safe Against Spyware (Short Paper)," in IEEE Symposium on Security and Privacy, May 2006.
8. Omar Zakaria, Toomaj Zangooei, Mohd Afizi Mohd Shukran, "Enhancing Mixing Recognition-Based and Recall-Based Approach in Graphical Password Scheme", IJACT, Vol. 4, No. 15, pp. 189–197, 2012.
9. Haichang Gao, Xiyang Liu, Sidong Wang, Honggang Liu, Ruyi Dai, "Design and analysis of a graphical password scheme", in Proceeding(s) of the 2009 Fourth International Conference on Innovative Computing, Information and Control, IEEE Computer Society, pp. 675–678, 2009.
10. Tetsuji Takada, Takehito Onuki, Hideki Koike, "Awase-E: Recognition-based Image Authentication Scheme Using Users' Personal Photographs", in Proceeding(s) of Innovations in Information Technology, pp.1–5, 2006.
11. Almuairfi Sadiq, Veeraraghavan Prakash, Chilamkurti Naveen, "A novel image-based implicit password authentication system (IPAS) for mobile and non-mobile devices". Journal of Mathematical and Computer Modelling, Elsevier, 2012.
12. Muhammad Daniel Hafiz, Abdul Hanan Abdullah, Norafida Ithnin, Hazinah K. Mammi; "Towards Identifying Usability and Security Features of Graphical Password in Knowledge Based Authentication Technique", IEEE Explore, 2008.
13. A. P. Sabzevar, A. Stavrou, "Universal Multi-factor authentication using graphical pass-words", In Proceeding(s) of the IEEE International Conference on Signal Image Technology and Internet Based Systems, pp. 625–632, 2008.

14. Kameswara Rao, Sushma Yalamanchili, "Novel Shoulder-Surfing Resistant Authentication Schemes using Text-Graphical Passwords", International Journal of Information and Network Security (IJINS), Volume 1, Issue 3, 2012.
15. M. Kameswara Rao, P. Aparna, G. Avinash Akash, K. Mounica, "A Graphical Password Authentication System for Touch Screen Based Devices", International Journal of Applied Engineering Research, Volume 9, Issue 18, 2014.

A Conceptual Framework for Big Data Implementation to Handle Large Volume of Complex Data

Manas Kumar Sanyal, Sajal Kanti Bhadra and Sudhangsu Das

Abstract Globally industries, businesses, people, government are producing and consuming vast amounts of data on daily basis. Now-a-days, it's become challenging to the IT world to deal with the variety and velocity of large volume of data. To overcome these bottlenecks, Big Data is taking a big role for catering data capturing, organizing and analyzing process in innovative and faster way. Big Data software and services foster organizational growth by generating values and ideas out of the voluminous, fast moving and heterogeneous data and by enabling completely a new innovative Information Technology (IT) eco-system that have not been possible before. The ideas and values are derived from the IT eco-system based on advanced data-analysis on top of the IT Servers, System Architecture or Network and Physical objects virtualization. In this research paper, authors have presented a conceptual framework for providing solution of the problem where required huge volume of data processing using different BIG data technology stack. The proposed model have given solution through data capturing, organizing data, analyzing data, finally making value and decision for the concern stakeholders.

Keywords Big data · Apache hadoop · NoSQL · MapReduce · R programing

M.K. Sanyal (✉) · S.K. Bhadra · S. Das
Department of Business Administration, University of Kalyani, Kalyani, India
e-mail: manas_sanyal@rediffmail.com

S.K. Bhadra
e-mail: sajal.bhadra@gmail.com

S. Das
e-mail: iamsud@gmail.com

© Springer India 2016
S.C. Satapathy et al. (eds.), *Information Systems Design and Intelligent Applications*, Advances in Intelligent Systems and Computing 433,
DOI 10.1007/978-81-322-2755-7_47

455

1 Introduction

Gartner defines: "Big data is neither a technology in itself, nor is it a distinct and uniquely measured market of products or vendor revenue. It is a technology phenomenon happened due to the rapid rate of data growth, complex new data types and parallel advancements in technology stake that have created an IT ecosystem of software and hardware products that are useful for its users to analyze the collected data to extract more values and granular levels of insight" [1, 2].

Big Data is not just handling with huge *Volume* of data. It has 3 more dimensions (Fig. 1), Variety, Velocity and Value [3, 4]. *The Variety* is because social media websites had led to the explosion of unstructured data sources like web pages, logs, video streams, audio-video files, blog entries and social media posts, text messages and email. *The Velocity* at which new and updated stream of data is flowing into the system is incessant and at Real-time. *The Values or insights to derive* out of huge amount of variety of data captured with high velocity. Important information or value can be hidden among irrelevant data. It's the biggest challenge for enterprises to identify value, out the ocean of data; it is generating or receiving from outside world.

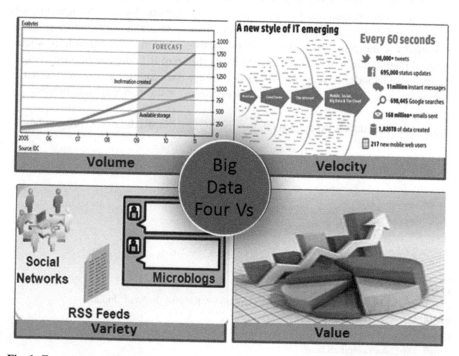

Fig. 1 Four versus of big data

Analysis of Big data can provide insights to the senior management that they would earlier not be privy to and hence enable them to take smarter, decisive and faster decisions in a world where change itself has changed [5].

Managers now can easily review inventory levels, sales pattern and customer demands on daily basis to decide which products are highly demanded or most profitable or which products are required to be procured when and in which quantity. The solution enables chain limit losses by scheduling price reductions to move perishable items prior to spoilage, effectively lowering losses on perishable goods, which are approximately 35 % of the chain's products. Stores can adjust quickly as the government's price settings on staple foods fluctuate, and the company can compile sales tax data 98 % faster than before. These improvements resulted in a 30 % increase in revenue and a US$7 million boost in profitability for the company [6].

As per the recent survey conducted by SAS among the business leaders [7], it was found that 82 % of them think that Big Data solutions are very, somewhat or extremely important for their business to achieve competitive advantages and 84 % of them have either implemented Big Data solutions (17 %) or will going to implement Big Data solutions within next 5 years (as in Fig. 2).

Oracle conducted a survey on 333 North American executives to identify their organizations' readiness and capability to extract innovative ideas from the existing system to improve its operations and explore new opportunities and new revenue. It was found that 94 % of organizations are capturing and processing more business data today than 2 years back and 93 % of them believe that their organization is losing revenue due to inability of effective leveraging of the collected data from the existing system. On average the estimated lost opportunity is to be around 14 % of annual revenue [8]. Big Data solutions can play a major role to overcome these issues faced by the enterprises to sustain in this competitive world.

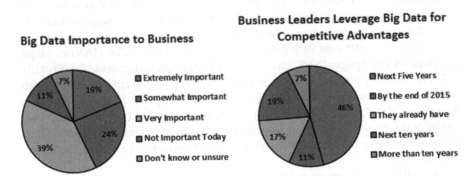

Fig. 2 Big data importance to staying competitive

2 Literature Review

In today's globally integrated, time-sensitive environment, right data is crucial across the entire business value chain, including merchandising and inventory management, distribution, marketing, sales, service, returns, finance and more. All of these must be optimized under continual directives to become leaner, more cost effective and increasingly profitable. So the question for many organizations remain, "How do we efficiently utilize these information to gain competitive advantage?" [9].

IBM's Big Data @ Work survey identifies one of the clear picture that the most organizations are currently in their initial thinking stages related to Big Data Adoption [10]. As per Gartner Report in 2012, the retail industry has the highest implementation rate of Big Data solutions. Williams Sonoma is using Big Data applications to capture information like customer purchase history, their likings and disliking, preferences etc. and after processing these data in Big Data applications, it identifies the targeted customers and products and sending discounts and offers related to specific product which has increased its huge sale [11]. Wal-Mart, the market giant retailer, uses Big Data Solutions to increase its customers, sales and business volumes. They are using mobile navigation system to identify the location information of the customers and send discount or offers related massages depending on customer's previous shopping styles and preference [2].

Various vendors are using various solution models to implement Big Data solutions for their clients. IBM uses Watson Analytics to offer cognitive, predictive, and visual analytics in an easy-to-use service to find a complete view of your business [12]. Splunk uses Splunk Enterprises for capturing, processing, and analyzing machine-generated big data on websites [13]. Kyar Nyo Aye, Ni Lar Thein proposed a big data solution, on open source and Hadoop MapReduce, Gluster File System, Apache Pig, Apache Hive and Jaql and compared the platform IBM big data platform and Splunk [14].

Amazon uses Amazon Web Services (AWS) as their big data solution tool [15]. Cloudera, market-leader in distributing of Hadoop software, uses "enterprise data hub" as big data solution that acts as the central point of data management within the organization [15]. HP uses its big-data-platform architecture called HAVEn. MapR uses Network File System (NFS) instead of HDFS to certain high data availability [15].

There are many vendors to provide Big Data solutions and are expert in various Big Data Implementation areas like Big Data Consulting and Solutions, Big Data Operations, Big Databases, Big Data Appliances, Big Data Storage, Big Data Analytics etc. ExpertOn Group conducted survey provided independent and neutral comparison among Big Data Hardware and Software vendors as well as the service providers [16].

As per their recent survey in 2015, IBM, PwC, Capgemini, SAP, T-System, HP, Steria Mummert are in Leader quadrant as Big Data Consulting and Solutions vendor whereas SAS, Teradata, Oracle, SAP, HP, Microsoft, Empolis, Software AG, Quick and Micro Strategy are placed in Leader quadrant as Big Data Analytics Software and Solutions (Fig. 3).

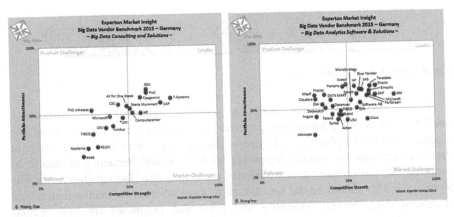

Fig. 3 Big data vendors market positioning as per ExpertOn group survey

3   Research Methodology

Big Data Solution Adoption is neither a software package/product implementation nor a hardware system installation. It is a complete data processing of an IT ecosystem where voluminous and varieties of data are produced or collected with rapid velocity. Authors have done extensive studies to understand the existing Big Data Solution approaches provided by the various vendors for various organizations depending on its business needs. Currently Big Data vendors are implementing Big Data Solution mainly for capturing and organize the data. Various brainstorming sessions were conducted along with the consultants from Big Data vendors to devise this new model for Big Data Solution. All the Big Data processing steps are explained clearly using appropriate diagrams and the required tools or technologies that can be used in the various data processing steps are mentioned in brief. Extensive Literature reviews were done based on the secondary data from various books, journals and research papers. Existing project implementation and consulting knowledge of authors in IT Systems helped to get the idea for designing the new Big Data Solution framework.

4   Conceptual Framework for Big Data Implementation

Authors have studied various Big Data solution approaches implemented so far for various organizations. Considering all the pros and cons of various Big Data Solutions, A uniform global framework is envisaged for Big Data Solutions which would be helpful for both the enterprises and vendors for their smooth implementation and execution of the business processes. The framework consists of four phases of data processing as shown in Fig. 4.

Fig. 4 Four-phases big data solution

All the business related data, in spite of their form i.e. Structured, Semi-Structured or Un-Structured, generated or collected at any velocity should be captured in the first phase called *Capture Data* phase. There are various types of databases that can be selected depending on the type and frequency of data to be captured. For unstructured data, distributed file system like Hadoop Distributed File System (HDFS) should be the best option to store whereas for semi-structure and structured data, NoSQL Database and Relational Databases are best options respectively to capture and store.

In *Organize Data* phase, Data is organized to eliminate redundancies and it can be partitioned logically to store in a relational database or a data warehouse. A programming paradigm called MapReduce is used to interpret and refine the data. In *Analyze Data* phase, the refined and organized data is fed into a relational database to perform further analysis, to search for performance issues and to improve business values. Various mining tools and analytical tools can be used in this phase. In *Values* and *Decision* phase, based on the analysis, it can be projected the results on a dashboard and can make business decisions.

All these four phases of data processing are not mandatory for an organization to adapt or implement as part of Big Data Solution and it depends on their business needs. For example, an organization generates huge amount of structured data on daily basis and has huge volume of business data, which is not possible to be accommodated in conventional relational databases, can implement first two phases (i.e. *Capture Data* and *Organize Data*) only and remaining two phases might not require immediately for its business need.

Most of the organizations, using IT Systems, must have their own existing systems for analyzing the collected data and extracting business values by preparing reports or publishing dashboards for top managements to take business decisions. They can implement first two phases to capture more data, in the form un-structure and semi-structure, to extract hidden business values and use their existing systems for Analyzing and Business Decisions if the managements think that the existing system is well intelligent to solve their purposes. Otherwise they can think of implementing all the phases of Big Data solutions. Thus it is organization's decision to choose which one to implement for their business needs.

As per our study, most of the organizations use relational database which is able to capture the structure data only. But various applications generate massive volumes of un-structure and semi-structure data that describe user behavior and application performance. Many organizations are unable to fully exploit these potentially valuable information because of their existing system are not intelligent enough to extract values from high volume un-structure and semi-structure data. Big Data implementation is the only solution to address all these above mentioned issues.

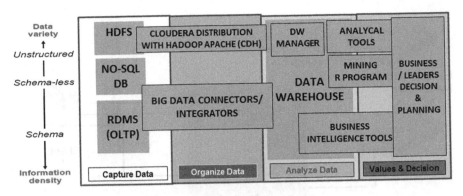

Fig. 5 Integrated big data solution framework

The conceptual framework is devised for Big Data Implementation for conducting the four phase of data processing as shown in Fig. 5.

Various tools and software are available for various phases. Big Data Implementation vendors and organizations can decide which one to implement depending on their business need and budget.

4.1 Capture Data

This phase captures or acquires data from the various source systems. The data may be of various types as discussed previously that structured, semi-structured and un-structured. Depending on the variety of data, various tools like Relational Database Management System (RDBMS), Not Only SQL (NoSQL) DB and Hadoop Distributed File System (HDFS) can be used to store data.

RDBMS RDBMS is a well-known schema-centric database where the collected data is stored into row-column construct, logically called Table. It is used to store high value, high density complex data in Tables using complex data relationships. ACID (Atomicity, Consistency, Isolation and Durability) is a set of factors that ensures that the reliability of the processed database transactions. There are well-known RDBMS available in the market like Oracle, DB2, SQL Server, MS Access, MySQL etc.

NoSQL DB NoSQL (Not Only SQL) database is a schema-less storage. There is no defined schema for data to store. Data is stored according to the data model of NoSQL database types, selected for the application. Various NoSQL databases have different associated data models like *Columnar* for Cassandra, *Key-value* for Voldemort and Oracle NoSQL database, *Document* for MongoDB, *Graph* for Neo4j etc. BASE (Basic Availability, Soft State and Eventual Consistency) ensures the data availability. NoSQL DB is useful to store high volume dynamic data with value. NoSQL Database stores all the major data types like voice, spatial, xml, document, text, numeric, video, image etc. Data is stored as byte arrays.

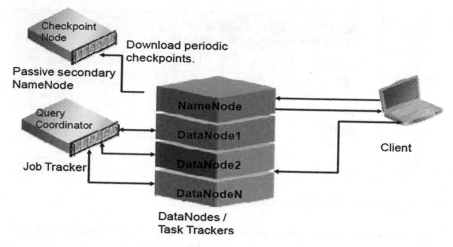

Fig. 6 Hadoop distributed file system cluster

HDFS Hadoop, developed by Apache, is a framework, operates by batch processing, for user applications running on large clusters built of commodity hardware. It is open-source software, written in Java, and it applies the MapReduce paradigm. It uses Hadoop Distributed File System (HDFS) to store and replicate data across multiple nodes. For example, in a MapReduce operation, Hadoop first divides the processing work in chunks and distributes the work among available servers (maps). Next it executes programmatic aggregation of data (that is reduces to address the business issue). Hadoop management and monitoring software determines which operations can and cannot be accomplished in parallel, based on dependencies. The default Hadoop storage engine is HDFS.

HDFS works on well-known master/slave architecture, where a single master *NameNode* is managing multiple *DataNodes* (slaves). A typical HDFS implementation consists of a dedicated system that executes only the *NameNode* software. Data is replicated and stored across multiple *DataNodes*. *Hadoop Client* is a terminal that initiates processing in HDFS (as in Fig. 6). *Job Tracker* hands out the job to *Task Trackers*. *Checkpoint Node* is a secondary *NameNode* in different system but with identical directory structure that can be imported to replace the primary *NameNode* if necessary. It periodically captures the snapshots of the *NameNode* directory, enabling faster recovery if the *NameNode* fails.

Considering the organization requirements depending on the factors like Data Volume, Real Time Information, Data Variety and Data Change, a combination of relational database, NoSQL database and HDFS can be used to capture data in the system.

4.2   Organize Data

Once data is captured and stored in Capture Data phase, it's required to refine the redundant data in Organize Data phase and partitioned into relational databases or data warehouse. There are various tools or software that can be used in this phase like Hive, Cloudera Distribution with Hadoop Apache and Big Data connectors and integrators.

Hive supports analyzing vast amount of data captured in Hadoop-compatible file systems such as HDFS, Amazon FS and soon. Hive uses a SQL-like query language called HiveQl to define and manipulate the data. It's a powerful tool that is widely used by Hadoop users to create normal tables and external tables to load delimited and semi-structured data.

Big data connectors are primarily used in the organize phase to refine the redundant data having high performance, security and efficiency. It includes Loader for Hadoop (LH), SQL Connector for HDFS (SCH), Data Integrator Application Adapter for Hadoop (DIAAH) and R Connector for Hadoop (RCH).

MapReduce is a programming model for batch data processing. It is designed to scale easily on Big Data and is particularly efficient in processing large data sets. MapReduce process is explained in details Fig. 7. In the *Map phase*, the Map tasks can be executed parallelly by generating the key-value pairs as the output. Next it's comes the Shuffle and Sort phase, where it partitions the output of previous phase using hash function and followed by sorting operation on the shuffled data. The main goal of partitioning is to ensure that all key-value pairs that share the same key should be placed into the same partition.

The final phase is the *Reduce phase*, where it assigns each partition to one Reducers running in parallel like Mappers. The number of reductions depends on the density of the data.

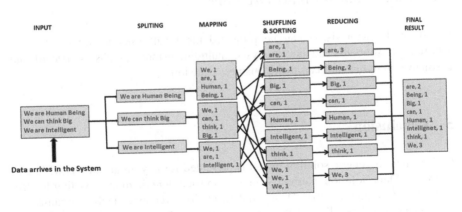

Fig. 7 MapReduce process example

4.3 Analysis Data

The main goal of this phase is to use various Analytical and Business Intelligence (BI) tools to extract value from the organized and refined data after feeding them to relational databases. Data Warehouses and DW managers played vital role in the regard. Various relational databases can be used for Data Warehouse depending on the cost and the complexity of data to be analyzed. Some well-known relational DBMS like Oracle from Oracle Corporation, DB2 from IBM and SQL server from Microsoft can be used.

There are various *Analytical tools, R programs* and *BI tools* to analyze the data from Data Warehouse to extract business values. Different vendors use different analytical tools depending on the organizations business. There are few well-known of Big Data Analytical tools available in the market like SAS, Actian Matrix and Actian Vector, Amazon Redshift service, Oracle Crystal Ball, Cloudera Impala, HP Vertica Analytics, IBM PureData System for Analytics, Microsoft SQL server 2012 Parallel Data Warehouse (PDW), SAP Hana and SAP IQ for business analytic.

R Program is an open-source statistical programming language and environment that provides a powerful graphical environment for the analyzed data visualization, several out-of-the-box statistical techniques and interactive GUI front ends for analyzing data.

There are few market leading Business Intelligence (BI) platforms available like SAP Business Objects, Oracle Hyperion BI tools, IBM Cognos Business objects, Microsoft SQL Server reporting, TIBCO Spotfire, MITS Distributed Analytics, Infor BI etc. These BI platforms can be implemented in any organization depending on needs.

4.4 Values and Business Decisions

Based on the analysis, it can be projected the results that includes enterprise reporting, dashboards, ad hoc analysis, multidimensional graphs, scorecard and predictive analytics and can make business decisions.

5 Conclusion

In this study, a generic solution model is presented for any organizations willing to adopt Big Data Solution. Currently there is no such generic model existing for Big Data Implementation. Big Data implementation vendors devise organization specific models depending on the organizations' business need. Any organizations and implementation vendors can use this conceptual framework to adapt the Big Data solution and leverage the utilities of this framework.

References

1. Akella, Janaki, Timo Kubach, Markus Löffler, and Uwe Schmid, *Data-driven management: Bringing more science into management*, McKinsey Technology Initiative Perspective, 2008.
2. Gartner Report on Big Data: August 2014.
3. Alan E. Webber, "B2B Customer Experience Priorities In an Economic Downturn: Key Customer Usability Initiatives In A Soft Economy," Forrester Research, Inc., Feb. 19, 2008.
4. "Analytics: The real-world use of big data", IBM Institute of Business Value, accessed Feb 11, 2012.
5. "Beyond the Hype of Big Data", CIO.com, October 2011, accessed Feb 11, 2012.
6. "Retail 2020: Reinventing retailing–once again." IBM and NewYork University Stern School of Business. January2012.
7. SAS 2013 Big Data Survey Research Brief, http://www.sas.com/resources/whitepaper/wp_58466.pdf.
8. Oracle Industries Scorecard http://www.oracle.com/us/industries/oracle-industries-scorecard-1692968.pdf.
9. Gleick, James, *The information:A history.A theory.A flood* (New York:Pantheon Books, 2011).
10. Schroeck, Michael, Rebecca Shockley, Dr. Janet Smart, Professor Dolores Romero-Morales and Professor Peter Tufano. IBM Institute for Business Value in collaboration with the Saïd Business School, University of Oxford. October 2012.
11. Kiron, David, Rebecca Shockley, Nina Kruschwitz, Glenn Finch and Dr. Michael Haydock. IBM Institute for Business Value in collaboration with MIT Sloan Management Review. October 2011.
12. IBM Analytic Tools http://www.ibm.com/marketplace/cloud/watson-analytics/us/en-us.
13. Splunk Big Data Tool http://www.splunk.com/en_us/products/splunk-enterprise.html.
14. Kyar Nyo Aye, Ni Lar Thein: A Comparison of Big Data Analytics Approaches Based on HadoopMapReduce, 2013.
15. Big Data Platform Comparisons http://www.informationweek.com/big-data/big-data-analytics/16-top-big-data-analytics-platforms/d/d-id/1113609?image_number=4.
16. Big Data Vendor Benchmark 2015 by Experton Group by Holm Landrock, Oliver Schonschek, Prof. Dr. Andreas Gadatsch.

Path Reliability in Automated Test Case Generation Process

Kavita Choudhary, Payal Rani and Shilpa

Abstract In software testing, the reliability of a path is an important factor and must be calculated to go forth in the testing process. This paper is the extended work (Cuckoo Search in Test Case Generation and Conforming Optimality using Firefly Algorithm, 2015) [1], previously test cases are generated using cuckoo search. The reliability is calculated with the help of control flow graphs and mathematical calculations for all the paths specified in flow graph (Integrating Path Testing with Software Reliability Estimation Using Control Flow Graph, 2008) [2]. We consider various test cases that are traversing different paths and accordingly compute reliability of the paths.

Keywords Reliability · Path testing · Test case generation · Control flow graph

1 Introduction

Reliability is the most important requirement for non-functional tasks which must be fulfilled in order to complete the phase of software testing. The probability that the system will work correctly under given environment condition for certain amount of time is reliability. Software reliability testing is performed to fulfill this requirement. In previous work, the successful generation and optimization of test cases for Quadratic classification problem using cuckoo search and firefly algorithm techniques are elaborated. This paper will consider the same problem and will calculate the reliability of the nodes of the software by considering the control flow graph.

K. Choudhary (✉) · Shilpa
ITM University, Gurgoan, Haryana, India
e-mail: kavitapunia@itmindia.edu

P. Rani
Banasthali University, Jaipur, India

© Springer India 2016
S.C. Satapathy et al. (eds.), *Information Systems Design and Intelligent Applications*, Advances in Intelligent Systems and Computing 433,
DOI 10.1007/978-81-322-2755-7_48

467

Cuckoo search is an algorithm which implements on the basis of the breeding behavior of cuckoo bird. The paper presents the pseudo code for the generation of test cases for Quadratic classification problem. The paper described one more algorithm the firefly algorithm which is based on flashing of fireflies to attract the mates. This algorithm is used for the optimization of the test cases generated by cuckoo search. The pseudo code presented in the paper can be used to check the optimality of any given set of test cases for a given problem. The pseudo code for cuckoo search and firefly algorithm is presented in Sect. 3. The categorization of the reliability of system is based on the estimation and prediction model. When for a given software historical data is used to analyze and estimate the reliability the software product, it is called as prediction model. The analysis and the estimation of current data from current efforts of development are known as estimation model. The reliability is estimated through the markov chain process. The process involves one of each path for each period of time. If the process starts from node 1, then for the next time it will move to node 2 with some probability and the probability is evaluated on the basis of current state. The state sequence generated in the course of time is markov chain. Process starts from root node with the highest probability and moves to next node with probability 1/5th.

2 Literature Review

Choudhary et al. [1] presents the two different algorithms for the test case generation and optimization. In the testing phase there is a requirement for generation of test case and after that optimization is done which test cases are optimal for the algorithm. Here, the paper discussed comparison between cuckoo search and firefly algorithm and results shows the coverage of every node is done on the set of test cases i.e. path coverage plays the important role. Gokhale et al. [3] put forwards a concept for the evaluation of the system architecture that is built on component reliability including architecture application. Architecture analysis is done on the basis of path or state approach. The estimation of the reliability through various executed path performed in approach of path based and on the other side, in state based a model of state space is joined to the control flow graph. Different issues encountered during implementation that is for results optimality and failure of interface to assess the system reliability. Huang et al. [4] suggested a concept for reliability assessment that is component based and solves 2 problems related to allocation of resources. Test results verify that the methods discussed here are able to solve problems regarding allocation of effort and also improves software on basis of its quality and its reliability. Trivedi et al. [5] presented that there is requirement for modeling concept that accounting the architecture and examine reliability via considering the components interaction, component's reliability and interface reliability accompanied by different components. The need for this concept is module identification that is modeling the system architecture, identification of behavior of failure and combination of failure with system architecture. With three different ways we can combine it i.e. path, state and by additive approach. The issues considered here for the component size definition, reliability of interface and probability of transition not examined. Huan et al. [6] proposed a framework from including path testing into the estimation of reliability of modular software systems. Here three methods are put forward for evaluation of reliability through sequence wise, branch wise and by loop. Test results showed that the reliability of path and software Reliability are highly related to it. Roberto et al. [7] presented an application to recognize the important ingredient so that allocate the best resource to those. The relation has been made of system reliability with the allocation of testing time. System enhances as reliability improves. Overall software reliability can be computed through individual component reliability and how many visits to these components. Hsu et al. [2] calculates the reliability of path and then this value is used to find estimated system reliability. Here, proposed process of Markov chain includes the control flow graph which represents the system reliability and also gives evidence to the theorem presentation for the structures of the program. This gives us various advantages including the ability to use the proposed method in the software testing phase. Hence it can be said that presented method is very convenient for assessment of reliability in systems that are architecture based. Fiondella et al. [8] proposed a module that will consider every component for the computation of software reliability. After that estimate the effort amount that is allotted to every component so that we can acquire the maximum reliability along with minimum effort. The reliability is conducted according to the relation between how much effort executed on component and their reliability. Main factor that impact on component reliability is the testing effort on this component. Effort can be calculated by complexity of technology used for implementation of component. Through reduction in specific component reliability and rate of fault detection, optimally we can achieve the spent effort. Gokhale et al. [9] proposed the approach for the assumption of system reliability that is on basis of architecture. The approach uses the measurements which come through the testing results for providing the parameters to the software model. The estimated value of reliability that is obtained using this approach will depend on the testing suit that is used. Hence should use the best possible test suite that suits the application. Xie et al. [10] put forward a process that predicts the reliability of software through defined model of reliability. There is a requirement of huge data for estimating the reliability. The growth of new large systems is carried in a way i.e. occurrence of already existing product is modified and checks the faults to consider the new parameters for the system. Huang et al. [11] initiated a concept in system reliability examination. The supposition is that correction can be easy applied on detected faults but that is wrong in reality. Fault detection considered as major issue and rectifying it is another issue. The paper discussed two different types of faults i.e. independent and dependant. The faults that are independent can be directly removed but if they are dependant faults then firstly have to remove the leading fault i.e. the consumption of huge time and higher cost it will take.

3 Reliability of Path from Root to Target Node

Here the Cuckoo search is used for the generation of test cases. The generated tests cases follow the path from root node to the target node. After following the path, the best solutions are optimized which are stored in the optimal solution repository and these are final set of the test cases. In this section, reliability of path from the calculated target node (using cuckoo search) is computed. The pseudo code for the test case generation using cuckoo search is as follows (Fig. 1):

3.1 Control Flow Graph Representing Transition Probability

The following flow graph comprises all conditions that is required for determination of roots and quadratic problem and also includes the reliability of node and transition probability. Firstly, the transition probability considered for node E is 1 and after that it is divided as per the path.

For the calculation of path reliability, N {E, 1, 2, 3, 4, 5, 6, 7, 8} is the complete node set in control flow graph and R defined as reliabilities of nodes. Let R1 be the

Fig. 1 Pseudo code for cuckoo search

```
            Pseudo code (Cuckoo Search)
 1.   Proc input: test Case Generator
 2.   TN[]<— Target nodes
 3.   Max_it <— TN.length
 4.   Pop_size <— ⌊Max_it/2⌋
 5.   i=1
 6.   initial_sol[Pop_size-1] = rand()
 7.   OSR<—0
 8.   WSR<—0
 9.   While(i!= Max_it) do
10.   ITN=TN[i-1]
11.   FV(ITN)=No of nodes covered
12.   Pop_S<—rand()
13.   Initial_sol[pop_size] <—Pop_S U Initial_sol[Pop_size-1]
14.   For j<—1 to Pop_size do
15.   FV(Initial_sol[j-1]) <— No of conditions satisfied
16.   If FV(Initial_sol[j-1])==FV(ITN)
17.   OS_TT<— Initial_sol[j-1]
18.   Go to step 21
19.   Else WS_TT<—Initial_sol[j-1]
20.   End if
21.   if(j=Pop_size)
22.   Pop_S<—rand()
23.   Initial_sol[j-1] <—Pop_S
24.   Else j=j+1
25.   End if
26.   Go to step 9
27.   i=i+1
28.   end while
29.   for i=1 to Max_it
30.   OSR= OSR U OS_TT
31.   WSR= WSR U WS_TT
32.   End_proc
```

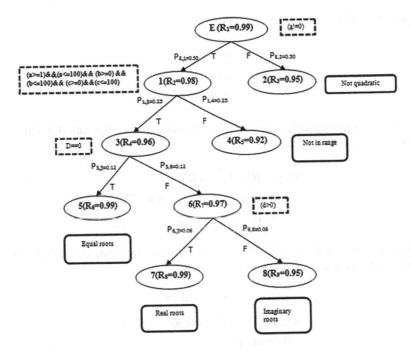

Fig. 2 Control flow graph

reliability of node E, R2 be the reliability of node 1, R3 of node 2, R4 of node 3, R5 of node 4, R6 of node 5, R7 of node 6, R8 of node 7, R9 of node 8. It determines that whether the current node is properly executed and then control is handled to the next node. P is considered as transition probability of edges and the $P_{i,j}$ represents the transition that has been taken from Node 1 to another. Pt_n represents the execution path from input module to the output module (Fig. 2).

The Control flow graph used here for the determination of nine modules where node E is the input module and node 2, 4, 5, 7, 8 are the output modules. To simplify the calculation, let the reliability of the modules be constants as follows:

$$R1 = 0.99 \quad R2 = 0.98 \quad R3 = 0.95 \quad R4 = 0.96 \quad R5 = 0.92$$
$$R6 = 0.99 \quad R7 = 0.97 \quad R8 = 0.99 \quad R9 = 0.95$$

Let the branching probabilities between the different modules be

$$P_{E,1} = 0.50 \quad P_{E,2} = 0.50 \quad P_{1,3} = 0.25 \quad P_{1,4} = 0.25$$
$$P_{3,5} = 0.12 \quad P_{3,6} = 0.12 \quad P_{6,7} = 0.06 \quad P_{6,8} = 0.06$$

Firstly generated random test case <5, 0, 95> that will execute on the path (EF). If the input node is E, then the execution will be start from this node. Therefore, transition probability for the node 1st will be considered as 1.

Now, reliability of path 1 is as follows:

$$R_{Pt1} = R_1^1 * R_3^{P_{E,2}}$$
$$= 0.99^1 * 0.95^{0.50}$$
$$= 0.96493264013$$

Now, we will consider test case <5, 44, 50> that traverses the path (ET,1F). Then reliability of path 2 is as follows:

$$R_{Pt2} = R_1^1 * R_2^{P_{E,1}} * R_5^{P_{1,4}}$$
$$= 0.99^1 * 0.98^{0.50} * 0.92^{0.25}$$
$$= 0.95983192137$$

Next test case will be <95, 95, 95> that follow the path (ET,1T,3T). Then reliability of this path is calculated as:

$$R_{Pt3} = R_1^1 * R_2^{P_{E,1}} * R_4^{P_{1,3}} * R_6^{P_{3,5}}$$
$$= 0.99^1 * 0.98^{0.50} * 0.96^{0.25} * 0.92^{0.12}$$
$$= 0.96044075392$$

Now, we consider test case <25, 25, 51> that traverses the path (ET,1T,3F,6T). The reliability of path 4 is as follows:

$$R_{Pt4} = R_1^1 * R_2^{P_{E,1}} * R_4^{P_{1,3}} * R_7^{P_{3,6}} * R_8^{P_{6,7}}$$
$$= 0.99^1 * 0.98^{0.50} * 0.96^{0.25} * 0.97^{0.12} * 0.99^{0.06}$$
$$= 0.96597694337$$

Lastly, we consider the test case <7, 10, 15> that traverses the path (ET,1T,3F,6F). Reliability of path 5 is calculated as:

$$R_{Pt5} = R_1^1 * R_2^{P_{E,1}} * R_4^{P_{1,3}} * R_7^{P_{3,6}} * R_9^{P_{6,8}}$$
$$= 0.99^1 * 0.98^{0.50} * 0.96^{0.25} * 0.97^{0.12} * 0.95^{0.06}$$
$$= 0.96358951371$$

4 Simulation Results

The results are shown in Table 1.

Table 1 Estimated path reliability (root to target node)

Iteration (i)	Target node	Path (root to target)	Random solution	Current cuckoo nest	Path (cuckoo nest elements)	Fitness value	OS_{Ti}	WS_{Ti}	Path reliability
i = 1	Node-2	(EF)	<95,95,95>	<5,0,95> <7,10,15> <95,95,95>	(EF) (ET,1T,3F,6F) (ET,1T,3T)	1 0 0	<5,0,95>	<7,10,15> <95,95,95>	0.9649
i = 2	Node-4	(FT,IF)	<5,44,50>	<7,10,15> <95,95,95> <5,44,50>	(ET,1T,3F,6F) (ET,1T,3T) (ET,1F)	1 1 2	<5,44,50>	<7,10,15> <95,95,95>	0.9598
i = 3	Node-5	(ET,1T,3T)	<25,25,51>	<7,10,15> <95,95,95> <25,25,51>	(ET,1T,3F,5F) (ET,1T,3T) (ET,1T,3F,6T)	2 3 2	<95,95,95>	<7,10,15> <25,25,51>	0.9604
i = 4	Node-7	(ET,1T,3F and T)	<3,4,5>	<7,10,15> <25,25,51> <3,4,5>	(ET,1T,3F,5F) (ET,1T,3F,5T) (ET,1T,3F,6F)	3 4 3	<25,25,51>	<7,10,15> <3,4,5>	0.9659
i = 5	Node-8	(ET,1T,3F,6F)	<70,70,70>	<7,10,15> <3,4,5> <70,70,70>	(ET,1T,3F,6F) (ET,1T,3F,5F) (ET,1T,3T)	4 4 2	<7,10,15>	<3,4,5> <70,70,70>	0.9635

5 Conclusion

Reliability for each node helps to estimate the overall path reliability. The reliability includes the testing of path in control flow graph which one is more efficient. The estimation of reliability is done through branch wise. Path coverage is an important aspect regarding the optimization. Mainly, the path coverage is directly proportional to the reliability. As excessive path is covered by the test cases the more will be chances of exactness in reliability. Table 1 represents the path reliability for the entire path from root to target node.

References

1. Choudhary K.: Cuckoo Search in Test Case Generation and Conforming Optimality using Firefly Algorithm. Proceedings of Second International Conference on Computer and Communication Technologies, IC3T 2015, Volume 1–3, Springer India (2015).
2. Hsu C. J. and Huang C. Y.: Integrating Path Testing with Software Reliability Estimation Using Control Flow Graph. Proceedings of the 2008 IEEE ICMIT, (2008).
3. Gokhale S. S.: Architecture-based software reliability analysis: Overview and limitations. IEEE Trans. Dependable and Secure Computing, vol. 4, pp. 32–40, (2007).
4. Lo J. H., Kuo S. Y., Lyu M. R., and Huang C. Y.: Optimal Resource Allocation and Reliability Analysis for Component-Based Software Applications. Proceedings of the 26th International Computer Software and Applications Conference (COMPSAC 2002), pp. 7–12, (2002).
5. K. Go. seva-Popstojanova and Trivedi K. S.: Architecture-based approach to Reliability Assessment of Software Systems. Performance Evaluation, vol. 45, pp. 179–204, (2001).
6. Hsu C. J. and Huan C. Y.: An Adaptive Reliability Analysis Using Path Testing for Complex Component-Based Software Systems. IEEE Transactions On Reliability, Vol. 60, No. 1, March (2011).
7. Roberto, Russo S. and Trivedi K. S.: Software reliability and testing time allocation-An architecture based Approach. IEEE Trans. Reliability, vol. 36, no. 3, pp. 322–337, June. (2010).
8. Fiondella L. and Gokhale S. S.: Optimal allocation of testing effort considering the Software Architecture. IEEE Trans. Reliability, vol. 61, no. 2, pp. 580–589, June. (2012).
9. Gokhale S. S., Wong W. E., Trivedi K. S. and Horgan J. R.: An Analytical Approach to Architecture Based Software Reliability Prediction. Proceedings of the 3rd International Computer Performance and Dependability Symposium (IPDS 1998), pp. 13–22, (1998).
10. Xie M., Hong G. Y. and Wohlin C.: Software Reliability Prediction incorporating Information from similar Projects. Journal of software and systems, vol. 49, No. 1, pp. 43–48, (1999).
11. Huang C. Y. and Lin C. T.: Software reliability analysis by considering fault dependency and debugging time lag. IEEE Trans. Reliability, vol. 55, no. 3, pp. 436–450, Sep. (2006).

A Study on Wii Remote Application as Tangible User Interface in Elementary Classroom Teaching

Mitali Sinha, Suman Deb and Sonia Nandi

Abstract Elementary teaching in classes has always been a challenge. The cognitive understanding of children (4–11 years of age) are very different from a matured person. Research and studies (Froebel, The pedagogics of the kindergarten, 2001 and Montessori, The Montessori method scientific pedagogy as applied to child education in "The Children's houses", 1992) [1, 2] have been carried out on children's behavior and problem solving ability which reveals that learning with actual physical objects produces better results than abstract representations. In this study, we have explored the benefits of TUI (Tangible User Interface) in children's thinking and learning process with Wii Remote. By providing both visual and physical representation, TUIs helps in reducing the cognitive load of thinking among children and increase their learning capabilities. A low-cost effective tool, "Nintendo Wii Remote" is rapidly finding a place in the field of learning technologies. Analysis conducted in this paper have shown the possibilities of exploring Wii Remote aiding in learning environment for children can significantly affect the learning outcome.

Keywords Tangible user interface · Wii remote

M. Sinha (✉) · S. Deb · S. Nandi
Computer Science and Engineering Department, NIT Agartala,
Agartala, India
e-mail: mitalisinha93@gmail.com

S. Deb
e-mail: sumandebcs@gmail.com

S. Nandi
e-mail: sonianandi90@gmail.com

1 Introduction

Children's cognitive thinking and learning process is very different from a matured person. Physical learning environment can enhance their development as they find relatively difficult to grasp knowledge from symbolic representation. Children's behavior and their approach to different learning environment have been studied for years by researchers [1, 2].

In elementary learning days children are actively indulged in playing activities like building blocks, shape puzzles etc. which helped them in developing various skills. Montessori (1912) explored the fact that children were taking more interest and attracted to learning environment that requires physical involvement which paved the way to the world of tangibility.

The idea of manipulating digital elements via physical objects have been introduced by Fitzmaurice et al. [3] which was followed by the pioneering work of Ishii and Ullmer [4] who came up with the term 'tangible bits' which removed the distinction between inputs and outputs by combining the digital and physical world. In this paper we will first give an overview of the early works of TUIs, specifically applicable for children's learning environment and then try to explore the possibilities of extending the Wii Remote's functionality for use in elementary classroom teaching.

2 Existing Works

The evolving field of tangible interfaces is getting preference in the educational domain. Resnick introduced "Digital Manipulatives" [5] in which physical objects were integrated with some computational ability that allowed children to explore concepts and enhance learning. MIT Media Lab Lifelong Kindergarten group provided a dimension to the involvement of TUI in education field with the work on Mindstorms by Papert [6]. The Lego/Logo construction kit from MIT Media Lab Lifelong Kindergarten group [7] links the construction kit with the Logo programming language by which children were able to use LEGO pieces along with sensors and motors and try to build newer and innovative machines and then writing programs to control those machines.

As mentioned earlier with the research work of Montessori [2] and Froebel [1] the hints of children's understanding of abstract concepts more easily with physical artifacts has been discovered. The Digital Montessori-inspired Manipulatives (MiMs) [8] were developed for children to grasp the more abstract concepts of dynamic behavior, probability, looping and branching. Mayer's cognitive theory of multimedia learning [9] highlighted the fact that involving the visual and verbal effects to form into coherent mental models or representation may result in quick learning. Padding system [10] were proposed for children with intellectual disabilities where a number of tiles were designed that interact with games specially designed for children with intellectual disabilities through the process of which it was found that combining the visual and verbal effects helped the children in developing their skills. However there is a need of exploring this domain of TUI enhancing memorability of children, but it is still a strong and appealing area for research.

The primary dependent variables in children's learning is the need of entertainment and engagement. TUIs for children's learning move around the concept of preserving the entertainment factor along with learning. Storytelling became a promising application of TUI for helping children with literature and engaging them constantly. Storymat by Ryokai and Cassell [11] is an example of such application where a play mat with RFID technology is implemented with the RFID-tagged toys being moved upon it. Children also find it more attractive and entertaining while visiting interactive museums that provides them physical interactions with the object like the Waltz dice game in the Vienna Haus der Music (Museum of Sound) automatic sounds are produced in rolling a dice and Todd Machover's Brain Opera where sound is generated in response to movement, touch and voice.

The tangible interfaces like I/O Brush [12], Story Mat [11] have been developed to engage children in different learning activities along with entertainment. Comparing with these technologies, the Wii Remote has immersed as cost effective tool whose functionality has been extended and explored in children's learning.

The WiiInteract [13] was developed to provide a tangible interface to the children linking digital manipulations to real life experiences. Here, the idea was to develop interactive interfaces for children using the Nintendo Wii Remote Controller for manipulation of the digital data. After their experiment and survey, it was stated that WiiInteract's applications resulted in more educational and efficient experiences.

In this [14] paper a health education based game was developed that also used Nintendo Wii Remote Controller as tangible device and performed experiments on group of students. A comparative study was carried out between two groups of students namely experimental group (participate in the activity with the use of Wii Remote) and control groups (participate in the activity without the use of Wii Remote) and their outcomes of learning was studied. Their study resulted in children of experimental groups using Wii Remote were having more learning outcomes related to fun factor and the level of interest, however the improvement in memorability needs to be explored further. And finally concluded that benefits of using TUIs in learning environment is quite acceptable and promising.

In this [15] paper a design is proposed to embed tangibility into multimedia learning of preschoolers. Their case studies provided a positive result on the learning outcomes of the preschoolers in terms of usability, enjoyment and feasibility.

3 Motive Behind the Study

With the advancement of new technologies, it is time to rethink on the concept of conventional teaching learning environment at elementary level. In the quest of finding an effective way of embedding technology in education, TUIs are extensively studied. At early age, children learn naturally and intuitively. Introducing them to formal learning methods may result in poor learning outcome with the lack of interest towards learning. Introduction of TUIs can provide an enhanced learning environment where children can learn intuitively with fun and enjoyment. Taking into consideration, the cost, reachability and usability of existing tangible technologies, the Wii remote has immersed as an effective solution with its low cost and varying functionality.

4 Methodology

4.1 Component Structure

A Japanese company Nintendo Co. Ltd. released a gaming console Wii in 2006 that competed with the other existing gaming consoles (for example, Microsoft's Xbox 360 and Sony's PlayStation 3). The main attraction of the Wii was in the motion-control magic of it's remote popularly known as Wii Remote. Along with the use of a pointing device, the Wii Remote also provides movement detection in three dimensions. A built-in accelerometer tracks the movement and an IR sensor detects the position. In 2009, Nintendo released Wii motion plus that gave the traditional Wii Remote with a gyroscope sensor to complement the accelerometer which resulted in more accurate motion detection (Fig. 1).

Fig. 1 Wii remote [16]

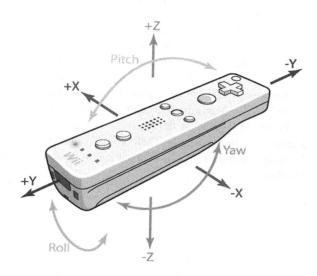

4.2 Work Flow

With the in-built motion sensing capability the Wii Remote interact and manipulate items on screen by gesture recognition providing tangibility. This made Wii Remote an effective tool to be used as an interacting and pointing device (Fig. 2).

In Fishkin's analysis of TUIs [14], the TUIs were characterized by five levels of metaphors namely none, noun, verb, noun and verb, full which helped in comparing the physical and the virtual manipulation. Wii Remote can be categorized as a TUI depending on the fact that it resembles most of these metaphors such as it can resemble the "noun and verb" metaphor in games where the Wii Remote itself is used as the object and performs movements like swinging (for example, The Legend of Zelda). In the above sections on different innovative tangible technologies, most of them restricts the user to perform whole body movement while a good number of tangible technologies are concerned with the cognitive and mental development of children. However the invent of technologies like Wii increased the domain space for TUIs. Gestures are considered to be playing an important role in communication and expressing intuitively. Wii provides an environment for the users to act intuitively through gestures providing a whole body movement. Although the Wii was originally developed for gaming purpose, eventually it's application was extended towards learning systems.

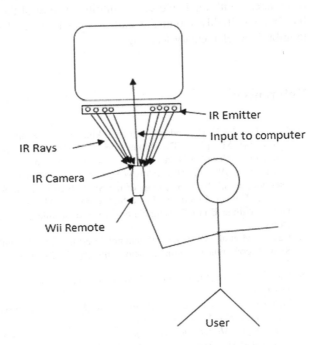

Fig. 2 Wii remote interaction with screen

5 Outcome

The study of deployment of tangible technologies in education resulted in enhancing the thinking and innovative ability of the children in learning abstract concepts. It also provides some evidences in increasing the memorability of children introducing a high rate of entertainment factor in the learning process.

The advent of Wii Remote opens a new area of introducing children to the world of Tangible User Interface. Due to low cost Wii Remote can be easily deployed in learning process even in rural areas limiting the implementation cost of the system within $100.

6 Conclusion

The role of tangible user interfaces in developing different learning skills in children is studied. A brief outline of the existing tangible techniques are explored that enhance thinking and innovative ability, enhancing memorability in children, learning abstract concepts and finally its role in maintaining enjoyment and engagement along with learning. A study on Nintendo Wii Remote as a tangible device is carried out which ought to provide a positive result on learning outcomes of children. With the low cost and intuitive nature of the Wii Remote controller, it has become highly promising and acceptable area of exploring the benefits of tangibility in elementary learning.

References

1. Froebel F (2001) The pedagogics of the kindergarten. Kindergarten Messenger.
2. Montessori M (1992) The Montessori method scientific pedagogy as applied to child education in "The Children's houses". Frederick A. Stokes Co., New York.
3. Fitzmaurice GW, Ishii H, Buxton WAS (1995) "Bricks: laying the foundations for graspable user interfaces," In: Proceedings of the SIGCHI conference on human factors in computing systems, Denver, Colorado, United States, pp 442–449.
4. Ishii H, Ullmer B (1997) Tangible bits: towards seamless interfaces between people, bits and atoms.
5. Resnick M et al. (1998) Digital manipulatives: new toys to think with. In: Proceedings of the SIGCHI conference on human factors in computing systems, New York, NY, USA, p 281–287.
6. Papert S (1980) Mindstorms: children, computers, and powerful ideas. Bas, Cambridge, MA USA.
7. M. Resnick, S. Ocko, and S. Papert, "LEGO, Logo, and Design," Children's Environments Quarterly 5, No. 4, 14–18 (1998).
8. R. E. Mayer, *Multimedia Learning*, Cambridge University Press, Cambridge, UK, 2009.
9. Amal Dandashi, Abdel G K, Sawsan Saad, Zaara Barhoumi, Jihad AlJaam,Abdulmotaleb El Saddik, Research Article, "Enhancing the cognitive and learning skills of children with

intellectual Disability through Physical activity and edutainment games" International Journal of Distributed Sensor Networks, 2014.

10. Ryokai K, Cassell J (1999) StoryMat: a play space for collaborative storytelling. In CHI 99 extended abstracts on human factors in computing systems, New York, NY, USA, pp 272–273.

11. Ryokai K, Marti S, Ishii H (2004) I/O brush: drawing with everyday objects as ink. In: Proceedings of the SIGCHI conference on human factors in computing systems, New York, NY, USA, p 303–310.

12. Jee Yeon Hwang and Ellen Yi-Luen Do, "WiiInteract: Designing Immersive and InteractiveApplication with a Wii Remote Controller", 15th International Conference on Computer Games: AI, Animation, Mobile, Interactive Multimedia, Educational and Serious Games, July 28–31, Galt House Hotel, Louisville, Kentucky, USA.

13. Jeng Hong Ho, Steven ZhiYing Zhou, Dong Wei, and Alfred Low," Investigating the Effects of Educational Game with Wii Remote on Outcomes of Learning", Z. Pan et al. (Eds.): Transactions on Edutainment III, LNCS 5940, pp. 240–252, 2009. c_ Springer-Verlag Berlin Heidelberg 2009.

14. Kenneth P. Fishkin, "A taxonomy for and analysis of Tangible interfaces", pers Ubiquit Comput, 2004.

15. C. Tsong, T. Chong, Z. Samsudin: "Tangible Multimedia: A case study for bringing tangibility into multimedia learning", TOJET, vol 11 issue 4–October 2012.

16. http://www.google.co.in/imgres?imgurl=http://www.osculator.net/doc/_media/faq:prywiimote.gif.

Enhanced Understanding of Education Content Using 3D Depth Vision

Sonia Nandi, Suman Deb and Mitali Sinha

Abstract Perspective representation of three dimensional physical object and scientific content printed makes an educational content easily understandable by the learner. Creation of abstract content on any subject for teaching purpose is a hard task for the teachers. In this regard, for an enhanced teaching learning experience, a very popular gaming tool named kinect tried in a different way to make the students understand the educational content in an easy and interesting way. In this paper, it is tried to visualize the perspective view of the two dimensional printed content on book in a real life scenario. It is tried out to generate equivalent educational content as interesting as gaming content for better understanding and encouraging student and to grasp knowledge in an efficient way.

Keywords Kinect · Game based learning · Depth vision of kinect · Gesture · Exer learning · 3D depth vision

1 Introduction

From the inception of civilization, learning was the inevitable part which made the human take part into evolution. But learning in natural way is limited and only serves the basic properties of human. For betterment of living, special trainings are required and these trainings can be acquired faster and in efficient manner from experienced persons.

S. Nandi (✉) · S. Deb · M. Sinha
National Institute of Technology, Agartala, India
e-mail: sonianandi90@gmail.com

S. Deb
e-mail: sumandebcs@gmail.com

M. Sinha
e-mail: mitalisinha93@gmail.com

© Springer India 2016
S.C. Satapathy et al. (eds.), *Information Systems Design and Intelligent Applications*, Advances in Intelligent Systems and Computing 433,
DOI 10.1007/978-81-322-2755-7_50

483

Human starts learning with the growth of life. They learn through their natural activities like playing, social interaction etc. This natural method of learning is called Unschooling [1] or natural learning. Gradually children start learning special skills from their parents and then in formal way from the professionals like teachers.

Specialized training and learning method need special attention which is not always given to the children or students by the professionals. People are affinities to games much more than any formal learning as it involves the person in more interactive manner in learning mechanism. To maximize the benefit of learning in less time, introducing game based learning is a better option in present time.

The primary requisite for proper learning is attention, motivation, pleasure and skill. In monotonous traditional learning, we only require constant attention but pleasure and skill sometimes ruled out whereas on the contrary the games like video game and other interactive games give pleasure and skill along with feedback or score. In this paper we tried to incorporate the concept of gaming and natural learning for specialized classroom teaching and learning process. Activities involving physical movement make the children more attained in learning method. Using kinect, we can motivate the children to learn by making interactive and reciprocative loop. For example, when a task is done, a slightly difficult task is given to child. On the other side, if he/she fails to complete the task, a task of same difficulty level is given to the child and if he fails repeatedly, an easy task is given again to motivate him. This concept of involving game-based exercises always involves the student in the thought process of learning which is an added advantage and the environment becomes dynamic.

Learning becomes monotonous if the physical movement is restricted. In this regard, other difficulties like obesity and different physical disorder also adds up. Introducing kinect for learning involves physical movement while learning can reduce the problems and enhance the learning system as physical activities and attention will be in higher scale.

2 Using Depth Vision of Kinect

Depth vision is the capacity to visualize an object in a 3 dimension space- width, height and depth by calculating the distance between the object and the eye. In kinect, we have the depth sensor camera.

2.1 Depth Sensor Technology

According to [2], images of depth sensor is same as that of ordinary or RGB camera, but the pixel of depth sensor image represents the distance between the sensor and the solid object point instead of representing color.

The main idea behind the depth sensor of kinect is that the infrared projector emits IR patterns continuously and the CMOS camera also called IR depth sensor of kinect captures the IR image simultaneously. The IR pattern consists of a number of dots. These dots in the pattern are used to calculate the position of each pixel in the x, y, and z coordinate. The process of examining the pattern is called structured light [3]. When the IR pattern strikes the object, the information of the surface and the depth is calculated. In this way, kinect constructs the depth diagram of an object. This depth diagram is done with the help of a hardware called Prime Sense [4] which is inside the kinect.

We can also recognize the forms of non-verbal communication like gestures with the help of this depth diagram of kinect. Gestures involve hand gestures, skeleton tracking, speech recognition and hand recognition.

2.2 Kinect

Kinect is mainly built by the Microsoft for gaming purpose. But now it has gained popularity in many other fields like medical, education, researches, etc.

Kinect is a "hands-free control device" which can sense the body movements and voice. It consists of depth sensor, infrared emitter, color camera, LED, a set of four Microphones with everything packed inside a plastic box.

According to [5], kinect can scan an area of 80–300 cm in the horizontal direction; it can scan an angle of 57°, in the vertical direction it can scan up to 43° whereas diagonally it can scan 70° (Figs. 1 and 2).

Fig. 1 Diagram of kinect [9]

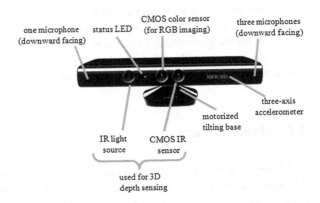

Fig. 2 Viewing angle of
kinect

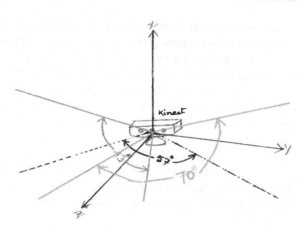

3 Kinect as a Learning Tool

In recent days, kinect has got a wide exposure in the field of education. It can be
used for learning in elementary education as well as in advanced education system.
Though, kinect is mainly introduced for learning purpose but the gaming content
can be replaced with educational content and experiment content. We can also
introduce kinect in regular classes where a large number of people work together in
a single platform in competitive manner. The mental capacity of the students will
also be enhanced with the physical movement.

Generally, in elementary learning institutions, training or learning is not interesting for a long time if it is only theoretical based or less involvement of movements is there. But the learning method utilizing kinect allow a pupil to learn along
with physical movement which boosts up his interest in gaining knowledge.

To motivate someone, instant scoring is an important factor. But in learning instant
scoring is not so much evident as after a long span of time, learning evaluation takes
place. For building continuous interest instant scoring is very important which can be
done with the help of learning technology using kinect. Using kinect based application people can learn by using different gestures, poses defined by the application.

3.1 Hand Gesture

In different learning applications based on kinect, pupil waves their hands or makes
different postures in front of kinect. In an e-learning classroom [2], pupil raises their
hand in the class where a kinect is there to monitor the class. The kinect then
focuses only that remote student so that the professor can interact with him/her
easily like the interaction between the student and the teacher in the real time
classroom environment.

4 Exer-Games to Exer-Learning

Exer-games mean games involving exercises i.e. physical movements. In kinect based application, a person can play games which involved exercises i.e. making different body movements like waving hands, nodding head, jumping etc. which keeps them fit as well as entertained. People irrespective of ages are always interested in gaming. Gaming also requires less training where self-induced learning and training mechanisms are used by an individual. By analyzing this, researchers thought of transforming exer-games into exer-learning.

Exer-learning can motivate an individual to learn as it involves physical movement along with learning. It will enhance not the physical ability but it will also help to get rid of different diseases like cardio-vascular disorder, obesity, diabetes because of continuous sitting in the classroom or before a machine. It will also help to build a continuous interest in learning. Exer-learning can also help the children with disabilities to improve their problem solving ability (Fig. 3).

5 Related Works

Many applications for game-based learning have been developed utilizing kinect as the major component. Kinems [6] build game based learning software that will help the children having autism, dyslexia, ADHD etc. by increasing the Childs' motivation, concentration, and interest. The software consists of a game named "Cuckoo" which will make them understand the analog clock. In this game the task is to set the analog clock by moving his hands in front of kinect. Another game "Bilisus game" can increase the analysis power of the child. In this, the child has to differentiate between greater and smaller cluster of dots. This has to be done without counting as the time provided is less.

Kinect angles [7] is a kinect educational game which is used to give the children an idea of percentages, fractions and angles. Kinect time helps the pupils to get an idea of timings.

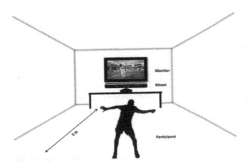

Fig. 3 Exer-gaming versus exer-learning [8, 10]

Jumpido [8] is another educational application which is used in many schools to incorporate kinesthetic learning. It consists of a series of primary school math app which will motivate the pupil to learn math in an interesting way.

6 Application Setup

All the previously mentioned exer-learning tools provide different platforms for the kids to understand the concept of mathematics in an easy and interesting way. In this paper, we tried to explore the possibility of exer-games to understand trigonometry concepts in an interesting way. There is an avtar in the application, which will resemble as the player in the application. It imitates the movements of the player. For developing such environment, we have used Microsoft Kinect.

To introduce kinect in the classroom, first we have to place kinect in the classroom in such a way that it can cover the whole body of the student. That is the student should be within a certain angle of 57° in horizontal direction. And a screen should be there in front of the students where they can visualize the trigonometry objects, shapes.

6.1 Application

In this application, there will be a tree in the screen. And if the individual comes closer to the kinect sensor, the tree will get zoomed and if he goes away from the sensor, tree will get zoomed out. With the help of this he can get an idea of the distance between his body and the tree. Again if he projects his hands in front of the tree by making an angle, it will show the angle between his hands. The moment he will move towards or backward of the sensor with his hands projected the angle between hands and the distance between the tree and the body changes (Figs. 4 and 5).

Fig. 4 As the user moving towards the sensor, the user is getting a closer view of the tree on the screen

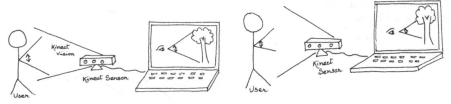

Fig. 5 As the user changes the gap between his hands, the angle between the hands got changed which can be seen on the screen

6.2 Interaction

User can interact with the application by hand gestures and by walking in front of the kinect. To start the game, the user will have to follow the guidance in which a particular pose will be defined to start the game. In this application, one student will perform the experiment, and the other students of the class can view the experiment on the screen.

7 Comparison with Conventional Methods

In the conventional method, most of the time students find it difficult to understand the trigonometry problems for example, height and distance problems. Conducting the experiment with 20 students in a class, it is seen that the student answered the given problems in less time when the proposed method is used.

8 Conclusion

Using exer-learning technologies, it is seen that students are getting motivated to learn in a greater extent. They can understand the height and distance problems in a very easy way. For developing an effective game based learning system, the educationalist and developers must work together.

From Fig. 6 it is seen that students learning outcome is significantly growing by usage of this system. Till up to the present work, the system is limited only for one person as a candidate. In future we will be trying to incorporate more than one child at a time in this learning system using kinect. The proposed methodology in this paper is not only limited for learning, it can also be extended for entertainment and regular aerobics for health care.

Fig. 6 Unsorted students
performance with time

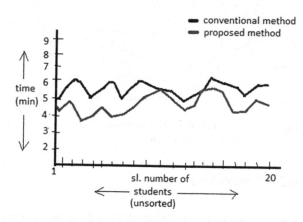

References

1. Wikipedia, https://en.wikipedia.org/wiki/Unschooling.
2. Balaji Hariharan, Padmini. S, Uma Gopalakrishnan in 2014 "Gesture Recognition Using Kinect in a Virtual Classroom Environment" IEEE.
3. Wikipedia, https://en.wikipedia.org/wiki/Structured_light.
4. "How kinect works" by John MacCormick.
5. Adi Sucipto, Agung Harsoyo, Pranoto Hidaya Rusmin in 2012 International Conference on System Engineering and Technology September 11–12, 2012, Bandung, Indonesia "Implementation of Gesture Recognition on Aquarium Application".
6. Kinems, www.kinems.com.
7. Kinect angles, https://drenton72.wordpress.com/2012/05/23/kinect-angles-v2-4/.
8. Jumpido, www.jumpido.com/.
9. http://buildsmartrobots.ning.com/profiles/blogs/building-a-kinect-based-robot-for-under-500-00.
10. http://journal.frontiersin.org/article/10.3389/fnhum.2014.00469/full.

AHP-Based Ranking of Cloud-Service Providers

Rajanpreet Kaur Chahal and Sarbjeet Singh

Abstract The paper presents an approach to rank various Cloud-Service providers (CSPs) on the basis of four parameters, namely, Performance, Availability, Scalability and Accuracy, using Analytic Hierarchy Process (AHP). The CSPs were monitored by the cloud storage company named Nasuni and tests were conducted to evaluate the CSPs for the four parameters mentioned above. This paper makes use of the data provided by Nasuni to propose a method for ranking various CSPs. AHP has been selected for this purpose because it uses the foundations of mathematics and psychology to enable one to make complicated decisions. Instead of recommending a correct decision, AHP provides the decision makers with an opportunity to select the option that is most befitting to their goals.

Keywords Ranking of CSPs · Analytic hierarchy process (AHP) · Cloud computing

1 Introduction

Cloud computing can be considered as the biggest emerging trend in Information Technology. It is based on the premise "why buy when you can rent". It provides on-demand services to its clients who pay for the services based on their usage. It gives the clients a chance to use the resources provided by the Cloud Service Provider (CSP) in exchange of remuneration.

R.K. Chahal (✉) · S. Singh
Computer Science and Engineering, UIET, Panjab University,
Chandigarh, India
e-mail: rajan_chahal@yahoo.com

S. Singh
e-mail: sarbjeet@pu.ac.in

© Springer India 2016
S.C. Satapathy et al. (eds.), *Information Systems Design and Intelligent Applications*, Advances in Intelligent Systems and Computing 433,
DOI 10.1007/978-81-322-2755-7_51

Cloud Computing has been defined by National Institute of Standards and Technology as "a model for enabling convenient, on-demand network access to a shared pool of configurable computing resources that can be rapidly provisioned and released with minimal management effort or service provider interaction" [1].

Cloud Computing has a vast potential and in order to tap in the advantages provided by it, various companies have ventured into the cloud computing business. With a huge choice of CSPs offering the services, the clients often have to choose one CSP over others. The purpose of this paper is to propose a method to help clients decide which CSP to choose. Section 2 of the paper consists of related work, Sect. 3 explains the proposed method with results and Sect. 4 concludes the paper.

2 Related Work

Saurabh, et al. have proposed an AHP based mechanism to rank the Cloud Service Providers using the SMICloud framework which provides accuracy in QoS determination and selection of cloud service by the users [2]. Zibin et al. have proposed a ranking method called QoS Ranking Prediction which employs the past experiences of other users. The experiments have proved this approach to perform better than other rating approaches and the famous greedy method [3]. To decrease the computational demands, Smitha et al. have proposed a broker-based approach where the cloud brokers maintain the data of the cloud service providers and rank them accordingly, keeping in view the request made by the customer [4]. Saurabh et al. have suggested a mechanism to prioritise the cloud services on the basis of quality measurement of each of the distinctive cloud services [5]. Ioannis et al. have addressed the concern of sheer uncertainty and vagueness in ascertaining comparison between various cloud services against the user requirements and specifications by proposing a framework that allows for a unified method of multi objective assessment of cloud services with precise and imprecise metrics [6]. Arezoo et al. have proposed a Weight Service Rank approach for ranking the cloud computing services. The approach uses QoS features and derives its strength from the point that it uses real-world QoS features [7]. Zia et al. have proposed a parallel cloud service selection technique wherein QoS history of cloud services is evaluated using MCDM process during different periods of time. The cloud services are ranked for each time period and then these results are combined to arrive at an overall ranking of cloud services [8]. Praveen and Morarjee have proposed a personalized ranking prediction structural framework known as cloud rank. It has been developed to forecast quality of service ranking without requiring any need of real time feeds or real world services invocation from the consumers or users [9].

3 Explanation of the Process with Results

The proposed process ranks the CSPs on the basis of data collected from the tests conducted by Nasuni [10]. Tests were conducted to monitor the CSPs on the basis of Performance, Availability, Scalability and Accuracy on a scale of 1–5, where 1 stands for very low and 5 stands for very high. The definitions of the parameters are given in Sect. 3.1.

3.1 Dataset

Nasuni has been monitoring CSPs since 2009. The detailed results of the Nasuni's evaluation of the CSPs for their Performance, Availability, Scalability and Accuracy can be found in [10]. These parameters have been measured using the following criteria:

Performance: It is measured by the read, write and delete speed of the service. Faster the speed, greater is the performance. This parameter measures the CSPs' ability to tackle thousands of write, read and delete speeds for files of various sizes. Nasuni ran the test on numerous instances of testing machine using various non-serial test runs so as to minimize the possibility of the results being affected by external network issues.

(i) Write Speed—As far as write speed is concerned, Nasuni's evaluation showed CSP3 to be the best write performer followed by CSP1 and CSP4. There is not much difference in the write performance of CSP1 and CSP4. CSP5 and CSP2 lag behind in their write performance with a large margin. Using these results, CSP1 has been assigned rank 4, CSP2 has rank 1, CSP3 has rank 5, CSP4 has rank 3 and CSP5 has rank 2, where rank 5 is the best and rank 1 is the worst.

(ii) Read Speed—In read speed, CSP3 again tops the chart followed by CSP4 and CSP1. Whereas in write speed, CSP1 performed slightly better than CSP4 to claim the second spot, in read speed CSP4 has outperformed CSP1 to claim the second position. CSP5 and CSP2 have improved read speed than write speed, but still they lag far behind the other three CSPs. Assigning values 1–5 as explained in part (i), CSP1, CSP2, CSP3, CSP4 and CSP5 have ranks 3,1,5,4 and 2 respectively.

(iii) Delete Speed—Comparing the delete speed of all the CSPs, CSP3 retains its top spot in delete performance as well. CSP1 has outperformed CSP4 in delete speed to come at second position while CSP4 has third position. CSP2 and CSP5 again lag far behind others. So, the ranks assigned to CSP1, CSP2, CSP3, CSP4 and CSP5 are 4,2,5,3 and 1 respectively.

Taking the average of the ranks assigned in (i), (ii) and (iii), we arrive at the ranks of the CSPs for their overall performance on the basis of write, read and delete speed. These ranks are: CSP1 has rank 3.6, CSP2 has rank 1.3, CSP3 has rank 5, CSP4 has rank 3.3 and CSP5 has rank 1.6.

Analyzing these results, we note that CSP3 is the best performer in terms of the read, write and delete operations. CSP1 and CSP4 compete with one another to claim the second spot. Similarly, CSP2 and CSP5 are the contenders for the lowest position. The conflict has been resolved by taking into account the performance in all the three operations and calculating the overall performance ranks. According to overall ranks, CSP3 emerges as the winner, followed by CSP1, CSP4, CSP5 and CSP2 in that order.

Availability: It is measured on the basis of response time of the service. Lower the response time, greater is the availability. The test for this parameter includes monitoring each CSP's response time to write, read and delete request at sixty-second interval. Reading and deleting a file ensures that each CSP is responsive to entire data for whole time and not just the recent cached data. The test enables one to evaluate the response time of the CSPs as well as their reliability and latency.

Hence, analyzing the results for response times of the CSPs mentioned in [10], CSP3 emerges as the best in terms of availability and CSP2 as the worst. Ranks assigned to the CSP1, CSP2, CSP3, CSP4 and CSP5 on the basis of availability are 4,1,5,2 and 3 respectively.

Scalability: It is measured by the ability of the service to handle increased workloads. Scalability is measured using the variations in the performance of the services when load is increased. The ability of a service to perform consistently under increased loads often acts as the Achilles heel of a system. So, the test for this parameter includes measuring the performance of the CSPs when the number of objects is increased to millions.

Lower the variation in performance with increased load, higher is the scalability. Nasuni's results have shown CSP1 to be the best in terms of scalability with CSP5 being the worst. The ranking achieved thereby is: CSP1 has rank 5, CSP2 has rank 3, CSP3 has rank 4, CSP4 has rank 2 and CSP5 has rank 1.

Accuracy: It is measured on the basis of read and write errors. Lower the percentage of errors, higher is the accuracy.

(i) Read Errors—Nasuni's results show that CSP3 did not commit any error during the read requests and hence is at the top spot. CSP3 is followed by CSP5, CSP1, CSP2 and CSP4 in that order. Ranking of CSPs on the basis of read errors is: CSP1 has rank 3, CSP2 has rank 2, CSP3 has rank 5, CSP4 has rank 1 and CSP5 has rank 4. As mentioned earlier, rank 5 is assigned to the best performer and rank 1 is assigned to the worst performer.

(ii) Write Errors—Nasuni's tests reveal that CSP3 commits no errors in writing attempts also and hence retains its top position. However, CSP3 is joined by CSP1 and CSP2 with zero write errors as opposed to their performance during read attempts. CSP5 and CSP4 have significantly improved their performance compared to the reading attempts. Ranking achieved on the basis of write errors is: CSP1, CSP2 and CSP3 all have rank 5, CSP4 has rank 3 and CSP5 has rank 4.

Averaging the rankings for read and write errors, we calculate the ranks of CSP1, CSP2, CSP3, CSP4 and CSP5 for accuracy as 4, 3.5, 5, 2 and 4 respectively.

It is clear that CSP3 is the best in terms of accuracy. After combining both reading and writing errors, there is a tie between CSP1 and CSP5. CSP2 follows closely with a slight difference. The overall ranks for performance, availability, scalability and accuracy are used for further analysis to rank the CSPs from best to worst. The dataset obtained after following the above procedure is shown in Table 1.

Table 1 contains the data that we will be using in further analysis.

3.2 Steps

After obtaining the Performance, Availability, Scalability and Accuracy values as explained in the previous part, the further analysis is carried out using the Analytic Hierarchy Process (AHP) technique. The whole process is explained in the steps below:

Step 1: From the values given in Table 1, find the maximum value in each column and term it as MAX(Parameter). So the values obtained are: MAX(Performance) = 5, MAX(Availability) = 5, MAX(Scalability) = 5, MAX(Accuracy) = 5

Step 2: A normalized matrix P is created using the values of Table 1 and MAX (Parameter). The normalized values thus obtained are termed as NORM (Parameter). P can be created using (1), where x = 1,2,3,4,5.

$$NORM(Parameter)_{CSPx} = \frac{Parameter(CSPx)}{MAX(Parameter)}. \qquad (1)$$

The NORM(Parameter) values thus obtained are shown in Table 2.

Table 1 Dataset for ranking CSPs based on auditor data

	Performance	Availability	Scalability	Accuracy
CSP1	3.6	4	5	4
CSP2	1.3	1	3	3.5
CSP3	5	5	4	5
CSP4	3.3	2	2	2
CSP5	1.6	3	1	4

1 Very low, *2* low, *3* average, *4* high, *5* very high

Table 2 Normalized matrix P

	Performance	Availability	Scalability	Accuracy
CSP1	0.72	0.8	1	0.8
CSP2	0.26	0.2	0.6	0.7
CSP3	1	1	0.8	1
CSP4	0.66	0.4	0.4	0.4
CSP5	0.32	0.6	0.2	0.8

Step 3: Relative importance of each parameter to the goal is evaluated. Our goal is to select the best CSP. Accordingly, an importance number is assigned to each parameter. These numbers are assigned according to Saaty [11]. The Saaty scale is given in Table 3.

The importance number for each parameter is represented as IN (Parameter).

Since auditor is a neutral third party monitoring the CSPs, it cannot favor one parameter over another. For auditor, all the parameters are equally important for ranking the CSPs from best to worst. Therefore, the importance number assigned by the auditor to all the parameters is 1, which signifies equal importance of all the parameters for our goal (refer Table 3). The importance numbers for the four parameters in the case of auditor are given in Table 4.

Table 3 Saaty scale for importance of parameters [11]

Importance number	Definition
1	Equal importance
2	Weak
3	Moderate importance
4	Above moderate
5	Strong importance
6	Above strong
7	Very strong
8	Very, very strong
9	Extreme importance

Table 4 Importance numbers

Parameter	Importance number
Performance	1
Availability	1
Scalability	1
Accuracy	1

Table 5 Importance matrix M

	Performance	Availability	Scalability	Accuracy
Performance	1	1	1	1
Availability	1	1	1	1
Scalability	1	1	1	1
Accuracy	1	1	1	1
Sum	4	4	4	4

Step 4: Using the importance numbers assigned to each parameter, an importance matrix M is created according to (2). The values in the matrix M are referred to as IMP(Parameter$_{mn}$). The importance matrix M is shown in Table 5.

$$IMP(Parameter_{mn}) = \frac{IN(Parameter_m)}{IN(Parameter_n)}. \tag{2}$$

where Parameter$_m$, Parameter$_n$ = Performance, Availability, Scalability, Accuracy

IMP(Parameter$_{mn}$) = Importance of Parameter m compared to Parameter n.

After calculating the IMP(Parameter$_{mn}$) for all the parameters, a column-wise Sum is calculated for each parameter in matrix M. This value is referred to as SUM(IMP(Parameter$_{mn}$)).

Step 5: A weight matrix W is constructed using the IMP(Parameter$_{mn}$) and SUM (IMP(Parameter$_{mn}$)) values according to (3), where W (Parameter$_{mn}$) = weight of Parameter$_m$ compared to Parameter$_n$.

$$W(Parameter_{mn}) = \frac{IMP(Parameter_{mn})}{SUM(IMP(Parameter_{mn}))}. \tag{3}$$

The final weights for each parameter, represented as W(Parameter), are then obtained using (4), where x = number of parameters (i.e. 4).

$$W(Parameter) = \frac{\sum W(Parameter_{mn})}{x}. \tag{4}$$

The value of weights obtained for all parameters in our case is 0.25.

Step 6: A rank matrix R is constructed using the weights of the parameters and the normalized values calculated in matrix P. The value in each cell of matrix R is known as Rank(Parameter). Matrix R is shown in Table 6.

$$Rank(Parameter) = W(Parameter) * NORM(Parameter). \tag{5}$$

Table 6 Rank matrix R

	Performance	Availability	Scalability	Accuracy
CSP1	0.18	0.2	0.25	0.2
CSP2	0.065	0.05	0.15	0.175
CSP3	0.25	0.25	0.2	0.25
CSP4	0.165	0.1	0.1	0.1
CSP5	0.08	0.15	0.05	0.2

Table 7 Ranks of the CSPs based on auditor data

CSPs	Rank
CSP1	0.83
CSP2	0.44
CSP3	0.95
CSP4	0.465
CSP5	0.48

Step 7: Finally, the CSPs are ranked by summation of Rank(Parameter) values.

$$Rank(CSP) = \sum_{\substack{Parameter_m = Reliability, \\ Performance, Security, \\ Usability}} Rank(Parameter_m). \quad (6)$$

The final ranks of the CSPs thus obtained are shown in Table 7.

The CSPs are ranked in the order CSP3 > CSP1 > CSP5 > CSP4 > CSP2.

The ranks obtained through AHP technique are shown in Fig. 1. After taking into account the behavior of the five CSPs for Performance, Availability, Scalability and Accuracy, we have arrived at the ranking where CSP3 is the best option, CSP1 is the second best, followed by CSP5, CSP4 and CSP2.

Fig. 1 Final rank of CSPs

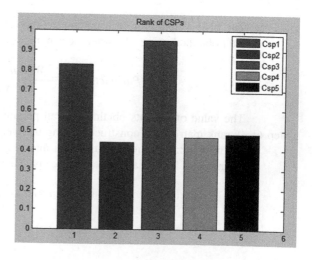

4 Conclusion and Future Scope

In this paper, we have proposed an AHP-based ranking of cloud service providers based on the tests conducted by a neutral third party, which can be referred to as an auditor. This approach provides a fair and unbiased ranking of CSPs based objectively on the results of the tests. It also has the advantage of incorporating both mathematics and psychology to make a decision of selecting the best option out of the available ones keeping in mind the goal and preference of the parameters to the overall goal. Plus, the model is scalable as both the parameters and the CSPs can be increased in number as and when required. In future, we would like to evaluate the performance of our approach with respect to other mechanisms.

References

1. Liu, F., Tong, J., Mao, J., Bohn, R., Messina, J., Badger, L., Leaf, D.: NIST Cloud Computing Reference Architecture. Recommendations of the National Institute of Standards and Technology, Special Publication 500-292, Gaithersburg (2011).
2. Garg, S.K., Versteeg, S., Buyya, R.: SMICloud: A framework for comparing and ranking cloud services. In: Proc. Fourth IEEE International Conference on Utility and Cloud Computing, pp. 210–218, doi:10.1109/UCC.2011.36 (2011).
3. Zheng, Z., Wu, X., Zhang, Y., Lyu, M.R., Wang, J.: QoS ranking prediction for cloud services. In: IEEE Transactions on Parallel and Distributed Systems, vol 24, no 6, pp 1213–1222, doi:10.1109/TPDS.2012.285 (2013).
4. Sundareswaran, S., Squicciarini, A., Lin, D.: A brokerage-based approach for cloud service selection. In: Proc. IEEE Fifth International Conference on Cloud Computing, pp. 558–565, doi:10.1109/CLOUD.2012.119 (2012).
5. Garg, S.K., Versteeg, S., Buyya, R.: A framework for ranking of cloud computing services. In: Future Generation Computer Systems, vol. 29, no. 4, pp. 1012–1023, doi:10.1016/j.future.2012.06.006 (2013).
6. Jahani, A., Khanli, L.M., Razavi, S.N.: W_SR: A QoS based ranking approach for cloud computing service. In: Computer Engineering and Applications Journal, vol. 3, no. 2, pp. 55–62, ISSN: 2252-5459 (2014).
7. Patiniotakis, I., Rizou, S., Verginadis, Y., Mentzas, G.: Managing imprecise criteria in cloud service ranking with a fuzzy multi-criteria decision making method. In: Service-Oriented and Cloud Computing, pp. 34–48, doi:10.1007/978-3-642-40651-5_4 (2013).
8. Rehman, Z., Hussain, O.K., Hussain, F.K.: Parallel cloud service selection and ranking based on QoS history. In: International Journal of Parallel Programming, vol. 42, no. 5, pp. 820–852, doi:10.1007/s10766-013-0276-3 (2014).
9. Kumar, G.P., Morarjee, K.: Ranking prediction for cloud services from the past usages. In: International Journal of Computer Sciences and Engineering, vol. 2, no. 9, pp. 22–25, E-ISSN: 2347-2693 (2014).
10. Nasuni: The State of Cloud Storage, Nasuni Corporation, Massachusetts (2014).
11. Saaty, T.L.: Decision making with the analytic hierarchy process. In: International Journal of Services Sciences, vol. 1, no. 1, pp. 83–98, doi:10.1504/IJSSCI.2008.017590 (2008).

4 Conclusion and Future Scope

References

Hardware Prototyping for Enhanced Resilient Onboard Navigation Unit

Archana Sreekumar and V. Radhamani Pillay

Abstract Dependability of safety critical system by fault tolerant design approach use redundant architecture to tolerate hardware faults. Navigation, Guidance and Control units of onboard computers in Indian satellite launch vehicles rely on hot standby dual redundancy. Effective use of computational resources is desirable in such applications where weight, size, power and volume are critical. Resource augmentation based on task criticality can achieve an increased slack margin which further is used for software and transient fault handling and improved system performance. In this paper, design and development of a hardware prototype with fault injection on an LPC 1768 ARM processor, for validating and testing the fault tolerant resource augmented scheduling of onboard computers is presented. The resource augmented system with added flexibility has been evaluated for improved performance and superior management of faults. The system provides better slack margin and resource utilization which leads for tolerating increased number of transient and software faults.

Keywords Safety critical system · Augmented dual redundancy · Hardware prototype and testing · Fault injection · Fault recovery

1 Introduction

Automation and technological advancement has caused an increased need for dependency in safety critical systems like medical equipments, avionics, automotive, etc. Onboard computers in satellite launch vehicles using hot standby dual redundancy controls and maintains overall operational functionality of the launch vehicle. The redundant computational resources can be effectively utilized for improved fault tolerance and enhanced performance. An earlier work rooted in

A. Sreekumar (✉) · V. Radhamani Pillay
Amrita Vishwa Vidyapeetham, Coimbatore, India
e-mail: archanasreekumar2007@gmail.com

© Springer India 2016
S.C. Satapathy et al. (eds.), *Information Systems Design and Intelligent Applications*, Advances in Intelligent Systems and Computing 433,
DOI 10.1007/978-81-322-2755-7_52

501

criticality based task allocation in such redundant systems increases the slack margin available. Secondary alternative tasks for handling software faults, recovery approaches for transient fault handling can be incorporated into the augmented system while maintaining allowable safety margin.

In this paper, a fault injection model for testing the fault tolerance algorithm for an augmented Navigation unit of the NGC system has been implemented. Fault tolerant recovery mechanisms have been implemented for permanent hardware and software faults, transient hardware and software faults. An algorithm has been developed and implemented in LPC1768 ARM Cortex M processors for prototype testing. Performance evaluation for permanent hardware faults and testing of the frame work for better resilience to transient faults on such a frame work with task dependencies and tight synchronization allows a measure of the system behavior in this domain.

2 Literature Survey

Real time systems are used to control safety critical systems such as patient monitoring system, aircrafts, launch vehicles, etc. Krishna discusses about real time scheduling which ensures the meeting of hard deadlines in the system despite processor and transient failures [1]. Four different scheduling paradigms are introduced by Stankovic and Ramamritham in [2] like static table driven scheduling, where tasks are scheduled offline.

The satellite launch vehicle onboard computer architecture and design based on fault tolerant methods are put forth by Basu [3]. Different fault tolerant designs and approaches for tolerating hardware faults in avionics systems are discussed in [4]. Software assisted recovery for transient faults in hard real time systems are put forth in [5]. A study on insertion of checkpoints within software in an aerospace domain application is discussed by Leach [6]. Punnekkat et al. [7] provides an exact schedulability test for fault tolerant system with checkpoints employed for fault tolerance.

An algorithm for utilizing the redundancy in onboard computer and obtaining more slack for extra computations is introduced in [8]. Concept of combining resource augmented task allocation for safety critical system and real time scheduling techniques like RM and EDF scheduling are discussed in [9]. The concept of adaptive fault tolerance in a commonly used hardware redundancy technique in a cruise control multi processor system has been detailed in [10]. A hardware prototype with LPC2148 has been used for testing and evaluating the performance of adaptive fault tolerance mechanism for permanent hardware faults in a resource augmented cruise system has been discussed in [11].

A methodology based on fault injection for system level dependability analysis of multi processor system is put forth in [12]. A fault simulation method based on software implemented fault injection for dependability evaluation in safety critical systems has been given in [13].

3 Background Study

Tasks in real time systems have critical deadlines and missing any deadline leads to catastrophic consequences. To avoid this, redundancy techniques are in use, but these techniques can lead to power and cost increase. So techniques to complement this redundancy and effectively utilize redundant resources are important. Fault tolerant scheduling techniques in homogeneous and heterogeneous systems ensure that the system will meet task deadlines with reduced redundancy needs in the system [1]. The resources need to be pre-allocated for a task to meet its deadline, hence leading to static table driven scheduling used in hard real time systems [2]. Predictability is maintained in static table driven scheduling where a schedule is determined such that the tasks meet the timing constraints and precedence constraints, etc.

Onboard computers exercises over all control on the vehicle during flight and perform computation related to navigation, guidance and control [3]. Hot standby dual redundancy is employed for onboard computers for overcoming failures without any outage. The computing cycles in the system can be mainly of two types, major computation cycles and minor computation cycles. The architecture of onboard computer system consists of parallel buses, serial buses, software modules, etc. Triple modular redundancy, dual redundancy are commonly used fault tolerant methods for permanent hardware faults and watch dog timers can be used for detection of these faults [4]. Most of the processor failures are due to transient faults, the effects of such faults can be minimized by using recovery approach methods. For recovering from the transient faults it is not necessary to schedule an alternate task, instead a backward recovery can be exercised to recover from previous saved correct state [5]. Primary backup approach is a commonly used software fault tolerance technique, where the code is executed by the processor and output is checked using acceptance tests [1]. The acceptance test checks are done to verify whether the results are within a range and timing constraints are satisfied. The backup copies are executed only when the acceptance test fails and these backup copies will be lighter than primary copies. The correct values or important states of the system after each acceptance test will be saved into the memory [6]. These checkpoints help in fault tolerance by rolling back the system to previously saved state.

A new scheme for improving the efficiency of onboard computers can be obtained by resource augmented scheduling which is based on task criticality [8]. The system will have more functional capability during fault free operation while system performs in an safe minimal operation during fault condition. Combining resource augmented task allocation along with real time scheduling techniques like Rate Monotonic scheduling (RM) and Earliest Deadline First (EDF) provide better programming flexibility and effective utilization of resources [9]. A scheduling method combining dynamic best effort and dynamic planning based scheduling along with resource management has been used for scheduling periodic and aperiodic task in cruise control system [10].

Fault injection methods can be used for assessment of system dependability and helps in studying the behavior of the system. Methods for evaluating the effects of faults not only on the system output but also on the internals of system during overall execution are explained in [12]. A software fault injection model for an anti lock braking system has been developed and analyzed in [13]. The methodology combines structural and functional models for achieving higher accuracy.

4   Approach

4.1   System Model and Computation Cycles

Navigation, Guidance and Control unit in onboard computers of the Indian satellite launch vehicle has a hot standby dual redundant architecture. The cross-strapping connections present in the system help in detection and synchronization of functions. Primary chain and redundant chains have navigation, guidance and control units independently, where navigation unit alone has been considered and represented by N1 in primary chain and N2 in redundant chain respectively. Bidirectional bus between the two units enables the communication and exchanges of health information between the units.

Watch dog timer helps in detection of faults in each of functional units. A periodic health check has been done during every minor cycle in each unit in the dual chain system. When a permanent fault occurs in a functional unit the succeeding units and redundant unit become aware of the fault with bidirectional and cross-strapping connections.

4.2   Task Model

Navigation unit determines the position and velocity of the vehicle [14]. The output from navigation units have been given to the guidance unit and are also saved in a global memory.

Navigation tasks shown in Table 1, denoted by ζ_1, ζ_2 and ζ_3, are executed in every major cycle of hyper period 500 ms. The tasks are periodic and include critical (C), non-critical (N) and optional task (O). Optional tasks included may consists of ζ_{NO1} which has a time period of 500 ms and computation time of 70 ms, second optional task ζ_{NO2} of period 500 ms and processing time of 70 ms. Optional tasks include functions like faster control loops or other complementary functions which can be dropped when situation demands [15]. Optional tasks can be system dependent or system independent, dropping a system independent task causes a minimal effect in the system performance [8].

Table 1 Task table

Task	Criticality	Execution time (units)	Time period (units)
ζ_1	C	15	50
ζ_2	N	98	500
ζ_3	N	99	500

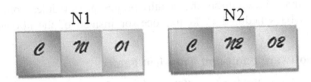

Fig. 1 Task allocation-resource augmented dual redundant system

4.3 Task Allocation

Task allocation in a dual redundant Navigation system—the critical allocated and simultaneously scheduled in both the processors while the non-critical tasks are shared and scheduled in both the processors. The subscripts 1 and 2 indicate primary chain and secondary chain respectively. Optional tasks enhance the performance of the system, like faster control loops, specialized computations and can be scheduled on both the processors.

Figure 1 shows the task allocation in an augmented Navigation unit with the optional tasks allocated to the available unused computational resources.

4.4 Fault Tolerance in Augmented Framework of Navigation Unit

The augmentation due to the task allocation paradigm gives more slack time and an algorithm for effective use of this accrued slack time for fault tolerance and system performance improvements have been discussed in earlier work done by the same author [16, 17]. The algorithm integrates approaches for tolerating permanent hardware and software faults, transient hardware and software faults with real time scheduling in a resource augmented framework.

Permanent Hardware Fault

The system provides with a flexibility to select different modes of operation. By default system starts in mode 3 operation where N1 processor schedules critical task (C_1), non-critical task (N_1) and optional tasks (O_1) while N2 processor schedules critical tasks (C_2) and non-critical tasks (N_2). The slack time available in N2 processor has been more compared to N1 processor, the amount of non-critical tasks scheduled in N2 has been greater than N1. Occurrence of permanent hardware fault in any one of the processors alerts the other processor; this causes the healthy processor to select a particular mode of operation.

Permanent hardware fault in any of the processor in dual system causes the healthy processor to enter into mode 2 or mode 1 operation. In mode 2 operation, the healthy processor discards the non-critical task that is to be scheduled in fault processor and continues scheduling the optional tasks allocated in the healthy processor. In mode 1 execution, the healthy processor schedules non-critical tasks which have to be scheduled in fault processor instead of the allocated optional tasks.

Permanent Software Fault and Transient Faults

The tasks have been divided into different blocks; after each block an acceptance test has been scheduled to identify permanent software faults and transient software and hardware faults. The test checks for the timing and value of the outputs generated and correct output which passes the test has been saved in the local memory of the processor. Slack times accrued by augmentation have been distributed between the tasks based on the execution time of the complete task [16]

Permanent software faults have been recovered by recovery block method, where a secondary alternative of the failed block has been executed in the same processor. Transient faults may result in changes in flags and bit flips, these errors can be overcome by task recovery approaches. When a transient fault affects the block for the first time, a backward recovery approach has been selected and the task re-executes from the previously saved state. When the transient fault still prevails, a forward recovery approach has been selected and the correct states for execution next block of task have been loaded from the memory. The fault tolerance approaches used for overcoming software and transient faults utilizes the augmented distributed slack time.

4.5   Performance Metrics

The performance metrics like utilization (U), slack margin (S) and mean time to recover (MTTR) has been evaluated during run time. The slack time made available is based on the total execution time of all tasks in the system. Utilization is the summation of ratio of the execution time of individual task to the time period of the task and it projects how effectively a task set utilizes processor resources. Mean time to recover gives the time system takes to recover from the failure.

5 Hardware Implementation and Testing

LPC1768 ARM Cortex-M3 processor has been used for rapid prototyping of the Navigation functional unit of the onboard computers.

The experimental setup consists of two processors and an array of LEDs for indicating scheduling of tasks (Fig. 2). The inputs given for the system consists of sensing task which periodically checks the status of bidirectional communication bus, control tasks which controls the operation of the system based on the inputs and actuating tasks. LEDs have been used to indicate the execution of the current task and LCD displays the current status (modes, fault condition, utilization and slack margin) over a hyper period. For practical real time experimentation a time scaling has been resorted to.

A blue LED built into the board indicates the health status of the processors. Switches have been interfaced into the evaluation board for injecting the faults and selecting the mode of operation. Power supply failure has been injected as a permanent hardware fault into the processor. Permanent hardware fault injection model has been given in Fig. 3. The processors N1 and N2 exchange synchronization and health signals using two bidirectional connections. LCD displays during fault and fault free operation are shown in Fig. 4

In the permanent software fault injection model (Fig. 5), the tasks have been divided into blocks and a LED array has been used to indicate primary task blocks and another LED array indicates the execution of secondary alternative for each blocks. Switch has been employed for injecting the fault and this cause an interruption in timing of the executing task and processor switches to another non-desirable function. This function has been indicated by employing a buzzer for a period of 3 time units. White LED indicates an acceptance test which detects software error occurred and timing breaks. Once a software error has been detected processor schedules corresponding secondary alternative task. The slack time allotted between each task decreases based on the execution of secondary alternatives blocks of the corresponding task blocks.

Fig. 2 Transient fault injection prototype

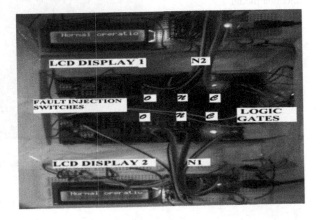

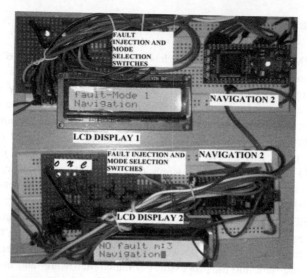

Fig. 3 Hardware fault injection model

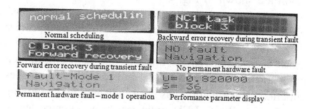

Fig. 4 LCD displays during different operational conditions

Fig. 5 Software fault injection model

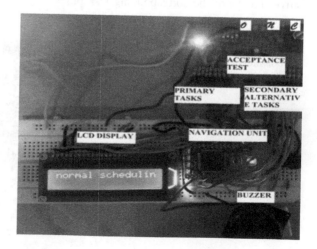

Transient faults—both hardware and software transients have been considered. The software transient faults have been implemented by interrupting the execution of the program and changing flags while hardware interrupts are implemented by logic gates. The hardware transient fault changes the logic value that has to be available in the system after execution of each task block (Fig. 2). On detection of transient fault, the task block has been re-executed and performs the next acceptance test. If transient fault affects the re-executed task block, a forward recovery approach has been undertaken. LCD displays the current status of the system during recovery approaches.

The system outputs displayed on desktop—task execution timings and performance metrics during hardware fault injection for verification by the user. The performance of the system has been evaluated and tested for 15 experimental runs of the hardware prototype and at different instants of fault injection.

6 Results

Permanent hardware fault injection model has been evaluated for different time instances of fault injections. Permanent hardware fault replicating power line failure has been injected in both secondary and primary processors, and utilization (U), slack time (S) and mean time to recover (MTTR) available after fault injections are evaluated.

Performance of the system over several fault injection simulations has been considered. Table 2 indicates the utilization and slack time during fault and free condition in hardware fault injection model. During fault condition, utilization and slack time available in the healthy processor have been taken into account. During fault free operation, average value of the utilization and slack time of both the processors have been considered. For a given fixed work load and system conditions, the utilization of healthy processors increase by 15.8 % and slack time decreases by 48 %. More slack time available during fault free operation can be effectively utilized by other performance improving tasks and also for fault tolerance. The mean time to recover (MTTR) from a permanent hardware fault has been obtained as 12.7 time units.

Average total mean time to recover from faults when one, two, three and four occurring over a hyper period of time are considered for transient fault injection model and software fault injection model. Many experimental runs have been

Table 2 Utilization and slack time for permanent hardware fault injection model

Parameter	Fault free condition		Fault condition
	Mode 1	Mode 2	
Utilization (U)	0.875	0.82	0.755
Slack time (S) (time units)	25	36	49

Table 3 Mean time to recover (MTTR) from permanent software fault and transient faults

No. of faults	Mean time to recover (MTTR)— permanent software faults (time units)	Mean time to recover (MTTR)— transient faults (time units)
1	9.85	8.4
2	14.3	27
3	18	35
4	14	48

considered for obtaining the results and the values are given in Table 3. The faults are injected over a hyper period of 265 time units under constant load and system conditions. The instants of fault injection are generated according random Poisson distribution in case of transient faults and permanent hardware faults, while an exponential distribution is considered for permanent software faults.

The average of the total mean time to recovery from 4 transient faults occurring over a hyper period is less with respect to overall slack time available (49 time units). Here, in the system uniform checkpointing combined with acceptance tests has been considered. The mean time to recover depends on the instances of checkpoint and acceptance test.

7 Conclusion

Fault tolerance algorithms for tolerating permanent hardware and software faults, transient hardware and software faults have been tested and evaluated with fault injection models. The task test of the Navigation unit of NGC framework has been selected and performance of the system has been evaluated for different instants of fault injection. Better slack time and utilization are available in an augmented frame work of Navigation unit for a given work load. The mean time to recover indicates the effectiveness of the system to tolerate one and more number of faults. A prototype for resilient fault tolerant framework for a resource augmented navigation unit has been developed to tolerate increased number of transient and software faults. Future work can include optimizing the checkpoints and design a more optimistic approach for fault detection where faults are assumed to be detected if they occur at the initial intervals of the task execution. A pessimistic approach has been employed in this work because the fault occurrences and detection are assumed to occur at the end of task blocks. This approach can be extended to a generic m-redundancy framework with considerable benefits to fault tolerance mechanisms. Optimal checkpointing in recovery approaches can improve the accuracy of the mean time to recovery (MTTR).

References

1. Krishna C M.: Fault tolerant scheduling in homogenous real time systems. ACM Computing Surveys (CSUR). vol. 46. Issue 4 (2014).
2. Krithi Ramamamritham, John Stankovic: Scheduling algorithms and operating system support for Real-Time systems. In: proceedings of IEEE. vol. 82 (1994).
3. Basu et al, A fault tolerant computer system for Indian satellite launch vehicle programme, Sadhana, vol. 11, pp. 221–231 (1987).
4. Hitt E F., Mulcare D.: The Fault tolerant Avionics. Digital Avionics Handbook, 2nd Edition. Cary R Spitzer. CRC press LLC (2007).
5. Basu D., Paramasivam R.: Approach to software assisted recovery from hardware transient faults for real time systems Computer Safety. Reliability and Security. In: 19th International Conference SAFECOMP 2000, vol. 1943, pp. 264–274 (2000).
6. Leach R J.: Setting checkpoints in legacy code to improve fault-tolerance. Journal of Systems and Software, vol. 81 (6), pp. 920–928 (2007).
7. Punnekkat S et al.: Analysis for Checkpointing for Real-Time systems. The International Journal of Time-Critical Computing Systems, vol. 20, pp. 83–102 (2001).
8. Radhamani Pillay et al.: An improved redundancy scheme for optimal utilization of onboard Computers. In: IEEE INDICON 2009, India (2009).
9. Radhamani Pillay V et al.: Optimizing Resources in Real-time Scheduling for Fault Tolerant Processors. In: 2010 1st International Conference on Parallel Distributed and Grid Computing (PDGC) (2010).
10. Annam Swetha et al.: Design, Analysis and Implementation of Improved Adaptive Fault Tolerant Model for Cruise Control Multiprocessor System. IJCA, vol. 86, (2014).
11. Annam Swetha et al.: A Real-Time Performance Monitoring Tool for Dual Redundant and Resource Augmented Framework of Cruise Control System. IJCA, vol. 94, (2014).
12. Antonio Miele.: A fault-injection methodology for the system-level dependability analysis of multiprocessor embedded systems. Microprocessors and Microsystems vol. 38, pp. 567–580, (2014).
13. Dawid Trawczyski., Janusz Sosnowski.: Fault injection testing of safety critical applications. Technical transaction on computer science and information systems, pp. 87–107, (2010).
14. F. Stesina, Design and Verification of Guidance, Navigation and Control systems for Space application, Porto Institutional repository, (2014).
15. Robert Ian Davis.: On exploiting spare capacity in hard real time systems. University of York, (1995).
16. Archana Sreekumar, Radhamani Pillay V.: An Increased Resilient Fault Tolerant Framework for Navigation, Guidance, Control system with Augmented Resources. International journal of Embedded Systems (IJES), unpublished (2015).
17. Archana Sreekumar., Radhamani Pillay V.: Fault Tolerant Scheduling with Enhanced Performance for Onboard computers- Evaluation. In: 4th International Conference on Frontiers in Intelligent computing: Theory and Applications 2015, NIT Durgapur, unpublished (2015).

Web Data Analysis Using Negative Association Rule Mining

Raghvendra Kumar, Prasant Kumar Pattnaik and Yogesh Sharma

Abstract Today era is combination of information and communication technology (ICT), everyone wants to share and store their information through the internet, so there is huge amount of data is searched every day, there is lots of web data is collected in every seconds and with the help of web usage mining, we can discover useful pattern from the web databases. For analyzing this huge amount of web data, we required one of the useful concepts is web site managements. In which we discover the useful pattern, discover or analyzing the useful information from the web database. Here we used the concept of negative association rule mining for analyzing the web log files, for finding the strong association between the web data's.

Keywords Data mining · Web data mining · Association rule mining · Negative association rule mining · Data analysis

1 Introduction

In applying sequence learning models to Web-page recommendation, association rules and probabilistic models have been commonly used. Some models [1–3], such as sequential modeling, have shown their significant effectiveness in recommendation generation. In order to model the transitions between different Web-pages in Web sessions, Markov models and tree-based structures are strong candidates.

R. Kumar (✉) · Y. Sharma
Faculty of Engineering and Technology, JNU, Jodhpur, Rajasthan, India
e-mail: raghvendraagrawal7@gmail.com

Y. Sharma
e-mail: yogeshsharma@gmail.com

P.K. Pattnaik
School of Computer Engineering, KIIT University, Bhubaneswar, India
e-mail: patnaikprasant@gmail.com

© Springer India 2016
S.C. Satapathy et al. (eds.), *Information Systems Design and Intelligent Applications*, Advances in Intelligent Systems and Computing 433,
DOI 10.1007/978-81-322-2755-7_53

513

2 Proposed Work

In this proposed work shall use the apache web server log file [4–10] of a financial web site user having more than 16 million hits in a month's time [11]. A web server log contains the following information

1. Number of Hits
2. Number of Visitors
3. Visitor Referring Website
4. Visitor Referral Website
5. Time and Duration
6. Path Analysis
7. Visitor IP Address
8. Browser Type
9. Cookies

Data of a typical web server is shown in following sample data of the first raw of the log file:

10.32.1.43[10/Nov/2013:00:07:00]"GET/flower_store/product.screen?product_id= FL-DLH-02   HTTP/1.1"20010901"http://mystore.splunk.com/flower_store/category. screen?category_id=GIFTS&JSESSIONID=SD7SL1FF9ADFF2"'"Mozilla/5.0 (X11; U; Linux i686; en-US; rv: 1. 8. 0. 10) Gecko/20070223 Cent OS/1.5.0.10-0.1.el4.centos Firefox/1.5.0.10" 4361 3217.

This work shall use the basic steps of the web usage mining [12] with modifications to improve the performance and accuracy of the extracted information. For improvements, work will be done in following modified steps [13]:

1. Data Collection—From a server log of a website
2. Data Pre-processing—Formatting data from the server log obtained. The server log contains huge information in each record which is separated and stored in lists with rows and columns. Some redundant information is removed and data rows are ordered on the basis of the date and time of hit.
3. Pattern Discovery—In this step I will apply the following (but not to max) association rules:

    I. If Webpage X is hit then Webpage Y is also hit by the user
    II. If Webpage X is hit then user moves away from the site by clicking an external link
    III. If a user comes on site once then he visits it again

4. Pattern Analysis—From the association rules applied in step 3 data shall be analysed to find the support and confidence for each rule and filtering of records shall be done on basis of low support or confidence. Knowledge will be produced for the web admin in readable format for future use.
5. Result Generation—Each phase will be measured for time taken and time complexity to show that the performance of the proposed system is high.

Comparatives shall also be produced to show the accuracy of the produced patterns in respect of the base work.

Proposed work will be implemented using C#.NET Windows and Server Log Dataset Extracted from a live server from a financial website. The work will be done in following steps: first step is Interface Development, in which C#.net windows/JAVA Swing forms will be used to develop the interface. Various relevant navigational buttons will be included in it which allows traversing with validations. Second step is Loading Server Log Data Set in which, Server Log dataset is a collection of server messages generated when a server starts and stops. The details such as login, logout, system shutdown, errors and other information's will be stored in it by the server machine during the normal course of operations on the server. These will be used for clustering of message using association rules. Third step is Data Pre-processing, in data pre processing file handling will be used for loading the server log file in project and will be applied data pre-processing on the loaded data. In data pre-processing stopping and stemming will be applied to filter data for important words. Stopping will be used to remove all special characters, remove any words with length less than or equal to 3 characters and removing any words which are prepositions. Stemming will be applied to remove words with having similar meaning and different tense of English. Fourth step is negative Association Rule Mining, in this phase; various characteristics of the server log file will be used for creating negative association rules. These rules shall be applied to create clusters. And last phase analysis phase in which we analyze our results with the help of graphs; Results will be calculated by using the clusters and formulas for calculation of performance and accuracy of the system. Graphs will be drawn using the various results as calculated for all accuracy, precision, recall and specificity.

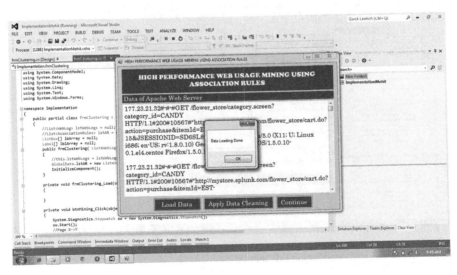

Fig. 1 Shows the process of loading of web data

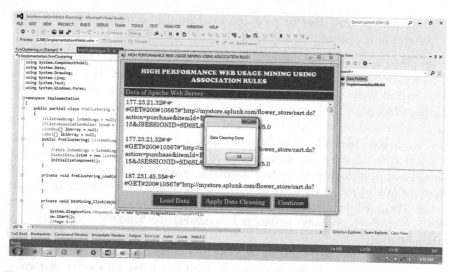

Fig. 2 Shows the process of data cleaning

The proposed algorithm of this paper is shows the steps, how we are analyzing our web data. And Fig. 1, shows the process of loading of web data, Fig. 2 shows the process of data cleaning after the loading of web data is done, Fig. 3 shows the data analyzing after applying the negative association rule mining, Fig. 4 shows the data analysis, and Fig. 5 shows the final output of our process, in which its shows the negative support count, negative confidence and negative lift or importance.

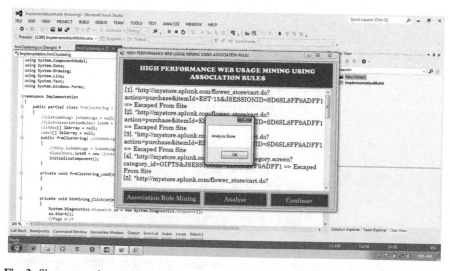

Fig. 3 Shows negative association rule mining

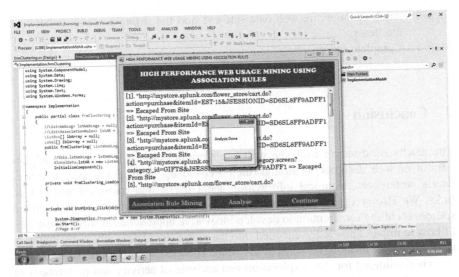

Fig. 4 Shows the data analysis

Algorithm

Step 1: Collect the useful Data from the web files or weblogs.

Step 2: Load the Web data into in our Data analyzer.

Step 3: After data loading of data apply data cleaning.

Step 4: After applying the data cleaning apply the association rule mining for data analysis

Step 5: After the data analysis calculates the negative support, negative confidence and negative lift or importance.

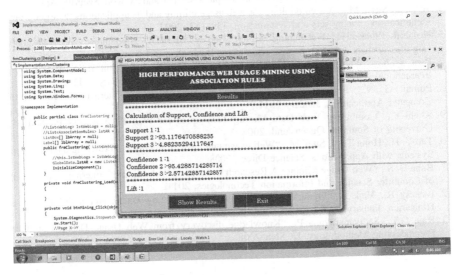

Fig. 5 Shows the final output of our process

Step 6:   Analyze the entire web data sets and find the strong association between
          the items.
Step 7:   Complete the process

3   Conclusion

The results we discover from the web usage mining can be used by web administrator or web designer to arrange their website by determining system errors,
user's preferences, technical information about users, and corrupted and broken
links. We also recognize that large amount of information stored in unstructured
sources is calling for attention to develop innovative approaches to solve challenges
of how to impart knowledge encoded into the unstructured data efficiently and how
to explore more meaningful ways to utilize the knowledge. This work has important
aspects considered for data exploration and analysis of activity and preferences of
users and researchers. In future this work can be further extended for applying
clustering methods using association rules drawn efficiently in this work.

References

1. Guandong Xu., "Web Mining Techniques for Recommendation and Personalization", Victoria University, Australia, March 2008.
2. Bamshad Mobasher., "Data Mining for Web Personalization", LCNS, Springer-Verleg Berlin Heidelberg, 2007.
3. Krishnamoorthi R., Suneetha K. R., Identifying User Behavior by Analyzing Web Server Access Log File", International Journal of Computer Science and Network Security, April 2009.
4. Arya, S., Silva, M., "A methodology for web usage mining and its applications to target group identification", Fuzzy sets and systems, pp. 139–152, 2004.
5. Kosala R., Blockeel H., 'Web mining research: a Survey", SIGKDD Explorations, 2, pp. 1–15, 2000.
6. Log files formats, http://www.w3c.org, Access Date: Dec. 2013.
7. Pani S. K., "Web Usage Mining: A Survey on Pattern Extraction from Web Logs", International Journal of Instrumentation, Control & Automation, January 2011.
8. Purohit G. N., "Page Ranking Algorithms for Web Mining", International Journal of Computer Applications, January 2011.
9. Khalil F., "Combining Web Data Mining Techniques for Web Page Access Prediction", University of Southern Queensland, 2008.
10. Tao Y. H., Hong T. P., Su Y.M., "Web usage mining with intentional browsing data", Expert Systems with Applications, Science Direct, 2008.
11. Babu D. S., "Web Usage Mining: A Research Concept of Web Mining", International Journal of Computer Science and Information Technologies, 2011.
12. Iváncsy R., Vajk I., "Frequent Pattern Mining in Web Log Data", Act a Polytechnic a Hungarica, January 2006.
13. Raju G.T., Sathyanarayana P., "Knowledge discovery from Web Usage Data", Complete Preprocessing Methodology, IJCSNS 2008.

Cloud Based Thermal Management System Design and Its Analysis

Namrata Das and Anirban Kundu

Abstract We are going to propose an advanced architecture sensing real time temperature of a particular location for transmitting the data to a cloud database. Current data have been analysed based on previously recorded data. If any abnormal data is observed, then the system produces an alarming message to the concerned authorities. Analytical data guide users to solve real time problems observing anomalies in the system.

Keywords Heat sensor · Microcontroller · Distributed database · Network · Cloud · Cloud thermal manager

1 Introduction

1.1 Overview

A network is a collection of computers or other hardware components that are interconnected through some communication channels which allow to share resources and information. Data is transferred in form of packets from source to destinations. Devices present in network that originate, route and terminate the data are typically known as network nodes.

A wireless sensor network (WSN) is distributed having autonomous sensors for monitoring physical and/or environmental conditions in a spatial way. Each sensor network node has several parts, such as radio transceiver with an internal antenna, connection to an external antenna, a microcontroller, a battery, etc. These networks can adopt any network topologies [1].

N. Das (✉)
Netaji Subhash Engineering College, Kolkata 700152, India
e-mail: namratacse@gmail.com

A. Kundu
Innovation Research Lab, Kolkata 711103, West Bengal, India
e-mail: anik76in@gmail.com

© Springer India 2016
S.C. Satapathy et al. (eds.), *Information Systems Design and Intelligent Applications*, Advances in Intelligent Systems and Computing 433,
DOI 10.1007/978-81-322-2755-7_54

1.2 Cloud Preliminaries

Cloud computing is an abstract level of concept which relies on sharing resources for accomplishing consistency and economies of scale throughout the network having resemblance to a utility. In cloud computing, a single server could be accessed by multiple users in order to retrieve and/or update data without having any license for different applications. Now-a-days, people are shifted from dedicated hardware to a shared cloud infrastructure as per demand basis. Availability of high-capacity cloud networks using low-cost computers and other related resources have led to a growth in cloud computing. It exhibits adoption of hardware virtualization, service-oriented architecture, and autonomic and utility computing [2]. Companies could be either scaled up based on increased computing needs, or, scaled down if demand is decreased. Cloud computing is the outcome of evolutions and further adoptions of several existing methodologies. Target of cloud computing is to allow users over the distributed networks having zero expertise of the specified system [3, 4]. Virtualization is useful in cloud computing scenario. Cloud computing is based on the concepts of Service-oriented-Architecture [5].

1.3 Related Works

Researchers have proposed varieties of theories on thermal management in cloud based system networks. Sensor network has been discussed in [6]. Users have found inconveniences with increase of temperature using specific devices having high power consumption. Researchers have done survey on various approaches to handle heat management in several devices of the network [7]. Authors have designed an interactive temperature management system in which users are allowed to supply inputs for managing the heat dissipation and are able to operate the applications as per their requirements [8].

It has become indispensable to lower power and cost of cooling of cloud data centers due to increment of data and varieties of data analyses. Thermal impact of input/output access patterns on data storage systems has been investigated in [9]. A comparison of cooling cost of storage systems has been done by different data placement schemes. Researchers have shown in [10] that unexpected events, like strategic attacks, wrong configurations, failures due to high server density of high performance computing data-centers, lead to thermal anomalies creating impact on total cost of data center operations. Researchers have described several challenges of data center thermal management [11]. Authors have established a new practical environment modeling for mobile robots in unknown circumstances [12].

2 Proposed Work

2.1 Overview

In this paper, a thermal management architecture has been proposed based on cloud scenario for taking real time temperature of a particular location and further transmit it to a server database for storing purpose. Users' are capable of enquiring particular record(s) based on temperature, time, and date. An interface network handles users' queries. This interface network is connected to the distributed database of cloud management system. A managerial node is present in each database for analysing users' queries to fetch records from concerned databases as referred in Fig. 1.

2.2 Hardware Components Used

In our proposed system, we have used a hardware device to sense the temperature and transmit it to the server interface. LM35 is used as a heat sensor in our approach. LM35 has an output voltage proportional to temperature having scale factor 0.01 V/°C. LM35 does not require any peripheral calibration and maintains an accuracy of ±0.4 ° C at room temperature and ±0.8 °C over a range of 0 to +100 °C [13]. The sensor self-heating causes less than 0.1 °C temperature rise in still air. ADC is used to convert the analog signal to corresponding digital code. AT89C51 is used as the microcontroller chip. The AT89C51 is a low-power, high-performance CMOS 8-bit microcomputer with 4 kbytes of Flash Programmable and Erasable Read Only Memory (PEROM) [14]. The device is manufactured using Atmel's high density non-volatile

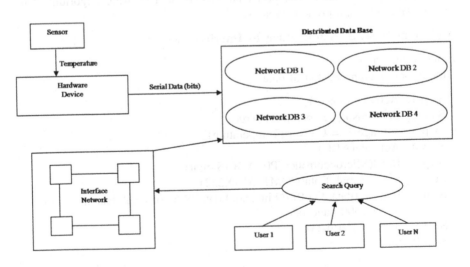

Fig. 1 Proposed architecture

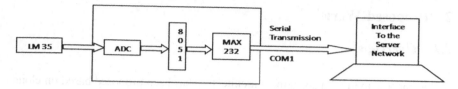

Fig. 2 Transmission of a signal from heat sensor to a computer

memory technology and is fully compatible with the industry standard MCS-51 instruction set and pin out. It has four ports namely port 0, port 1, port 2, and port 3. We have used port 0 as the data output port. MAX232 is used for transferring data using serial communication to the server interface. The MAX232 is an IC which converts signals from RS-232 serial port to signals appropriate for using in TTL compatible digital logic circuits [15]. The MAX232 is a typical driver and/or receiver converting Receiving Signal (RX), Transmitting Signal (TX), Clear-To-Send (CTS) and Request-To-Send (RTS) signals. Proposed hardware model has been depicted in Fig. 2.

2.3 Procedure

In Algorithm 1, temperature has been sensed by the concerned sensor such as LM35. Further, analog signal has been transformed into digital using ADC chip. Digital data is sent to serial port of interface to the server using COM1 via P_0 port of the 8051 microcontroller. MAX232 chip is responsible for this type of transmissions. 9 bit (8 bit for actual signal/data and 1 bit for stop-bit) serial data has been received. The transmission uses 9600 baud rate, and the entire data is being kept in the system buffer. The port data have been read using Algorithm 2 for storing data into appropriate databases.

Algorithm 1: Send_Temperature_to_Interface_Server

Input: Temperature
Output: Serial Transmission of data

    Step 1: Activate(Sensor)
    Step 2: Set Sensor(Temperature)=True
    Step 3: Voltage V = Convert(Temperature)
    Step 4: Activate(ADC)
    Step 5: Read(Microcontroller_P0, ADC(Signal))
    Step 6: Send_Serial_Data(COM1, MAX232)
    Step 7: Receive(data, 9600) //9 bit serial data (8 bit original data and 1 stop bit) using 9600 baud rate
    Step 8: Save_buffer(data)
    Step 9: Stop

Algorithm 2: Store_Database

Input: 9 bit serial data
Output: Data storage in real-time within distributed database

Step 1: Read(Buffer_9bit)
Step 2; Split(Data_8bit, Stop_1bit)
Step 3: Set Data = Data_8bit
Step 4: Read(System_Date)
Step 5: If (System_Day == "Sunday")

    Step 5.1: If (System_Time >= "06:01:00" && System_Time <= "12:00:00")
    Store(Data) = Database("Sunday","06:01 to 12:00")
    Step 5.2: If (System_Time >= "12:01:00" && System_Time <= "18:00:00")
    Store(Data) = Database("Sunday", "12:01 to 18:00")
    Step 5.3: If (System_Time >= "18:01:00" && System_Time <= "24:00:00")
    Store(Data) = Database("Sunday", "18:01 to 24:00")
    Step 5.4: If (System_Time >= "00:01:00" && System_Time <= "06:00:00")
    Store(Data) = Database("Sunday", "00:01 to 06:00")

Step 6: If (System_Day == "Monday")

    Step 6.1: If (System_Time >= "06:01:00" && System_Time <= "12:00:00")
    Store(Data) = Database("Monday", "06:01 to 12:00")
    Step 6.2: If (System_Time >= "12:01:00" && System_Time <= "18:00:00")
    Store(Data) = Database("Monday", "12:01 to 18:00")
    Step 6.3: If (System_Time >= "18:01:00" && System_Time <= "24:00:00")
    Store(Data) = Database("Monday", "18:01 to 24:00")
    Step 6.4: If (System_Time >= "00:01:00" && System_Time <= "06:00:00")
    Store(Data) = Database("Monday", "00:01 to 06:00")

Step 7: If (System_Day == "Tuesday")

    Step 7.1: If (System_Time >= "06:01:00" && System_Time <= "12:00:00")
    Store(Data) = Database("Tuesday", "06:01 to 12:00")
    Step 7.2: If (System_Time >= "12:01:00" && System_Time <= "18:00:00")
    Store(Data) = Database("Tuesday", "12:01 to 18:00")

Step 7.3: If (System_Time >= "18:01:00" && System_Time <= "24:00:00")
Store(Data) = Database("Tuesday", "18:01 to 24:00")

Step 7.4: If (System_Time >= "00:01:00" && System_Time <= "06:00:00")
Store(Data) = Database("Tuesday", "00:01 to 06:00")

Step 8: If (System_Day == "Wednesday")

Step 8.1: If (System_Time >= "06:01:00" && System_Time <= "12:00:00")
Store(Data) = Database("Wednesday", "06:01 to 12:00")

Step 8.2: If (System_Time >= "12:01:00" && System_Time <= "18:00:00")
Store(Data) = Database("Wednesday", "12:01 to 18:00")

Step 8.3: If (System_Time >= "18:01:00" && System_Time <= "24:00:00")
Store(Data) = Database("Wednesday","18:01 to 24:00")

Step 8.4: If (System_Time >= "00:01:00" && System_Time <= "06:00:00")
Store(Data) = Database("Wednesday", "00:01 to 06:00")

Step 9: If (System_Day == "Thursday")

Step 9.1: If (System_Time >= "06:01:00" && System_Time <= "12:00:00")
Store(Data) = Database("Thursday", "06:01 to 12:00")

Step 9.2: If (System_Time >= "12:01:00" && System_Time <= "18:00:00")
Store(Data) = Database("Thursday", "12:01 to 18:00")

Step 9.3: If (System_Time >= "18:01:00" && System_Time <= "24:00:00")
Store(Data) = Database("Thursday", "18:01 to 24:00")

Step 9.4: If (System_Time >= "00:01:00" && System_Time <= "06:00:00")
Store(Data) = Database("Thursday", "00:01 to 06:00")

Step 10: If (System_Day == "Friday")

Step 10.1: If (System_Time >= "06:01:00" && System_Time <= "12:00:00")
Store(Data) = Database("Friday", "06:01 to 12:00")

Step 10.2: If (System_Time >= "12:01:00" && System_Time <= "18:00:00")
Store(Data) = Database("Friday", "12:01 to 18:00")

Step 10.3: If (System_Time >= "18:01:00" && System_Time <= "24:00:00")
Store(Data) = Database("Friday", "18:01 to 24:00")

Step 10.4: If (System_Time >= "00:01:00" && System_Time <= "06:00:00")
Store(Data) = Database("Friday", "00:01 to 06:00")

Step 11: If (System_Day == "Saturday")

Step 11.1: If (System_Time >= "06:01:00" && System_Time <= "12:00:00")
Store(Data) = Database("Saturday", "06:01 to 12:00")

Step 11.2: If (System_Time >= "12:01:00" && System_Time <= "18:00:00")
Store(Data) = Database("Saturday", "12:01 to 18:00")

Step 11.3: If (System_Time >= "18:01:00" && System_Time <= "24:00:00")
Store(Data) = Database("Saturday", "18:01 to 24:00")

Step 11.4: If (System_Time >= "00:01:00" && System_Time <= "06:00:00")
Store(Data) = Database("Saturday", "00:01 to 06:00")

Step 12: Goto Step 1

3 Experimental Analysis

In our proposed system, several circumstances have been modelled in real-time. Sensor data is being updated having a fixed interval. In this section, distinct graphs are being demonstrated to exhibit cloud system performances having 1 s time interval. Temperature graph of harddisk has been shown in Fig. 3a with respect to time in case of interface server of the cloud. Figure 3b exhibits the condition of the

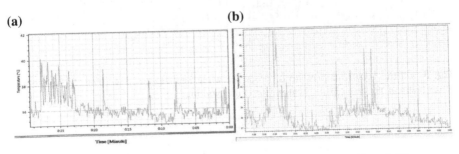

Fig. 3 **a** Time versus temperature of harddisk while heat sensor is working. **b** Time versus temperature of harddisk while heat sensor is not working

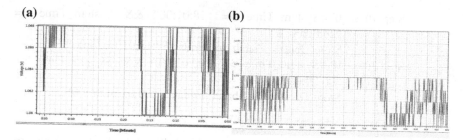

Fig. 4 **a** Time versus voltage of harddisk while heat sensor is working. **b** Time versus voltage of harddisk while heat sensor is not working

server while the sensor is not activated. Typically, temperature is shown in degree centigrade (°C). Figure 4a shows the voltage response in respect of time of the interface server's harddisk. Figure 4b exhibits non-working condition of heat sensor based section of the cloud using temperature graph. Figure 5a, b show how the processor of the interface server is responding with time respectively. Figure 6a describes how much the processor of the interface server is loaded with respect to time. Figure 6b shows the processor load in case of idle condition of heat sensor. Figure 7 exhibits the voltage changes with respect to temperature change of LM35 which is the heat sensor used in our network. The voltage raise of LM35 is 0.01 V per °C. Figure 8 describes the Sensor data and its storage locations in case of active conditions of the sensor within our cloud. In our system, there are seven databases for seven days in a week. Each database is again splitted into four sub-databases according to system time with 4 h interval.

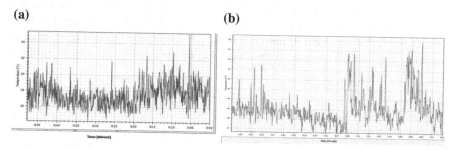

Fig. 5 **a** Time versus temperature of processor while heat sensor is working. **b** Time versus temperature of processor while heat sensor is not working

(a)   **(b)**

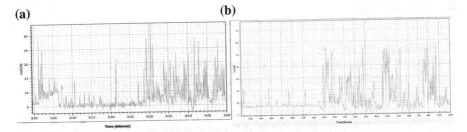

Fig. 6 a Time versus load of processor while heat sensor is working. **b** Time versus load of processor while heat sensor is not working

Fig. 7 Temperature versus voltage change of LM35 (heat sensor)

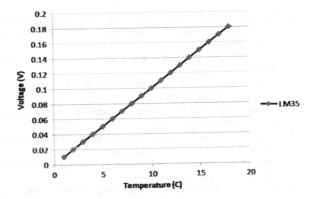

Fig. 8 System time versus storage location while heat sensor is working

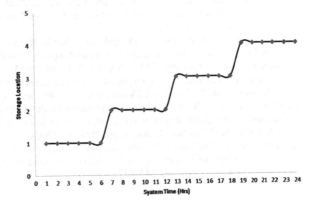

4   Conclusion

In this paper, we are have proposed an advanced architecture sensing real time temperature of particular location for transmitting data to cloud. Heat sensor has been used for collecting real-time data within cloud for analytical purpose. We have monitored conditions of harddisk within interface server and processor. The system performance comparisons have been done using heat sensors. Temperature and voltage of harddisk have changed in different conditions. Change in temperature and load on processor of interface server are monitored.

References

1. Santra, S., Acharjya, P. P. (2013), "A Study And Analysis on Computer Network Topology For Data Communication," *International Journal of Emerging Technology and Advanced Engineering*, vol. 3, no. 1.
2. "Cloud Computing: Clash of the clouds," *The Economist*, 2009-10-15, Retrieved 2009-11-03 (2009).
3. "Gartner Says Cloud Computing Will Be As Influential As E-business," *Gartner.com*, Retrieved 2010-08-22 (2010).
4. Gruman, G., "What cloud computing really means," *InfoWorld*, 2008-04-07, Retrieved 2009-06-02 (2008).
5. Bouley, D. "Impact of Virtualization on Data Center Physical Infrastructure," *Green Grid*, APC by Schneider Electric, white paper 27 (2010).
6. Estrin, D., Govindan, R., Heidemann, J., "Embedding the internet," *Communications of ACM*, vol. 43, no. 5, pp. 39–41 (2000).
7. Bhagwat, A., Prasad, G. R., "Approaches to Thermal Management in Handheld Devices: A Survey," *International Journal of Computer Trends and Technology*, vol. 21, no. 2 (2015).
8. Ambrose, J. A., "HEATSMART: An Interactive Application aware Thermal Management Framework for MPSoC Embedded Systems," *2011 6th International Conference on Industrial and Information Systems*, August, Sri Lanka (2011).
9. Jiang, X., Alghamdi, M. I., Assaf, M. M. A., Ruan, X., Zhang, J., Qiu, M., Qin, X., "Thermal Modeling and Analysis of Cloud Data Storage Systems," *Journal of Communications*, vol. 9, no. 4, pp. 299–311 (2014).
10. Lee, E. K., Viswanathan, H., Pompili, D., "Model-Based Thermal Anomaly Detection in Cloud Datacenters," *2013 IEEE International Conference on Distributed Computing in Sensor Systems*, May, Cambridge, MA, USA, pp. 191–198 (2013).
11. Schmidt, R. R., Cruz, E. E., Iyengar, M. K., "Challenges of data center thermal management," *IBM Journal of Research and Development*, vol. 49, no. 4/5, pp. 709–723 (2005).
12. Park, S. H., Lee, B. H., "Practical environment modeling based on a heuristic sensor fusion method," *2004 IEEE International Conference on Robotics and Automation*, New Orleans, LA, USA, April 26–May 1, vol. 1, pp. 200–205 (2004).
13. http://www.facstaff.bucknell.edu/mastascu/elessonsHTML/Sensors/TempLM35.html.
14. http://www.tutorial.freehost7.com/MC89c51/feature-and-description-of-89c51.htm.
15. http://en.wikipedia.org/wiki/MAX232.

ICC Cricket World Cup Prediction Model

Avisek Das, Ashish Ranjan Parida and Praveen Ranjan Srivastava

Abstract The paper aims to predict the winner of the Cricket World Cup by taking into consideration the various factors which plays an important role in deciding the final outcome of a game. Of the several factors, five has been taken into consideration. These are whether team wins the toss or not, whether the team bats first or not, whether the match is a day match or day/night match, whether the team is playing in its home ground or away from home and at what round of the tournament the match has been played. This paper has used the method of Analytic Hierarchy Process (AHP) to compare the different parameters and come to the final conclusion.

Keywords Analytic hierarchy process · Cricket world cup · Winner prediction

1 Introduction

Started around the 13th century, cricket has become a very popular game played worldwide. There are numerous types of statistical analyses that can be applied to various sports—both information rich games like basketball and baseball as well as less information rich games like curling [1]. However, cricket hasn't seen much of work done on it in terms of analytical modeling as in some other sports like basketball.

Cricket being a game of uncertainty, draws a lot of speculation about who would win. The world cup being one of the greatest cricket festivals, every country want their team to win the match. From People taking the help of astrologers to worshipping their favorite God, there has been numerous supernatural ways of predicting the winner of World Cup. Like Paul, the octopus who supposedly predicted

A. Das (✉) · A.R. Parida · P.R. Srivastava
IIM Rohtak, Rohtak, Haryana, India
e-mail: avisek_3@yahoo.co.in

A.R. Parida
e-mail: ashishavenue88@gmail.com

P.R. Srivastava
e-mail: praveen.ranjan@iimrohtak.ac.in

© Springer India 2016
S.C. Satapathy et al. (eds.), *Information Systems Design and Intelligent Applications*, Advances in Intelligent Systems and Computing 433,
DOI 10.1007/978-81-322-2755-7_55

the results of association football matches, there have been predictors in World Cup Cricket too. A Robot made by University of Canterbury's robot, Ikram, also predicted the results of the world cup cricket only by seeing the Flags of the country playing the world cup [2]. But the winner of the cricket world cup cannot be left to prediction by supernatural powers. All the factors should be properly analyzed before coming into any conclusion [3].

This paper applies analytics to this problem. It seeks to understand what had determined World Cup winners in the past and whether those factors are likely to impact the outcome of this tournament in order to establish the prediction. Though there are many factors that influence the outcome of a cricket match this paper looks at five major factors which would have a considerable impact on the win or loss of the game.

The paper primarily focuses on predicting the winner of the Cricket World Cup 2015 which started from February 2015. The paper is written in the month of February 2015 when the game has just begun.

Some salient points of the paper are:

- The paper predicts the winner of the matches and gives the probability of each of the teams to win. It is a prediction model.
- Data used for the reference are of the world cups from 1987 to 2011. 8 teams— India, Sri Lanka, West Indies, New Zealand, South Africa, Pakistan, Australia and England has been taken into consideration and new teams have not been taken into consideration for the lack of enough data for analysis.
- This prediction model uses AHP [4, 5] to predict the winning probability of a team using various decision factors.
- The Paper has taken Decision factors as whether team wins the toss or not, whether the team bats first or not, whether the match is a day match or day/night match, whether the team is playing in its home ground or away from home and at what round of the tournament the match has been played. The group matches are clubbed into one category while super six, semi-finals and finals has been clubbed into Non Group matches (NG). The winning Probability of the team is calculated on the basis of decision factors.

The data for the decision factors required to perform AHP has been collected and collated from The Home of Cricket Archives [6].

2 Decision Making Using AHP (Analytical Hierarchy Process) Modeling

AHP is a multi-criteria decision making method which was created to optimize choice making when one is confronted with a blend of subjective, quantitative, and at times clashing factors. AHP has been extremely successful in making confounded, frequently irreversible choices [3].

Choice making has been innately mind boggling when numerous variables must be weighed against competing needs. One of the cutting edge tools created in the most recent 30 years used to survey, organize, rank, and assess choice decisions is the AHP developed by Saaty [4, 5].

The understanding of decision makers is utilized by AHP to break up problems into hierarchical structures. The number of levels combined with the decision maker's model determines the complexity of the problem [3].

3 Data

The outcomes of all the world cup matches from 1987 to 2011 between the above mentioned 8 teams has been considered. The no-result and tie matches and matches played with other lower ranked (in the ICC rankings) teams were ignored for simplicity. The research focused only on the 5 major parameters which have a considerable effect on deciding the final outcome of the game. The parameters were decided with the aid of a previous research done by Bandulasiri [7]. The parameters are Toss Winner, First Innings Batting Team, Day/Night or Day Match, Venue and Round of the tournament. The significance of each parameter is discussed below.

1. **Toss**—Nobody can deny the importance of toss in a cricket match. In many a cases it has been seen that winning the toss becomes an important factor in deciding the winner of the match. In Day and Night matches it is more so because of the dew factor which comes in giving advantage to the batting side.

2. **Batting first**—Batting first has a lot influence on the game, since the total runs scored by the team batting first becomes the target for the second team. The team batting first plays under no pressure. It also tries to make full use of the pitch to its advantage. The wearing of the pitch towards the second half of the match also helps the batting first team when they bowl and subsequently use slow and spin bowlers.

3. **Day and Night or Day Match**—Batting under floodlights is always a difficulty because, when compared with itself, a team will always do better chasing in daylight than under artificial lights.

4. **Home ground**—Teams that play at home have relative advantage because of various factors like home pitches, weather and the crowd.

5. **Round**—Even the round of the tournament has a significant effect in the outcome of the matches. The pressure on the teams rises to an altogether different level once the knockout rounds start. Keeping this in mind we have included this factor too in our analysis. Group matches is denoted by 'G' and non-group matches (super-six, quarterfinals, semifinals) by 'NG'.

3.1 Sample Data

Table 1 represents data for one of the teams (West Indies). Each row in the table represents a World Cup match. In the match represented in the first row, West Indies had lost the coin toss and didn't bat first. It was a Day and Night match, not

Table 1 West Indies data

Winning	Winning coin toss	Batting first	Day and night	Home	Round	Opponent
No	No	No	D and N	Away	G	AUS
Yes	No	No	Day	Away	G	Ind
No	No	Yes	Day	Away	G	Eng
No	Yes	No	D and N	Away	G	Eng
No	No	Yes	Day	Away	G	NZ
Yes	Yes	No	Day	Away	G	NZ
No	YES	No	Day	Away	G	NZ
No	No	Yes	Day	Away	NG	NZ
No	Yes	No	Day	Away	G	SA
yes	Yes	yes	Day	Away	G	SA
Yes	Yes	Yes	D and N	Away	G	SA
No	Yes	No	Day	Home	NG	SA
No	No	Yes	D and N	Away	G	SA
No	No	No	D and N	Away	G	SA
No	Yes	Yes	Day	Away	G	Pak
Yes	Yes	Yes	Day	Away	G	Pak
Yes	Yes	No	Day	Away	G	Pak
No	No	No	Day	Away	G	Pak
Yes	No	Yes	Day	Home	G	Pak
No	Yes	Yes	Day	Away	NG	Pak
Yes	No	Yes	Day	Away	G	Pak
No	Yes	No	Day	Away	G	Ind
No	No	No	Day	Away	G	Ind
Yes	No	No	Day	Away	G	AUS
No	No	No	Day	Away	NG	AUS
No	No	Yes	Day	Away	G	AUS
No	Yes	No	Day	Home	NG	AUS
Yes	No	Yes	Day	Away	G	SL
Yes	No	No	Day	Away	G	SL
Yes	No	Yes	Day	Away	G	SL
No	No	No	D and N	Away	G	SL
No	Yes	No	Day	Home	NG	SL
No	No	Yes	D and N	Away	G	Eng
No	No	Yes	Day	Away	NG	Eng
No	No	No	D and N	Away	G	Eng

held in West Indies and it was a group match with Australia. The outcome of the match being West Indies was defeated. Similarly other rows can be explained.

Similarly data for other teams were obtained. Using these data and some basic assumptions, the AHP structure [4, 5] was constructed as is described below.

4 Solution

4.1 Assumptions

1. Outcome of cricket is dichotomous i.e. matches results as either Win or Loss has only been considered for analysis.
2. Out of the 14 teams playing this 2015 World Cup, top 8 teams have only been considered—top 4 from each group.
3. Pressure conditions on the teams can be divided into two categories—pressure in the group (G) matches and in the non-group (NG) matches.

4.2 Methodology

AHP [4, 5] has been followed to predict the match outcomes and come up with the winner of the tournament.

Step 1: First the values of the parameters were converted into binary values as is depicted in Table 2.

Table 3 shows a part of the data (first five rows of data from Table 1) for West Indies:

Table 4 shows Table 3 transformed into binary values:

Table 2 Binary representation	Y	1	Winning game	Wct	1	Winning coin toss
		0	Losing game		0	Losing coin toss
	Bf	1	Batting first	Dn	1	Day and night match
		0	Batting second		0	Day match
	H	1	Home game	R	1	Group match
		0	Away Game		0	Non-group match

Table 3 Part of West Indies data

Winning	Winning coin toss	Batting first	Day and night	Home	Round
No	No	No	D and N	Away	G
Yes	No	No	Day	Away	G
No	No	Yes	Day	Away	G
No	Yes	No	D and N	Away	G
No	No	Yes	Day	Away	G

In the first column **WCT-Y** means the team (in this case West Indies) won the coin toss and **W-Y** means the team won the match. Thus **WCT-Y, W-Y** is 1 if the team won the coin toss as well as won the match and it is 0 if either is false. In the first five rows (of Table 3) there is no such combination for the value to be 1. In the second column, **WCT-Y, W-N** means whether the team won the coin toss and lost the match. In the 4th row it can be seen from Table 3 that the team lost the match but won the coin toss. So the 4th row in Table 4 is 1. Similarly other columns of Table 4 can be explained.

Step 2: Using the transformed table (Table 4), the sum of each column was obtained and tabulated as follows:

Table 5 depicts a matrix structure which shows the number of matches won (W) or lost (L) for a 'Yes' value of each of the five parameters. After winning the coin toss, West Indies has won 5 matches and lost 9 matches as is obtained from Table 1. These values are represented in the first data column of Table 5. Similarly other columns can be explained. The last row is the sum of the above two rows i.e. West Indies has won coin toss a total of 14 times.

Step 3: After that AHP normalized matrix was constructed as:

The formula used for prioritizing Wct over Bf is $\Sigma Wct / \Sigma Bf$ [4, 5]. ΣWct is 14 and ΣBf is 16 from Table 5. Thus the 2nd cell is obtained as 14/16 = 0.875. Similarly all other cells have been filled up (Table 6).

Step 4: The individual matrices were created for each parameter. For **Wct** it was obtained as.

This is obtained from Table 5. The 2nd cell value is obtained as 5/9 = 0.555556 (Ratio of No. of matches won to No. of matches lost for matches in which West Indies won the coin toss) (Table 7).

Table 4 Binary table

WCT-Y, W-Y	WCT-Y, W-N	BF-Y, W-Y	BF-Y, W-N	D and N-Y, W-Y	D and N-Y, W-N	H-Y, W-Y	H-Y, W-N	R-G, W-Y	R-G, W-N
0	0	0	0	0	1	0	0	0	1
0	0	0	0	0	0	0	0	1	0
0	0	0	1	0	0	0	0	0	1
0	1	0	0	0	1	0	0	0	1
0	0	0	1	0	0	0	0	0	1

Table 5 Team statistics matrix

	Wct	Bf	Dn	H	R
W	5	7	1	1	12
L	9	9	7	3	16
	14	16	8	4	28

Table 6 AHP normalized matrix

	Wct	Bf	Dn	H	R
Wct	1	0.875	1.75	3.5	0.5
Bf	1.142857	1	2	4	0.571429
Dn	0.571429	0.5	1	2	0.285714
H	0.285714	0.25	0.5	1	0.142857
R	2	1.75	3.5	7	1

Table 7 Parameter matrix

Wct	W	L
W	1	0.555556
L	1.8	1

Step 5: Similarly, data for all other teams were calculated.

Step 6: Using the above data, AHP analysis was performed and winning probabilities of the teams were obtained.

The AHP structure for West Indies is:

Figure 1 represents five parameters with their weightages and win and loss probabilities under each parameter for West Indies. The winning probability for West Indies is obtained as 0.2*0.357 + 0.228*0.438 + 0.114*0.125 + 0.057*0.25 + 0.401*0.429 = 0.372 [4, 5].

The winning probabilities of all the teams under study are:

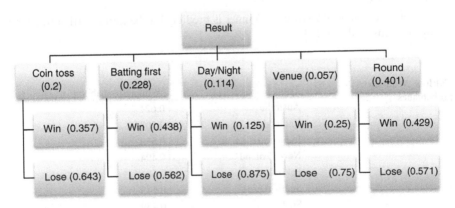

Fig. 1 AHP structure for West Indies

Step 7: However to find out the winning probabilities of each encounter, the probabilities obtained from AHP were further adjusted with another probability factor. This factor was obtained from past records of head to head encounters.

The final tabulation for winning probabilities is:

If Australia and Sri Lanka have played each other 7 times in World Cup with Australia winning 6 of those, the factor for Australia will be 6/7 = 0.857142857 and that for Sri Lanka will be 1 − 6/7 = 0.142857143.

Now, chances of Australia winning in a match against Sri Lanka is

0.621*0.857142857 = 0.532285714 (from Table 8) and chances of Sri Lanka winning the same match is 0.0627 (Table 9).

Step 8: The chosen 8 teams have already been divided into two Pools for this year's World Cup. In the group stage each team is to play with one another. Using the predefined group structure given by ICC and the calculations of chances we get the following two tables (Tables 10 and 11).

- From Pool A, Australia tops the list. However the other 3 teams are at equal points. In reality it is calculated through NRR but for simplicity this paper considers the team with highest overall winning probability to pass through to the next round. In this case it is England (0.508) (*Since this paper has been written after the 2015 Cricket World Cup it is a known fact that it was New Zealand who reached the semifinals and not England. However according to this paper it is England who reach the semifinals. This can be attributed to two factors—(1) not considering the other three teams who were in Pool A and (2) not considering the current form of New Zealand which has been discussed later*).
- The outcome from Pool B is straightforward. The two teams qualifying into next round are South Africa and defending champions, India with South Africa topping the group.

Step 9: Now, the next round will be the semi-finals with the matches as:

- Australia versus India—Winner is predicted to be Australia (0.4347) against India's 0.1599
- England versus South Africa—Winner is predicted to be South Africa (0.2715) against England's 0.254

Table 8 Winning probabilities

Team	Winning probability
Australia	0.621
England	0.508
India	0.533
New Zealand	0.494
Pakistan	0.603
South Africa	0.543
Sri Lanka	0.439
West Indies	0.372

Table 9 Factorized winning probabilities

	Australia	England	India	New Zealand	Pakistan	South Africa	Sri Lanka	West Indies
Australia	1	0.443571429	0.4347	0.414	0.3105	0.46575	0.532285714	0.276
England	0.145142857	1	0.254	0.1905	0.254	0.254	0.3048	0.423333333
India	0.1599	0.2665	1	0.228428571	0.533	0	0.228428571	0.304571429
New Zealand	0.164666667	0.30875	0.282285714	1	0.1235	0.329333333	0.1976	0.247
Pakistan	0.3015	0.3015	0	0.45225	1	0	0.603	0.1809
South Africa	0.13575	0.2715	0.543	0.181	0.543	1	0.362	0.362
Sri Lanka	0.062714286	0.1756	0.250857143	0.2634	0	0.146333333	1	0.146333333
West Indies	0.206666667	0.062	0.159428571	0.186	0.2604	0.124	0.248	1

Table 10 Pool A table

Pool A	New Zealand	Sri Lanka	Australia	England	Total wins
New Zealand		0	0	1	1
Sri Lanka	1		0	0	1
Australia	1	1		1	3
England	0	1	0		1

Step 10: Final will be between **Australia** and **South Africa** and the winner according to this model is predicted to be **Australia** (0.46575) against South Africa's 0.13575.

(*If it would have been New Zealand and South Africa in the semifinals, according to this paper New Zealand would have reached the finals with a winning probability of 0.329 over South Africa's 0.181 (from Table 9), thus validating the model depicted in this paper*).

5 Discussion

This paper tries to build a statistical model to predict the winner of a cricket tournament; here it is the World Cup. It uses AHP for the same which is one of the most successful tools used for multivariate decision analysis. Other methods like logistic regression have been found to be unsuccessful due to scarcity of data.

The pros and cons of this model are:

Pros

- The model gives a probabilistic winning score for the teams. Thus apart from the winner we can also get an idea about the relative margin by which one team is expected to win over another.
- The model proposed being based on AHP, the costs are low when compared to other methods like econometrics, cost-benefit programming [8].

Cons

- Problem Structuring—In case of AHP modeling, structuring is very important. A different structure may lead to different final outcome. In this paper, a change of Pool structure has the potential to change the ranking of the teams.

Table 11 Pool B table

Pool B	India	South Africa	West Indies	Pakistan	Total wins
India		0	1	1	2
South Africa	1		1	1	3
West Indies	0	0		1	1
Pakistan	0	0	0		0

- "Rank Reversal" problem—Whenever there is any addition or deletion of any criteria or choices into the initial set of options, the likelihood scores of the teams may change. For instance if we would replace one of the parameters with some other which might seem relevant, the whole ranking structure of the teams may change [8, 9].

6 Conclusion and Future Work

The outcome of a cricket match is dependent on a large number of factors. Among them the 5 important ones taken into consideration are winning the toss, batting first/second, playing during the day or night, having the home advantage or not and the pressure on the team to perform which is determined by whether the match is a group match or non-group match. The other important factors which could have been considered for a better prediction are:

- Current form of the team, say for last 10 matches
- Difference in the ICC ratings of the two encountering teams

Still, this model is pretty successful in coming out with fairly accurate results. The model depicts the importance of each of the chosen factors on the teams individually. For e.g. in case of West Indies the relative importance of Round as a factor is the highest (0.401) and that of Venue is the lowest (0.057) which means that West Indies as a team is very much sensitive to the stage of the tournament in which they are playing and indifferent on the ground conditions.

References

1. Schumaker, Robert P., Osama K. Solieman, and Hsinchun Chen.: Sports data mining: The field. Sports Data Mining. Springer US, 2010. 1–13.
2. Ikram the robot predicts Cricket World Cup winners. http://blogs.canterbury.ac.nz/insiders/2015/02/10/ikram-robot-predicts-cricket-world-cup-winners/.
3. Melvin Alexander, Social Security Administration, Baltimore, MD: Decision-Making using the Analytic Hierarchy Process (AHP) and SAS/IML (2012).
4. Saaty, T.L.: A Scaling Method for Priorities in Hierarchical Structures, Journal of Mathematical Psychology, 15, 234–281 (1977).
5. Saaty, T.L.: The Analytic Hierarchy Process, McGraw-Hill, New York (1980).
6. Home of Cricket Archives, www.cricketarchive.com.
7. Ananda Bandulasiri, Ph.D: Predicting the Winner in One Day International Cricket, Journal of Mathematical Sciences & Mathematics Education, Vol. 3, No. 1 (2008).
8. Hartwich, F.: Weighing of Agricultural Research Results: Strength and Limitations of the Analytical Hierarchy Process (AHP) (1999).
9. Warren, J.: Uncertainties in the Analytic Hierarchy Process (2004).

Towards Distributed Solution to the State Explosion Problem

Lamia Allal, Ghalem Belalem and Philippe Dhaussy

Abstract In the life cycle of any software system, a crucial phase formalization and validation through verification or testing induces an identification of errors infiltrated during its design. This is achieved through verification by model checking. A model checking algorithm is based on two steps: the construction of state space of the system specification and the verification of this state space. However, these steps are limited by the state explosion problem, which occurs when models are large. In this paper, we propose a solution to this problem to improve performance in execution time and memory space by performing the exploration of state space in a distributed architecture consisting of several machines.

Keywords Model checking · State explosion problem · Formal methods · State compression

1 Introduction

Model Checking is a verification technique based on exhaustive exploration of system states seeking behaviors that do not satisfy its specification. A model checker can be viewed as a black box that accepts as input a system and a property expressed on this system and returns as output a response indicating whether the

L. Allal (✉) · G. Belalem
Department of Computer Science, Faculty of Exact and Applied Sciences,
University of Oran 1 Ahmed Ben Bella, Oran, Algeria
e-mail: allal.lamia@gmail.com

G. Belalem
e-mail: ghalem1dz@gmail.com

P. Dhaussy
Lab-STICC UMR CNRS 6285 ENSTA Bretagne, Brest, France
e-mail: philippe.dhaussy@ensta-bretagne.fr

© Springer India 2016
S.C. Satapathy et al. (eds.), *Information Systems Design and Intelligent Applications*, Advances in Intelligent Systems and Computing 433,
DOI 10.1007/978-81-322-2755-7_56

property is verified or not. The algorithms implemented include two steps, a construction of state space of the system and a traversal of this space for errors. The state space is represented as a graph which describes all possible evolutions of the system. The nodes of the graph represent the states of the system and the edges represent transitions between these states [1].

The advantage of model checking verification is the accuracy of the response [2]. However, this technique suffers from a known as the state explosion problem [3, 4]. This problem occurs when the state space to explore is large and can't be explored by algorithms due to a lack of resources (execution time and memory space) because the memory required to perform the exploration is higher than the space memory contained in the machine. That's the reason for why the size of the systems is generally small. This article is placed in the context of verification of systems by Model Checking. We present a solution to the state explosion problem. This article is divided into four parts. The second section presents some solutions that address the state explosion problem. The third section is devoted to the presentation of the proposed approach. The fourth section presents a comparative study between our solution and some solutions. The fifth section is devoted to the development of a scenario using the proposed approach. We finish with a conclusion and some perspectives for future work.

2 Some Related Work

There are several research in the field of Model Checking. In this section, we present some solutions that have been proposed to address the state explosion problem. These are based on different methods and data structures. Each solution tries to improve performance in execution time and memory space. Solutions differ depending on the architecture used and on the verification done (on-the-fly, partial order, symbolic).

2.1 Symbolic Model Checking

It is based on the representation of states and transitions of the system by sets, this means that the states are manipulated by packets instead of being considered one by one [5]. Symbolic model checking [6] is based on the overall iterative calculation of states, the most widely used method for the statements of representation uses binary decision diagrams (BDD). BDD were used to encode Boolean functions. This technique can be expensive in memory because the representation tree shape generates a lot of redundancy. Several contributions have been made based on symbolic model checking, in [7], the authors proposed a model checking algorithm based on BDD structures for the verification of temporal logic mu-calculus. The experiments were performed on a single synchronous pipeline.

In [8], the authors proposed a solution that extends the model checker NUSMV, the properties to verify the model are in RTCTL logic. The proposed algorithm is used to check these properties on Kripke structures. The extension of the model checker is temporized, it means that each transition is labeled by the possible duration for this transition.

2.2 On-the-Fly Model Checking

This technique [9] consists of manipulating graphs as an initial state and a successor function. It is based on the DFS algorithm for the construction of the state space. Mukherjee et al. [10] presented a solution to the state explosion problem. This approach is based on the storage of states on their compressed form. The proposed algorithm is sequential. The first state (initial state) is stored explicitly, the other states are stored in their compressed form into hash tables. The stored information is the difference (D) between the old state and the new state. It is represented by the following expression: D = newState-oldState. When a state is generated, it must be compared with all treated states. However, compressing and comparing it with states stored in difference form might lead to an error owing to multiple states having the same difference form. Therefore it is essential to revert a stored state before comparing. To do that, they must be under their explicit form. Therefore, all states are reconstructed, it means that all states will be under their explicit form.

The solution described in [11] allows a distribution of state space exploration during the verification of models by model checking using SPIN [12, 13]. Each node has a set V that contains the explored states and a queue U to store the unvisited states.

2.3 Partial Order Reduction

System verification is based on an exhaustive exploration of the state space. To remedy to the state explosion problem many approaches have been proposed. These methods are called partial reduction methods. It can eliminate the equity interleaving produced by independent transitions in competing systems. In [14] some changes has been done to avoid the redundant interleaving sequence for an execution.

In [15], both techniques on the fly and partial order are combined in order to have better performance.

3 Our Contribution: State Space Compression (SSC)

The state space exploration is the main step for verification and validation of models. This allows for a construction of state space during verification by model checking. The main problem with this method is the state explosion problem occurring when memory space is insufficient to perform a full state space exploration.

Our solution is based on the solution presented in [10] where each state is stored in a compressed form into hash tables. In this solution, when a state is generated, compressed states explored and stored in memory are not reconstructed for comparisons. Another state that is generated is not directly stored in memory, a test on the state is required (done by a verification on the hash table). If the state has already been explored, then move to the next state, otherwise store this state. When a state is processed, its successors are generated and stacked in a stack. This process is repeated until all states have been processed. The activity diagram shown in Fig. 1 shows the execution steps of the proposed approach.

The algorithm (*Distributed state space exploration*) shows the execution steps of the proposed approach. The second instruction indicates whether the termination was detected by looking at nodes' stacks. If empty, the exploration ends, otherwise, process the next states.

```
program Distributed state space exploration
    begin
      State-generation();
      while(termination-not-detected)
          if(visited-state)
              State-popped();
              Termination-verification();
          else
              State-storage();
              Successors-generation();
          endif
      endwhile
    end.
```

3.1 State Generation

This step is used to extract a state for model in order to apply the appropriate treatment. When a state is explored, it is stored in memory to avoid duplication. This process is repeated until all states are explored.

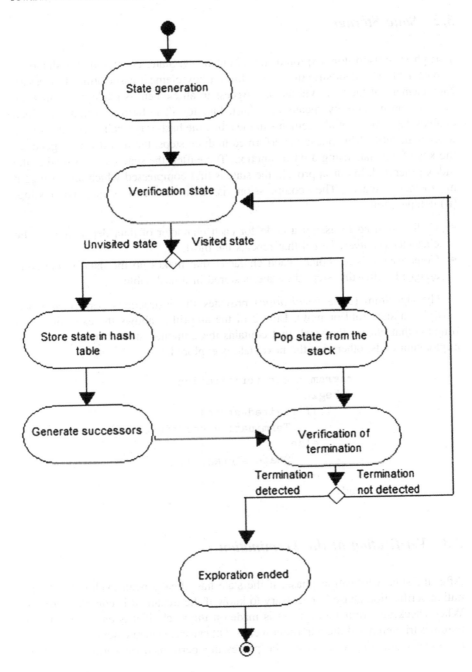

Fig. 1 Execution phases of the solution

3.2 State Storage

This phase aims to store explored states in memory. States are stored in hash tables. A hash table is a data structure that allows a key element association. It accesses each element of the table via its key. Access to an element is made by turning the key in a hash value by means of a hash function. The hash is a number which enables the location of all elements in the table, the hash is typically the index of the item in the table. This phase is performed in three steps, the first one is to generate the key of the state using a hash function. Thereafter, the state can be stored at the index generated. In our approach, the state is first compressed which allows a gain in memory space. The compression is done in two steps: (i) coding, (ii) compression.

- Coding is used to assign a code for each character of data depending on the character (*number*, *letter*) that repeats the most.
- Compression is to replace each character that makes up the data by the code assigned. After this step, the state is stored in a hash table.

The algorithm (*State verification*) provides the instructions performed when verifying a state (old or new). Line 2 of the algorithm allows the execution of a function that checks the stack that contains the untreated state. If it is empty, the exploration ends, otherwise the next state is explored.

```
program State verification
    begin
        if(visited-state)
            Termination-verification();
        else
            State-storage();
        endif
    end.
```

3.3 Verification of the Termination

After the state verification phase, if the state has already been explored, a termination verification phase is necessary to know if the treatment is completed or not. When checking termination, a test is made on the stack, if it is empty, the termination is triggered and the exploration ends. Otherwise, generate next state and redo the same work. Figure 2 shows the processing performed on verification of the

Fig. 2 Verification of the termination

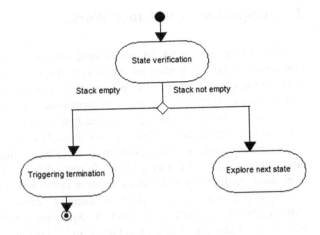

termination. The algorithm (*Termination verification*) shows the processing performed at verification termination. Line 2 of the algorithm is a statement that can perform a test to check if there are unexplored states.

```
program Termination verification
    begin
        if(Stack-is-not-empty)
            State-verification();
        else
            Termination-triggered();
        endif
end.
```

3.4 Successors Generation

This phase is performed in two steps and represents a treatment that can generate new states when other states were treated. When a state is generated, it is stored in physical memory. After that, its successors are generated to be stored too. The generated states are added to the set of unprocessed states. This set is represented by a stack structure. This operation is repeated until perform an exhaustive state space exploration.

3.5 Stop Exploration

This phase represents the last step of the state space exploration. Termination is detected by the stack containing the states to be treated. When it is empty, there is no longer any state to explore. In this case, the termination is detected and the exploration phase is complete.

4   Comparison with Other Works

In this section, we focus on positioning our approach to some solutions. The comparison based is performed on seven metrics: Data structures, performance improvement in execution time and memory space, architecture used, state duplication, state distribution, collisions, compression and decompression.

The data structures used in [10] are hash tables used for storing states. The states with the same index are stored in tables through linked lists.

In [16], the authors used a localization table that defines whether a state is old or new and hash tables for storing states. In our solution, two types of structures have been used, hash tables and a stack to store unprocessed states.

The experiments were performed using different architectures. In [16], the authors relied on a parallel algorithm. A sequential algorithm was used in [10]. The state space exploration has been achieved in [11] using a distributed algorithm. Our solution is based on a distributed architecture based on a number of worker processes.

There is no duplication of states in [10] because a verification is performed at each state generation to check whether the state has been explored or not. There is no risk of collision states in [10] because each state is stored at the generated index. Collisions can occur in [16] for two states having the same cell number and a different key. In the proposed approach, for each state, a unique index is generated (in hash table), so collisions cannot occur. In [10], states are compressed before being stored. In the proposed solution, data compression is used to reduce the memory required to perform the state space exploration. State decompression (state reconstruction) is not performed to verify whether a generated state has been explored or not.

5   Description of a Scenario

In this section, we execute an example representing the different steps of the contribution by using a model consisting of 8 states: s_1, s_2, s_3, s_4, s_5, s_6, s_7, s_8 (see Fig. 3) using the proposed approach.

The exploration begins by generating the initial state s_1. s_1 is compressed and stored in the hash table.

Then, from Fig. 3, the successors of s_1 are generated: s_2 and s_3. A Verification is made on the hash table (a set of known states) to know if these states are old or new. In this case, s_2 and s_3 are compressed and added to this set and to the set of unprocessed states. After that, the state s_3 is explored. All its next configurations are generates, s_7. s_7 is new, it's added to the hash table and to the set of unprocessed

Fig. 3 Model consisting of 8 states

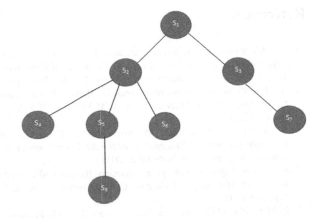

states. s_7 has no successors. Then, the successors of s_2 are generated: s_4, s_5, s_6. the states s_4, s_5 and s_6 are new, therefore, they are compressed and added to the hash table and to the set of unprocessed states. s_4 and s_6 have no successors, s_5 has one successor s_8, it's compressed and stored in the hash table and in the set of unprocessed states. s_8 has no successors. The set of unprocessed states is empty, therefore the termination is triggered and the exploration is completed.

6   Conclusion

Model checking is a technique based on three concepts: A model system to verify, a specification as a property of the system and algorithms to check whether the design meets its specification. This technique suffers from the state explosion problem occurring when systems become too large.

In this article, we presented a new solution to this problem. Our approach aims to improve the performance on execution time and memory space by using data compression and avoiding decompression step that allows a gain in time. Using a single hash function, we can know whether a state is old or new compared to the solutions presented in [16] where the test is done using a bloom filter or a hash function and a map function. The use of the bloom filter is not reliable because the returned result may be a false positive. In addition, collisions can occur in [16] when the map function returns the same state for two different states. Currently, we are implementing this approach within a distributed architecture. We plan in a near future to study a function to perform an effective partitioning of states on the nodes in a distributed architecture.

References

1. S. Christensen, L. M. Kristensen, and T. Mailund. A sweep-line method for state space exploration. In *Proceedings of the 7th International Conference on Tools and Algorithms for the Construction and Analysis of Systems*, TACAS'2001, pp. 450–464, April 2001.
2. M. Kwiatkowska, G. Norman, and D. Parker. Prism: Probabilistic model checking for performance and reliability analysis. *ACM SIGMETRICS Performance Evaluation Review*, 36 (4):40–45, 2009.
3. E. M. Clarke, W. Klieber, M. Nováček, and P. Zuliani. Model checking and the state explosion problem. In *Proceedings of the 8th Laser Summer School on Software Engineering*, volume 7682, pp. 1–30, September 2011.
4. R. Pelánek. Fighting state space explosion: Review and evaluation. In *Proceedings of the 13th on Formal Methods for Industrial Critical Systems*, volume 5596 of *FMICS'08*, pp. 37–52, September 2008.
5. E. M. Clarke, O Grumberg, S Jha, Y Lu, and H Veith. Progress on the state explosion problem in model checking. In *Informatics—10 Years Back. 10 Years Ahead*, pp. 176–194, London, UK, UK, 2001. Springer-Verlag.
6. K. L McMillan. *Symbolic Model Checking: An Approach to the State Explosion Problem*. PhD thesis, Pittsburgh, PA, USA, 1992. UMI Order No. GAX92-24209.
7. J. R. Burch, E. M. Clarke, K. L. McMillan, D. L. Dill, and L. J. Hwang. Symbolic model checking: 1020 states and beyond. In *Proceedings of the Fifth Annual IEEE Symposium on Logic in Computer Science*, pp. 428–439, Philadelphia, PA, 1990. IEEE.
8. N. M Arkey and Ph. Schnoebelen. Symbolic model checking of simply-timed systems. In *Formal Techniques in Real-Time and Fault Tolerant Systems*, FTRTFT 2004, volume 3253, pp. 102–117, Grenoble, France, 2004. LSV, ENS Cachan, Springer Berlin Heidelberg.
9. G. Bhat, R. Cleaveland, and O. Grumberg. Efficient on the fly model checking for ctl. In *Proceedings of the 10th Annual IEEE Symposium on Logic in Computer Science*, LICS'95, 1995.
10. A. Mukherjee, Z.Tari, and P. Bertok. Memory efficient state-space analysis in software model-checking. In *Proceedings of the Thirty-Third Australasian Conference on Computer Science*, volume 102 of *ACSC'10*, pp. 23–32, January 2010.
11. F. Lerda and R. Sisto. Distributed-memory model-checking with spin. In *Proceedings of the 5th and 6th International SPIN Workshops on Theoretical and Practical Aspects of SPIN Model Checking*, volume 1680, pp. 22–39, July 1999.
12. P. Dhaussy, J. C. Roger, and F. Boniol. Reducing state explosion with context modeling for model-checking. In *Proceedings of the 13th International Symposium on High Assurance Systems Engineering*, HASE, pp. 130–137, November 2011.
13. G. J. Holzmann. The model checker spin. *IEEE Transactions on Software Engineering*, 23 (5):279–295, May 1997.
14. G. J. Holzmann and D Peled. An improvement in formal verification. In *Proceedings of the 7th IFIP WG6.1 International Conference on Formal Description Techniques VII*, pp. 197–211, 1995.
15. D Peled. Combining partial order reductions with on-the-fly model-checking. In *Proceedings of the 6th International Conference on Computer Aided Verification*, CAV'94, pp. 377–390, 1994.
16. R. T. Saad, S. D. Zilio, and B. Berthomieu. Mixed shared-distributed hash tables approaches for parallel state space construction. In *Proceedings of the 10th International Symposium on Parallel and Distributed Computing*, ISPDC'11, pp. 9–16, July 2011.

Real Time Bus Monitoring System

Jay Sarraf, Ishaani Priyadarshini and Prasant Kumar Pattnaik

Abstract The real time bus monitoring system may be designed to serve as a tracking system for the frequent local bus travelers using a GPS (Global positioning system) device and GPRS (General packet radio service) system. This paper focuses on system that help passengers locate the current location of the buses and expected arrival time of the buses to their nearest bus stop. The location and ETA (Estimated Time of Arrival) will be shown on the mobile app and can also be received through SMS (Short Messaging Service). The location can also be tracked by the network administrator through a web application which will keep the complete location history of the busses.

Keywords Real time tracking · Gps tracking · GPS · GIS

1   Introduction

The real time bus monitoring system uses GPS (Global Positioning System) to identify the current location of the buses. The location calculated by the GPS is in the form of latitude and longitude. The latitude and longitude are directed to the database server through GPRS service. It chooses GPRS network for superiority in transmitting speed, its forever on-line character and most importantly the reasonable cost [1]. The location data are stored into the database server and then it is retrieved on the map server to locate and display the location to users in the graphical user interface.

J. Sarraf (✉) · I. Priyadarshini · P.K. Pattnaik
School of Computer Engineering, KIIT Univeristy, Bhubaneswar, India
e-mail: jaysarraf596@gmail.com

I. Priyadarshini
e-mail: ishaanidisha@gmail.com

P.K. Pattnaik
e-mail: patnaikprasantfcs@kiit.ac.in

2 Related Work

There have been numerous research and implementation for vehicle tracking system using GPS and GSM services but the accuracy of GPS has always been a questioned because of the up lift of Selective Availability [2]. The accuracy of the GPS could be around 5–15 m in a densely populated area because major satellites signals are always reflected and blocked by large buildings [3]. There are several technologies available for cellular phone positioning either network based or mobile station based [4]. These technologies used the signal attenuation, angle of arrival, time of arrival, time difference of arrival and time advanced to locate unmodified network phone [5]. In TDMA (Time Division Multiple Access) time for the signal to reach mobile phones is known so this information can be used to calculate location information. However the positioning technology of GPS and GSM is mature enough for position accuracy and available for civil use.

3 Motivation

Bus passengers often face difficulties in finding the bus position or waiting for a long time at the bus stop for the arrival of bus. They may be unable to decide whether to wait for the bus or hire a cab/rickshaw due to which they are often late to work. Students are late for classes. A small survey was taken within the students and employees of a University about their opinion on the current transportation system and the following conclusion was made:

- 80 % of the passengers reported delay to their destination as they had to wait for the buses instead of walking or hiring a cab.
- 98 % of the passengers confirmed that knowing the location of the buses could help them decide whether to wait or walk.
- 95 % of the passengers reported that finding location with the help of a mobile device may be beneficial.

If passengers have the easiest way to search for and locate the bus they are willing to travel it would help them take accurate decision whether to wait for the bus or just hire the cab. This Real time bus monitoring system may provide passengers the service to find the bus through a mobile device. In the proposed system, the buses may be integrated with individual GPS devices to constantly send the location to a specified server where the location and time will be stored in the database. The location may be well retrieved through a smart phone application or short messaging service.

4 Project Description and Goals

The goal of this project is to provide assistance to the bus passengers by helping them to locate the bus position and view the calculated ETA of the bus with respect to its nearest bus stop. The bus location and the ETA will be displayed on a smart phone application. In case the passenger does not have a smart phone then they can track the bus location through SMS (Short messaging service) [6].

The proposed system may work to achieve the following goals:

- To provide accurate location of the buses to the passengers with the accurate ETA
- To be the cost effective regarding the hardware and efficient regarding the power performance.
- Easily installable in every bus.
- Low power consumption.
- To cut the service subscription charges as provided by third party location based service providers.
- Data integrity protection as the application might be hosted by the university based servers and may be accessible to the respective network administrator only within the university wide network area.

5 Proposed Architecture

In the above architecture as shown in the Fig. 1, the bus is installed with the GPS tracking device which is enabled with the GSM/GPRS module. The GPS device may obtain spatial data from its nearest available satellites. The location may be

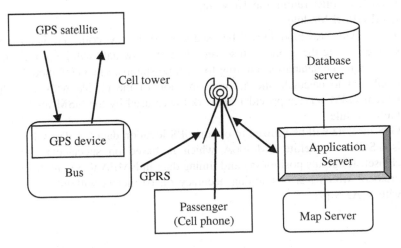

Fig. 1 Architectural diagram of real time bus monitoring

determined considering the nearest point from the bus position and the road portion using Map Matching Method. The time and distance required for arrival at the bus stop is given by the formula [7]

$$distance = \min\left(\sqrt{(P - P_i)^2 + (Q - Q_i)^2}\right)$$

where, min is the minimum distance between the bus station tag and bus reader, (P, Q) is the GPS coordinates of the reader in the bus and (Pi, Qi) is a point on the tag segment.

The location may be then sent to the database server using the GSM/GPRS module. The application server linked to the database server will read the location input from the database. Hence the location is displayed on the map using the map server.

6 Components Used

6.1 Hardware Utilities

The hardware components such as microcontroller, GSM/GPRS module, GPS module will be used during the development of this system which are discussed below:

A. Microcontroller

The micro-controller is the central controller for the entire unit. The model employed in the project is the Arduino Duemilanove, which is an open source electronics prototyping board based on the Atmel ATMega328 8 bit micro-controller running at 16 MHz.

B. GSM/GPRS Module

The GSM/GPRS module will be used to send the raw collected data from the GPS device to the application server where the raw data will be processed and stored into the database according to the time stamps. The GSM module can be employed to calculate the accurate location of the device with the help of current GSM service provider network tower used by the GSM module.

C. GPS Module

GPS module will be utilized to obtain GPS location data [8]. The GPS module is USGlobal satellite EM-406A which is based on spectacular SiRF Star chipset. It outputs positioning and timing data in NMEA 0183 and SiRF binary protocol which has positioning accuracy of 10 meters without and 5 meters with WAAS [9].

6.2 Server Utilities

A Software will be designed to display graphical location of the buses with their ETA. The software incorporating Microsoft Visual Studio (a high level programming language tool) will be accessible through the website address and as well as through mobile application installed in the passengers phone.

The server utilities are toolkits such as Microsoft visual studio and SQL Server to store location of the buses. The web application is designed to track the location from web interface and a mobile application is used for the users to track the location using the bus numbers.

A. Web Application

The real time bus monitoring web application will display the current location of the buses as well as the total distance and route covered by individual buses to keep the track of their record.

The real time bus monitoring system web application will be showing the current bus location using the bus numbers through an interface (Fig. 2).

B. Mobile Application

The real time bus tracking mobile application will help passengers find current location of the bus according to the bus number or the bus name provided (Fig. 3).

The mobile application will be designed to support open source operating systems so that it can be available to all users widely.

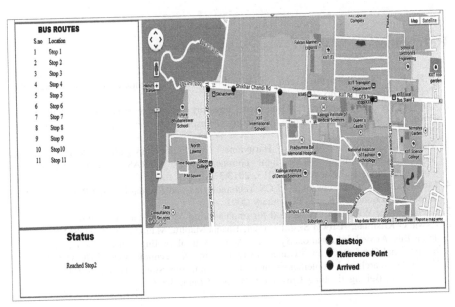

Fig. 2 Partial preview of the real time bus monitoring system web application

Fig. 3 Mobile application interface

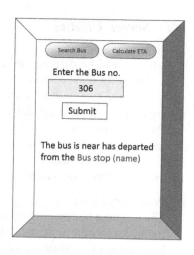

7 Conclusion

With the implementation of the project a complete track of the buses can be maintained around the city through the web application. The application will be hosted on the university's web server which will reduce the cost of subscription charges provided for the tracking services. This will also protect the integrity of the location data of the busses as the servers will be accessible only through the local domain's network.

References

1. Y.H. Pan, "Development Trend of Wireless Mobile Communication", Communication Technologies, vol. 4, no. 2, pp. 55–57, 2006.
2. Ochieng, W. Y., & Sauer, K. Urban road transport navigation: Performance of the Global Positioning System after selective availability. *Transportation Research Part C*, 10, 171–187, 2002.
3. Madoka Nakajima and Shinichiro Haruyama, "New indoor navigation system for visually impaired people using visible light communication", EURASIP Journal on Wireless Communications and Networking 2013, 2013:37.
4. Motorola, Inc., "Overview of 2G LCS Technnologies and Standards", 3GPP TSG SA2 LCS Workshop, London, UK, 11–12 January 2001.
5. Svein Yngvar Willassen, "A method for implementing Mobile Station Location in GSM".
6. Aswin G Krishnan, Ashwin Sushil Kumar, Bhadra Madhu, Manogna KVS, "GSM Based Real Time Bus Arrival Information System", Amrita School of Engineering, Coimbatore, India, International Conference on Advances in Engineering & Technology-2014 (ICAET-2014).
7. H.Niu, "Research and Implementation of Bus Monitoring System Based on GIS", Bachelor dissertation, Beijing Jiaotong University, Beijing, China, 2008.

8. Mrs. Swati Chandurkar, Sneha Mugade, Sanjana Sinha, Megharani Misal, Pooja Borekar, "Implementation of Real Time Bus Monitoring and Passenger Information System", IJSRP, Volume 3, Issue 5, May 2013.
9. Ahmed El-Rabbany, "Introduction to GPS: the global positioning system", ARTECH House, INC, 2002 pp. 117–127.

High Performance DFT Architectures Using Winograd Fast Fourier Transform Algorithm

Shubhangi Rathkanthiwar, Sandeep Kakde, Rajesh Thakare,
Rahul Kamdi and Shailesh Kamble

Abstract This paper presents area and latency aware design of Discrete Fourier Transform (DFT) architectures using Winograd Fast Fourier Transform algorithm (WFFT). WFFT is one of the Fast Fourier algorithms which calculate prime sized DFTs. The main component of DFT architectures are Adders and Multipliers. This paper presents DFT architectures using Winograd Fast Fourier Algorithm with Carry Look Ahead Adder and add/shift multiplier and also with Semi-complex Multipliers. In this paper, different prime size DFTs are calculated using polynomial base WFFT as well as conventional algorithm. Area and latency are calculated in Xilinx synthesizer. Polynomial WFFT include Chinese Remainder theorem which increases complexity for higher orders. This paper mainly focuses on prime size 5-point and 7–point WFFT architectures, implemented in Verilog and simulated using Xilinx ISE 13.1. Each sub module is designed using data flow style and finally top level integration is done using structural modeling. DFT architecture has wide range of applications in various domain includes use in Digital Terrestrial/Television Multimedia Broadcasting standard.

Keywords DFT · WFFT · Conventional base algorithm

S. Rathkanthiwar (✉) · S. Kakde · R. Thakare · R. Kamdi · S. Kamble
Y.C.C.E., Nagpur, India
e-mail: svr_1967@yahoo.com

S. Kakde
e-mail: sandip.kakde@gmail.com

R. Thakare
e-mail: rdt2909@gmail.com

R. Kamdi
e-mail: rahulkamdi19@gmail.com

S. Kamble
e-mail: shailesh_2kin@rediffmail.com

559

© Springer India 2016
S.C. Satapathy et al. (eds.), *Information Systems Design and Intelligent Applications*, Advances in Intelligent Systems and Computing 433,
DOI 10.1007/978-81-322-2755-7_58

1 Introduction

Many applications required miniaturization in area and power which provides opportunities for signal processing researchers to optimize design in terms of different parameters which contributes to area and latency reduction. Discrete Fourier transform (DFT) is used for transforming time domain samples into frequency domain samples. Digital signal processing is one of the promising areas in the field of ULSI. Conventional method based algorithm required N^2 multipliers and $N^2 - N$ adders for calculating N point DFT which required large area and more complex in structure for the implementation also required more time delay unit. Polynomial base Winograd Fast Fourier Transform algorithm can be implemented to reduce arithmetic operations in terms of multipliers unit with less complex in structure [1, 2]. There are many different approaches present leading to the optimization in various parameters. Novel pipeline approach of Winograd FFT results in less power dissipation and circuit area compared with other approaches which are present in house estimation tools [3]. The core part of TDS-OFDM system, 3780 point FFT can be decomposed in different ways to achieve best performance in terms of signal to noise ratio and cost of Hardware [4, 5]. One of the approaches of WFFT in which multiplication operations is replaced by shift, addition and subtraction which results in very high speed and power efficiency [6]. Iteratively proposed DFT architecture reduced approximately 45 % multiplications and required less hardware resources [7]. In long size WFFT applications pipelined architecture would cost more power and area compare to the memory based design [8]. Floating point implementation of DFT architectures become inexpensive in terms of area [9]. Delay reduction at post layout simulation is observed in 64-point FFT architecture [10].

2 Previous Work

Winograd FFT algorithm can be used to compute a DFT faster than the other FFT algorithms as multiplication operation took longer time for computation as compared to the addition or subtraction operations and WFFT required lesser number of multipliers. The Cooley-Tukey [11] Fast Fourier transform algorithm is one of the high speed algorithm still it requires $N \log_2 N$ multiplications for N point DFT while Prime factor algorithm also known as Good Thomas algorithm (1958) [12] required $N \log_2 N$ multiplications. In 1976, Winograd [13] developed an algorithm known as Winograd Fast Fourier transform algorithm which requires more number of addition or subtraction operations but only approximately N multiplication for smaller size and more than N for higher order of N point DFT. This paper is organised as follows. Section 3 explains Discrete Fourier Transform (DFT) by conventional/Definition Based approach. Section 4 deals with the Winograd FFT while Sect. 5 shows Verilog implementation results of 5 and 7 point DFT and also some comparative result analysis of 2, 3, 5 and 7 point DFT and also area comparison of 2 and 3 point DFT. Finally Sect. 6 concludes the paper.

3 Discrete Fourier Transform

The DFT is used to convert a finite discrete time domain sequence x(n) into a frequency domain sequence denoted by X(K). The N point DFT is mathematically given as

$$X(k) = \sum_{n=0}^{N-1} x(n) W_N^{nk} \qquad (1)$$

where, k varies from 0 to N − 1.

W_N^{nk} is the Twiddle factor defined as multiplicative constant used to recursively combine smaller discrete Fourier transform and mathematically given as

$$W_N^{nk} = e^{-\frac{2\pi i}{N} nk} \qquad (2)$$

The input samples can also be considered as a complex numbers but usually real numbers are preferred for complexity issues and the output coefficients obtain may also be complex. From Eq. (1), it is clear that for calculation of N point DFT required N^2 multipliers and $N(N-1)$ adders i.e. consider example of 3 point DFT, requirement of multipliers become nine and adders six [14]. For implementation purpose it becomes inconvenient for such a large multipliers unit requirement for small size DFTs. Hence, Fast Fourier algorithms are employed to tackle such conditions.

4 Winograd Fast Fourier Transform

There are many Fast Fourier algorithms are present namely Cooley and Tukey [11], Good Thomas algorithm [12] etc. Those algorithms are easy to implement but multipliers unit requirement is N log N or even simplified N/2 log N while in case of Winograd FFT it becomes approximately N. Winograd FFT mainly consist of two major parts Rader algorithm and Winograd small convolution algorithm.

4.1 Rader Algorithm

Rader algorithm is used the concept of primitive root to calculate base. If n is positive integer then integers between 1 and n − 1 which is co-prime and should contain all the terms considered as Primitive root.

$$k = r^m \mod N \tag{3}$$

where, m = 0, 1,....N − 2, r is primitive root, N is size of DFT.

Equation (3) represents all terms associated with size of DFT. We will always get one less term than size of DFT. Firstly DC component is calculated as x(0) then after that cyclic convolution of X(k)−x(0) is mathematically calculated by following formulae.

$$\left\{ \left[x(r^0 \mod N), x(r^1 \mod N), \ldots x(r^{N-2} \mod N) \right] - \oplus x(0) \right\}$$
$$\left[W_N^{r^0 \mod N}, \ldots \ldots W_N^{r^{N-2} \mod N} \right] \tag{4}$$

Equation (4) can also represent in matrix form for simplification in calculations. Rader algorithm is the first step of Winograd FFT in which cyclic convolution is used.

Thus, N point DFT is converted into N − 1 point cyclic convolution.

4.2 Winograd Small Convolution Algorithm

Winograd small convolution algorithm is polynomial base algorithm which used cyclic convolution with minimal numbers of multipliers and adders with reduction in computational complexity. Cyclic convolution is performed between Impulse samples and input time domain samples which are converted into N − 1 point samples by Rader algorithm. Consider unit impulse response for most accurate result of input time samples. Let us consider h(p) = {h_0, h_1, ...h_{N−1}} be the impulse response and x(p) = {x_0, x_1, ...x_{N−1}} be the input time samples then cyclic convolution is given by

$$S(p) = h(p) \times (p) \mod (p^{N-1}) \tag{5}$$

The following are different steps which are used to calculate Winograd small convolution algorithms.

Step 1: choose a polynomial m(p) with the degree higher than h(p) and x(p). Split that polynomial into small prime polynomial such as

$$m(p) = m^{(0)}(p)m^{(1)}(p)\ldots m^{(k)}(p) \tag{6}$$

where, k varies from 0 to N − 2.

Step 2: calculate $M^{(i)}(p) = m(p)/m^{(i)}(p)$ where $m^{(i)}(p)$ are small prime polynomials. Then used Chinese remainder theorem for calculating $n^{(i)}(p)$ by assuming value of $N^{(i)}(p)$ such that the following equation satisfy.

$$M^{(i)}(p)N^{(i)}(p) + m^{(i)}(p)n^{(i)}(p) = 1 \tag{7}$$

Step 3: calculate the impulse and input time domain coefficients.

$$h^{(i)}(p) = h(p) \bmod m^{(i)}(p) \tag{8}$$

$$x^{(i)}(p) = x(p) \bmod m^{(i)}(p) \tag{9}$$

where i = 0, 1, ...k

Step 4: compute $s^{(i)}(p)$ by the following equations

$$s^{(i)}(p) = h^{(i)}(p)x^{(i)}(p) \bmod m^{(i)}(p) \tag{10}$$

From Eq. (10) we get the different equations for $s^{(i)}(p)$. Using those equations requirement of number of multipliers and adders cleared.

Step 5: again by simplifying and solving the equations for $s^{(i)}(p)$ by following formula

$$S(p) = \sum_{i=0}^{k} s^{(i)}(p)N^{(i)}(p)M^{(i)}(p) \bmod m^{(i)}(p) \tag{11}$$

Equation (11) is the final equation of algorithm in which number of multipliers and adders can no further reduced. Winograd Fast Fourier Transform algorithm combines Rader algorithm and Winograd small convolution algorithm. Input time domain samples converted into cyclic convolution by Rader algorithm then proceed further with Winograd small convolution algorithm to finally reduce computational complexity [15].

5-point and 7-point signal flow structures are shown in Figs. 1 and 2. Multiplicative coefficients which are used in respective signal flows are evaluated in following Table 1. The value of u can be considered as $u = -2\pi/N$, where N is size of WFFT.

5 Verilog Implementation Results

We designed 5-point and 7-point WFFT module is simulated in software Xilinx ISE 13.1. Structural coding of the complete architecture is done by Verilog language. Simulation result of test bench can be obtained after

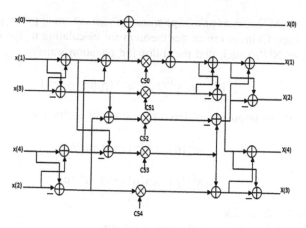

Fig. 1 5-point WFFT signal flow graph

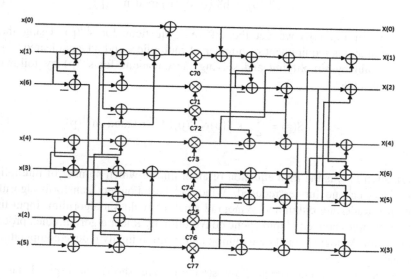

Fig. 2 7-point WFFT signal flow graph

Table 1 Comparison between WFFT and DFT

Fast Fourier Transform (FFT)				
N-point	Definition based method		Winograd FFT	
N	Multipliers	Adders	Multipliers	Adders
5	25	20	5	17
7	49	42	8	36

successful synthesis of the module. Various results of simulations are shown in Figs. 3 and 4.

Tables 2 and 3 shows area and delay synthesize results for 2-point and 3-point DFTs using conventional as well as Winograd algorithm.

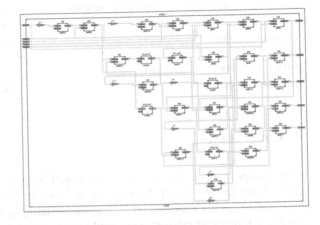

Fig. 3 RTL schematic of 5-point WFFT

Fig. 4 Simulation waveform for 5-point FFT

Table 2 Synthesis results of WFFT and DFT

Family: Spartan 3			
Target device: XA3S200			
2-point FFT architecture			
	Conventional algorithm	Winograd algorithm using add/shift multiplier	Winograd algorithm using semi-complex multiplier
No. of slices	23	19	19
No. of 4 input LUT's	38	33	33
IOB's	32	32	32
Delay (ns)	20.16	17.58	15.17

Table 3 Synthesis results of WFFT and DFT

Family: Spartan 3			
Target device: XA3S200			
3-point FFT Architecture			
	Conventional algorithm	Winograd algorithm using add/shift multiplier	Winograd algorithm using semi-complex multiplier
No. of slices	101	81	79
No. of 4 input LUT's	162	139	142
IOB's	48	48	48
Delay (ns)	32.95	27.99	28.25

6 Conclusion

In this paper, latency and area of 2-point and 3-ponit DFT Architectures using conventional and Winograd algorithm is as shown in Tables 2 and 3 which clearly shows reduction in Delay and area. When considering performance Winograd FFT algorithm, it becomes efficient compared to conventional definition based approach. High performance is achieved in DFT architecture in terms of area and latency. Overall Winograd FFT is one of the best algorithms for smaller sized DFTs as well as larger sized DFTs when it fits into divide and conquers approach.

References

1. F.Qureshi, M. Garrido and O. Gustafsson, "Unified architecture for 2, 3, 4, 5 and 7-point DFTs based on winograd Fourier transform algorithm", IET, Electronics letter, 28 Feb 2013 vol. 49 pp. 348–349.
2. G.A. Sathishkumar and Dr. K. Boopathy bagan, "A hardware implementation of winograd Fourier transform algorithm for cryptography", Ubiquitous computing and communication journal, 2010, volume 3 number 4.
3. Adem Coskun, Izzet Kale, Richard C. S. Morling, Robert Hughes, Stephen Brown and Piero Angeletti, "The design of low complexity low power pipelined short length winograd Fourier transform", IEEE, 2014, 978-1-4799-3432-4/14.
4. He Jing, Li Tianyue and Xu Xinyu, "Analysis and performance comparison of 3780 point FFT processor architectures", IEEE, 2011, 978-1-4577-0321-8/11.
5. Florent Camarda, Jean Christophe Prevotet and Fabienne Nouvel, "Implementation of a reconfigurable fast Fourier transform application to digital terrestrial television broadcasting", IEEE, 2009, 978-1-4244-3892-1/09.
6. Malcolm D. Macleod, "Multiplierless winograd and prime factor FFT implementation", IEEE, September 2004, signal processing letters, vol. 11, no. 9.
7. Jinan Leng, Lei Xie, Huifang Chen and Kuang Wang, "Optimized iterative WFTA method for the 3780 point FFT scheme", IEEE 2011, 978-1-61284-307-0/11.

8. Chen-Fong Hsiao and Yuan Chen, "A generalized mixed-radix algorithm for memory-based FFT processors", IEEE transactions on circuits and systems—ii: express briefs, January 2010, vol. 57, no. 1.
9. Aniket Shukla and Mayuresh Deshmukh. "Comparative Study Of Various FFT Algorithm Implementation On FPGA", International Journal of Emerging Trends in Signal Processing Volume 1, Issue 1, November 2012.
10. Amit Kumar Pathak, Ravi Mohan, Sumit Sharma, "Flexible Digital Design for DSP, FFT Macros Design & Implementation", International Journal of Advanced Research in Electronics and Communication Engineering (IJARECE) Volume 2, Issue 1, January 2013.
11. Cooley J. and Tukey J., "An algorithm for the machine calculation of complex Fourier series", Math. Comput., 1965, 19, pp. 97–301.
12. Burrus C.S. and Eschenbacher P., "An in-place, in- order prime factor FFT algorithm", IEEE Trans. Acoust. Speech Signal Process, 1981, 29, (4), pp. 806–817.
13. S. Winograd, "On computing the discrete Fourier transform", Nat. Acad. Sci. USA, 1976, 73, (4), pp. 1005–1006.
14. Prathamesh Vinchurkar, Dr. S.V. Rathkanthiwar, Sandeep Kakde," HDL Implementation of DFT Architectures using Winograd Fast Fourier Transfrom Algorithm",5th IEEE International Conference on Communications and Network (CSNT) 2015, Gwalior, India, 978-1-4799-1797-6/15 ©IEEE.
15. U. Meyer-Baese, "Digital signal processing with field programmable gate array", Third edition, Springer, 2007.

Dynamic Voltage Restorer Based on Neural Network and Particle Swarm Optimization for Voltage Mitigation

Monika Gupta and Aditya Sindhu

Abstract Dynamic Voltage Restorer (DVR) is one of the most widely implemented power devices used to mitigate voltage unbalance in the grid. The performance of DVR largely depends upon its control strategy, in which controller plays an important part. Literature has shown that the commonly used proportional integral (PI) and neural network (NN) controller have many inherent disadvantages including high total harmonic distortion (THD) and high delay time. In this paper, we have replaced the PI controller with a neural controller, whose weights are trained using Particle Swarm Optimization (PSO). A comparative analysis of the DVR performance is done in MATLAB SIMULINK environment for three controllers—PI, NN with back propagation and NN with PSO for 30 and 80 % voltage sag, 40 % voltage swell and unbalanced voltage (1-ϕ). The results obtained document that the hybrid neural controller with PSO has least distortions and is most robust of the three.

Keywords Neural controller · Dynamic voltage restorer · Neural network · Particle swarm optimization · Comparative analysis

1 Introduction

With an increased risk of damage to sensitive loads due to factors like overloaded, external conditions like weather and grid integration hazards like islanding, it is important to keep in mind appropriate safety devices regarding voltage transients and fluctuations. A DVR is one of the widely used load end installed power safety

M. Gupta (✉) · A. Sindhu
Department of Electrical & Electronics Engineering, Maharaja Agrasen
Institute of Technology, Delhi, India
e-mail: gupta.monika.219@gmail.com

A. Sindhu
e-mail: aditya.sindhu@hotmail.com

© Springer India 2016
S.C. Satapathy et al. (eds.), *Information Systems Design and Intelligent
Applications*, Advances in Intelligent Systems and Computing 433,
DOI 10.1007/978-81-322-2755-7_59

device whose function is to conveniently to restore or balance out the voltage across the load during periods of sag, swell or voltage transients, which may otherwise damage the sensitive load by producing subsequently large currents [1]. Voltage unbalance may also lead to shutdown of the device or cause tripping of protective devices like breakers and relays [1].

The industry standard solution to the voltage unbalance problems like sag and swell for the past few years is the DVR. A DVR injects the required corrective voltage waveform at the time of voltage unbalance. The control system of a DVR is quintessential to the performance of the DVR. The control mechanism of the DVR should be robust and adaptive to sudden changes in voltage. The most widely used controller today is the PI controller. The function of the controller is to input the error signal and give the required signal to the Pulse Width Modulator (PWM) controlled voltage source inverter when voltage imbalance occurs. It has however been observed that the PI controller has many inherent disadvantages including high THD values and high delay time [2–4].

The disadvantages of the PI controller can be overcome by using a neural network controller. A three layer neural network is designed and encoded to percept and store the load voltage levels, which are to be compared to a fixed threshold value. In this paper the neural weight training is done by PSO. This paper introduces a neural controller whose weights have been trained by the PSO and back propagation and compares the performance of the DVR with the traditional PI model and with the two neural controllers. PSO, being swarm based converges faster and is a more robust algorithm as compared to back propagation [5]. In this simulation we have coded the NN and PSO in a mfile and embedded this block in SIMULINK environment. This gives us more flexibility as compared to neural toolbox.

The paper organization is as follows. A brief theory of DVR operation and its control strategy is described in Sect. 2. Problem formulation, including the MATLAB SIMULINK model for the grid connected DVR and the corresponding simulation results are discussed Sect. 3. In Sect. 4 comparison of the different controllers is done followed by the Conclusion in Sect. 5.

2 DVR with the Proposed Controller

2.1 DVR

A DVR consists of a DC battery connected to a PWM controlled Inverter. The input to the inverter is a pulsed wave which is generated by the PWM generator. The PWM generator only produces pulses when it gets the signal from the DVR control system. The DVR output is connected by a special injection transfer to the part of the grid located at the load end, also called Point of Common Coupling or PCC. In this way, the DVR senses any increase or decrease in voltage levels

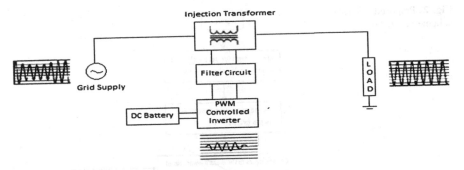

Fig. 1 Basic model of a grid connected DVR

through its control model and sends the signal to PWM controlled Inverter which provides the necessary corrected voltage during the time of voltage disturbance or transient. The basic model of the grid connected DVR is given in Fig. 1. The control mechanism of a DVR is responsible for recording and monitoring the voltage levels continuously. Whenever the voltage levels increase or decrease beyond a set voltage, a gated signal is send to the input of the PWM VSC Inverter which injects the corrective voltage into the grid.

2.2 Hybrid (NN and PSO) Controller

Particle swarm optimization (PSO) is an algorithm used for neuron weight training in a multi layered neural network. PSO is a swarm based robust algorithm which optimizes a problem by iteratively focusing on achieving the best output as compared to a set of desirable weights gbest and pbest [6, 7]. The swarm particles are then initiated from a random position and made to move in the search space. Their movement in search space is governed by velocity (1) and position (2) equations.

$$v(t+1) = wv(t) + c_1 r_1 [\hat{x}(t) - x(t)] + c_2 r_2 [g(t) - x(t)] \tag{1}$$

$$x(t+1) = x(t) + v(t+1) \tag{2}$$

The proposed controller scheme for the DVR using PSO is given in Fig. 2. We start by identifying the output (setpoint for signal generation) and input (error) signals for the neural network. Following this, certain neural parameters like epoch, samples and weights were set and neurons initialized governed by a set formulas in each layer. The error was then measured and if the measured error margins were large, new gbest and pbest values of weights were calculated and the swarm of weights was updated.

Fig. 2 Proposed controller
scheme using PSO

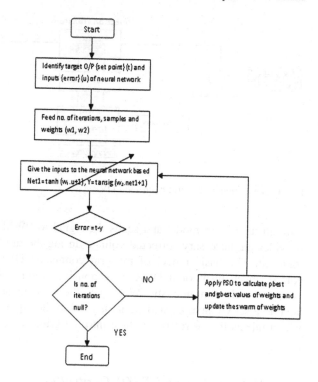

3   Problem Formulation and Simulated Results

The model for grid connected DVR was implemented in MATLAB SIMULINK environment as shown in Fig. 3a.

The model consists of a 250 V, 15 MVA generator supplying power to an isolated micro grid, supplying power to an RLC load of 60 kW, 20 Mvar. Bus Bars B1 and B2 act as voltage measurement blocks. The Voltage Source (VSC) inverter is connected to the grid via a 100 V/100 V, 15 MVA three winding Injection transformer. The point of connection of the Injection transformer with the grid is the Point of Common Coupling or the PCC. The power supplied to the VSC Inverter comes from a 700 V DC battery. The input to the VSC Inverter comes from a PWM Generator which is itself fed signals from the output of the PI/NN Controller block, after undergoing Clarke's transformation.

Figure 3b shows the control system of the DVR implemented in the MATLAB SIMULINK environment. The input voltage levels are compared to V_{ref}, a standard voltage reference level. Both the blocks undergo Clarke's transformation and are then compared. The error is then sent to the PI/NN controller which is responsible for detecting any voltage imbalance and sending the required signal to the input of the PWM Generator. For this simulation, the three controllers (PI,

(a)

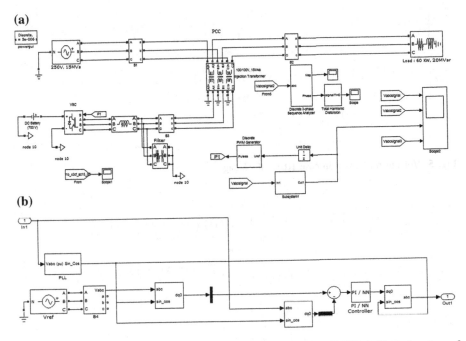

(b)

Fig. 3 a MATLAB SIMULINK implementation for a grid connected DVR. **b** Control system of the DVR

NN + BP, NN + PSO) were tested individually and the performance of the DVR for each case was observed.

The simulation results for voltage mitigation (NN + PSO controller) are divided into 3 cases—sag, swell and unbalanced phase. In Case 1(a), shown in Fig. 4 we have given 30 % sag to the simulation between 0.2 and 0.4 s. The same is repeated with 80 % sag in Case 1(b), depicted in Fig. 5. Figure 6 shows 40 % Swell (Case 2) and Fig. 7 depicts voltage mitigation in case of an unbalanced phase (phase R) between 0.2 and 0.6 s. In all the cases, the first plot corresponds to the generator voltage, second displays the load voltage and third shows the DVR output.

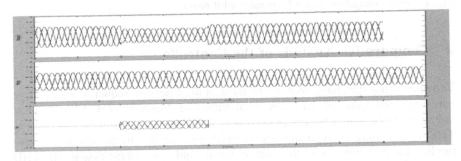

Fig. 4 Voltage mitigation in case 1(a)—30 % sag

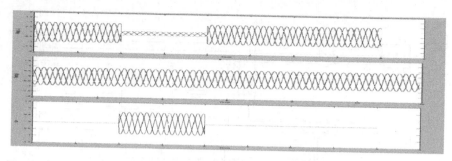

Fig. 5 Voltage mitigation in case 1(b)—80 % sag

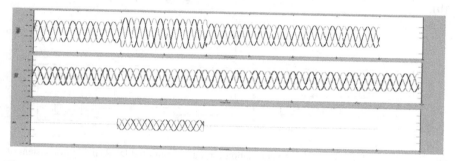

Fig. 6 Voltage mitigation in case 2—40 % swell

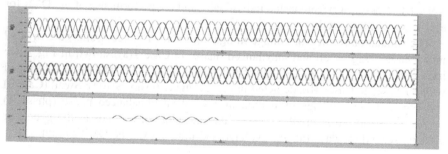

Fig. 7 Voltage mitigation in case 3—unbalanced R phase

4 Comparative Analysis of the Controllers

The three controllers have been on the basis of their positive sequence load voltage waveform and value of maximum total harmonic distortion (THD) for the load voltage waveform after the compensation by the DVR. Total harmonic distortion or THD is a measure of the total harmonic distortion present in the system and also predicts the non linearity of the system. For a stable and robust system, the THD should be as less as possible. Figures 8, 9 and 10 correspond to the load voltage waveforms for PI, NN with BP and NN with PSO controllers respectively. Figures 11, 12 and 13 show the plot of THD for the three controllers (in the same order).

From Table 1 and the plots for THD it is evident that the hybrid NN PSO controller has the least THD and the PI controller has maximum THD. From the above results and plots it can thus be sufficiently concluded that hybrid neural network controller with PSO is the most accurate and most robust out of the three simulated controllers.

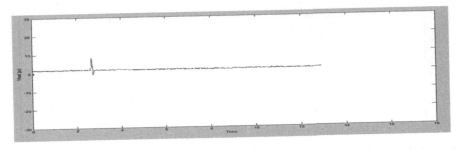

Fig. 8 Plot for load voltage for PI controller

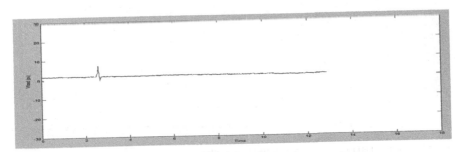

Fig. 9 Plot for load voltage for NN + BP controller

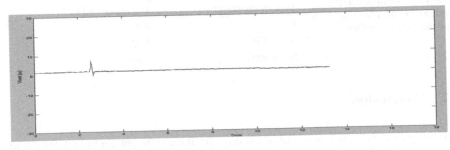

Fig. 10 Plot for load voltage for NN + PSO controller

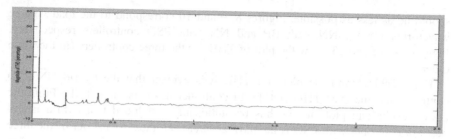

Fig. 11 Plot for THD in case of PI controller

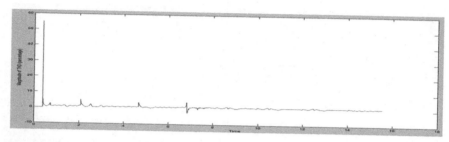

Fig. 12 Plot for THD in case of NN + BP controller

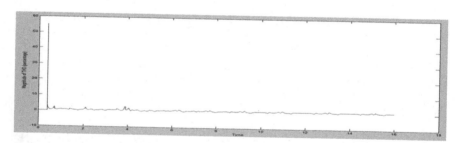

Fig. 13 Plot for THD in case of NN + PSO controller

Table 1 Maximum THD values (%) measured for each of the three controllers

Type of controller	Maximum THD (%)
PI	9.08
NN + BP	5.42
NN + PSO	3.54

5 Conclusion

In this paper, a comparative analysis is done for three different controllers for a grid connected DVR. A DVR connected via injection transformers to an active supply system as a part of simulation in MATLAB SIMULINK environment and voltage fluctuations namely sag, swell and unbalanced phase transient was given in the system for a particular duration of time, during which the voltage mitigation action of the DVR was also observed. Three different controllers—PI, neural network controller with back propagation and neural controller with PSO were implemented individually and the performance of the controllers was compared using maximum THD values and load voltage plots. The results prove that hybrid neural network controller with PSO is the most accurate and most robust controller out of the three.

References

1. Awad, H.; Svensson, J.; Bollen, M.: Mitigation of Unbalanced Voltage Dips Using Static Series Compensator. In: IEEE Trans. On Power Elec., Vol. 19, No. 13. (2004).
2. Choi, S.S.; Li, B.H., Vilathgamuwa, D.M.: Dynamic voltage restoration with minimum energy injection. In: IEEE Trans. Power Syst., vol. 15, no. 1, pp. 51–57. (2000).
3. IEEE recommended practices and requirements for harmonic control in electrical power system. IEEE Standard 519-1992. (1992).
4. Sankaran, C.: Power quality. CRC PRESS. (2002).
5. Kennedy, J., Eberhart, R.C.: Particle Swarm Optimization. In: The IEEE International Conference On Neural Networks, Vol. 4, pp. 1942–1948. (1998).
6. Mendes, R., Kennedy, J., Neves, J.: The fully informed particle swarm: simpler, maybe better. In: IEEE Transactions on Evolutionary Computation, 8, pp. 204–210.
7. Van den Bergh, F.: Analysis of particle swarm optimization. Ph.D. thesis, University of Pretoria, South Africa. (2002).

Analysis of Norms in Adaptive Algorithm on Application of System Identification

Sarthak Panda and Mihir Narayan Mohanty

Abstract System identification is an important area in signal processing research. It aims to retrieve the system's unknown specifications from its output only. This technology has a wide variety of applications in engineering and control, industrics, as well as medical fields. Typically, the identification of models expressed as mathematical equations. Linear, Non-Linear, Non parametric and Hybrid models are few deciding factors on which different techniques for System Identification relies on. In this paper, we discuss in detail the LMS algorithm and NLMS algorithm. In particular, various types of norms are included in LMS algorithm and the NLMS algorithm is modified according to the norms. Considering different norms in LMS algorithm we have analyzed the application of System identification. Also, it has been verified for both linear and non-linear models. Finally, for non-linear system identification based on Wilcoxon norm has been proposed. The results as well as the comparison show that the Wilcoxon norm is one of the better norms than others and is applied for System identification. The results show its efficacy.

Keywords System identification · LMS algorithm · NLMS · Norms · Wilcoxon norm

1 Introduction

System identification has wide application in solving a range of problems. Hence it occupies an important place in solving the major engineering issue. In the past years Mathematical system theory has been evolved into a strong scientific course of vast applicability, which gives the idea about analysis and synthesis of the systems.

S. Panda (✉) · M.N. Mohanty
ITER, S 'O' A University, Bhubaneswar, Odisha, India
e-mail: sarthak216@rediffmail.com

M.N. Mohanty
e-mail: mihirmohanty@soauniversity.ac.in

© Springer India 2016
S.C. Satapathy et al. (eds.), *Information Systems Design and Intelligent Applications*, Advances in Intelligent Systems and Computing 433,
DOI 10.1007/978-81-322-2755-7_60

579

Linear algebra, complex variable theory and theory of ordinary linear differential equations are being designed based on linear operator are the best techniques for systems to be identified. Stability property plays a vital role in System identification. So, design techniques for dynamic systems are designed by taking system's stability in consideration. For linear time invariant systems, stability conditions has been developed over past decades and according to this designing methods have been adapted for these systems. In order to system-by-system basis stability of non-linear systems can be developed.

Otherwise we can say, System identification is to estimate a *black box* or *grey box* model of a dynamic system based on observing input-output. According to Zadeh's statement in 1962, System Identification is *the determination on the basis of input and output, of a system within a specified class system, to which the system under test is equivalent (in terms of a criterion).* Obtaining systematic information from experimental observations is a key aspect of any scientific work. The obtained information leads to the formulation of a model of the system under consideration. Such a model is some form of pattern that explains the observed experimental results and allows for predictions of future system responses to be made. It leads to System identification. This job can be performed by using certain algorithms which are adaptable. Some of them are Least Mean Square (LMS) Algorithm, Normalized Least Mean Square (NLMS) Algorithm, Wilcoxon Least Mean Square (WLMS) Algorithm and many more. A saving in Computational time can be saved using System identification as compared to physical modeling, since only the input signal to the system and output signals from the system are of concern [1, 2].

Modeling is used to characterize a real process and study its behavior. The set of processes in a system determines the behavior of the system. Sometimes a dynamical model can be difficult to obtain due to the complexity of the process.

2 Related Literature

System identification has been performed using adaptive algorithms as well as neural network methods. Also modification of these techniques is achieved by researchers, though there is the future scope [3–8]. Many works in this area have been attempted by many researchers since some decades. In such scenario, LMS algorithm gains popularity among all the algorithms. Further, the modification of this LMS algorithm has been used in various applications. In 2013, Ghauri et al. used three types of adaptive filters. These filters were applied to identification purpose of the unknown system using LMS, NLMS and RLS algorithms. They found that NLMS and RLS have more computational complexity than LMS and NLMS is the normalized version of LMS. RLS shows its efficacy more than the other two though it is a complex algorithm. Based on Least Mean Square Error (LMSE) all these algorithms worked and filter weights were recursively revised as to bring output equal to the desired signal [3].

In [4], authors have proposed a new adaptive algorithm to improve the performance. Some of the algorithms are variable step-size Least Mean Square (VSSLMS) algorithm and Zero-attracting (ZA) VSSLMS, used when the system is in sparse condition. They observed that when the sparsity of the system reduces the enforcement of the ZA-VSSLMS algorithm degrades. Again, they proposed weighted zero-attracting (WZA)-VSSLMS algorithm. When the sparsity of the system falls, it functions worthier than the ZA-VSSLMS. In 2009, Majhi et al. used Wilcoxon norm as the cost function and proposed a new robust WLMS (Wilcoxon least mean square) algorithm. When outliers were present in the training signal Wilcoxon Least Mean Square algorithm was helpful for identification of all zero plants, which exhibit the remarkable performance over the least Mean Square algorithm. Again in 2013, when outliers were present authors approached the system identification issue by using Wilcoxon based LMS. Also the result was compared with the conventional LMS. In addition to it, the error was studied for the deviation factor. Dash et al. have verified the application of Wilcoxon norm in linear system identification in initial stage that follows for non-linear system. In the next stage, Sign-Sign Wilcoxon norm based approach has been verified for the same. Finally they have modified to a Variable Step-Size Sign-Sign Wilcoxon technique for both linear and non-linear system identification and compared with above two techniques. They observed that the designed technique is robust against outliers in the desired signal and simultaneously the convergence rate is higher than Wilcoxon norm based approach [5–7].

3 Methods of System Identification

Figure 1 shows the standard block diagram for system identification. The input is enforced to the plant as well as to the adaptive model that is parallel to the unknown system. $W(n)$ is the impulse response of the linear section of the plant, followed by non-linearity (NL). Then White Gaussian noise is added to non-linear output and the noise can be denoted as $N(n)$. The desired output $D(n)$ is then compared with the estimated output $Y(n)$ to generate the error $E(n)$, and the adaptive algorithm was applied for updating the weights of the model [8].

The system has chosen as Linear and Non-linear system and are described as follows:

(i) Linear system

A linear system has been considered as

$$D = A \times (\sin(2\pi ft + \emptyset))$$

where, D = Desired signal; A = Amplitude; f = Frequency; t = Time period; $\emptyset$ = Phase.

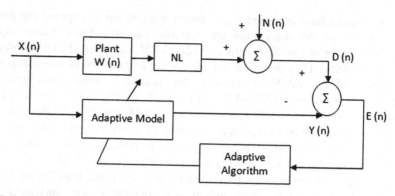

Fig. 1 Block diagram of non-linear system identification

(ii) Non-Linear system

Again a Non-linear system is described as

$$D = A \times (\sin(2\pi ft + \emptyset)) + B \times (\sin^2(2\pi ft + \emptyset)) + C \times (\cos^3(2\pi ft + \emptyset))$$

where, D = Desired signal; A, B, C = Amplitudes; f = Frequency; t = Time period $\emptyset$ = Phase.

Then these two linear and non-linear systems are proposed to three different types of adaptive algorithms (LMS, NLMS, and WLMS) and based on Mean Square Error efficient one is chosen.

Least Mean Square (LMS) Algorithm

In adaptive systems coefficients can be found by using Least Mean Squares (LMS) algorithm. Then the error signal can be produced which is the difference between the desired and the actual signal. It is based on gradient descent method in which the filter is adapted, at the current time instance based on the error. The weight update relation is derived from [1, 2] as:

$$W(n+1) = W(n) + 2\mu E(n)X(n) \tag{1}$$

The impulse response of the unknown system can be found by taking the difference between the desired signal $D(n)$ and the estimated output $Y(n)$ which is used to find the tap weights. And, $W(n)$ is the existing value of the weight and $W(n+1)$ is the updated value. A positive constant μ (learning rate) is taken which controls the rate of convergence and the stability of the system. μ has to be chosen smaller than $1/2\lambda m$, where λm is the largest Eigen value of the correlation matrix R. The training sequence needs to be longer if and only if learning rate μ is small.

Normalized Least Mean Square (NLMS) Algorithm

The Normalized LMS algorithm is the customized version of LMS algorithm. The weight updation relation of NLMS algorithm can be given by [2]:

$$W(n+1) = W(n) + \mu \cdot E(n) \cdot \frac{X(n)}{\delta + \|X(n)\|^2} \tag{2}$$

where, $\mu(n) = \mu/\delta + \|X(n)\|^2$. In NLMS algorithm δ is time-varying step size. The step size can boost the convergence rate of the adaptive filter. The NLMS algorithm has a higher converging rate as compared to the LMS algorithm with a higher residual error.

Different Norms Associated with LMS Algorithm

The following norms are considered to evaluate the LMS algorithm as NLMS algorithm application in System identification [9–12].

Euclidean Norm

Let an Euclidean space considered as n-dimensional space be A^n, the length of the vector $X = (X_1, X_2, \ldots, X_n)$, can be captured by the relation

$$\|X\| = \sqrt{X_1^2 + X_2^2 + \cdots + X_n^2} \tag{3}$$

Manhattan Norm or Taxicab Norm

$$\|X\|_1 = \sum_{i=1}^{n} |X_i| \tag{4}$$

As per the name "taxicab" it can be defined by the distance a taxi has to go in a rectangular manner to get from the origin to the point x.

p-Norm

Let $q \geq 1$ is a real number.

$$\|X\|_q = \left(\sum_{i=1}^{n} |X_i|^q \right)^{\frac{1}{q}} \tag{5}$$

For $q = 1$ we get the Manhattan norm, for $q = 2$ we get Euclidean norm, and when q reaches ∞ the p-norm can be called as maximum norm.

Maximum Norm

$$\|X\|_\infty = \max(|X_1|, |X_2|, \ldots, |X_n|) \tag{6}$$

The set of vectors, whose Maximum norm is a constant h then the surface of a hypercube can be formed with edge length $2h$.

Wilcoxon Norm

A norm called Wilcoxon norm which is robust in nature is chosen as the cost function in the proposed model. The weights of the model are updated using standard LMS, which steadily reduces the norm. Wilcoxon Norm of a vector can be defined by, a score function as in [6, 7]. If the error vector of lth particle at tth generation can be denoted as $[e_{1, l}(t), ..., e_{K, l}(t)]^T$ for K input samples. Then the errors are sorted in an increasing order from which the rank $r(e_{K, l}(t))$ each lth error term is obtained. The score related with each rank of the error can be expressed [7] as:

$$a(i) = \rho(u) = \sqrt{12}(u)$$
$$= \sqrt{12}\left(\frac{i}{K+1} - 0.5\right) \tag{7}$$

where, K is a positive number. And the rank related to each term can be denoted by $(1 \leq i \leq K)$. Now, at tth generation of each lth particle the Wilcoxon norm is then calculated as:

$$C_l(t) = \sum_{i=1}^{N} a(i)e_{i,l}(t) \tag{8}$$

The training procedure using Least mean square Algorithm will continue till $C_l(t)$, which is known as the cost function in Eq. (8) decrease to the minimum value.

4   Result and Discussion

By taking learning rate μ as 0.085 and step size δ as 0.01 when it is being simulated we get the following observations. The testing analysis is achieved in presence of Gaussian noise of SNR 20 dB.

It has been seen that in non-linear system Normalized LMS converges faster than LMS and we can observe from the output that error effectively is minimized in case of NLMS more. In Figs. 2 and 3 mean squared error and root mean square error has been compared between LMS and NLMS in non-linear system and by observing both MSE and RMSE plots approximately in 10th iteration NLMS starts to converging while LMS takes 10 more iterations. In Fig. 2 when the process starts initially the error was at 10 dB and after 300 iteration the error finally reached to −27 dB in LMS while in NLMS after 300 iteration error reduces to −48 dB (Figs. 4 and 5).

Again by taking same value of $\mu = 0.085$ and $\delta = 0.01$ and SNR = 20 dB different norms were applied in NLMS algorithm. On comparing the above given norms we observed that approximately from 10th iteration Euclidean norm is converging and initially the error was 10 dB and after completion of 300 iteration the error reduces to −38 dB. In p-norm convergence started from 17th iteration and

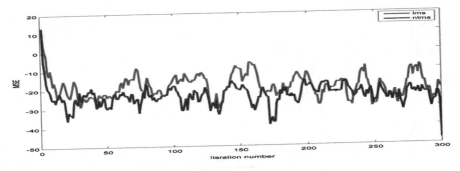

Fig. 2 MSE in LMS and NLMS algorithm in non-linear system

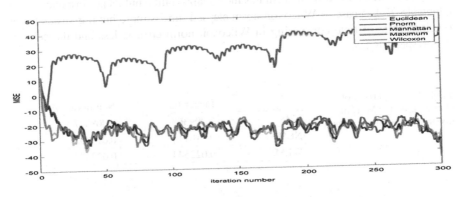

Fig. 3 RMSE in LMS and NLMS algorithm in non-linear system

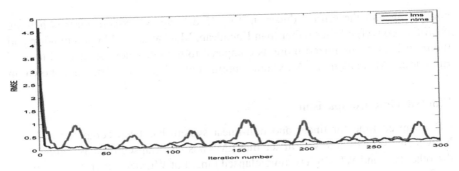

Fig. 4 MSE of different norms in NLMS algorithm

the error reduces to −34 dB similarly in Manhattan norm at 15th iteration we can observe the convergence and the error reduce to −35 dB. Maximum norm gives us the poor result as we have considered the maximum absolute value of the input signal. But while implementing the Wilcoxon norm from 5th iteration it starts

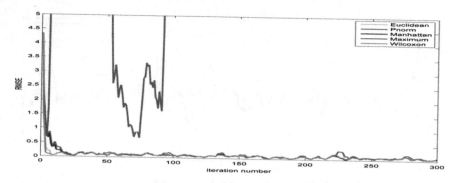

Fig. 5 RMSE of different norms in NLMS algorithm

converging and the error reduces up to −45 dB. So, Wilcoxon norm is helping NLMS to converge faster rather than Euclidean, Manhattan, p, Maximum norm and the error is also minimized more as compared to the other norms. Elapsed time of Euclidean, Manhattan, p, Maximum norm and Wilcoxon norm are shown in Table 2.

Elapsed Time Comparison

CPU elapsed time for linear and non-linear system has been compared in LMS, NLMS and WLMS and we can see that LMS has less elapsed time as compared to the other two and WLMS has more elapsed time, but WLMS is giving us the better performance (Table 1).

Table 2 compares the CPU elapsed time between the different norms. Form the above comparison maximum norm has more elapsed time but its performance is not adaptable. But in case of Wilcoxon norm elapsed time is more but it gives us better result in spite of time consuming in Wilcoxon norm error is less and the convergence speed is fast.

Table 1 Comparison of elapsed time for MSE in linear and non-linear system

	Linear (s)	Non-linear (s)
LMS	0.009368	0.008047
NLMS	0.010118	0.019993
WLMS	0.022544	0.022378

Table 2 Comparison of elapsed time for MSE in different norms

Different norms	Elapsed time (s)
Euclidean	0.023776
P	0.006998
Manhattan	0.006262
Maximum	0.005049
Wilcoxon	0.022653

5 Conclusion

In both Linear and Non-linear system for small input vector LMS algorithm is more convenient but its performance decays for large input vector. For large input computational complexities increases for LMS. Its efficiency reduces because observing the desired output will take more time in LMS. NLMS is perfect for real time application, because NLMS has higher convergence rate than LMS. LMS shows better results non-real time applications such as system identification. From Figs. 6 and 7 we can conclude that among LMS, NLMS and WLMS, WLMS is converging more faster as compared to the other two in both linear and non linear case. Since the convergence rate of the discussed techniques are very fast it can be applied to the fast varying system. Convergence rate of WLMS algorithm is much better than LMS and NLMS algorithm.

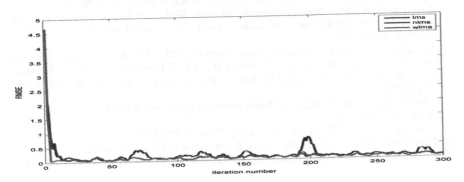

Fig. 6 RMSE of LMS, NLMS, WLMS in a non-linear system

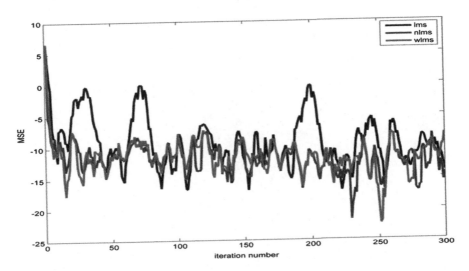

Fig. 7 MSE of LMS, NLMS, WLMS in a non-linear system

References

1. B. Widrow and S. D. Stearns, 'Adaptive Signal Processing' 2nd Edition, Pearson Education.
2. Simon Haykin, 'Adaptive Filter Theory', Prentice Hall, 2002, ISBN 0-13-048434-2.
3. Sajjad Ahmed Ghauri, Muhammad Farhan Sohail, "System Identification using LMS, NLMS and RLS", 2013 IEEE Student Conference on Research and Development (SCOReD), 16–17, pp. 65–69, December 2013, Putrajaya, Malaysia.
4. Mohammad Shukri Salman, Mohammad N. S. Jahromi, Aykut Hocanin and Osman Kukrer, "A zero-attracting variable step-size LMS algorithm for sparse system identification", 2012 IX International Symposium on Telecommunications (BIHTEL) October 25–27, 2012, Sarajevo, Bosnia and Herzegovina.
5. Mihir Narayan Mohanty, Badrinarayan Sahu, Prasanta Kumar Nayak, Laxmi Prasad Mishra, "Non-linear Dynamic System Identification using FLLWNN with novel Learning Method", SEMCCO, Book Chapter in Book title Swarm, Evolutionary, and Memetic Computing, LNCS 7677, pp. 332–341, © Springer-Verlag Berlin Heidelberg 2013.
6. B. Majhi, G. Panda and B. Mulgrew, "Robust identification using new Wilcoxon least mean square algorithm", ELECTRONICS LETTERS 12th March 2009 Vol. 45 No. 6.
7. Sidhartha Dash, Mihir Narayan Mohanty, Member, IEEE, " A Comparative Analysis for WLMS Algorithm in System Identification", Lecture Notes on Information Theory Vol. 1, No. 3, September 2013.
8. Sidhartha Dash, Mihir Narayan Mohanty, Member, IEEE, "Variable Sign-Sign Wilcoxon Algorithm: A Novel Approach for System Identification", International Journal of Electrical and Computer Engineering (IJECE) Vol. 2, No. 4, August 2012, pp. 481–486, ISSN: 2088-8708.
9. Rudin, Walter (1964). Principles of Mathematical Analysis. New York: McGraw-Hill. p. 151. ISBN 0-07-054235-X.
10. Eugene F. Krause (1987). Taxicab Geometry. Dover. ISBN 0-486-25202-7.
11. Deza, Elena; Deza, Michel Marie (2009). Encyclopedia of Distances. Springer. p. 94.
12. Erwin Kreyszig, 'Advanced Engineering Mathematics', JOHN WILEY & SONS, INC., 2006.

SKT: A New Approach for Secure Key Transmission Using MGPISXFS

Kamal Kumar Gola, Vaibhav Sharma and Rahul Rathore

Abstract Cryptography is the concept used to enhance secure communication between two parties; these parties can be two persons in same building or two persons in different organizations of the world. Cryptography is said to be 100 % secure but every time attacker try to break the magic of cryptography. One loose point in cryptography is key transmission. It is most sensitive transmission in the field of cryptography. Many Techniques are proposed time to time to secure key transmission. Some techniques have shown good result up to some extent but not at all fully secured. In this paper we have proposed a key transmission technique to enhance security, Confidentiality and Integrity. We have described detailed algorithm to perform key transmission. Also we have compared the proposed algorithm with existing algorithms.

Keywords Secret key · Hash function · Permutation · Substitution · XOR operation · MD5 algorithm · Gray code

1 Introduction

In public key Cryptosystem, key pair of private and public is used. Public key is known to all. Private Key is private to the organization. It can be generated by organization itself or by any third party organization i.e. Key Distribution center Private Key must be secured from attackers and hackers. Public key cryptosystem

K.K. Gola (✉) · V. Sharma · R. Rathore
Department of Computer Science and Engineering,
Teerthanker Mahaveer University, Moradabad, India
e-mail: kkgolaa1503@gmail.com

V. Sharma
e-mail: vaibhavaatrey@gmail.com

R. Rathore
e-mail: rahul.rathore15@gmail.com

© Springer India 2016
S.C. Satapathy et al. (eds.), *Information Systems Design and Intelligent Applications*, Advances in Intelligent Systems and Computing 433,
DOI 10.1007/978-81-322-2755-7_61

plays a great role in cryptography to enhance confidentiality, integrity, authentication and digital signature. For secure communication encrypted data must be secure apart from it one important issue is to secure the key. Each time attacker tries to attack the encrypted data is of no use until he/she is not having the original key. There has been some security policy for key management and distribution. Keys should be maintained in such a manner that only legitimate user can get the access of his/her key. If any unauthorized user tries to access, it will not be able to access the key. A security policy inhabits the definition of the possible threats to the system.

Proposed technique ensures that one should not the access to the key in any way. This technique not only ensures the security of the key it will also ensure the integrity of the key. Because there might be a chance when attacker fails to de-code the key he/she might change the key. In such case this technique will help to know that whether the received key is the original one or someone changed the key during transmission. Proposed approach works on most of the security issues that might appear in between sender and receiver while sharing a key. With this approach one can minimize the risk of compromising the key during transmission. If somehow they tried to get the key they can easily access the message. Our objective in this paper is to secure the key during transmission. In this paper we are proposing a technique to secure the key transmission in addition to enhance the integrity and data Confidentiality.

2 Literature Review

A large variety of symmetric encryption are Data Encryption Standard [1], Triple DES [2], the Rivest Cipher. The renowned types of symmetric encryption are Data Rivest Cipher (RC4) [3] and the Advanced Encryption Standard (AES). All the three algorithms are block cipher techniques while RC4 is a stream cipher. All these techniques need to use a Key management technique for securing their data transfer.

In [4] the author analyzed the problem of secure data transmission and slow data transmission speed; to overcome from this problem they proposed a technique, which is the combination of encryption and data compression. The compression reduce the size of data and increase the data transfer rate, then the compressed data will be encrypted by using the private key encryption technique, to provides security during the communication process.

In [5] the author proposed an algorithm for efficient key generation and key transfer protocol for group communication in which, KGC is responsible to distribute group keys inform to all group members in a secure manner, the keys are maintained forward and backward confidentially whenever a new member join or left the group. Algorithm explained in [5] relies on NP class.

In [6] the author proposed a secure key data transfer technique by appending the ideas of cryptography and steganography. The author stated that encryption and decryption can be done by using symmetric asymmetric and hashing algorithms. To resolve these issues they used DCT, 2DDCT and FFT transformation and DES for cryptography. DES is used to encrypt the data and some part of the data is hidden in an image by 2DDCT and rest of the data will generate two secret keys for high security.

In [7] the author proposed an algorithm to reduce the risk of attacks on key transmission and keep it integrated and confidential. The authors used the SHA 256 to provide integrity and confidentiality during key transmission. That used 64 bit permutation table as well as 4 * 15 bit substitution table.

We have enhanced the work proposed in [7] by using 128 bit permutation along with 4 * 64 bit substitution table. By considering complex permutation table and substitution the proposed technique result is more secure, maintains more data Integrity and Confidentiality.

Proposed Algorithm can be used in any of the above cipher algorithm to securely transmit the data. The goal of our research work is to provide an extreme secure environment by appropriately securing secret key for both secret key transmission and public key algorithms.

3 Proposed Model for Secure Key Transmission

See Fig. 1.

4 Proposed Algorithm

4.1 Key Process at Sender Side

Step-1 First the sender selects a key to encrypt the message.

Step-2 Convert the key into its equivalent decimal number like a to z will denote by 0–25. And A–Z denote by 26–51, 0–9 will be expressed as 52–61.

Step-3 Now convert the numbers into its binary number with eight bits. If the total number of bits in key are less than 512 bits then padding of 11111111, will be done to make a total of 512 bits. During decryption if the receiver get 11111111 in the data then this has to be discarded at this stage.

Step-4 Now calculate the hash value of step-3 using MD5 algorithm and convert the hash value into its equivalent binary number with four bits.

Step-5 Now append all the bits of step-3 and step-5, after that divide all the bits into five parts, each part having 128 bits.

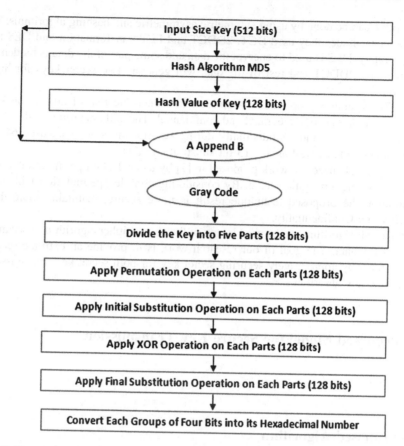

Fig. 1 Proposed model for secure key transmission

Step-6   Convert all the bits of each part into its gray code. After that apply the permutation operation on each part according to the given Table 1 and copies all the entries row wise from the table for all parts.

Step-7   Apply the substitution operation on each part of setp-6 according to the substitution Table 2.

Step-8   Now apply the XOR operation given below.

Step-9   Now append the bits of Part-1 first (Step-8) then Part-A, Part-B, Part-C and Part-D.

Step-10   Now leave the first two bits and last two bits to make a group of six bits and apply final substitution operation on each group of six bits according to the given final substitution table [7]. After final substitution place the first and last two bits at their same location and arrange the bits into group of four bits and convert that into its hexadecimal number which is the final encrypted key.

Table 1 Permutation table

0	2	4	6	8	10	12	14
1	3	5	7	9	11	13	15
16	18	20	22	24	26	28	30
17	19	21	23	25	27	29	31
32	34	36	38	40	42	44	46
33	35	37	39	41	43	45	47
48	50	52	54	56	58	60	62
49	51	52	55	57	59	61	63
64	66	68	70	72	74	76	78
65	67	69	71	73	75	77	79
80	82	84	86	88	90	92	94
81	83	85	87	89	91	93	95
96	98	100	102	104	106	108	110
97	99	101	103	105	107	109	111
112	114	116	118	120	122	124	126
113	115	117	119	121	123	25	127

4.2 Key Process at Receiver Side

Receiver can receive the Key using the reverse process from Step 10 to Step 1. (As explained in Key Process at Sender Side).

Note: To find the reverse of initial and final substitution, Receiver calculate 1st and last digit value in equivalent decimal and consider it as row values. Now search in table for row value to find the middle 4 digits and get the column number of that cell. For example: 001111, it forms row value 1 and actual value is 7. Now according to Substitution Table [7] Column value is 11. Now copy 1st and last digit of 6 bit number and place the binary equivalent of Column value i.e. 001111 will become 010111.

5 Implementation

Step-1 Let the key is

"KamalkumargolalGulistaKhan2VaibhavSharma3RahulRathore4"

Step-2 Converted key into its equivalent decimal number and then convert the key into its binary equivalent of eight bits and append the bits to make it 512 bits. 00100100000000000001100000000000000101100100100000 10100000011000000000000010001001000000000111000001011000000 000001101010010000000010100000010110000100000010010000100 100010100001001000001000100110110001011110000000000010000

Table 2 Initial substitution table

Outer bits	\	Middle six bits of the key																			
		0	1	2	3	4	5	6	7	8	9	10	11	12	13	14	15	16	17	18	19	20
0		0	1	3	5	7	9	11	13	15	17	19	21	23	25	27	29	31	33	35	37	39
1		2	4	6	8	10	12	11	16	18	20	22	24	26	28	30	32	34	36	38	40	42
2		62	60	58	56	54	52	50	48	46	44	42	40	38	36	34	32	30	28	26	24	22
3		63	61	59	57	55	53	51	49	47	45	43	41	39	37	35	33	31	29	27	25	23
21		23	24	25	26	27	28	29	30	31	31	32	33	34	34	35	36	37	38	39	39	40
41		45	47	49	51	53	55	57	59	61	61	63	2	4	4	6	8	10	12	14	14	16
44		48	50	52	54	56	58	60	62	2	0	1	3	5	5	7	9	11	13	15	15	17
20		16	14	12	10	8	6	4	2	63	63	61	59	57	57	55	53	51	49	17	17	45
21		17	15	13	11	9	7	5	3	1	1	0	60	60	60	56	56	54	52	50	50	48
41		43	45	46	47	48	49	50	51	52	52	53	54	55	56	57	58	59	60	61	62	63
18		22	24	28	30	32	34	36	38	40	40	42	44	46	46	50	52	54	56	58	60	62
19		23	25	29	31	33	35	37	39	41	41	43	45	47	49	51	53	55	57	59	61	63
43		39	37	33	31	29	27	25	23	21	21	19	17	15	13	11	9	7	5	3	1	0
46		42	38	36	34	32	30	28	26	24	24	22	20	18	16	14	12	10	8	6	4	2

000000100000111000000000001010100101100000001110000000000
100010000110000000000001101100010101100000000000011100010
100000010110010101100000000000100110000011100001110001000
100000100001101111111111111111111111111111111111111111111
11111111111111111111111111111111111111111111111111111111

Step-4   Calculated hash value of step-3 using MD5 algorithm and its binary
          equivalent of four bits.
          4A 16 B9 B9 26 D9 D4 9F BE 5F 59 99 5D 19 DE EA 01001010000
          10110101110011011100100100110110110011101010010011111101111
          1001011111010110011001100101011101000110011101111011101010

Step-5   Now append all the bits of step-3 and step-5, after that divide all the bits
          into five parts, each part having 128 bits.

          Part-1   0010010000000000000011000000000000000101100100100000
                    10100000011000000000000010001001000000000111000010
                    110000000000011010100100000

          Part-2   0001010000001011000010000001001000010011000101000
                    0010000010001001101100010111100000000000001000000000
                    01000001110000000000010101

          Part-3   0010110000000111000000000000100010000110000000000001
                    10110001010110000000000000011100010100000001011001010
                    11000000000001001100000111

          Part-4   00001110000100010000010000011011111111111111111111111
                    111111111111111111111111111111111111111111111111111111
                    111111111111111111111111111

          Part-5   01001010000101101011100110111001001001011011001110
                    10100100111111011111001011110101100110011001010111
                    01000110011101111011101010

Step-6   Convert the each Parts of Step-5 into gray code. After that apply the
          permutation operation and copies the entries row wise.

          Part-1   0101000001100000001100000000000001111010010011000 1
                    10011011000000000001000000101110000100100000100 1110
                    0000100000011111000011 0100

          Part-2   0011001101100010101000110010010100111011 01000110010
                    10010011001011110011000110100100000100000000 10000010
                    00000100101000001100000111

          Part-3   01110000010000101000001000000101101100000000000011
                    0011100110110100000000000001010110011011000101111 10
                    00011000000011100001000010

          Part-4   00100010000101011001011000100010000000000000000000
                    00000000000000000000000000000000000000000000000000
                    000000000000000000000000000

Part-5 01110010101101111100010010111011110011000111011011
11000011011000100010010011100110000001111111110110
0011010111010010110101011

Step-7 Apply the substitution operation on each part of setp-6 according to the Initial Substitution Table 2.

Part-1 00010000010000000101111000000000011111010100101011
01001010000000000010000011011111011000000011011011
10001111100111100101100110

Part-2 01101001010001001011101101001101011110010000110001
00100010010110101110011001101111100000000010000011
10000111101111101100010001

Part-3 01100000000010011111000000011011110010000000000010
01111011010101000000000000010100110110100010010010
10010000000110111000000100

Part-4 01000100010110111010000010000100000000000000000000
00000000000000000000000000000000000000000000000000
0000000000000000000000000

Part-5 01100100100100111111001010001011111000100110111011
10000010110000000100011000100111110101000010111010
10110101010001011100101111

Step-8 Apply XOR operation on Step-7 according to the Fig. 2 and copy Part-1 from the step-7.

Part-A 01111001000001001110010101001101000001000100011010
01101000010111010110001101011000011000000001011000
000010000010000010011101

Part-B 00011001000000000001110101000000000000001000110000
00010011000010010100011010011001010101000101111010
100110000011011000111001

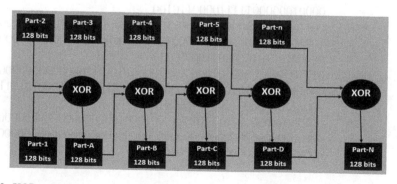

Fig. 2 XOR operation

Part-C 0101101100101101110011010000001000000000001000110000
0001001100001001010000110100110010101010001011111010
10011000001110110001110011

Part-D 0011111110111110001111111000100111100010001010010011
1001000111001001000001010110101101111110000000001000
001011010111111011011100

Part-1 0001000001000000010111100000000001111101010010100011
0100101000000000001000001101111101100000011011011
10001111100111100101100110

Step-9 Now append the bits of Part-1 first (Step-8) then Part-A, Part-B, Part-C and Part-D. After that leave the first two bits and last two bits to make a group of six bits and apply final substitution operation on each group of six bits according to the given final substitution Table 2 **00**011110011
11000011111000000000001111100100010111110111101111100000000
01110000101101001111100011101100010000001110011110111111000
00111000111100001111101110001110111100000100101000011011
11010001010010001010010011011011000010101000000000100000100
00010011011100001110100000000001000100100000000000000000001
11111000010101111110100100001101100000010110000110111001101
11111110010110101000101110101011101100101100010111000001001
00000000110010000000001001000000010011111110111111011100100
00011000110000100000110101000011111010000011011011100101000
101111100011100001110101111101100011101111010011110010110
01100100101000000110010011001001111101101001111000

Step-10 After final substitution place the first and last two bits at their same location and arrange the bits into group of four bits and convert that into its hexadecimal number which is the final encrypted key. **1E787C007C8 BEF780385A7C762073DF071E1F71DE0CA1DE8A4526D854020826 E1D004480000FC57F486C0B0DCBFCB51757658B82006402404FEF DC831841A87D06DCA5F1C3AF63BD3CACC940C993ED3C**

6 Conclusions

This Paper presents a secure key transmission algorithm to enhance security, data integrity and confidentiality. This algorithm can be particularly used for private key distribution. This algorithm uses MD5 for message digest or hash value. The proposed algorithm used permutation and substitution table which were not previously defined to make it more secure. It is more secure than some previously defined algorithm by number of reasons as it contains 512 bits key. We apply here complex operation of permutations, substitution, Gray codes and XOR operation to make it more secure. Along with sender side key transmission algorithm this paper also explains the receiver side algorithm to recover the key. As compared to previously known algorithm [7] this techniques is showing more secure way to transfer key, to maintain integrity and have data confidentiality.

References

1. Prashanti. G, Deepthi. S, Sandhya Rani. K, "A Novel Approach for Data Encryption Standard Algorithm", International Journal of Engineering and Advanced Technology (IJEAT), Volume-2, Issue-5, June 2013.
2. T. J. Watson, Don B. Johnson, Stephen M. Matyas, "Triple DES Cipher Block Chaining with Output Feedback Masking on Coppersmith", IBM Journal of Research and Development, Volume 40, Number 2, 1996 pp. 253–261.
3. Ritu Pahal Vikas kumar, "Efficient Implementation of AES", International Journal of Advanced Research in Computer Science and Software Engineering, Volume 3, Issue 7, July 2013.
4. Ajit Singh and Rimple Gilhotra,International Journal of Network Security & Its Applications (IJNSA), Vol.3, No.3, May 2011. doi:10.5121/ijnsa.2011.330558.
5. R. Velumadhava Rao, K Selvamani, R Elakkiya, Advanced Computing: An International Journal (ACIJ), Vol.3, No.6, November 2012 doi:10.5121/acij.2012.361083.
6. S.G. Gino Sophia, K. Padmaveni, Linda Joseph, International Journal of Engineering Research and Applications (IJERA) ISSN: 2248–9622 www.ijera.com Vol. 3, Issue 1, January -February 2013, pp.1492–1496 1492.
7. Vaibhav Sharma, Kamal Kumar Gola, Gulista Khan, Rahul Rathore, 2015 Second International Conference on Advances in Computing and Communication Engineering, Key Sharing Technique To Preserve Integrity And Confidentiality, 978-1-4799-1734-1/15 © 2015 IEEE, doi:10.1109/ICACCE.2015.48.

A Comparative Study of Different Approaches for the Speaker Recognition

Kanaka Durga Returi, Vaka Murali Mohan
and Praveen Kumar Lagisetty

Abstract A Comparative Study of different Approaches for the Speaker Recognition is presented in this paper. In this study speaker speech signal is normalized. This normalized signal for the different function are tested with some of the parameters and compared with the original signal. Some of them are tested in this study those are Hamming function; Gaussian function; Blackman Harris function; Bertlett Hanning function; Chebyshev function; Kaiser function; Hann function and Parzen function. All these functions are tested are compared in this paper.

Keywords Speaker recognition · Signal · Normalization

1 Introduction

Speaker recognition is an emerging field and has been studied from several decades. This is the process to recognize a person based on the information incorporated in the signals of the speech. Biometric based authentication measures individuals' unique physical or behavioral characteristics. The fingerprints and scans of the retina are additional dependable ways of user identification. Biometric authentication has some key advantages over knowledge and token based authentication techniques. Now a days in many applications speaker recognition is utilized as biometric for differentiate the information services on computers. Speaker recognition recommends the capability to restore or augment the individual recognition numbers and passwords by means of impressions that cannot be stolen or misplaced.

K.D. Returi
Department of Computer Science & Engineering, GITAM Institute of Technology, GITAM University, Visakhapatnam, AP, India

V.M. Mohan (✉) · P.K. Lagisetty
Department of Computer Science & Engineering, TRR College of Engineering, Inole, Patancheru, Hyderabad, TS, India
e-mail: murali_vaka@yahoo.com; vakamuralimohan@gmail.com

© Springer India 2016
S.C. Satapathy et al. (eds.), *Information Systems Design and Intelligent Applications*, Advances in Intelligent Systems and Computing 433,
DOI 10.1007/978-81-322-2755-7_62

599

Speaker recognition is a popular technology which has applicability in various fields like tele-banking, telephone shopping, voice dialing, services related to database access, information services, security control over confidential areas, voice mail, remote access to computers, forensic applications, etc. The advancement of automatic speaker recognition technology is one of the most important fields that combine the areas like speech science and pattern recognition. Human beings have been trying to allow the computer to recognize natural speech from the time of invention of a computer. Speaker recognition technology brings together the subjects like computer science, voice linguistics, signal processing, and intelligent systems. It is an important subject not only in the field of research but also as a useful application. Especially in day to day life, speaker and speech recognition systems are being used very commonly.

Speaker recognition provides a convenient means of establishing the identity of a person based on his voice. An important scientifical issue is a suitable and user pleasant interface for Human computer interaction. The widespread computer input interface is key board or mouse and a visual display for output. It is usual for humans to imagine such type of interface with computers. Speech recognition systems permit ordinary people to speak to the computer to retrieve information. The automatic recognition of speaker and speech recognition are very strongly related. As the goal of speech recognition system is to recognize the spoken words, on the other hand the goal of speaker recognition is to identify personality of the speaker through extraction, recognition and characterization contained in speech signal.

The most important communication technique among human beings is speech; at the time of development of the computer, people are demanding to allow computer to recognize the normal speech. Speaker recognition is a technique which is having close associations among computer science, intelligence systems, voice linguistics, signal processing, etc. Now a days it is understandable with the aim of speakers that can be recognized from their voices during real time system point of view. By focusing on the technology, so many companies are facilitating today's speaker recognition as well as speaker identification systems which can be find out modern implementations and forecast the future developments.

2 Literature Review

Speaker recognition is one of the most speedily developing areas now-a-days. A lot of work has been done previously in the same area by several models and methods some of them are Tobias Herbig et al. [1] presented speaker identification and speech recognition models to achieve speech characteristics to obtain good recognition rate. Marco et al. [2] reported the relevance of microphone array process and system robustness to improve the performance by using hidden Markov models. Selami and Bilginer [3] developed a common vector approach for speaker recognition with Fisher's linear discriminant analysis and Gaussian mixture models. Shung Yung Lung [4] presented a system for the recognition of speaker by using wavelet transform and vector of the expression by kernel analysis. Shung [5] developed a thresholding method for noise reduction by using Gradient-based adaptive learning algorithms. Avci and Avci [6] reported a speaker identification system by using speech signal. Avci [7] reported the combination of the feature extraction and an approach with wavelet packet analysis of the speech signal. Hamid et al. [8] presented speech representations by using wavelet filter-banks with discrete wavelet transform and the performance is analyzed with neural networks. Leandro et al. [9] reported a genetic algorithm for the representation of the speech by using wavelet decomposition and classification. Muhammad and Asif Zahoor [10] proposed a new and innovative idea based on fractional signal processing through least mean square algorithm for modeling and design of the system. Boashash et al. [11] described a FetMov detection by utilizing the time frequency signal processing approach. It also provided measurements for significant challenges of the signal processing. Carvalho et al. [12] presented a method for the identification of the errors through various signal processing approaches.

3 Speaker Recognition System

Speaker recognition is a method of mechanically recognizing the individuals depending upon the information integrated in speech signals. In general speaker recognition system deals with identification of the speaker at the end. Speech recognition is different from language recognition since these concepts deal with recognition of speech (i.e. the words that are spoken) and recognition of language deals with recognizing the language in which the words or sentences are spoken. The final goal of speaker identification has two steps: the essential goal is to recognize the speaker irrespective of what is being said. The potential duty of the recognition is to fix whether the speech belongs to a maintained individual. Generally speaker recognition is the technique of automatically recognizing a speaker. The speaker is established and found the quality of the speech wave in its order. The task is to confirm the uniqueness of the person from his voice. This procedure involves only binary choice of the claimed characteristics (Fig. 1).

Fig. 1 Generic speaker recognition model

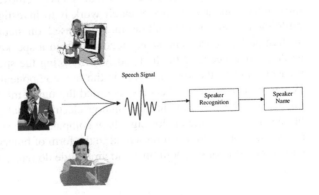

3.1 Real Time Applications

This technique is applicable in Military Security Systems, Forensic Services, Voice Dialing and Mail, Tele Banking, Access and Services of the Database and Information, Computers Security through Remote Access and Control for confidential information areas. Nowadays, this approach is being utilized for Telephone Shopping. The stored data in computers are protected in different ways; coding and decoding by utilized text or voice password for protecting data files. Entire computer is protected directly at the booting stage. Speaker recognition system has many potential applications such as

Security Control Speaker recognition systems can be used for law enforcement. They can help identify suspects. Some security applications employ sophisticated techniques to check whether a speaker is present where that particular speaker is supposed to be.

Telephone Banking Access to bank accounts may be voice controlled. Such systems may want to verify whether the authorized person is trying to access the accounts, private and personal details. Intelligent machines may be programmed to adapt and respond to the user.

Information retrieval systems Participants in conferences or meetings may be identified by special machine technology. Automatic transcriptions containing a record of who said what can be also obtained from large quantities of audio information if such machine technology is used in conjunction with continuous recognition of the speech systems.

Speech and Gender recognition system Speaker recognition systems are usefully employed by speech recognition systems. Many speaker independent speech recognizers are already using gender recognition system for improving the performance.

4 Methodology

In the speaker recognition system, the captured signal is first edited through the pre-processing stage, which consists of silence removal and normalization. The most important idea of this research work is to investigate the various options in building a speaker recognition method based on neural network for obtaining system performance. The signal recorded from a speaker is an analog signal. The analog signal needs to be digitized, while storing the speech to a computer. So the measurement of the speech signal is the first component of the speech processing. When the voices of speakers are recorded through a microphone, the analog signal pressurizes the air after that the analog electric signal passes through the microphone. To process the analog signals in computers, the signals need to be converted into digital form and it is represented in the form of binary digits 0 and 1. An analog signal is continuous in both time and amplitude domains, where as a digital signal is

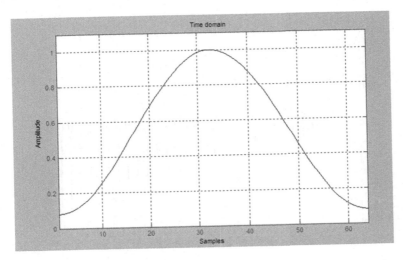

Fig. 2 Original signal in time domain

discrete in both domains. MATLAB was used to develop a model system in this paper. The original signal of "female" in time domain is shown in Fig. 2 and this figure shows the original signal function in terms of time domain.

The speech signals contain silence or noise areas. The silence signal contains no information and it is useless for identification of the speaker. So, keeping the silence signal increases the size of the processing signal and it takes more time and more space when retrieving the features. The actual speech segments of the signal part are utilized for recognition. For silence removal there are two ways namely physical way and processing way. If there is a small data base of speech signal physical way is feasible and time saving. In this normalization process the sampling function is symmetric and the Length of the function is 64. With this parameters several function are utilized for the normalization of the signal. Some of them are tested in this study those are Hamming function; Gaussian function; Blackman Harris function; Bertlett Hanning function; Chebyshev function; Kaiser function; Hann function and Parzen function. All these function s are tested are compared in this paper.

5 Results and Discussion

The signal function is Normalized with the above functions and its results are shown below.

Figure 3 shows that the Hamming function is utilized for the normalization within the frequency domain and It also given that the Leakage Factor is 0.03 %; Relative Sidelobe Attenuation is −42.5 dB; Mainlobe width (−3 dB): 0.039.

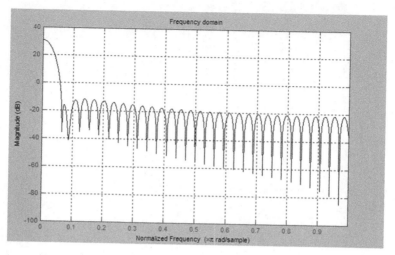

Fig. 3 Hamming function

Figure 4 shows that the Gaussian Function is utilized for the normalization within the frequency domain and It also given that the Leakage Factor is 0.01 %; Relative Sidelobe Attenuation is −44.1 dB; Mainlobe width (−3 dB): 0.043.

Figure 5 shows that the Blackman Harris Function is utilized for the normalization within the frequency domain and It also given that the Leakage Factor is 0 %; Relative Sidelobe Attenuation is −92 dB; Mainlobe width (−3 dB): 0.0586.

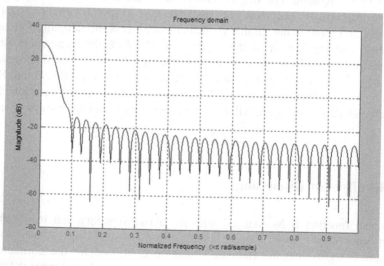

Fig. 4 Gaussian function

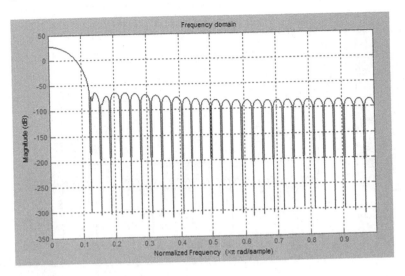

Fig. 5 Blackman Harris function

Figure 6 shows that the Bertlett Hanning is utilized for the normalization within the frequency domain and It also given that the Leakage Factor is 0.03 %; Relative Sidelobe Attenuation is −35.9 dB; Mainlobe width (−3 dB): 0.04297

Figure 7 shows that the Chebyshev Function is utilized for the normalization within the frequency domain and It also given that the Leakage Factor is 0 %; Relative Sidelobe Attenuation is −100 dB; Mainlobe width (−3 dB): 0.0547.

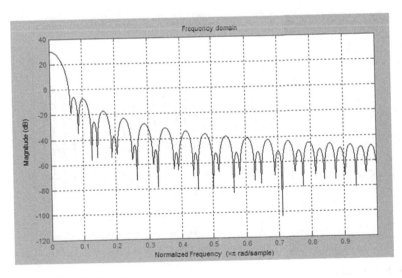

Fig. 6 Bertlett Hanning function

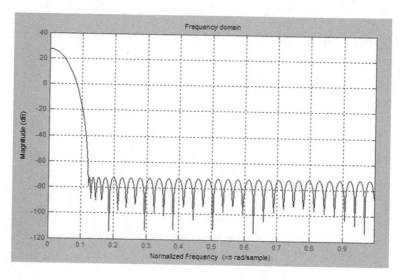

Fig. 7 Chebyshev function

Figure 8 shows that the Kaiser Function is utilized for the normalization within the frequency domain and It also given that the Leakage Factor is 8.36 %; Relative Sidelobe Attenuation is −13.6 dB; Mainlobe width (−3 dB): 0.0273 with β = 0.5.

Figure 9 shows that the Hann Function is utilized for the normalization within the frequency domain and It also given that the Hann: Leakage Factor is 0.05 %; Relative Sidelobe Attenuation is −31.5 dB; Mainlobe width (−3 dB): 0.04297.

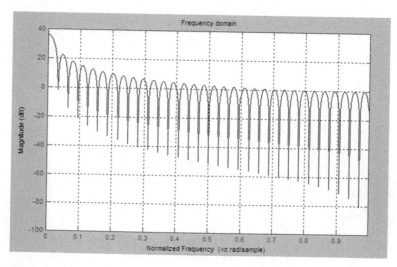

Fig. 8 Kaiser function

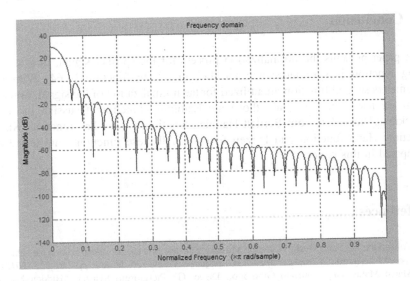

Fig. 9 Hann function

Figure 10 shows that the Parzen Function is utilized for the normalization within the frequency domain and It also given that the Leakage Factor is 0 %; Relative Sidelobe Attenuation is −53.1 dB; Mainlobe width (−3 dB): 0.054688.

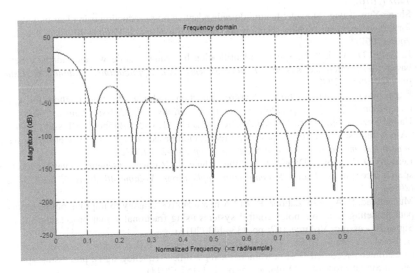

Fig. 10 Parzen function

6 Conclusion

This paper presents the normalization process for the speaker signal, in this sampling function is symmetric and the Length of the function is 64. With this parameters several function are utilized for the normalization of the signal. Some of them are tested in this study those are Hamming function; Gaussian function; Blackman Harris function; Bertlett Hanning function; Chebyshev function; Kaiser function; Hann function and Parzen function. All these function s are tested are compared in this paper.

References

1. Tobias Herbig., Franz Gerl., Wolfgang Minker, "Self-learning speaker identification for enhanced speech recognition" *Computer Speech & Language, Volume 26, Issue 3, pp 210–227 (2012)*.
2. Marco Matassoni., Maurizio Omologo., Diego, G., Piergiorgio Svaizer, "Hidden Markov model training with contaminated speech material for distant-talking speech recognition" *Computer Speech & Language, Volume16, Issue 2, pp 205–223 (2002)*.
3. Selami Sadıç., M. Bilginer Gülmezoğlu, "Common vector approach and its combination with GMM for text-independent speaker recognition" *Expert Systems with Applications, Volume 38, Issue 9, pp 11394–11400 (2011)*.
4. Shung-Yung Lung, "Improved wavelet feature extraction using kernel analysis for text independent speaker recognition" *Digital Signal Processing, Volume 20, Issue 5, pp 1400–1407 (2010)*.
5. Shung Yung Lung "Wavelet feature domain adaptive noise reduction using learning algorithm for text-independent speaker recognition" *Pattern Recognition, Volume 40, Issue 9, pp 2603–2606 (2007)*.
6. E. Avci., D. Avci, "The speaker identification by using genetic wavelet adaptive network based fuzzy inference system" *Expert Systems with Applications, Volume 36, Issue 6, pp 9928–9940 (2009)*.
7. Engin Avci, "A new optimum feature extraction and classification method for speaker recognition: GWPNN" *Expert Systems with Applications, Vol. 32, Issue 2, pp 485–498 (2007)*.
8. Hamid Reza Tohidypour., Seyyed Ali Seyyedsalehi., Hossein Behbood., Hossein Roshandel "A new representation for speech frame recognition based on redundant wavelet filter banks" *Speech Communication, Volume 54, Issue 2, pp 256–271 (2012)*.
9. Leandro D. Vignolo., Diego H. Milone., Hugo L. Rufiner, "Genetic wavelet packets for speech recognition" *Expert Systems with Applications, Volume 40, Issue 6, pp 2350–2359 (2013)*.
10. Muhammad, S, A., Asif Zahoor, R, M "A new adaptive strategy to improve online secondary path modeling in active noise control systems using fractional signal processing approach" *Signal Processing, Volume 107, pp 433–443 (2015)*.
11. B. Boashash., M.S. Khlif., T. Ben-Jabeur., C.E. East., P.B. Colditz "Passive detection of accelerometer-recorded fetal movements using a time–frequency signal processing approach" *Digital Signal Processing, Volume 25, pp 134–155 (2014)*.
12. Vítor Carvalho., Michael Belsley., Rosa M. Vasconcelos., Filomena O. Soares "Yarn periodical errors determination using three signal processing approaches" *Digital Signal Processing, Volume 23, Issue 5, pp 1427–1440 (2013)*.

A Time Efficient Leaf Rust Disease Detection Technique of Wheat Leaf Images Using Pearson Correlation Coefficient and Rough Fuzzy C-Means

Dhiman Mondal and Dipak Kumar Kole

Abstract In agricultural sector diagnosis of crop disease is an important issue, since it has a marked influence on the production of agriculture of a nation. It is very essential to diagnose disease in an early stage to control them and to reduce crop losses. This paper presents a time efficient proposed technique to detect the presence of leaf rust disease in wheat leaf using image processing, rough set and fuzzy c-means. The proposed technique is experimented on one hundred standard diseased and non-diseased wheat leaf images and achieved 95 and 94 % success rate respectively depending on most three dominated features and single most dominated feature, Ratio of Infected Leaf Area (RILA). The three most dominated features and single most dominated feature are selected out of ten features by the Pearson correlation coefficient. A significant point of the proposed method is that all the features are converted into size invariant features.

Keywords Rough-fuzzy-C-Means · Pearson correlation coefficient · RILA

1 Introduction

The world's most widely cultivated and important crop is wheat. Since Green Revolution, India achieved a remarkable position in the production of wheat. In wheat production, India holds third position in the world and has produced near about 95,000 Metric Ton in the year 2014 [1]. Generally the experts in the area of agriculture use a naked eye method to identify the diseases and takes decision by detecting the changes in leaf color and other related symptoms like shape change,

D. Mondal (✉) · D.K. Kole
Department of CSE, Jalpaiguri Government Engineering College,
Jalpaiguri, India
e-mail: mondal.dhiman@gmail.com

D.K. Kole
e-mail: dipak.kole@gmail.com

© Springer India 2016
S.C. Satapathy et al. (eds.), *Information Systems Design and Intelligent Applications*, Advances in Intelligent Systems and Computing 433,
DOI 10.1007/978-81-322-2755-7_63

609

texture change etc. But this method is time consuming, requires a lots of efforts, impractical for large fields. Red spot or fungal disease detection using eigen feature regularization and extraction has been addressed by Gurjar and Gulhane [2] with a success rate of 90 %. Meunkaewjinda et al. have discussed grape leaf disease diagnosis based on color extraction [3]. The concept of ANN is used to classify the diseased leaf image using color based, texture based and shape based features in [4, 5]. In [6], Jaware et al. developed a Fast method for detection and classification of plant diseases based on RGB image by creating color transformation structure of RGB leaf images. Different detection and diagnosis of plant leaf disease using integrated image processing approaches are discussed in [7]. The concept of color based, texture based and shape based features are used to extract features and to classify leaf diseases of wheat, grape in [8–10]. Detection of downy mildew disease detection of grape leaves using the concept of fuzzy set theory has been suggested by Kole et al. [11].

In this paper, a time efficient method is discussed for the detection of leaf rust disease in wheat leaf. Initially ten features are chosen depending leaf rust diseased wheat leaf images. Then all ten features are transformed into size invariant features by mapping individual feature range into a common normalized feature range [0, 1]. After that a set of dominated features is obtained based on the Pearson correlation coefficient method. The proposed technique is experimented on one hundred standards diseased and non-diseased wheat leaf images with the dominated features set containing three features which gives us 95 % success rate taking 302.768 s execution time. The same experiment is done using single most dominated, RILA feature which gives 94 % success rate talking 33.21 s execution time.

2 Preliminaries

2.1 Rust Diseases of Wheat

The most common wheat diseases are leaf blight, common root rot, cottony snow mold, foot rot, strawbreaker, Leaf (brown) rust, Powdery mildew etc. Rust disease in wheat is one of the most devastating diseases. Out of the three rust diseases in wheat the most common is called leaf (brown) rust. Its symptoms are isolated uredinia on upper leaf surface and rarely on leaf sheaths. The other two kind of rust disease are stripe (yellow) rust and stem (black) rust.

2.2 Features Used in the Proposed Techniques

For features extraction based on texture of the image, a gray level co-occurrence matrix is used. The co-occurrence matrix $C(i, j)$ counts the number of co-occurrence of pixels with gray-levels i and j respectively, at a given distance d. The matrix is given by:

$$C(i,j) = cord \begin{cases} ((x_1,y_1),(x_2,y_2)) \in XY \times XY \\ f(x_1,y_1) = i, f(x_2,y_2) = j, \\ (x_2,y_2) = (x_1,y_1) + (d\cos\theta, d\sin\theta) \\ 0 < i,j < N \end{cases} \quad (1)$$

where d is the distance defined in polar coordinates (d, θ) with discrete length and orientation. θ takes values 0, 45, 90, 135, 180, 225, 270 and 315. $Cord\{\}$ represents the number of elements presents in the set.

The features considered from the gray-level image are Mean, Standard Deviation, and Entropy.

RILA (Ratio of Infected Leaf Area) Feature of Wheat Leaf Image The amount of infected area of each leaf image has been calculated by using segmentation and R, G, B plane of each individual image, and have been described in Sect. 3.1.1.

3 Proposed Technique

The proposed technique consists of two phases, which is shown in the Fig. 1. First phase is the training phase and the second phase is the testing phase. The first two stages of each phase are common, which are Image Acquisition and Pre-Processing. In the first phase, subsequent stages are Feature Extraction, Dominating Feature Selection and Supervised Learning.

Fig. 1 Flow chart of proposed method

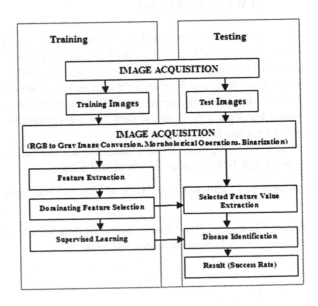

Table 1 Feature number with respective feature name

Feature no.	Feature name	Feature no.	Feature name
1	R-plane mean	6	G-plane entropy
2	R-plane standard deviation	7	B-plane mean
3	R-plane entropy	8	B-plane standard deviation
4	G-plane mean	9	B-plane entropy
5	G-plane standard deviation	10	RILA

The last three stages of second phase are Selected Feature Values Extraction, Disease Detection and the success rate as a Result. Different stages are being discussed in the following subsections. In Image acquisition stage, the images are captured by using high resolution color camera with a fixed background to avoid the background objects and other noises. And in the pre-processing stage, all the image background is converted to white and some morphological operations are applied on the images before converting them into binary images.

3.1 Feature Extraction

Ten feature values, as mentioned in Table 1, are extracted from each leaf image of the training samples. Out of 10 features, 9 features are calculated as mentioned in Sect. 2.2. The last feature RILA is calculated as follows.

3.1.1 The RILA (Ratio of Infected Leaf Area) Feature Calculation of Wheat Leaf Image

To calculate the value of RILA feature, at first, the actual leaf area of the RGB image is calculated by ignoring the background pixels. Then from the binary image of the RGB image, background pixels are searched and the corresponding pixel value in the R, G and B-plane of the RGB image are changed to a value 255. After that, in each plane of the resultant RGB image, those pixels are counted having values less than 255. And the total count is divided by three times the actual leaf area to get the RILA feature value.

The following Algorithm 1 describes the method to calculate the most dominated RILA feature value of wheat leaf image.

Algorithm 1: Calculating RILA of Wheat Leaf Image

Input: Binary and RGB image corresponding to wheat leaf image.
Output: RILA feature of wheat leaf image

Step1: Calculate the actual leaf area from RGB leaf image.

Step2: In the binary image check each pixel whether it is 1 (image pixel) or 0 (background pixel).

Step3: If it is background pixel then initialize value of that pixel to 255 corresponding to each plane of the RGB image, i.e., R-plane, G-plane and B-plane, otherwise do not change the pixel value.

Step4: Calculate the total number of pixel having values less than 255 in each plane of the RGB image.

Step5: Divide the total number of pixel having value less than 255 as calculated in step4 by three times the actual area of leaf image as calculated in *Step1*. And store the ratio of the infected leaf area

3.2 Pearson Correlation Coefficient Based Dominated Feature Set Selection

The feature-class correlation value in Pearson Correlation Coefficient ranges from −1 to +1. The negative correlation indicates that high value of one feature is associated with low values of class label and vice versa. The feature-class correlation value (FCCV) or degree of dependence indicates the following status:

$0.9 \leq |FCCV| \leq 1$ Correlation is high

$0.8 \leq |FCCV| < 0.9$ Correlation is moderate

$0.51 \leq |FCCV| < 0.8$ Correlation is low

$|FCCV| < 0.5$ Correlation is very low

where, $|FCCV|$ indicates the absolute value of FCCV.

After calculating feature-class correlation coefficient corresponding to each of the 10 features, those features having feature-class correlation absolute value greater than 0.8, are selected as the dominating features. And the procedure selects three features: G-plane mean, B-plane mean and RILA as the dominating features.

3.3 Detection of Leaf Rust Disease

In this stage, a list of diseased and non-diseased image of wheat leaf is taken as input. Then the feature values corresponding to dominating feature set is calculated. Then a feature matrix of size $n \times m$ is generated, where n is the number of input images and m is the number of dominating features and normalized the feature matrix in [0, 1] range.

3.3.1 Detection of Leaf Rust Disease Using the Single Most Dominated Ratio of Infected Leaf Area (RILA) Feature

Since RILA is the most dominating feature, an experiment has been done to detect or classify the input images based on this feature. In this experiment first, a training data set, consisting of 50 diseased and non-diseased wheat leaf images are taken into account. Then RILA feature value of wheat leaf images is calculated to decide the threshold value of RILA feature through experiment. After obtaining the threshold value of RILA feature, training data set along with another 50 wheat leaf images consisting of diseased and non-diseased images are considered for testing. As non-disease leaf image should have minimum or no infected leaf area, so the RILA feature value of disease free images will be less with respect to diseased images. Here those images whose RILA feature value is less than the pre determined threshold value is considered as disease free and those images have greater than or equal to the pre determined threshold value are considered as diseased images. The range of feature this value is 0–1. The curve between RILA versus success rate is plotted from 0 to 1 stepping 0.05 intervals along the RILA value. A RILA value is selected as Best Threshold Value of RILA feature, for which the success rate is maximum. In this experiment on wheat leaf images, a value, 0.675 is obtained as the best threshold value of RILA feature.

3.3.2 Detection of Leaf Rust Disease Using Dominating Feature Set and Rough Fuzzy C-Means

In the training phase of this method, a training data set consisting of 50 diseased and non-diseased wheat leaf images are taken into account. Taking the feature values of training data corresponding to the selected dominating feature set, a feature matrix is generated. The feature matrix is transformed into the normalized feature matrix in the range [0, 1], as because the range of individual feature is different from others. Then the normalized feature matrix and the two initial cluster centers chosen randomly are used as input to the proposed Algorithm 2 based on rough fuzzy C-means algorithm [12]. At the end of the training phase, the final updated cluster centre values are considered as standard cluster centre and used in the testing phase for detecting leaf rust disease in wheat leaf images. In the testing phase, training data set of 50 wheat leaf images and another 50 diseased and non-diseased wheat leaf images are considered. These 100 wheat leaf images and the standard cluster centre are used to detect the leaf rust disease in wheat leaf images.

Proposed Algorithm for Detection of Leaf Rust Disease in Wheat Leaf

Fuzzy C-means, as proposed in [13, 14], partitions n patterns $\{x_k\}$ into c clusters allowing partial membership ($u_{ik} \in [0, 1]$) of patterns to clusters. The membership function u_{ik} is defined in the Eq. (2) and each cluster center is updated according to the Eq. (3) considering each cluster as rough set [15].

$$u_{ik} = \frac{1}{\sum_{j=1}^{c} \left(\frac{d_{ik}}{d_{jk}}\right)^{\frac{2}{m-1}}} \quad (2)$$

$$v_i = \begin{cases} w_{low} \dfrac{\sum_{x_k \in \underline{B}X_i} u_{ik}^m x_k}{\sum_{x_k \in \underline{B}X_i} u_{ik}^m}, & if \ \underline{B}X_i \neq \emptyset \wedge BNX_i \neq \emptyset \\[2ex] \dfrac{\sum_{x_k \in BNX_i} u_{ik}^m x_k}{\sum_{x_k \in BNX_i} u_{ik}^m}, & if \ \underline{B}X_i = \emptyset \wedge BNX_i \neq \emptyset \\[2ex] \dfrac{\sum_{x_k \in \underline{B}X_i} u_{ik}^m x_k}{\sum_{x_k \in \underline{B}X_i} u_{ik}^m}, & otherwise \end{cases} \quad (3)$$

Algorithm 2: Detection of Leaf Rust disease in Wheat Leaf

Input : Feature values corresponding to a set of *n* number of wheat leaf images to detect Leaf Rust disease
Output: Identification of diseased and non-diseased wheat leaf images

Step1: **for** l=1:MAX_ITERATION
Step2:   Assign Initial Cluster Centroids v_i for C clusters
Step3:   Compute membership grade (u_{ik}) by using equation (2) for C clusters and N patterns
Step4:   **for** each object (pattern) x_k in the dataset **do**
Step5: $u_{ik} \leftarrow$ the maximum membership grade for pattern k
Step6:   **for** each $j \in \{1,2,....,C\}$ $j{\neq}i$ **do**
Step7:   **if** $d_{ik} - d_{jk} < threshold$ **then**
Step8:   Assign x_k to both upper approximations $x_k \in \bar{B}X_p$, $x_k \in \bar{B}X_j$ and x_k
                       can not be a member of any lower approximation
Step9:   **else** assign x_k to the lower approximation $x_k \in \underline{B}X_i$
Step10:   **end if**
Step11:   **end for**
Step12:   **end for**
Step13:   Find out the objects (patterns) in the boundary region
Step14:   **If** boundary is not empty **then**
Step15:   **for** each object in the boundary do the following
Step16:   find out the maximum membership grade and associated cluster number (z)
Step17:   Add the object to the lower approximation of z.
Step18:   Remove the object from upper approximation of that cluster
Step19:   **end for**
Step20:   **end if**
Step21:   compute new cluster centroids v_i according to equation (3)
Step22: **end for**
Step23: Output $\underline{B}X_p$, $\bar{B}X_i$ for each cluster $i \in \{1,2,....,C\}$

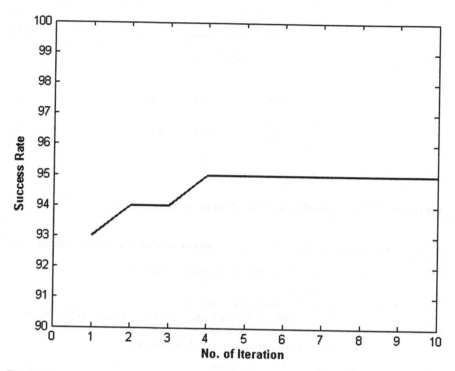

Fig. 2 Success rate versus number of iterations

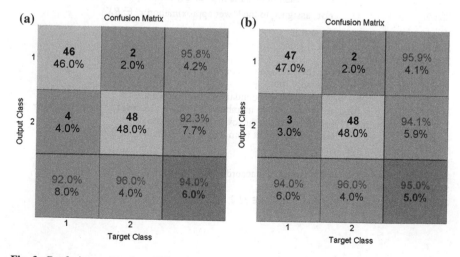

Fig. 3 Confusion matrix for **a** RILA feature based. **b** Dominated feature set based classification

Table 2 **a** RILA feature based classification, **b** dominated feature set based classification

Technique used	No. of feature (feature name)	Success rate (%)	Execution time (s)
a	1 (RILA)	94	33.206
b	3 (G-plane mean, B-plane mean, RILA)	95	302.756

4 Experimental Results

The experiment is done on the system having the configuration, Intel(R) Core(TM) i5-3470 CPU @3.20 GHz, 3.19 GHz 2 GB RAM, windows XP SP2, MATLAB R2009b. During the experiment, we first find out feature-class correlation values corresponding to each feature. Correlation Value (Feature No.): 0.834475 (7), 0.813863 (4), 0.642705 (1), 0.024001 (8), −0.07812 (5), −0.18496 (2), −0.66851 (9), −0.71131 (6), −0.7641 (3), −0.86406 (10). The most dominated and the set of dominated features with respect to absolute threshold value 0.8 are 10 and {4, 7, 10}. The experiment has been done on 100 standard diseased and non-diseased wheat leaf images. In case of most dominated feature, RILA, the best threshold value 0.675 is determined. The RILA feature based detection technique gives 94 % success rate in 33.206 s execution time. The dominated feature set using Rough Fuzzy C-Means technique gives 95 % success rate in 302.756 s execution time.

5 Conclusion

This paper presents time efficient technique to detect the presence of leaf rust disease in wheat leaf. The Single most dominating feature based detection technique gives 94 % success rate in 33.206 s execution time, whereas three most dominating feature set based rough fuzzy C-means technique gives 95 % success rate in 302.756 s execution time. Although, the success rate of second method is increased by 1 % but the time taken by the second method is much higher than the first one (Figs. 2 and 3; Table 2).

Acknowledgments The authors would like to thank Dr. Amitava Ghosh, ex-Economic Botanist IX, Agriculture dept., Govt. of West Bengal, for providing wheat leaves images with scientist's comments.

References

1. http://www.indexmundi.com.
2. Ajay A. Gurjar, Viraj A. Gulhane. "Disease Detection On Cotton Leaves by Eigen feature Regularization and Extraction Technique", IJECSCSE, Vol.1,No. 1, pp 1–4, (2012).

3. A. Meunkaewjinda, P. Kumsawat, K. Attakitmongcol and Sri kaew. "Grape leaf disease detection from color imagery using hybrid intelligent system", 5th. International Conference on Electrical Engineering/Electronics, Computer, Telecommunications and Information Technology, Vol.1, Krabi, pp. 513–516,(2008).
4. Libo Liu, Guomin Zhou. "Extraction of the Rice Leaf Disease Image Based on BP Neural Network", CiSE 2009, pp. 1–3, (2009).
5. K. Muthukannan, P. Latha, R. Pon Selvi and P. Nisha. "Classification of Diseased Plant Leaves using Neural Network Algorithms", ARPN Journal of Engineering and Applied Sciences, Vol. 10, No. 4, pp. 1913–1919, (March 2015).
6. Tushar H Jaware, Ravindra D Badgujar and Prashant G Patil. "Crop disease detection using image segmentation", Proceedings of NCACC'12, pp:190–194, (2012).
7. Diptesh Majumdar & et al, "Review: Detection & Diagnosis of Plant Leaf Disease Using Integrated Image Processing Approach", International Journal of Computer Engineering and Applications, Volume VI, Issue-III, pp. 1–16,(June 2014).
8. Yuan Tian; Chunjiang Zhao; Shenglian Lu; Xinyu Guo, "SVM-based Multiple Classifier System for recognition of wheat leaf diseases," in World Automation Congress (WAC), 2012, vol., no., pp.189–193, (24–28 June 2012).
9. Mr. Hrishikesh, P. Kanjalkar, Prof. S.S. Lokhande. "Feature Extraction of Leaf Diseases", International Journal, IJARCET, Volume 3, Issue 1, pp. 1502–1505, (January 2014).
10. Diptesh Majumdar & et al, "Application of Fuzzy C-Means Clustering Method to Classify Wheat Leaf Images based on the presence of rust disease", FICTA 2014, pp. 277–284, (November 2014).
11. Dipak K. Kole & et al, "Detection of Downy Mildew Disease present in the Grape Leaves based on Fuzzy Set theory", ICACNI 2014, Volume 1, Springer Smart Innovation, Systems and Technologies Volume 27, 2014, pp 377–384, (June 2014).
12. Rafael Falcon & et al, : "Rough Clustering with Partial Supervision" In: "Rough Set Theory: A True Landmark in Data Analysis", Studies in Computational Intelligence, Vol 174, Springer-Verlag Berlin Heidelberg, pp. 137–161, (2009).
13. Z. Pawlak, Rough Sets: Theoretical Aspects of Reasoning About Data, Kluwer Academic Publishers, Boston, (1991).
14. J.C. Bezdek, Pattern Recognition with Fuzzy Objective Function Algorithms, Plenum Press, New York (1981).
15. G. Peters.: Some refinements of rough k-means clustering, Pattern Recognition, Volume 39, Issue 8, pp. 1481–1491, (August 2006).

Task Scheduling Algorithms with Multiple Factor in Cloud Computing Environment

Nidhi Bansal, Amit Awasthi and Shruti Bansal

Abstract Optimized task scheduling concepts can meet user requirements efficiently by using priority concepts. Increasing the resource utilization and reducing the cost, both are compulsory factors to be compromise in task scheduling algorithms of cloud computation for executing many tasks. With updating the technology many new features in cloud computing introduced such as fault tolerance, high resource utilization, expandability, flexibility, reduced overhead for users, reduced cost, required services etc., this paper discussed task scheduling algorithms based on priority for virtual machines and tasks. This algorithm performs good results with balance the load, but it's not effective with cost performance. Secondly comparative study also has been done in this paper between various scheduling algorithms by CloudSim simulator.

Keywords Cloud computing · Task scheduling · Cost · Load balancing

1 Introduction

In these days all the application runs via virtual machines and resources are allocated to all virtual machines. All applications are independent, unique in technology and has no connection to each other. Resources are compromised on every event or activity performed with all individual units of products and service.

N. Bansal (✉)
Department of Computer Science and Engineering, Vidya College of Engineering, Meerut, India
e-mail: nidhi18jul@gmail.com

A. Awasthi
Department of Computer Science and Engineering NITTTR, Chandigarh, India
e-mail: awasthi.amit1989@gmail.com

S. Bansal
Department of Electronic & Communication, Shobhit University, Meerut, India
e-mail: shruti_mrt@rediffmail.com

© Springer India 2016
S.C. Satapathy et al. (eds.), *Information Systems Design and Intelligent Applications*, Advances in Intelligent Systems and Computing 433,
DOI 10.1007/978-81-322-2755-7_64

The concept of scheduling is performing very important role in cloud computing with requirement of users in the market. 'Mapping the tasks' is the basic concept of scheduling. This is a necessary condition in successful working of cloud as many factors must be examined for useful scheduling. The feasible resources should be appropriate for execution in task scheduling (Tables 1, 2, 3 and 4).

The mechanism of task scheduling can not only satisfy to the user, but also increase the utilization for resources [1]. Load balancing factor must be calculated to acquire more resource utilization. The process of load balancing is distributing the load between various nodes to enhance utilization of resource and task response time while also neglecting the condition where nodes are fully occupied while many other distinct nodes are free or performing for limited work. Load balancing assure that processors in the setup or all node in the rooted network connection does approximately the uniform amount of execution at any present of time.

To get the complete cost of every user's applications, all individual service of resources (like Processor cost, Internal memory used, Input/Output cost, etc.) need to be calculated. When the complete cost of all resources has been identified, factual cost and output dissection depend on it can be retrieving, related to all of the traditional concepts of scheduling. Traditional concept leading the use of absolute tasks for users and exceeds the overheads in applications of cloud computation. It may be true that any distinct tasks may not the reason of exceeding costs for

Table 1 FCFS Vs VM-Tree

Cloudlets	FCFS	VM_Tree
50	5668.944	5894.572
70	5512.491	5602.98
100	5486.416	5794.572

Table 2 FCFS Vs PSO

Cloudlets	FCFS	PSO
50	5668.944	2473.44
70	5512.491	2864.571
100	5486.416	2929.76

Table 3 QoS of FCFS

Cloudlets	FCFS	QoS
50	5668.944	4481.248
70	5512.491	4298.72
100	5486.416	4389.984

Table 4 FCFS Vs ABC

Cloudlets	FCFS	ABC
50	5668.944	2473.44
70	5512.491	2864.571
100	5486.416	2929.76

resources in traditional way. The result is that exceeds in estimate and raises the cost. To compete towards market, some organization has had to reduce the cost or prices of expansive items or products. But they have been capable to get huge mark-ups on less expensive tasks. Minimization in cost or price has raised marketing for individual units/item but degrades the complete strength and good mark-ups on specific tasks or product didn't equal the down in the confine output of expensive items [2].

However, Load Balancing and Allocation Cost are the primary issues for task scheduling algorithms of user's applications in cloud computation. Presently the complete cost and proper resource utilization of the scheduling issue has encouraged researchers to recommend multifarious cost related task scheduling algorithms. More advanced algorithm designed on these factors are introduced by innovators or researchers such as Activity Based Cost, Particle Swarm Optimization, DLA (Double Level Priority) and Balancing the Load [1–3, 6–13] etc.

This paper proposes an implementation view for scheduling techniques with simulated outcome in CloudSim3.0 simulator by taking the factor cost and balancing the load.

The remaining part of this research paper is implemented as followed: Sect. 2 explains/methodology. Section 3 derives the implementation by simulation and analysis. Section 4 shows the conclusion.

2 Various Task Schedulings

This research paper examined with traditional method i.e. First Come First Serve and optimized scheduling methods i.e. VMT, PSO, QoS, ABC. Brief introduction for every scheduling is described here:

Generally First Come First Serve algorithm is take it as by default scheduling to explain any traditional concept of scheduling because there is no priority for any parameter and its very simple in implementation without any error.

Now comes to optimized or prioritized scheduling concepts. First optimized method is explained a tree form hierarchy named Virtual Machine_Tree for proper running of input tasks with including the concept of priority for machines and for cloudlets/tasks also. Depth First Search is modified according to scheduling concept to get effective output [4].

Second optimization technique introduces an algorithm which relates the method of small position value (SPV). Particle Swarm Optimization (PSO) is used to reduce the execution cost for task scheduling in cloud computation [3].

QoS-driven is the third method used in this paper that proposed a concept with considering many task attributes like privileges of application, expectation, length of executed task and the time awaiting in series to execute and sort applications by the priority [5].

ABC (Activity Based Costing) is the fourth one that introduces an optimized method based on priority in terms of benefit for SP (service provider). The

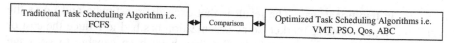

Fig. 1 Basic model for implementing task scheduling

traditional concept of task processing cannot fulfill the user's requirement. Activity-based method measures the every event or activity cost for all objects and the outcome is better than traditional method in Cloud Computation [2].

2.1 Methodology

Figure 1 At step first, all optimized methods and traditional scheduling algorithm have been compared using simulator CloudSim3.0. Comparison shows that optimized algorithms (priority concepts) are always perform better than the traditional methods. Load balancing and allocation cost parameter are calculated in these comparisons. At the final step, outcome of this research paper is very able to find an effective technique that executes or performs good to get more resource utilization and reduce cost.

3 Simulation

CloudSim3.0 simulator is used to simulate the all these task scheduling algorithms explained above. To compare the effective performance under many distinct parameters. An open environment is considered with two host node to implement the scheduling techniques with thirty autonomous tasks. It can be dynamically changed throughout the simulation. This simulation mainly focuses on the load balancing and cost factors between these scheduling methods.

To estimate the effective performance of scheduling methods, datacentres, virtual machines and many cloudlets based on user's requirement are created in simulator. Now time to schedule the tasks based on all entire scheduling methods for example, virtual machine tree [4], particle swarm optimization [3] and QoS-driven [5] activity based costing [2].

3.1 Performance Metrics

We have evaluated the scheduling techniques using allocation cost and load balancing metrics and compared with traditional scheduling algorithm. Parameters are:

(1) Size of virtual machine is 10,000 with 512 memory allocation, 250 instruc-
    tions per seconds, 100 BandWidth.
(2) Length of task is 40,000 and the file size is 300.
(3) Memory allocated for host is 16,384, 1,000,000 for Storage, 10,000
    BandWidth.

The structure designed for the CloudSim tool is includes two datacenters, thirty
virtual machines, two hosts in each DC and 4 Processing Element (PE) or CPU
cores for each host. Implementation has been done using 50 tasks to analyze the
algorithms for much number of tasks/cloudlets. As the cloudlets (applications) are
submitted by the user it is the task of the cloud broker (Behalf of client, Cloud
broker works and search the best virtual machine to execute the application, the VM
is selected by measuring the various parameters, for example size, bandwidth, cost)
to allocate those tasks to the VM and then Virtual Machine Manager selects the host
on which this VM should be worked based on the allocation policy of virtual
machines. When VM is assigned to the host then VM starts for execution.
Every VM has a virtual processor called PE (processing element) in CloudSim.
The VM can have much processors or process elements which simulates the real
multi-core CPUs.

Cost is measured by:

$$Cost = datacenterhost.costPerStorage * vm.size + datacenterhost.costPerRam * vm.ram$$
$$+ datacenterhost.costPerBw * vm.bw$$
$$+ datacenterhost.costPerMips * (vm.mips * vm.numberOfPes);$$

Load balancing is measured by:

$$AL = VmL/n$$

AL represents average load.
VmL represents Load of virtual machines is calculated by the load average of the
cloudlets that execute on it.
n represents number of virtual machines.

3.2 Simulation and Results

3.2.1 Load Balancing Parameter

With designing the environment, Fig. 2 illustrates 10–50 input tasks and five virtual
machines are taking for calculate the load balancing parameter in traditional method

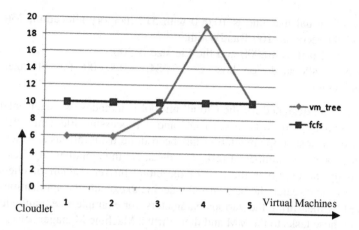

Fig. 2 Comparison with load balancing factor between VM_Tree and FCFS method

and optimized method. By introducing the figure, much number of input cloudlets runs on that machines have high memory, instructions and bandwidth in VM_tree optimization algorithm. While load are same for all VM's in FCFS algorithm.

3.2.2 Cost Parameter

With designing the environment, 10–100 cloudlets and thirty virtual machines are taking and calculate the cost factor with four all optimized methods illustrated below.

Figure 3, explains the comparison between VM_Tree and FCFS scheduling algorithms with allocation cost metrics against the number of cloudlets. It identifies that allocation cost is increased in VM_Tree optimized scheduling algorithm.

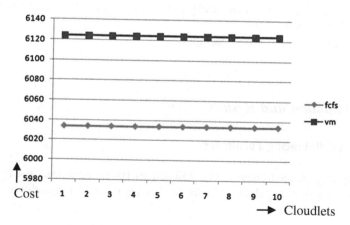

Fig. 3 Variation in cost factor between VM_Tree and FCFS scheduling

Hence proved that VM_Tree task scheduling method is perform good in load balancing factor only.

This approach can also be justified by rest of the algorithms with cost parameter only.

Figure 4, illustrates the comparison between PSO and FCFS task scheduling algorithm with parameter allocation cost against the number of cloudlets.

Figure 5, measuring the allocation cost is against the number of cloudlets and shows the comparison between QoS and FCFS (First Come First Serve) task scheduling algorithm. An optimized method i.e. Qos gives better performance from the traditional one.

Figure 6, illustrates the comparison between ABC and FCFS task scheduling algorithm based on cost against the number of cloudlets. It clearly identifies that allocation cost is reduced in optimized scheduling algorithm.

Results of this study proved that all the optimized task scheduling algorithms are very efficient.

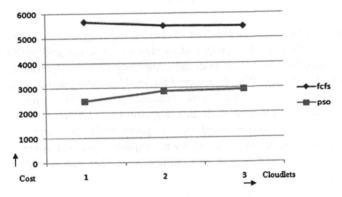

Fig. 4 Variation in cost factor between PSO and FCFS scheduling

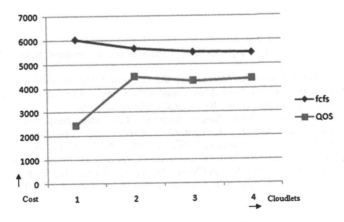

Fig. 5 Variation in cost factor between QoS and FCFS scheduling

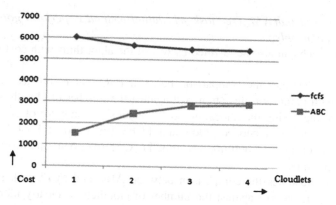

Fig. 6 Allocation cost variation in ABC and FCFS scheduling

4   Conclusion and Future Work

The traditional way of task scheduling in cloud computing is to execute or schedule the task is very difficult. For proper utilization of resources and optimal solution with task scheduling algorithms in cloud computing is very important according to its hike fame day by day. With considered the load balancing and cost parameter for virtual machine tree optimized task scheduling algorithm, proved that cost parameter is not so efficient with this algorithm. Apart from this result other comparisons are also implemented in this paper with cost parameter in cloud computing environment. Many more users' requirements will consider in future, such as resource reliability and availability to further enhance the scheduling techniques.

References

1. Yiqiu Fang, Fei Wang and Junwei Ge, "A Task Scheduling Algorithm Based on Load Balancing in Cloud Computing" Springer, Lecture Notes in Computer Science Vol. 6318, 2010.
2. Qi Cao, Zhi-Bo Wei and Wen-Mao Gong, "An Optimized Algorithm for Task Scheduling Based On Activity Based Costing in Cloud Computing", IEEE The 3rd International Conference on Bioinformatics and Biomedical Engineering 2009.
3. Lizheng Guo, Shuguang Zhao, Shigen Shen and Changyuan Jiang, "Task Scheduling Optimization in Cloud Computing Based on Heuristic Algorithm", ACADEMY PUBLISHER, Journal of Networks, Vol. 7, No. 3, pp. 547–553, 2012.
4. Raghavendra Achar, P. Santhi Thilagam, Shwetha D, Pooja H, Roshni and Andrea, "Optimal Scheduling of Computational Task in Cloud using Virtual Machine Tree", Third International Conference on Emerging Applications of Information Technology, IEEE 2012.

5. Xiaonoan Wu, Mengqing Deng, Runlian Zhang and Shengyuan, "A task scheduling algorithm based on QoS-driven in Cloud Computing", First International Conference on Information Technology and Quantitative Management, Procedia Computer Science, Vol. 17, pp. 1162–1169, ELSEVIER 2013.
6. R. Buyya and M. Murshed, "GridSim: A Toolkit for the Modelling and Simulation of Distributed Resource Management and scheduling for Grid Computing", The Journal of Concurrency and Computation: Practice and Experience (CCPE), Vol. 14, No. 13–15, Wiley Press, November 2002.
7. Calheiros, R.N., Ranjan R, De Rose, C.A.F. and Buyya R., "CloudSim: A Novel Framework for Modeling and Simulation of Cloud Computing Infrastructures and Services" in Technical Report, GRIDS-TR-2009-1, Grid Computing and Distributed Systems Laboratory, The University of Melbourne, Australia, 2009.
8. Rajkumar Buyya, Rajiv Ranjan and Rodrigo N. Calheiros, "Modeling and Simulation of Scalable Cloud Computing Environments and the CloudSim Toolkit: Challenges and Opportunities", High Performance Computing & Simulation, pp. 1–11, June 2009.
9. GAO Zhong-wen; ZHANG Kai, "The Research on cloud computing resource scheduling method based on Time-Cost-Trust model" IEEE 2012 2nd International Conference on Computer Science and Network Technology (ICCSNT).
10. Cory Janssen : http://www.techopedia.com/definition/23455/first-come-first-served-fcfs.
11. Wang Zong jiang; Zheng Qiu sheng, "A New Task Scheduling Algorithm in Hybrid Cloud Environment" IEEE 2012, (CSC) International Conference on Cloud Computing and Service Computing.
12. Hu Wu, Zhuo Tang, Renfa Li, -A Priority constrained scheduling strategy of Multiple Workflows for Cloud Computing‖, ISBN 978-89-5519-163-9 ICACT 2012.
13. Shachee Parikh and Richa Sinha, "Double Level Priority based Optimization Algorithm for Task Scheduling in Cloud Computing" International Journal of Computer Applications Vol. 62, No. 20, 2013.

The State of the Art in Software Reliability Prediction: Software Metrics and Fuzzy Logic Perspective

S.W.A. Rizvi, V.K. Singh and R.A. Khan

Abstract Every day a bulk of software are developed by industries to fulfill the customer and user requirements. Definitely, it has increased the facilities but on the other hand it also increase the probability of errors, faults, failures and also the complexity in the system that subsequently reduces the understandability of the software, make the software more error prone, highly complex and less reliable. As reliability in software based systems is a critical issue, its prediction is of great importance. In this paper, the state of the art in Software Reliability prediction has been presented with two perspectives; Software Metrics and Fuzzy Logic. The overall idea of the paper is to present, analyze, investigate and discuss the various approaches as well as reliability prediction models that are based on either reliability relevant metrics or Fuzzy Logic or both. At the end, paper presents a list of critical findings identified during literature review of various prediction models.

Keywords Software reliability · Fuzzy logic · Software metrics · Software defects

1 Introduction

The computer system is the loco that drive the scientific investigation, business decision making and engineering problem solving. In the modern scenario computerization and electronic system contain a significant presence of software.

S.W.A. Rizvi (✉)
Department of CSE, BBDU, Dr. Bhimrao Ambedkar University, Lucknow, India
e-mail: swabbasrizvi@gmail.com

V.K. Singh
Department of IT, BBDNITM, Dr. Bhimrao Ambedkar University, Lucknow, India
e-mail: vksingh@bbdnitm.ac.in

R.A. Khan
Department of IT, Dr. Bhimrao Ambedkar University, Lucknow, India
e-mail: khanraees@yahoo.com

© Springer India 2016
S.C. Satapathy et al. (eds.), *Information Systems Design and Intelligent Applications*, Advances in Intelligent Systems and Computing 433,
DOI 10.1007/978-81-322-2755-7_65

629

Each sector whether it is transportation, telecommunication, military, industrial process, home appliances, entertainment offices, aircrafts or even wristwatches is highly influenced by computer and internet. Today, the software drives a huge spectrum of activities from elementary education to genetic engineering [1]. With the beginning of the computer era, computers are bringing drastic changes to human life in all respects. Dependences of everyone over the computers have increased and at the same time internet has reduced the distance between geographical boundaries all over the globe. Even sitting at home anyone can manage and handle their responsibilities and assignment through the internet [2]. Such dependence on computers creates a pressure on the industry to develop more user friendly software and consequently increases the probability of commenting errors, faults and making the software less reliable [3].

In general software reliability is defined as 'how well the software meets its requirements' and also 'the probability of failure free operation for the specified period of time in a specified environment' [4–6]. The literature exposes many unpleasant happenings related to failure of software in the health and defense sectors [2–4, 7, 8] due to that many people lost their lives. One of the major causes behind all these misshaping is the presence of unreliable software. This paper is organized as follows; Sect. 2 presents the survey of metrics based software fault and reliability prediction, while fuzzy logic based approaches and models are presented in Sect. 3. Section 4 provide summary of critical findings identified during literature review. Section 5 presents some directions emerging out of the critical finding, while the paper concludes in Sect. 6.

2 Metric Based Software Fault and Reliability Prediction

In the last two decades a number of software defect prediction models has considered different categories of software metrics such as traditional software metrics, process metrics and object oriented software metrics [9–12]. However, considering all the software metrics whether they are traditional or object oriented or process metrics, to predict software defects have some flaws, like expensive application and processing cost, presence of less significant metrics in the considered combination, presence of computationally complex metrics, correlation between some of the metrics in the considered combination and so on. In a study [13] author has shown that the appropriate selection of software metrics plays a crucial role in improving the defect prediction as well as prediction accuracy of the proposed model or approach.

In a similar research Li et al. [14, 15] proposes a framework to choose the software engineering measures that are the best software reliability predictors. The studies further highlight that using these reliability measures during different stages of software development lifecycle may result in the development of more reliable products. Studies [16, 17] highlight size of the software as the main factor influencing the number of residual defects in software. Fenton and Neil in [18] said that most of the software fault prediction models are based on size and complexity metrics in order to predict the residual defects. In [19] predictive models were proposed that include a functional relationship of program error measures with metrics related to the complexity of the software. Through the regression analysis, authors had explored the association between software complexity and the residual defects. The study further concluded that the residual defects in the software can be predicted through software complexity or software size metrics. One of the researches [20] also focuses on software size metrics as a measure of the intrinsic complexity of the software. Maa et al. [21] analyze the capability of requirement metrics for predicting software defect during design phase. A software reliability predictor was built using requirement and design metrics in the early stage of the SDLC. The study compares six machine learning algorithms against the require- ments based metrics, design metrics and their combination to analyze the ability of metrics for software defect prediction.

In [22] Okutan and Yildiz have used Bayesian networks to know how software metrics are related with defect proneness? The study had used metrics data from the promise repository. Beside the software metrics available in promise repository, authors defined two more metrics NOD (based on the number of developers) and LOCQ (related to the quality of source code). On the basis of the results from the experiments, the study had inferred that, on defect proneness, the three metrics CBO, WMC, and LCOM were not as effective as RFC, LOC, and LOCQ. Besides that the study had also highlighted that the two metrics NOC and DIT had very restricted effect and are undependable.

Mohanta et al. [23, 24] proposed a fault model, based on bottom-up approach, for early prediction of the reliability of object-oriented software through the architectural design metrics. Initially a probabilistic approach, Bayesian Belief Network was adopted to predict the software reliability of the components. Subsequently reliability of the system was predicted on the basis of component's reliabilities along with their usage frequencies. In [10] Catal et al. highlighted the method-level metrics as the most dominant metrics in fault prediction research area. Radjenovic et al. [9] in a systematic literature review highlighted that, the most frequently used object-oriented metrics are the metrics proposed by Chidamber and Kemerer. Besides that the study also inferred that comparing on the usage fre- quency, the object-oriented metrics were used twice than the traditional process based metrics or the metrics based on source code. As long as fault prediction is concerned, object-oriented metrics have been proved to be more successful than the conventional metrics based on size and complexity of the software.

From the above it could be easily inferred that the software metrics are playing a prominent role in ensuring the software reliability as well as the quality of the software. The only thing that it to be taken care of, is the selection of appropriate and suitable combination of software metrics in the context of the development phase where the reliability is to be predicted.

3 Fuzzy Logic Based Software Defect and Reliability Prediction

Looking at the literature it can be noticed that conventional or statistical calculations are considering only accurate computations and ignoring the fuzziness and uncertainties. In fact the term fuzzy denotes imprecision, approximate reasoning, uncertainty and noisy environment [25]. The concept of fuzzy sets was introduced by Zadeh [26] to represent vagueness in linguistics as a mathematical way. It can be considered a generalization of classical set theory. Many researchers had contributed [27–32] in the area of reliability prediction using fuzzy logic. Pandey and Goyal [30] have proposed an early fault prediction model that is based on software metrics and developing organization's CMM level. Total eight metrics was identified and used with fuzzy system to predict the defects at the end of each phase of software life cycle and thereafter overall reliability of the software product. The study has considered the fuzzy profiles of various metrics in different scale, but has not explained the criteria used for developing these fuzzy profiles. Yadav et al. [31] present a model that predicts the number of residual faults before testing stage. The study had used the software metrics along with fuzzy logic to predict the remaining defects of the software that are expected during testing or when the software would be actually used.

In the Ph.D. submitted at Florida Atlantic University [33], Zhiwei Xu had studied the fuzzy logic in Software Reliability Engineering, along the different dimensions. The focus of his study was on how fuzzy based expert system could be used for early risk assessment? In [34] authors had developed a model for reliability improvement that makes use of data mining and software metrics. The model had exploited the potential of one of the data mining technique known as classification to categorize a software module as faulty or free from the factors that may cause faults. The study used ID3 algorithm to construct the decision tree. The information gained from decision tree helped to develop fuzzy rules and incorporated in fuzzy inference system. In the paper [35] Aljahdali explored the applications of fuzzy techniques to develop a SGRM (Software Reliability Growth Model). The proposed fuzzy model is a collection of linear sub-models those are joined together using fuzzy membership functions.

Recently, [32] developed a multistage fuzzy logic based model for residual fault prediction. In another study Yadav et al. [28], proposed an approach based on fuzzy logic to improve reliability. Fuzzy reasoning method is used by considering subjective information to generate crisp output. This crisp output is incorporated into a Bayesian framework to estimate the reliability. Khalsa [27] identified the high risks components early in the design phase. Fault prone modules are identified using the C&K metrics, while the defect density of the modules through MOOD metrics. Further, by the use of the fuzzy logic toolbox an algorithm was also suggested to identify the fault proneness and defect density of modules. In another fuzzy logic based study Yuan et al. [29] established a method for evaluation of software reliability using a Fuzzy-Neural hybrid network. The study proposed "Adaptive-Network based Fuzzy Inference System" (ANFIS) model to improve the accuracy of reliability prediction. The network was trained and verified that the ANFIS training speed is faster than the Radial Basis Function Network (RBFN).

In [36–38] yadav proposed a phase-wise model for defect prediction. Model is based on Fuzzy Logic and the nine metrics. At the end of the each phase of the SDLC the defect density indicator metric was predicted and subsequently used as an input for the next stage of the product development. The relevant metrics are judged as per linguistic terms and fuzzy techniques were applied in order to develop the model. Predictive accuracy of the model was validated using the data of twenty real software projects. The predicted defects of twenty software projects were found very near to the actual defects found in the testing phase. A new approach was introduced by Aljahdali et al. [39] using Fuzzy Logic and Normalized Root of Mean of the Square of Error (NRMSE) for software reliability prediction. This design of the fuzzy model was based on the Takagi-Sugeno (TS) fuzzy model. Their experimental work includes three different applications Real-Time and Control, Military and Operating System.

On the basis of above paragraphs it is evident that the Fuzzy Logic has proved its usefulness in capturing and processing subjective information in the early stages of software development. The key issue is how it is applied in making the software product more reliable.

4 Summary of Critical Findings

Following is the list of critical findings, identified during the above literature review of software faults and reliability prediction models:

1. In the absence of failure data during the early stages of product development appropriate software metrics may be a suitable mean of reflecting the expert knowledge regarding failure information and help in reliability prediction.
2. The unrealistic assumptions, smaller size of software testing data, probabilistic approach based prediction models and the fact that some measures cannot be defined precisely, are the key reasons that a fuzzy logic based approach should be consider for predicting the software defects.
3. To improve the defect prediction as well as prediction accuracy, appropriate selection of software metrics is very significant.
4. Predicting the software reliability early will be useful for both software engineers and managers since it provides vital information for making design and resource allocation decisions and thereby facilitates efficient and effective development process. Therefore, it is reasonable to develop models that more accurately predicts the number of faults/defects that are propagating undetected from one stage to the next. Such an effort would minimize or at least reduce future maintenance effort of the software also.

5. The fuzzy set theory has emerged as an alternative to capture the vagueness, uncertainty and imprecision present in the information. Therefore, in early stages, where the data is inadequate or is present in form of 'knowledge', use of fuzzy logic would be more appropriate. Any model based on Fuzzy Techniques help in the prediction of software residual defects.

6. It is evident from the literature that a significant number of models have been proposed by different researchers for estimation/prediction of software reliability in the past three decades, but these traditional reliability prediction models are neither universally successful in predicting the reliability nor generally tractable to users.

7. In the absence of failure data, while the software is in its early stages of development, reliability can be predicted in the light of those software metrics that are reliability relevant, maturity level of the developer and expert opinions. It can be noticed from some of the studies that software metrics along with the process maturity play a crucial role in early fault prediction. Therefore, it seems quite reasonable to integrate reliability relevant software metrics with process maturity for improved fault prediction accuracy.

8. A significant number of studies in the literature are concluding that number of residual faults are inversely proportional to the software reliability, means system reliability will be lesser as the number of residual faults (defects) in the system becomes more and vice-versa. Although there are a number of faults prediction models was proposed, but predicting faults in the absence of field failure data before testing phase are rarely discussed. Therefore, considering the role of residual faults in reliability prediction, more accurate models are needed to predict residual defects as early as possible.

9. It is not always feasible or necessary to predict the exact number of fault in a software module. Therefore some categorical measure is expected that helps to classify a module as faulty or free from the factors that may cause faults. Such measure will definitely help in improving the reliability.

10. In order to deliver reliable software, fuzzy techniques are playing a critical role. These techniques are emerging as robust optimization techniques that can solve highly complex, nonlinear, correlated and discontinuous problems. As the most of the early stage metrics are not very comprehensible and involve uncertainties. That's why fuzzy techniques have found useful in capturing and processing subjective data of these metrics. Therefore, fuzzy techniques are considered to be an appropriate tool in these situations.

5 Points to Ponder

This section of the paper pondering some suggestions on the basis of the literature review presented in previous sections.

1. In order to develop and deliver reliable software, timely prediction, identification and subsequent fixation of residual faults is of great significance.

2. In order to accomplish this task researchers are bound to depend on early stage measures. But generally in early stage of development sufficient objective data is unavailable, as most of the measures are subjective in nature and based on expert opinions.
3. Therefore, to handle such inherent subjectivity and uncertainty fuzzy techniques have come up as a reliable tool in capturing and processing subjective values of software metrics.
4. Only a few fuzzy based reliability models are present in the literature. But the main issue is the time and the stage of SDLC. These existing models are helping developers either by the end of coding stage or in the testing phase. This provides very late feedback to improve internal characteristics.
5. Therefore, there is an evolving need to develop a fuzzy based early reliability prediction model that guides the software developers before the start of the coding phase. Such early feedback and guidance allows them to make appropriate corrections, remove irregularities and reduce complexity in order to develop reliable quality software.

6   Conclusion

Specifically, it could be conclude that prediction of residual defects and software reliability is going to be a continuing challenge for many years to come. The critical review of models for predicting software fault and reliability prediction has shown that software metrics along with fuzzy logic techniques definitely prove to be a better approach in predicting the software reliability and residual defects. One important observation, that is being noticed that predicting reliability early in the development life cycle would provide software designers and developers an opportunity to appropriately alter the architecture of the software system, early in the development life cycle, for better reliability that leads to the overall reduction of future maintenance costs as well.

References

1. Lyu, M.R.: Handbook of Software Reliability Engineering. IEEE Computer Society Press, Los Alamitos, California (1996).
2. Dalal, S.R., Lyu, M.R., Mallows, C.L.: Software Reliability. John Wiley & Sons (2014).
3. Khan, R.A., Mustafa, K., Ahson, S.I.: Operation Profile-A key Factor for Reliability Estimation. University Press, Gautam Das and V. P. Gulati (Eds), CIT. pp. 347–354 (2004).
4. Arnold, D.N.: Two Disasters Caused by Computer Arithmetic Errors. http://www.ima.umn.edu/~arnold/455.f96/disasters.html.
5. Shooman, M.L.: Yes, Software Reliability can be Measured and Predicted. In: Proceedings of the 1987 Fall Joint Computer Conference on Exploring Technology, IEEE Computer Society, pp. 121–122 (1987).

6. Lyu, M.R.: Software Reliability Engineering: A Road Map. Future of Software Engineering. 153–170 (2007).
7. Yadav, D.K., Chaturvedi, S.K., Misra, R.B.: Early Software Defects Prediction Using Fuzzy Logic. International Journal of Performability Engineering. 8, 4, 399–408 (2012).
8. Lions, J.L.: ARIANE 5 flight 501 Failures-report by the Inquiry Board. http://www.di.unito.it/~damiani/ariane5rep.html.
9. Radjenovic, D., Hericko, M., Torkar, R., Zivkovic, A.: Software Fault Prediction Metrics: A Systematic Literature Review. Information and Software Technology, 55, 8, 1397–1418 (2013).
10. Catal, C., Diri, B.: A Systematic Review of Software Fault Predictions Studies. Expert System with Applications, 36, 4, 7346–7354 (2009).
11. Mizuno, O., Hata, H.: Yet Another Metric for Predicting Fault-Prone Modules. Advances in Software Engineering Communications in Computer and Information Science, Springer Berlin Heidelberg, 59, 296–304 (2009).
12. Catal, C.: Software Fault Prediction: A literature Review and Current Trends. Expert System with Applications, 38, 4, 4626–4636 (2011).
13. He, P., Li, B., Liu, X., Chen, J., Ma, Y.: An Empirical Study on Software Defect Prediction with a Simplified Metric Set. Information and Software Technology, 59, 170–190 (2015).
14. Li, M., Smidts, C.: A ranking of Software Engineering Measures based on Expert Opinion. IEEE Transaction on Software Engineering, 29, 9, 811–824 (2003).
15. Li, M., Smidts, C.: Ranking Software Engineering Measures Related to Reliability using Expert Opinion. In: Proceedings of 11th International Symposium on Software Reliability Engineering (ISSRE), SanJose, California, pp. 246–258 (2000).
16. Lipow, M.: Number of Faults per Line of Code. IEEE Transaction on Software Engineering, SE-8, 4, 437–439 (1982).
17. Gaffney, J.E.: Estimating the Number of Faults in Code. IEEE Transaction on Software Engineering, 10, 4, 141–152 (1984).
18. Fenton, N.E., Neil, M.: A Critique of Software Defect Prediction Models. IEEE Transaction on Software Engineering, 25, 5, 675–689 (1999).
19. Khoshgoftaar, T.M., Musson, J.C.: Predicting Software Development Errors using Software Complexity Metrics. IEEE Journal on Selected Areas in Communications, 8, 2, 253–261 (1990).
20. Fenton, N.E., Neil, M.: Predicting Software Defects in Varying Development Lifecycles using Bayesian Nets. Information and Software Technology, 49, 1, 32–43 (2007).
21. Maa, Y., Zhua, S., Qin, K., Luo, G.: Combining the Requirement Information for Software Defect Estimation in Design Time. Information Processing Letters, 114, 9, 469–474 (2014).
22. Okutan, Yildiz, O.T.: Software Defect Prediction using Bayesian Networks. Empirical Software Engineering, 19, 1, 154–181 (2014).
23. Mohanta, S., Vinod, G., Mall, R.: A Technique for Early Prediction of Software Reliability based on Design Metrics. International Journal of System Assurance Engineering and Management, 2, 4, 261–281 (2011).
24. Mohanta, S., Vinod, G., Ghosh, A.K., Mall, R.: An Approach for Early Prediction of Software Reliability. ACM SIGSOFT Software Engineering Notes, 35, 6, 1–9 (2010).
25. Magdalena, L.: What is Soft Computing? Revisiting Possible Answers. International Journal of Computational Intelligence Systems, 3, 2, 148–159 (2010).
26. Zadeh, L.: Fuzzy Sets. Information and Control, 8, 338–353 (1965).
27. Khalsa, S.K.: A Fuzzified Approach for the Prediction of Fault Proneness and Defect Density. In: Proceedings of the World Congress on Engineering,1, 218–223 (2009).
28. Yadav, O.P., Singh, N., Chinnam, R.B., Goel, P.S.: A Fuzzy Logic based Approach to Reliability Improvement Estimation during Product Development. Reliability Engineering and System Safety, 80, 1, 63–74 (2003).
29. Yuan, D., Zhang, C.: Evaluation Strategy for Software Reliability Based on ANFIS. In: IEEE International Conference on Electronics and Communications and Control (ICECC), pp. 3738–3741 (2011).

30. Pandey, A.K., Goyal, N.K.: Early Software Reliability Prediction. Springer, India (2013).
31. Yadav, D.K., Charurvedi, S.K., Mishra, R.B.: Early Software Defects Prediction using Fuzzy Logic. International Journal of Performability Engineering, 8, 4, 399–408 (2012).
32. Pandey, A.K., Goyal, N.K.: Multistage Model for Residual Fault Prediction. Early Software Reliability Prediction, Studies in Fuzziness and Soft Computing, Springer, 303, 59–80 (2013).
33. Zhiwei, Xu, Khoshgoftaar, M.T.: Fuzzy Logic Techniques for Software Reliability Engineering. Thesis (Ph.D.) Florida Atlantic University, Boca Raton (2001).
34. Pandey, A.K., Goyal, N.K.: Predicting Fault-prone Software Module Using Data Mining Technique and Fuzzy Logic. International Journal of Computer and Communication Technology, 2, 2–4, 56–63 (2010).
35. Aljahdali S., Sheta, A.F.: Predicting the Reliability of Software Systems Using Fuzzy Logic. In: Eighth International Conference on Information Technology: New Generations, IEEE, pp. 36–40 (2011).
36. Yadav, H.B., Yadav, D.K.: A Multistage Model for Defect Prediction of Software Development Life cycle using Fuzzy Logic. In: Proceedings of the Third International Conference on Soft Computing for Problem solving, Advances in Intelligent Systems and Computing, Springer India Publication, 259, pp 661–671 (2014).
37. Yadav, H.B., Yadav, D.K.: Early Software Reliability Analysis using Reliability Relevant Software Metrics. International Journal of System Assurance Engineering and Management, pp. 1–12 (2014).
38. Yadav, H.B., Yadav, D.K.: A Fuzzy Logic based Approach for Phase-wise Software Defects Prediction using Software Metrics. Information and Software Technology, 63, 44–57 (2015).
39. Aljahdali, S., Debnath, N.C.: Improved Software Reliability Prediction through Fuzzy Logic Modeling. In: Proceedings of the ISCA 13th International Conference on Intelligent and Adaptive Systems and Software Engineering, Nice, France, pp. 17–21 (2004).

An Indexed Approach for Multiple Data Storage in Cloud

Saswati Sarkar and Anirban Kundu

Abstract A cloud based data storage technique is going to be proposed in this research. Cloud based multiple data storage technique exhibits multiple data storage within a particular memory location using indexing. Proposed cloud based technique involves searching and storing data with less time consumption. Data analyzer, data transmission, and data acquisition in cloud have been introduced in this paper. Dynamic memory space allocation in cloud has been demonstrated. The paper introduces data searching and indexing techniques. Time and space analysis are represented graphically in this paper. Hit ratio in real-time scenario has been demonstrated in our work. Proposed cloud based multiple data Storage technique reduces memory access time. Comparisons have been shown for time difference realization.

Keywords Indexing · Hashing · Storage · Data searching · Cloud · Cloud based memory · Storage controller · Data acquisition · Data transmission

1 Introduction

Storage is a system maintaining data using electromagnetic or optical form. Data typically are being accessed by computerised processors. Storage system components are application, file system, I/O library, and Storage controller [1, 2]. Capacity of main memory is lower than secondary memory and it has high cost than secondary memory. The speed of cache memory is higher than primary memory as

S. Sarkar (✉)
Adamas Institute of Technology, West Bengal 700126, India
e-mail: ssam.saswati@gmail.com

A. Kundu
Netaji Subhash Engineering College, West Bengal 700152, India
e-mail: anik76in@gmail.com

S. Sarkar · A. Kundu
Innovation Research Lab, West Bengal 711103, India

© Springer India 2016
S.C. Satapathy et al. (eds.), *Information Systems Design and Intelligent Applications*, Advances in Intelligent Systems and Computing 433,
DOI 10.1007/978-81-322-2755-7_66

well as secondary memory. It has high cost and low capacity than primary memory and secondary memory [3].

The memory unit is essential component in digital computer. Memory unit is needed for storing information, data, instruction and program. Cache memory and main memory are the parts of memory unit.

Cache memory is a special high speed memory. It is a technique used to compensate speed mismatch between CPU and main memory using fast cache memory. Cache memory is small in size and high cost. There are two types of cache memories in use (L1 and L2) [4].

Volatile RAM is classified as static RAM and dynamic RAM. Non-Volatile ROM is classified as PROM, EPROM, and EEPROM [5].

Cloud based memory is faster and cheaper. It is not only cheaper to access memory than disk but also this cloud based memory is cheaper to access another computer's memory through the network. Cloud is typically a terminology used for describing distributed network using internet. It is an abstract concept used for internet based development and services [6].

Data searching is a technique to find the location of a given item or data within a collection of items. It is an efficient, scalable, and cheap procedure.

Data acquisition system (DAQ) maintains distinct sensors having related hardware and software. Data are transferred between two or more digital devices in case of data transmission [7, 8].

Cloud storage [9, 10] is a service model in which data is maintained, managed and backed up remotely using advanced techniques of networking using internet. Cloud has been a driving force to supply desired resources based on clients' choice for minimizing overall expenses along with virtualization [11].

2 Proposed Work

Main memory stores data and searching refers to the operation of finding location of particular data. Hashing and indexing are used for searching data. Index shows the exact position of a data within a particular memory location and calculates hit ratio as required in our proposed technique. The proposed cloud based technique store multiple data in a particular memory location using searching and indexing and counting number of hits and calculate hit ratio.

Consider, an array of size (4 * 10) having first room address "7 A00".
∴ Storage location of Data $a00$ = 7 A00. It is considered as cluster 1, and mathematically it is represented as 0 having binary value "000".
∴ Storage location of Data $a01$ = 7 A01. It is considered as cluster 2, and mathematically it is represented as 1 having binary value "001".
∴ Storage location of Data $a02$ = 7 A02. It is considered as cluster 3, and mathematically it is represented as 2 having binary value "010".

∴ Storage location of Data $a03 = 7\ A03$. It is considered as cluster 4, and mathematically it is represented as 3 having binary value "011".

∴ Storage location of Data $a04 = 7\ A04$. It is considered as cluster 5, and mathematically it is represented as 4 having binary value "100".

∴ Storage location of Data $a05 = 7\ A05$. It is considered as cluster 6, and mathematically it is represented as 5 having binary value "101".

∴ Storage location of Data $a06 = 7\ A06$. It is considered as cluster 7, and mathematically it is represented as 6 having binary value "110".

∴ Storage location of Data $a07 = 7\ A07$. It is considered as cluster 8, and mathematically it is represented as 7 having binary value "111".

Further data are stored in a similar fashion as shown above depending on the last three digits as follows:

Storage location of Data $a08 = 7\ A08$. It is considered as cluster 1, and mathematically it is represented as 8 having binary value "1000". Consider last three digits which are "000". Therefore, this value should be stored in cluster 1.

∴ Storage location of Data $a09 = 7\ A09$. It is considered as cluster 2, and mathematically it is represented as 9 having binary value "1001". Consider last three digits which are "001". Therefore, this value should be stored in cluster 2.

Data are stored in main memory (MM) from address 7 A00 to 7 A07. So, 8 frames are required. Total size of MM = 8 * 4 = 32 kB.

∴ Data a00 and a08 are stored in 7 A00 having space (frame) size of 4 kB. If $a00$ and $a08$ are of "long integer" data type (maximum), then it would take space of 20 (10 + 10) bytes. Two data are separated by ",", which is 1 byte of size. So, total size is 21(20 + 1)21 bytes in maximum (≤4 kB). Therefore, more than one data could be stored in one frame of main memory. Similarly, $a01$ and $a09$ are stored within same frame, and so on. Thus, cache memory access time is reduced, and hit ratio is increased in proposed technique.

2.1 Formation of Dynamic Memory Space in Cloud

As memory gets allocated at run time, dynamic memory space is being formed. The essential requirement of memory management is to provide way to allocate memory dynamically. Proposed dynamic memory space allows a program to obtain more memory space in real-time. It typically releases memory when no space is required. In our proposed approach, key values are allocated in a particular memory location in dynamic manner. Multiple key values are being allocated in same location(s) using delimiter. Memory space is being reduced. Figures 1 and 2 have shown memory space utilizations in existing approach and proposed approach respectively.

A00	A01	A02	A03	A04	A05	A06	A07	A08	A 09-address
aa	bb	cc	dd	ee	ff	gg	hh	ii	kk

Fig. 1 Memory space allocation

A 00	A 01	A 02	A 03	A 04	A 05	A 06	A 07-address
aa,ii	bb,kk	cc	dd	ee	ff	gg	hh

Fig. 2 Dynamic memory space allocation

2.2 Data Analyzer in Cloud

Data analyzer is important for data recovery in cloud. Data recovery is a process to handle data when the data is damaged and/or corrupted. It is also used to recover failed data. This is typically a process for evaluating data using analysis examining each component of data collected from various sources. In our approach, data replacement policy is initially not encouraged. All data are stored in main memory without using replacement or swapping.

2.3 Data Transmission in Cloud

In our proposed approach, more than one data are stored in one main memory location. When Cache memory is needed to search a particular data, the searching method is easy because in our proposed approach memory space is reduced. Data is transmitted from main memory to cache memory, only if hit is occurred.

2.4 Proposed Algorithms

Our proposed Algorithm 1 is used to store more data in a particular location of memory space of cloud and subsequently searching data from memory location and calculate hit ratio.

Algorithm 1: Calculate_HitRatio_of_Main_Memory

Input: Number of elements, Serial data, Base address of an array
Output: Hit ratio
Step 1: Three_Digit_Binary_Number = Convert (Last_Digit_of_Memory_Address)
Step 2: Check (Three_Digit_Binary_Number)
Step 3: If (Memory_Location (Three_Digit_Binary_Number) = = Empty) then
Step 4:   Store(Serial_Data, Number_of_Elements)
Step 5. Else If (Memory_Location (Three_Digit_Binary_Number) != Empty) then
Step 6:   Index (Number_of_Elements)
Step 7:   Store (Serial_Data, Number_of_Elements, Index)
                 //Index shows the exact position within a particular memory
        location
Step 8: Hit_Ratio = Search (Data)
Step 9: Stop

In this Algorithm 1, element, serial data and base address of an array are taken as inputs. Last digit of memory address is converted to three digits of binary number. Data are stored in a particular memory location one after another. If more than one data have same memory address, then all those data are stored into one frame. Finally, hit ratio has been calculated.

Algorithm 2: Index (Number_of_Elements)

Input: Number of elements stored within particular memory array, Data storage structure within same location
Output: Index position
Step 1: Search (Number_of_Elements)
Step 2: If (found) then
Step 3:   Generate (Index)
Step 4: Stop

In this Algorithm 2, elements and data storage structure are taken as inputs for searching elements. If the elements are found, then an index has been generated to find out particular index position.

Algorithm 3: Generate (Index)

Input: Kth key value
Output: Index

Step 1: Key_Value_Location = Remainder (Key_Value(K), Length_of_Array)
Step 2: If (more than one key values $\rightarrow$ same Key_Value_Location) then
Step 3: next_key_value_locations $\rightarrow$ $(A + 1^3)$, $(A + 2^3)$, $(A + 3^3)$, ...//A is first location
Step 4: Stop

In this Algorithm 3, key value is taken as input. Key value is divided by length of an array and calculates the remainder which is key value location. If more than one value has same location, then next key value location would be {$(A + 1^3)$, $(A + 2^3)$, $(A + 3^3)$, ...} where A is the first location.

3   Experimental Results

In Fig. 3, X-axis is denoted by time in seconds and Y-axis is denoted by load of main memory. Figure 3 is the real-time observations of main memory usage in respect of time, and it exhibits a little bit more usage of main memory while following our proposed approach. Figure 3 also shows the comparison of main

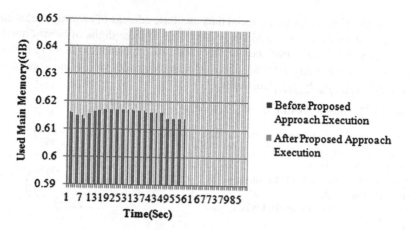

Fig. 3 Real-time observation of main memory load

memory load between "before proposed approach execution" and "after proposed approach execution". It means our approach does not require huge memory.

In Fig. 4, X-axis is denoted by time in seconds and Y-axis is denoted by load of secondary memory. Figure 4 is the real-time observations of secondary memory usage with respect to time, and it exhibits actual usage of secondary memory while following our proposed approach. This graph is the comparison of memory load between "before proposed approach executions" and "after proposed approach execution".

In Fig. 5, X-axis is denoted by time in seconds and Y-axis is denoted by total load of CPU. Figure 5 is real-time observations of CPU usage in respect of time. This graph exhibits that CPU is busy around 50 % while our proposed approach is being executed. Remaining 50 % of CPU is free to access by other programs.

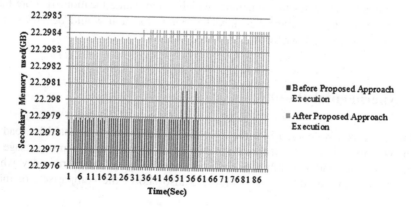

Fig. 4 Real-time observation of secondary memory load

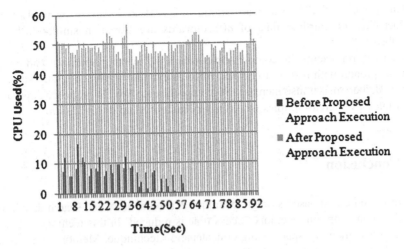

Fig. 5 Real time observation of CPU load

3.1 System Comparison

Existing approach

We have considered an array of size "4 * 10". Suppose first room address is "7 A00". Then, data is stored in "row major" order in main memory. Therefore, each room is considered as frame. Typically, size of one frame is 4 kB. So, total size of main memory is 40 kB considering only first row. Same concept has been applied for further rows of main memory of each server within the cloud.

Proposed approach

An array of size "4 * 10" has been considered. Suppose first room address is "7 A00". Data of a00 is stored in 7 A00 which is considered as "Cluster 1" or "000"

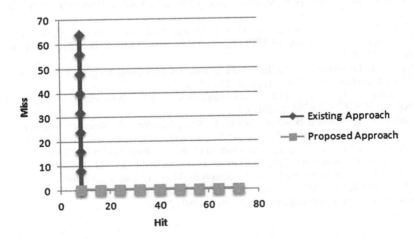

Fig. 6 System comparison graph

in binary. Data of $a01$ to $a07$ are stored in 7 $A01$ to 7 $A07$ respectively having distinct clusters. Further, data of $a08$ onwards are stored in similar fashion as described in Sect. 2.

Figure 6, represents the comparison between the existing approach and the proposed approach with respect to hit and miss. In existing approach, if key values are increased, then miss is also increased in a proportionate way. In proposed approach, if key values are increased, miss would not be increased up to certain limit.

4 Conclusion

The proposed cloud based storage technique is used to store and search data having less time consumption. Memory access time is reduced. In this technique, data are stored in less memory space using concatenation technique. Memory space is also reduced in proposed approach. Better results have been shown in our approach compared to existing storage technique.

References

1. Silberschatz, A., Galvin, P. B., Gagne, G.: Operating System Concepts.Wiley-India Edition, ISBN 978-81-265-2051-0 (2011).
2. Rumble, M. S., Kejriwal, A., Ousterhout, J.: Log-structured Memory for DRAM-based Storage. Stanford, ISBN 978-1-931971-08-9 (2014).
3. Guttman, A.: R-trees:a dynamic index structure for spatial searching. SIGMOD '84 Procedings of the 1984 ACM SIGMOD international conference on Management of data. Pages 47–57 (1984).
4. Ramasubramanian, N., Srinivas, V. V., Gounden, N. A.: Performance of cache memory subsystems for multicore architecture. International Journal of Computer Science, Engineering and Applications(IJCSEA), Vol.1, No.5 (2011).
5. Mano, M. M.:"Computer System Architecture," Pearson Edition, ISBN 978-81-317-0070-9 (2009).
6. Nazir, M. :Cloud Computing: Overview & Current Research Challenges. IOSR Journal of Computer Engineering (IOSR-JCE), ISSN: 2278-0661, ISBN: 2278-8727, vol. 8, no. 1, pp. 14–22 (2014).
7. Buyya, R.: Introduction to the IEEE Transactions on Cloud Computing. IEEE Transactions on Cloud Computing, vol. 1, no. 1, pp. 3–21 (2013).
8. Forouzan, A. B.: Data communication and networking. McGraw-Hill, ISBN-13:978-1-25-906475-3 (2013).
9. Reddy, V. K., ThirumalRao, B., Reddy, L. S. S., SaiKiran, P.: Research Issues in Cloud Computing. Global Journal of Computer Science and Technology, vol. 11, no. 11 (2011).
10. Chan W. K., Mei, L., Zhang, Z. : Modeling and testing of cloud application. IEEE Asia-Pacific Services Computing Conference. Singapore, December 7–11, pp. 111–118 (2009).
11. Gonzalez, N., Miers, C., Redigolo, F., Simplicio, M., Carvalho, T., Naslund, M., Pourzand, M.: A quantitative analysis of current security concerns and solutions for cloud computing. Journal of cloud computing.Springer Open Journal (2012).

Design New Biorthogonal Wavelet Filter for Extraction of Blood Vessels and Calculate the Statistical Features

Yogesh M. Rajput, Ramesh R. Manza, Rathod D. Deepali, Manjiri B. Patwari, Manoj Saswade and Neha Deshpande

Abstract World health organization predicts that in year 2012 there are about 347 million people worldwide have diabetes, more than 80 % of diabetes deaths occur in different countries. WHO projects that diabetes will be the 7th major cause leading death in 2030. Diabetic Retinopathy caused by leakage of blood or fluid from the retinal blood vessels and it will damage the retina. For extraction of retinal blood vessels we have invent new wavelet filter. The proposed filter gives the good extraction result as compare to exiting wavelet filter. In proposed algorithm, we have extract the retinal blood vessels features like area, diameter, length, thickness, mean, tortuosity, and bifurcations. The proposed algorithm is tested on 1191 fundus images and achieves sensitivity of 98 %, specificity of 92 % and accuracy of 95 %.

Keywords Wavelet filter · Retinal blood vessels · Area · Diameter · Length · Thickness · Mean · Tortuosity · Bifurcations

Y.M. Rajput (✉) · R.R. Manza · R.D. Deepali
Department of CS and IT, Dr. B. A. M. University, Aurangabad, MS, India
e-mail: yogesh.rajput128@gmail.com

M.B. Patwari
Institute of Management Studies & Information Technology, Vivekanand College Campus, Aurangabad, MS, India

M. Saswade
Saswade Eye Clinic, Aurangabad, MS, India

N. Deshpande
Guruprasad Netra Rugnalaya pvt. ltd, Aurangabad, MS, India

© Springer India 2016
S.C. Satapathy et al. (eds.), *Information Systems Design and Intelligent Applications*, Advances in Intelligent Systems and Computing 433,
DOI 10.1007/978-81-322-2755-7_67

1   Introduction

Diabetes, which can be characterized as a continuing increase of glucose level in the blood. Diabetes has become one of the most fast increasing health intimidations worldwide. Diabetic retinopathy is the highest common diabetic eye disease, occurs when blood vessels in the retina changes unusually. For extraction of diabetic retinopathy lesions the digital image processing techniques is widely used by the researcher. In proposed algorithm we have design new wavelet filter for extraction of retinal blood vessels. A wavelet is a localized function that can be used to capture informative, effective, and useful descriptions of a signal. If the signal is characterized as a function of time, then wavelets provide efficient localization in both time and either frequency or scale. Despite its short history, wavelet theory has established to be a powerful mathematical device for analysis and synthesis of signals and has found effective applications in a noteworthy diversity of disciplines such as physics, geophysics, numerical analysis, signal processing, biomedical engineering, statistics, and computer graphics. Number of basic functions that can be used as the mother wavelet for wavelet transformation. Since the mother wavelet produces all wavelet functions used in the transformation through translation and scaling, it governs the characteristics of the subsequent wavelet transform [1–7].

The authors proposed a method to support a non-intrusive analysis in current ophthalmology for early detection of retinal infections, treatment assessment or clinical study. This study emphasizes on the bias correction and an adaptive histogram equalization to enhance the retinal blood vessels. Formerly the blood vessels are extracted by probabilistic modelling that is improved by the expectation maximization algorithm. For evaluation these results STARE and DRIVE fundus image database is used [8]. The proposed method use the mathematical morphology and a fuzzy clustering algorithm with purification procedure. The proposed algorithm has tested on retinal images, and experimental results show that the algorithm is very effective for retinal blood vessels extraction [9].

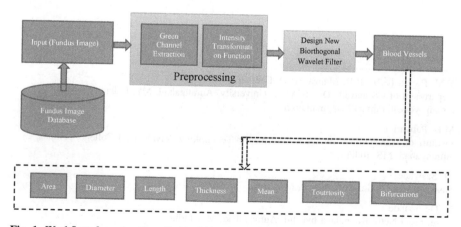

Fig. 1 Workflow for extraction of retinal blood vessels and calculate the statistical features

2 Methodology

The block diagram proposed system is depicted in the Fig. 1. Initially preprocessing is done for fundus image enhancement. For this preprocessing we have extracted the green channel and then apply intensity transformation function. Afterward design new wavelet filter for extraction of retinal blood vessels.

2.1 Biorthogonal Wavelet

The biorthogonal wavelet transform is invented of the decomposition process and the reconstruction process by using two different wavelets Ψ and $\tilde{\Psi}$. Ψ is used in the decomposition process, and $\tilde{\Psi}$ is used in the reconstruction method. Ψ and $\tilde{\Psi}$ are the dual and orthogonal to each other, and this association is called biorthogonal. There are two scale functions ϕ and $\tilde{\phi}$. in the above processes, these two scale functions are also dual and orthogonal. One is used in the decomposition process, and the second is used in the reconstruction process. So, there are four filters in biorthogonal wavelet transform. They are the decomposition low-pass filter $\{h_n\}$, the decomposition high-pass filter $\{g_n\}$, the reconstruction low-pass filter $\{\tilde{h}_n\}$ and the reconstruction high-pass filter $\{\tilde{g}_n\}$ [10].

2.2 Discrete Cosine Transform (DCT)

As for discrete cosine transform (DCT), we have

$$C^T(i,j) = \begin{cases} \frac{1}{\sqrt{N}}, & j = 0, & i = 0, 1, \ldots, N-1 \\ \sqrt{\frac{2}{N}}\cos\frac{j(2i+1)\pi}{2N}, & j = 1, 2, \ldots N-1, & i = 0, 1, \ldots, N-1 \end{cases} \tag{1}$$

$$H_i(j) = \sum_{i=0}^{N-1} C^T(i,k)C(k,j)F(k), \quad i,j = 0, 1, \ldots, N-1. \tag{2}$$

When Eq. (2) is applied to Eq. (1),

$$H_i(j) = \frac{1}{N}\left[F(0) + \sum_{k=1}^{N-1} 2\cos\frac{k(2i+1)\pi}{2N}\cos\frac{k(2j+1)\pi}{2N}F(k)\right]. \tag{3}$$

Because of $H_i(j) = H_j(i)$, we get

$$Q(n) = Q(-n), n - 0, 1, \ldots, N - 1. \tag{4}$$

Therefore, the frequency response of the system is

$$H(e^{j\omega}) = Q(0) + 2\sum_{k=1}^{N-1} Q(k)\cos(k\omega). \tag{5}$$

So, the system has strict zero phase characteristics and is an all phase filter.

2.3 Design of Biorthogonal Wavelet

2.3.1 Filter Coefficients Solver

The transfer function of Discrete Cosine Sequency Filter (DCSF) in DCT domain can be gotten with Eq. (5):

$$H(z) = Q(0) + \sum_{k=1}^{N-1} \left(Q(k)z^k + Q(-k)z^{-k} \right). \tag{6}$$

Obviously, $Q_{1/2}$ is consistent to the coefficients of each decomposition and reconstruction filter. Because of the strict zero-phase characteristic, we know $Q(k) = Q(-k)$. It means that coefficients of the biorthogonal wavelet transform must meet the requirement of symmetry

$$h_{2k-n} = h_n, \ g_{2k-n} = g_n, \ \tilde{h}_{2k-n} = \tilde{h}_n, \ \tilde{g}_{2k-n} = \tilde{g}_n. \tag{7}$$

In different wavelet transforms, $Q_{1/2}$ is consistent to different filters $\{h_n\}$, $\{g_n\}$, $\{\tilde{h}_n\}$ and $\{\tilde{g}_n\}$ the details are described as follows:

Decomposition filter: low-frequency $Q_L(k) = h_k$, input signal $x(n)$; high-frequency $Q_H(k) = g_{k+1}$, output signal $x(n+1)$.

Reconstruction filter: low-frequency $Q_L(k) = \tilde{h}_k$, input signal $x(n)$; high-frequency $Q_H(k) = \tilde{g}_{k+1}$, output signal $x(n+1)$.

Having transfer function of the system, the method for solving the coefficients of each filter is as follows:

(1) Firstly, the filter order is defined as N, in corresponding filters $\{h_n\}$, $\{\tilde{h}_n\}$, $\{g_n\}$, $\{\tilde{g}_n\}$, $N = \max(n) + 1$;
(2) If $Q_{1/2}$ is recognized, the filter parameter F can be obtained.

2.3.2 Design New Wavelet Filter Using MATLAB

Step 1: Create a biorthogonal wavelet of type 2
Step 2: Create the two filters linked with the biorthogonal wavelet and save them in a MAT-file.

$$Rf = [1/21/2];$$
$$Df = [7/8 \quad 9/8 \quad 1/8 \quad -1/8]/2;$$

Step 3: Add the new wavelet family to the pile of wavelet families.
Step 4: Display the two pairs of scaling and wavelet functions.
Step 5: We can now use this new biorthogonal wavelet to analyze a signal/image.

After extraction of retinal blood vessels, calculate its statistical features like area, diameter, length, thickness, mean, tortuosity and bifurcation points.

3 Result

For evaluation of this algorithm use some online databases and local fundus image database following Table 1 show the details of databases.

For extraction of retinal blood vessels we have proposed the biorthogonal wavelet filter by using Matlab software [11, 12]. After designing the new filter we compare the results with the existing filter such as symlet wavelet. Based on the statistical features like, area, diameter, length, thickness, mean, tortuosity and bifurcation points of blood vessels. We can say that the proposed filter is good as compare to the existing filter. Following Table 2 shows the features of retinal blood vessels by proposed biorthogonal wavelet filter (Figs. 2 and 3, Table 3).

Table 1 Fundus image database

Sr. no	Name of fundus database	Total images
1	SASWADE	500
2	STARE	402
3	DRIVE	40
4	Diarect DB 0	130
5	Diarect DB 1	89
6	HRF (Diabetic Retinopathy)	15
7	HRF (Glaucoma)	15
	Total	1191

Table 2 Features of retinal blood vessels by proposed biorthogonal wavelet filter (RRM)

Sr. no	Area	Diameter	Length	Thickness	Mean	Tortuosity	Bifurcation points
1	20	14	9.95	2	20	2	651
2	33	18	4.7	2	19	4	1434
3	27	17	8.5	2	19	2	677
4	31	18	6.42	2	19	4	148
5	22	15	5.66	2	20	3	309
6	40	20	5.71	2	20	2	619
7	74	27	8.26	2	20	1	509
8	37	19	6.32	2	19	2	186
9	38	20	5.86	2	20	3	404
10	42	21	6.44	2	20	3	933
11	37	19	7.55	2	19	1	205
12	78	28	8.16	2	20	3	426
13	26	16	5.5	2	20	5	907
14	48	22	6.24	2	20	3	418
15	104	32	9.95	2	20	2	651

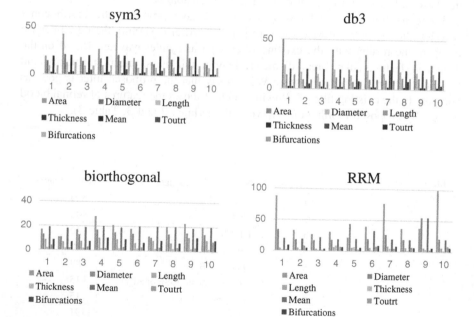

Fig. 2 Features of retinal blood vessels by existing symlet filter verses proposed new wavelet filter

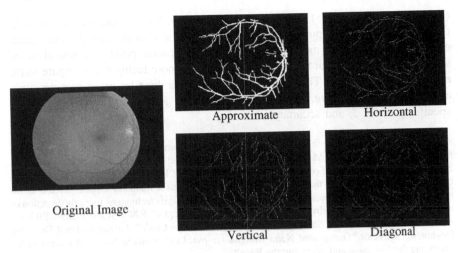

Fig. 3 Retinal blood vessels extraction using proposed wavelet filter (RRM)

Table 3 Features of retinal blood vessels by existing wavelet filter (symlet)

Sr. no	Area	Diameter	Length	Thickness	Mean	Tortuosity	Bifurcation points
1	17	13	6.71	2	19	2	649
2	15	12	7.67	2	19	4	1431
3	18	14	5.8	2	20	2	675
4	20	14	10.59	2	20	2	147
5	19	14	9.37	2	19	3	306
6	18	14	9.47	2	20	2	616
7	33	18	12.63	2	19	1	500
8	25	16	10.56	2	19	2	187
9	20	14	9.83	2	20	3	405
10	27	17	10.42	2	19	3	930
11	19	14	9.04	2	19	1	203
12	42	21	12.71	2	20	3	421
13	18	14	9.33	2	20	5	903
14	27	17	10.03	2	19	3	415
15	45	21	14.75	2	20	6	649

4 Conclusion

The result projected in this research article were obtained on SASWADE database and performance method were compared with STARE, DRIVE, DIARECT DB0, DIARECT DB1 and HRF database also. The features of retinal blood vessels which is extracted by the proposed "RRM" filter is compared with the existing "Symlet (sym3), daubechies (db3) and biorthogonal (bio3.3, bio3.5, and bio3.7)" wavelet filter for the validation purpose. And based on the statistical features (area, diameter, length, thickness, mean, tortuosity and bifurcation points) of retinal blood vessels, we conclude that the proposed filter gives more features as compare to the existing wavelet filters. The performance analysis is done by using receiver operating characteristic curve. The proposed algorithm achieves sensitivity of 98 %, specificity of 92 % and accuracy of 95 %.

Acknowledgments We are thankful to University Grant Commission (UGC) for providing us a financial support for the Major Research Project entitled "Development of Color Image Segmentation and Filtering Techniques for Early Detection of Diabetic Retinopathy" F. No.: 41 – 651/2012 (SR) also we are thankful to DST for providing us a financial support for the major research project entitled "Development of multi resolution analysis techniques for early detection of non-proliferative diabetic retinopathy without using angiography" F.No. SERB/F/2294/2013-14. Also thankful to Dr. Manoj Saswade, Director "Saswade Eye Clinic" Aurangabad and Dr. Neha Deshpande, Director "Guruprasad Netra Rungnalaya pvt. Ltd", Samarth Nagar, Aurangabad for providing the Database and accessing the Result.

References

1. Manjiri B. Patwari, Dr. Ramesh R. Manza, Dr. Manoj Saswade and Dr. Neha Deshpande, "A Critical Review of Expert Systems for Detection and Diagnosis of Diabetic Retinopathy", Ciit International Journal of Fuzzy Systems, February 2012, doi:FS022012001 ISSN 0974-9721.
2. Yogesh M. Rajput, Ramesh R. Manza, Manjiri B. Patwari, Neha Deshpande, "Retinal Blood Vessels Extraction Using 2D Median Filter", Third National Conference on Advances in Computing (NCAC-2013), 5th to 6th March 2013, School of Computer Sciences, North Maharashtra University, Jalgaon-425001 (MS) India.
3. Yogesh M. Rajput, Ramesh R. Manza, Manjiri B. Patwari, Neha Deshpande, "Retinal Optic Disc Detection Using Speed Up Robust Features", National Conference on Computer & Management Science [CMS-13], April 25–26, 2013, Radhai Mahavidyalaya, Auarngabad-431003(MS India).
4. Manjiri B. Patwari, Ramesh R. Manza, Yogesh M. Rajput, Manoj Saswade, Neha K. Deshpande, "Review on Detection and Classification of Diabetic Retinopathy Lesions Using Image Processing Techniques", International Journal of Engineering Research & Technology (IJERT), ISSN: 2278-0181, Vol. 2 Issue 10, October – 2013.
5. Manjiri B. Patwari, Ramesh R. Manza, Yogesh M. Rajput, Neha K. Deshpande, Manoj Saswade, "Extraction of the Retinal Blood Vessels and Detection of the Bifurcation Points", International Journal in Computer Application(IJCA), September 18, 2013. ISBN : 973-93-80877-61-7.
6. Manjiri B. Patwari, Dr. Ramesh R. Manza, Yogesh M. Rajput, Dr. Manoj Saswade, Dr. Neha K. Deshpande, "Personal Identification algorithm based on Retinal Blood Vessels Bifurcation", 2014 International Conference on Intelligent Computing Applications, 978-1-4799-3966-4/14 © 2014 IEEE, doi:10.1109/ICICA.2014.51.
7. Patwari Manjiri, Manza Ramesh, Rajput Yogesh, Saswade Manoj, Deshpande Neha, "Automated Localization of Optic Disk, Detection of Microaneurysms and Extraction of Blood Vessels to Bypass Angiography", Springer, Advances in Intelligent Systems and Computing. ISBN: 978-3-319-11933-5, doi:10.1007/978-3-319-11933-5_65. 2014.
8. Djibril Kaba, Chuang Wang, Yongmin Li, Ana Salazar-Gonzalez1, Xiaohui Liu and Ahmed Serag, "Retinal blood vessels extraction using probabilistic modelling", Health Information Science and Systems 2014, 2:2 doi:10.1186/2047-2501-2-2.

9. A. Kong, D. Zhang, M. Kamel, "An Automatic Hybrid Method For Retinal Blood Vessel Extraction", International Journal Appl. Math. Comput. Sci., 2008, Vol. 18, No. 3, 399–407. doi:10.2478/v10006-008-0036-5.
10. Baochen Jiang, et. al, "Implementation of Biorthogonal Wavelet Transform Using Discrete Cosine Sequency Filter", International Journal of Signal Processing, Image Processing and Pattern Recognition Vol. 6, No. 4, August, 2013.
11. "Understanding MATLAB" By Karbhari Kale, Ramesh R. Manza, Ganesh R. Manza, Vikas T. Humbe, Pravin L. Yannawar, Shroff Publisher & Distributer Pvt. Ltd., Navi Mumbai, April 2013. ISBN: 9789350237199.
12. "Understanding GUI using MATLAB for Students" By Ramesh Manza, Manjiri Patwari & Yogesh Rajput, Shroff Publisher & Distributer Pvt. Ltd., Navi Mumbai, April 2013. ISBN: 9789351109259.

Demand Side Management Using Bacterial Foraging Optimization Algorithm

B. Priya Esther, K. Shivarama Krishna, K. Sathish Kumar
and K. Ravi

Abstract Demand side management (DSM) is one of the most significant functions involved in the smart grid that provides an opportunity to the customers to carryout suitable decisions related to energy consumption, which assists the energy suppliers to decrease the peak load demand and to change the load profile. The existing demand side management strategies not only uses specific techniques and algorithms but it is restricted to small range of controllable loads. The proposed demand side management strategy uses load shifting technique to handle the large number of loads. Bacterial foraging optimization algorithm (BFOA) is implemented to solve the minimization problem. Simulations were performed on smart grid which consists of different type of loads in residential, commercial and industrial areas respectively. The simulation results evaluates that proposed strategy attaining substantial savings as well as it reduces the peak load demand of the smart grid.

Keywords Smart grid · Demand side management · Bacterial foraging optimization algorithm · Load shifting

1 Introduction

Smart grid is the integration of advanced communication technologies, sensors and control methodologies at each level (transmission and distribution level) of the power system, which supplies electric power in a smart and efficient manner to the

B. Priya Esther (✉) · K. Shivarama Krishna · K. Sathish Kumar · K. Ravi
School of Electrical Engineering, VIT University, Vellore 632014, Tamilnadu, India
e-mail: priyarafaela@gmail.com

K. Shivarama Krishna
e-mail: shivarama.krishna@vit.ac.in

K. Sathish Kumar
e-mail: kansathh21@yahoo.co.in

K. Ravi
e-mail: k.ravi@vit.ac.in

© Springer India 2016
S.C. Satapathy et al. (eds.), *Information Systems Design and Intelligent Applications*, Advances in Intelligent Systems and Computing 433,
DOI 10.1007/978-81-322-2755-7_68

657

customers [1, 2]. Majority of the existing demand side management strategies use dynamic programming [3] and linear programming [4, 5] which are system specific [3–7]. These strategies cannot handle the practical systems which consist of variety of loads. In smart grid, considerable part of generation is expected from the renewable energy sources which are intermittent in nature that leads to challenging task in making power dispatch operations from the grid. The existing algorithms such as heuristic based evolutionary algorithms have been developed to solve the problem [8–10]. There are six DSM strategies that can be applied in smart grid; these techniques are peak clipping, strategic conservation, strategic load growth, valley filling, flexible load shape and load shifting, shown in the following Fig. 1.

Both the peak clipping and valley filling technique employs direct load control to decrease the peak load demand and to build the off—peak demand respectively, whereas the load shifting technique is more efficient in shifting the loads during peak time. Strategic load growth and strategic conservation techniques are used to optimize the load.

1.1 Implementation of Bacterial Foraging Optimization Algorithm (BFOA) in DSM

The Bacterial foraging optimization algorithm (BFOA)is introduced by Passsino for the first time in the year 2002. It is mainly inspired by the chemotactic and foraging behavior of *Escherichia coli* (*E. coli*) bacterium. The bacteria are able to move towards the nutrient area and escapes from the poisonous area, by tumbling and smooth running. The four important mechanisms involved in BFOA are, one is Chemotaxis, other is Reproduction, third one is Elimination—dispersal and fourth one is swimming.

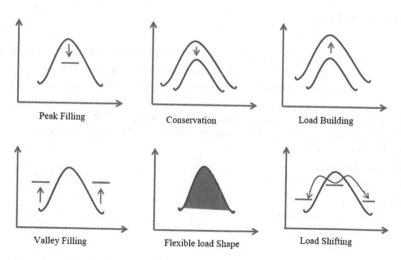

Fig. 1 Demand side management technique

In chemotactic step, if the bacterium finds new position in which nutrient medium higher than the existing position, then the bacterium takes one more step in that direction. This process is repeated until worst nutrient medium is reached. In reproduction step, the bacteria are arranged in the descending order based on the nutrient concentration acquired during the chemotaxis process. The first half of the population which have acquired enough nutrients will reproduce, that is each bacterium splits into two. The other half of the population will eventually die and they are eliminated from the population, while maintaining the initial population to be the same. The changes in the environment will affect the population and behavior of the bacteria; in order to analyze this phenomenon elimination dispersal step is introduced. In this step, a random number is assigned to each bacterium in the range of 0–1. If the random number value is less compared with the predetermined parameter value then it is survived, else it is eliminated from the environment. The fitness function and the equations related to the BFOA are listed below.

$$[\theta]^{i}[j+1,k,l] = [\theta]^{i}[j,k,l] + c[i]\frac{\Delta[i]}{\sqrt{\Delta^{i}[i].\Delta[i]}} \tag{1}$$

$\theta^{i}(j, k, l)$ is the expression of its bacterium at chemotactic step j, reproduction step k and elimination step l. This leads to a step of size $C(i)$ ith bacterium and Δ denotes a vector in the random direction whose elements lie in range $[-1, 1]$.

$$J_{cc}[\theta, P[j,k,l]] = \sum_{i=1}^{S} J_{cc}[\theta, \theta^{i}[j,k,l]] \tag{2}$$

where $J_{cc}(\theta, P(j,k,l))$ denotes the value of the objective function that should be added to the actual objective function in order to obtain a time varying objective function, S denotes the overall number of bacteria, p denotes the number of variables which should be optimized, that are present in each bacterium and

$$\theta = [\theta_1, \theta_2, \theta_3...\theta_p]$$

$$J[i,j,k,l] = J[i,j,k,l] + J_{cc}[\theta^{i}[j,k,l], P[j,k,l]] \tag{3}$$

$$Fitness = \frac{1}{1 + \sum_{t=1}^{24}[[PLoad[t] - Objective[t]]^2]} \tag{4}$$

The above fitness function is selected for BFOA, to achieve final load curve such that it should be very close to the objective load curve.

The flow chart for BFOA is shown in the following Fig. 2.

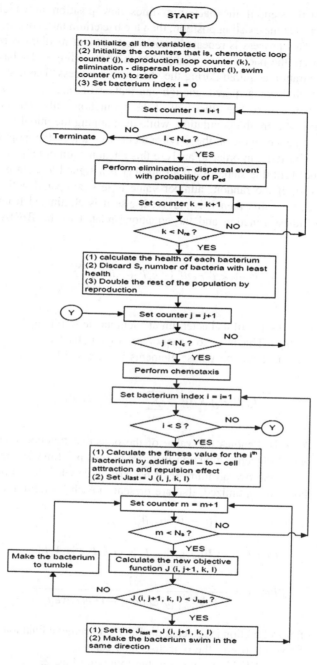

Fig. 2 Flow chart for BFOA

1.2 Problem Formulation

The load shifting technique has been implemented for DSM of smart grid, which is mathematically formulated in the following Eq. (5).
   Minimize

$$\sum_{t=1}^{N} (PLoad(t) - Objective(t))^2 \tag{5}$$

where
   PLoad (t) and Objective (t) denotes the actual consumption and objective curve at particular time t respectively.

$$PLoad(t) = Forecast(t) + Connect(t) - Disconnect(t) \tag{6}$$

where
   Forecast(t), Disconnect(t) and Connect(t) denotes the forecasted consumption, amount of loads disconnected and connected at particular time t respectively.
   Connect(t) is the combination of two parts of increment in load at particular time t, which is given in the below Eq. (7)

$$Connect[t] = \sum_{t=1}^{N} \sum_{k=0}^{n} [X_{kit} \cdot P_{1K}] + \sum_{l=1}^{j-1} \sum_{i=1}^{t-1} \sum_{k=1}^{D} X_{Ki[t-1]} P_{[1+l]k} \tag{7}$$

where
   X_{kit} denotes the number of type k devices which are moved in the time interval i to t,
   D denotes the number of device types,
   P_{1k} and $P_{(1+l)k}$ indicates the power consumption of the device of type k at time step 1 and (1 + l) respectively and j denotes the type k total duration of consumption.
   Disconnect(t) is the combination of two parts of decrement in load at particular time t, which is shown in the below Eq. (8)

$$Disconnect[t] = \sum_{q=t+1}^{t+m} \sum_{k=1}^{D} [X_{ktq} \cdot P_{1K}] + \sum_{l=1}^{j-1} \sum_{q=t+1}^{t+m} \sum_{k=1}^{D} X_{k[t-1]q} P_{[1+l]k} \tag{8}$$

X_{ktq} denotes the number of type k devices which are delayed in the time interval t to q and m denotes the maximum amount of delay.

The minimize function is subjected to the below constraints which are given in the below Eqs. (9, 10).

The devices that are moved should not have a negative value and it is given as follows

$$X_{kit} > 0 \, \forall i, j, k \tag{9}$$

The devices transferred should not be more than the devices that are accessible to control at particular time which is given as follows

$$\sum_{t-1}^{N} X_{kit} \leq Ctrlable[i] \tag{10}$$

where Ctrlable(i) represents the type k devices that are accessible to control at particular step time i.

2 System Used

To determine the effectiveness of proposed load shifting technique as DSM strategy for smart grid using BFOA, three different areas namely residential, commercial and industrial area with different types of devices have been considered. The complete network is operated at voltage level of 410 V, while the main grid resistance and reactance as 0.003 pu and reactance of 0.01 pu respectively. The length of the links in the residential, commercial and industrial micro-grids is 2, 3 and 5 km respectively. The wholesale energy prices and hourly forecasted load demand for residential, commercial and industrial micro-grid are listed in the Table 1.

The controllable devices and its data for commercial, residential and industrial areas are listed in the Tables 2, 3 and 4 respectively.

3 Results and Discussions

The DSM results of the residential, commercial and industrial areas are depicted in Figs. 3, 4 and 5 respectively.

The proposed DSM strategy has achieved the load consumption curve which is very close to the objective load curve. The proposed BFOA is effective in handling different type of loads, in residential, commercial and industrial areas (Tables 5 and 6).

Table 1 Forecasted load demand and wholesale energy prices

Time (h)	Wholesale price (ct/kWh)	Hourly forecasted load (kWh)		
		Residential micro grid	Commercial micro grid	Industrial micro grid
8–9	12.00	729.4	923.5	2045.5
9–10	9.19	713.5	1154.4	2435.1
10–11	12.27	713.5	1443.0	2629.9
11–12	20.69	808.7	1558.4	2727.3
12–13	26.82	824.5	1673.9	2435.1
13–14	27.35	761.1	1673.9	2678.6
14–15	13.81	745.2	1673.9	2678.6
15–16	17.31	681.8	1587.3	2629.9
16–17	16.42	666.0	1558.4	2532.5
17–18	9.83	951.4	1673.9	2094.2
18–19	8.63	1220.9	1818.2	1704.5
19–20	8.87	1331.9	1500.7	1509.7
20–21	8.35	1363.6	1298.7	1363.6
21–22	16.44	1252.6	1096.7	1314.9
22–23	16.19	1046.5	923.5	1120.1
23–24	8.87	761.1	577.2	1022.7
24–1	8.65	475.7	404.0	974.0
1–2	8.11	412.3	375.2	876.6
2–3	8.25	364.7	375.2	827.9
3–4	8.10	348.8	404.0	730.5
4–5	8.14	269.6	432.9	730.5
5–6	8.13	269.6	432.9	779.2
6–7	8.34	412.3	432.9	1120.1
7–8	9.35	539.1	663.8	1509.7

Table 2 Controllable devices and its data for commercial area

Device type	Devices hourly consumption (kW)			Number of devices
	1st hour	2nd hour	3rd hour	
Water dispenser	2.5	–	–	156
Dryer	3.5	–	–	117
Kettle	3.0	2.5	–	123
Oven	5.0	–	–	77
Coffee maker	2.0	2.0	–	99
Fan/AC	3.5	3.0	–	93
Air conditioner	4.0	3.5	3.0	56
Lights	2.0	1.75	1.5	87
Total	–	–	–	808

Table 3 Controllable devices and its data for residential area

Device type	Devices hourly consumption (kW)			Number of devices
	1st hour	2nd hour	3rd hour	
Kettle	2.0	–	–	406
Iron	1.0	–	–	340
Dish washer	0.7	–	–	288
Fan	0.20	0.20	0.20	288
Oven	1.3	–	–	279
Washing machine	0.5	0.4	–	268
Dryer	1.2	–	–	189
Vacuum cleaner	0.4	–	–	158
Frying pan	1.1	–	–	101
Blender	0.3	–	–	66
Rice-cooker	0.85	–	–	59
Hair dryer	1.5	–	–	58
Coffee maker	0.8	–	–	56
Toaster	0.9	–	–	48
Total	–	–	–	2604

Table 4 Controllable devices and its data for industrial area

Device type	Devices hourly consumption (kW)						Number of devices
	1st hour	2nd hour	3rd hour	4th hour	5th hour	6th hour	
Water heater	12.5	12.5	12.5	12.5	–	–	39
Welding machine	25.0	25.0	25.0	25.0	25.0	–	35
Fan/Ac	30.0	30.0	30.0	30.0	30.0	–	16
Arc furnace	50.0	50.0	50.0	50.0	50.0	50.0	8
Induction motor	100	100	100	100	100	100	5
DC motor	150	150	150	–	–	–	6
Total	–	–	–	–	–	–	109

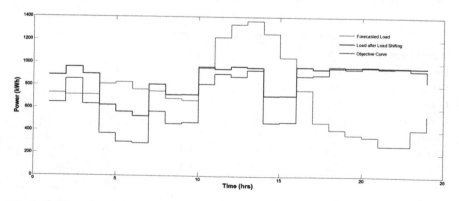

Fig. 3 DSM results of the residential area

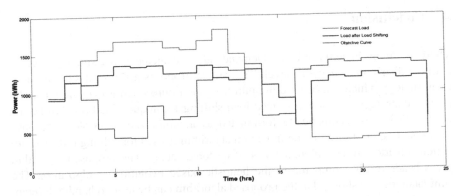

Fig. 4 DSM results of the commercial area

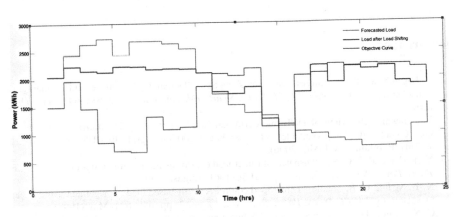

Fig. 5 DSM results of the industrial area

Table 5 Operational cost reduction using BFOA

Area	Cost without DSM ($)	Cost with DSM ($)	Percentage reduction (%)
Residential	2211.45	2047.63	7.4011
Commercial	3211.28	3020.60	5.9378
Industrial	5067.778	4556.34	10.091

Table 6 Peak demand reduction using BFOA

Area	Peak load without DSM (kW)	Peak load with DSM (kW)	Peak reduction (kW)	Percentage reduction (%)
Residential	1363.6	1106.3	18.869	7.4011
Commercial	1812.2	1462.5	19.296	5.9378
Industrial	2727.3	2338.6	14.252	10.091

4 Conclusion

There are many benefits of using Demand side management in a smart grid especially at distribution network level. This article proposes a Demand side management strategy which can be used in optimizing the future smart grid operations. The proposed strategy uses a generalized load shifting technique based on shifting the loads which is mathematically formulated as a minimization problem for optimization. Bacterial foraging optimization algorithm is used for solving this problem which considers three different types of customer areas. The proposed algorithm gives better results compared to the heuristic based evolutionary algorithm. The simulation results shows that the proposed algorithm can be used to handle different types of controllable devices in large quantity to achieve the objective of increased savings by reducing the peak load demand on the smart grid.

References

1. T. Logenthiran, Dipti Srinivasan and Tan Zong Shun,"Demand side management in smart grid using heuristic optimization," *IEEE Trans. Smart grid*, vol. 3, no. 3, pp. 688–694, September (2012).
2. P. Agrawal, "Overview of DOE microgrid activities," in *Proc. Symp. Microgrid*, Montreal, QC, Canada, [Online].Available: http://der.lbl.gov/2006microgrids_files/USA/Presentation_7_Part1_Poonumgrawal.Pdf, (2006).
3. Y. Y. Hsu and C. C. Su, "Dispatch of direct load control using dynamic programming," *IEEE Trans. Power Syst.*, vol. 6, no. 3, pp. 1056–1061, August (1991).
4. K.-H. Ng and G. B. Sheblé, "Direct load control-A profit-based load management using linear programming," *IEEE Trans. Power Syst.*, vol. 13, no. 2, pp. 688–694, May (1998).
5. C. N. Kurucz, D. Brandt, and S. Sim, "A linear programming model for reducing system peak through customer load control programs," *IEEETrans. Power Syst.*, vol. 11, no. 4, pp. 1817–1824, Nov. (1996).
6. A. I. Cohen and C. C. Wang, "An optimization method for load management scheduling," *IEEE Trans. Power Syst.*, vol. 3, no. 2, pp. 612–618, May (1988).
7. F. C. Schweppe, B. Daryanian, and R. D. Tabors, "Algorithms for a spot price responding residential load controller," *IEEE Trans. PowerSyst.*, vol. 4, no. 2, pp. 507–516, May (1989).
8. L. Yao, W. C. Chang, and R. L. Yen, "An iterative deepening genetic algorithm for scheduling of direct load control," *IEEE Trans. PowerSyst.*, vol. 20, no. 3, pp. 1414–1421, Aug. (2005).
9. T. Logenthiran, D. Srinivasan, and A.M. Khambadkone, "Multi-agent system for energy resource scheduling of integratedmicrogrids in a distributed system," *Electr. Power Syst. Res.*, vol. 81, no. 1, pp. 138–148, (2011).
10. T. Back, D. Fogel, and Z. Michalewicz, *Handbook of Evolutionary Computation*. New York: IOP Publ. and Oxford Univ. Press, (1997).

An Optimized Cluster Based Routing Technique in VANET for Next Generation Network

Arundhati Sahoo, Sanjit K. Swain, Binod K. Pattanayak
and Mihir N. Mohanty

Abstract Since last few years, research in the field of vehicular networking has gained much attention and popularity among the industries and academia. Intelligent approach for such technology is the challenge. In this paper, we have taken an attempt to optimize the routing algorithm for vehicular adhoc networking (VANET). Ant Colony Optimization (ACO) is an optimization technique and is applied based on clustering technique. To improve the safety factor and efficiency and to develop an intelligent transport system, it is highly conceptual with the wireless technology. It is a special type of MANET, because of the variation of routing protocols. Even if the protocols of MANET are feasible, they are not able to provide the optimum throughput required for a fast changing vehicular ad hoc network. Positions of the vehicles create the zone and the optimization is zone based. Ant Colony algorithm is combined with zone based clustering algorithm to improve the result. This approach combines the advantages of both the techniques, the ant colony algorithm as well as the zone based routing algorithm. Routing overhead has been compared between AODV, MARDYMO and TACR protocols and depicted in the graphical plots.

Keywords VANET · Routing protocol · Optimization · ACO

A. Sahoo
Research Scholar, Bhubaneswar, Odisha, India

S.K. Swain
Silicon Institute of Technology, Bhubaneswar, Odisha, India

B.K. Pattanayak · M.N. Mohanty (✉)
ITER, SOA University, Bhubaneswar, Odisha, India
e-mail: mihirmohanty@soauniversity.ac.in

© Springer India 2016
S.C. Satapathy et al. (eds.), *Information Systems Design and Intelligent Applications*, Advances in Intelligent Systems and Computing 433,
DOI 10.1007/978-81-322-2755-7_69

1 Introduction

VANET is used to provide communications between neighbour vehicles. At the same time, vehicles can communicate with fixed infrastructure on the roadside named as road side units (RSU). Many challenges are there to adopt the protocols those can serve in different topologies. VANETs represent an emerging, especially challenging class of MANETs. A new kind of Ad hoc network with an immense improvement in technological innovations is emerging these days known as VANET (Vehicular ad hoc network). Communication network must be wireless and made for Inter-Vehicular Communications (IVC) as well as Road-Vehicle Communications (RVC) in Mobile Ad Hoc Networks (MANETs). Two kinds of communication can be achieved to provide a list of applications like emergency vehicle warning, safety etc. One such communication is between various vehicles known as vehicle to vehicle (V2V) and the other type is among vehicles and roadside units (V2R). It is an important issue for supporting the smart intelligent transport system (ITS) in design of routing protocols for VANETs. The area of coverage with the architectural details remaining the same where a VANET has a larger coverage area than that of a MANET. The routing protocols designed for urban areas may not suitable for packet delivery in a sparse, and partially connected VANET, as reported in [1].

2 Related Literature

Many authors have been worked in this area, but some authors have compared the ad hoc routing protocol in various scenario of urban area. The objective was to build a robust communication network between mobile Vehicles for the safety [2].

A survey has been made [2] for routing protocols in VANET. Some of those belongs to mobicast, geocast, and broadcast protocol. Carry-and-forward is the key consideration for designing the routing protocols for VANETs. As a result, min-delay and delay-bounded routing protocols found attractive for VANET and was discussed. For classical routing Protocol, DSDV and DSR a routing algorithm MUDOR are simulated and analysed in [3]. Different metrics are used in [4] for dynamic culstors. Authors clustered minimum number of vehicles, equipped with IEEE 802.11p and UTRAN interfaces, were selected as vehicular gateways to link VANET to UMTS. Also they have studied two routing protocols known as Vehicle to Vehicle Communication and Vehicle to roadside communication, their merits/demerits. RSU is a fixed unit and has been modeled for VANET notes communication. It is discussed in [4, 5]. Reactive location based routing algorithm uses cluster-based flooding for VANETs, called location-based routing algorithm with cluster-based flooding [6]. In Compound clustering algorithm the position, velocity, acceleration and degree of a vehicle has taken into consideration but this algorithm improves the network stability but does not provide any mechanism to handle malicious vehicle [7]. A Direction based clustering algorithm has been considered for moving direction, but it has no provision for detecting vehicle behaviour [8]. Distance-Vector Routing (DSDV), DSR, and Position based routing algorithm, are not suitable for dynamic environment. AODV is said to be an on demand algorithm. The algorithm established the route as desired by source node [9–11]. Route established by AODV broken very frequently, if the network is with the mobile agents and dynamic by nature. DSDV is a table driven algorithm based on Bellman-Ford shortest path algorithm, but this needs regular update of routing tables, which requires small amount of bandwidth even when the network is idle. DSR algorithm also creates the route on-demand like AODV but this algorithm uses the mechanism of source routing. The routing overhead was achieved and this routing overhead is directly proportional to the path length as in [12–14]. it is designed for VANET, but DYMO protocol is well suited for MANET.

Although many researchers have been used and compared, the performance and simulation result of various routing techniques but we have used ANT Colony Optimization technique. It is the method to analyse the performance and overhead of different routing Protocol which is illustrated in Sect. 3.

3 Methodology

Network designers' choice is mostly clustering based model for vehicular adhoc network as clustering is an efficient resource as well as load balancing network. Routing can be effectively done in case of cluster based network for vehicles. Also it can provide the reuse of resources and also improve the capacity of VANET. In clustering process a cluster-head (CH) is created to store the informtion status of the members. Cluster-heads operate on two different frequencies to avoid teh inter and intra cluster communication. Frequency of re-affiliation should be minimized for highly dynamic network topologies. Choice of good clustering algorithm, there must be the reduction of cluster swap to improve the stability of the network [9, 10].

3.1 Cluster Creation Process

The Cluster Head collects data from any of the node and sends them to another cluster head. This type of solution provides less propagation delay with high delivery ratio. These may change their relative position on highways. The size and stability of clusters change abruptly, if lowest ID and node weight heuristics are used. It is found that the time consumed to generate a cluster is directly proportional to the number of clusters as shown in Fig. 1.

Fig. 1 Cluster creation time versus number of cluster

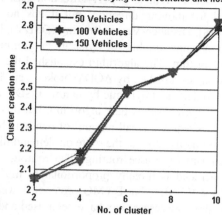

Cluster creation time with varying no.of vehicles and no. of clusters

3.2 Selection of Cluster Head

The algorithm makes a search for the available nodes in the cluster. The cluster contains two types of nodes.

(i) Vehicle nodes (ii) Roadside unit

Selection process of cluster-head should be more appropriate. We propose an appropriate cluster-head selection by taking the position and trust value of each vehicle into consideration.

For Roadside unit within the cluster, the algorithm made the choice of selection as cluster head, because of its better processing capability and static in nature. For such case, the algorithm is as follows.

(i) Select the slowest moving vehicle (S_{mv}) from the cluster. The coverage area can exist for along period.

(ii) Then calculate the trust value as per the selection of slow vehicle (T_{smv}) selected.

(iii) If selected trust value is greater than threshold trust value (TN_{th}), then select it as the cluster-head. Else again repeat for the slowest vehicle from step (1)

(iv) For the vehicle of same distance and trust value, the steps will be repeated from beginning.

By variation of number of clusters in different size, teh selection time is estimated. While the number of clusters is less, more time is requires for a high head selection irrespective of the number of nodes (Fig. 2).

Fig. 2 Time to select the cluster head versus no. of cluster

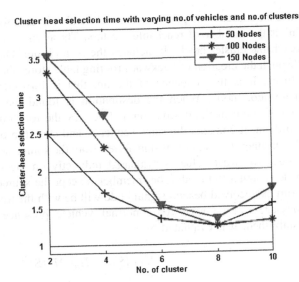

Cluster head selection time with varying no.of vehicles and no.of clusters

4 Optimized Routing Using ACO

Ant colony optimization algorithm (ACO) is an alternative choice for optimized routing. To find shortest paths from source to destination with a low cost overlay routing network the optimization is highly essential. ACO has been used for such purpose in this work. It is the stochastic decision based technique.

4.1 POSANT Routing Algorithm

Position based ant colony optimization technique is named as POSANT algorithm. It uses the location information to enhance its efficiency. It is able to find optimum routes in a given network which contains nodes. Ant Colony algorithm is combined with zone based clustering algorithm. It is difficult to store large amounting of routing network information in the clustering nodes, as the mobile notes have small memory. Since it is zone based optimized clustering algorithm, cluster head is also available in each zone. It is necessary to store the route information of the mobile notes and the information regarding zone boundary and other clusters.

4.2 Establishment of Route Using Optimization

To find the destination node the optimization using ACO has been used, subject to condition that both source and destination nodes are not in a same cluster. In this case the source node generates (n − 1) number of ants with unique sequence; where 'n' is the number of reachable clusters. Those ants are to be sent forward to the reachable cluster heads. It includes the own cluster. Once the ant moves forward, through the creation of backward routing table stores the identifier of the neighbour cluster. It is the sequence of the ant that is forwarded as the identifier of the destination node. When the destination is found the cluster head searches the identification of the destination and destroy the forward ant by creating the backward ant. It contains the same sequence and stored in backward routing tables. In this manner the message is transferred from the source node within the transmission range. The trust value is evaluated indirectly on the source node by sending a beacon signal to the all other members except the source. As a result the response from the received beacon signal node will be with the trust value of source. Beacon message can be used for abnormal vehicle presence, otherwise route can be established. It is evaluated by

$$IT_{CH}(S) = [\Pi_n(DT_n(S)]^{1/n} \tag{1}$$

where $IT_{CH}(S)$ = indirect trust of cluster-head (CH) on source node S. This can be calculated from indirectly given information by n neighbour nodes on node S and given to cluster-head (CH). The network threshold of indirect trust (ITN_{th}) can be evaluated using

$$ITN_{th} = \sum_{i=1}^{n-1} (IT_{vi})/n - 1 \tag{2}$$

where $ITvi$ = indirect trust of cluster-head on node i and n − 1 = number of nodes within the transmission range of cluster-head except CH (Fig. 3).

4.3 Variation of Transmission Range

Figure 4 illustrates the variation of transmission range in congested traffic with respect to present local density. K1, K2, K3 represents different value vehicle density at λ (Sensitivity of vehicle interaction) equal to 1/10, 1/7, 1/5. It is also observed that for lower value of density, transmission range is equal to maximum transmission range (MR) = 1000 (DSRC standard) which indicates less number of neighbour vehicles for a vehicle V. With the increase in density, transmission range is determined by the minimum value of MR and TR given by Eq. (2). K_{jam} is the maximum density at traffic jam.

Fig. 3 Cluster head
(CH) trust value calculation

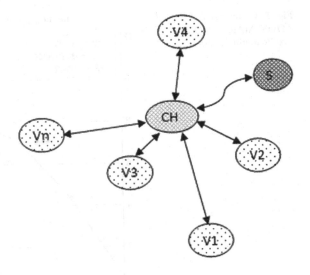

Fig. 4 Variation of
transmission range versus
estimated vehicle density

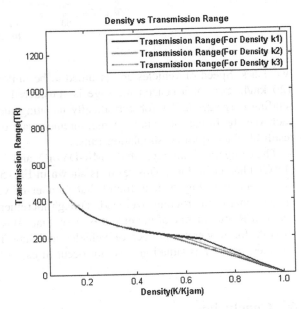

5 Result and Discussion

The proposed clustering and trust dependent ant colony routing method has been
simulated in MATLAB 2010a. We consider the VANET model described in
Sect. 3. The network has been simulated with 10–150 vehicles and position of those
vehicles has taken randomly along with some critical parameters for simulation
setup. Dimension of VANET size is assumed as 1500 * 1500 m of highway with

Fig. 5 Comparison of
AODV, MARDYMO and
TACR routing overhead

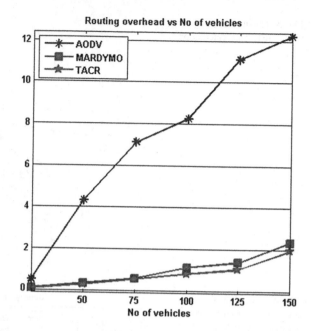

two lanes. Speed of vehicles are assumed to be uniform in between 60 km/h and 120 km/h. Size of a beacon message is considered as 200 bytes and interval of sending messages is 2 s for periodically updating the data record maintained by each vehicle. In the following our measurements are based on the averaging of the result obtained from 20 simulation runs.

The comparison among AODV, MR-DYMO, and Trust dependent ACO routing (TACR) has been done. This result is shown in Fig. 5.

As shown in Fig. 5, it is found that number of vehicles increased is directly proportional to the routing overhead. Though increment happend in MAR-DYMO and TACR, but the rate of increment is lesser than AODV. Further MAR-DYMO is suitable for moderate number of vehicles and fails in case of large number of vehicles, where this situation may not occur in case of TACR.

6   Conclusion

In this paper we have studied different routing Protocol initially. Further to optimize the routing ACO algorithm was used whose performance shows better result. It is also compared its overhead, the performance and overhead is more in AODV in comparison to MARDYMO and TACR. Other optimization approaches may be tried in future to enhance this piece of work.

References

1. Hannes Hartenstein and Kenneth P.Laberteaux. 'A Tutorial Survey on Vehicular Adhoc Networks', IEEE Communication Magazine, June 2008.
2. Pranav Kumar Singh, Kapang Lego, Themrichon Tuithung, "Simulation based Analysis of Adhoc Routing Protocol in Urban and Highway Scenario of VANET" International Journal of Computer Applications (0975–8887) Volume 12–No.10, January 2011, pp. 1–42.
3. Rakesh Kumar, 2Mayank Dave, "A Comparative Study of Various Routing Protocols in VANET", IJCSI International Journal of Computer Science Issues, Vol. 8, Issue 4, No 1, July 2011, pp 644.
4. Bijan Paul, Md. Ibrahim, Md. Abu Naser Bikas, ""VANET Routing Protocols: Pros and Cons", International Journal of Computer Applications (0975–8887) Volume 20–No. 3, April 2011, pp. 1–28.
5. T.D.C Little and A. Agarwal. 'An Information Propagation Scheme for VANETs', Proc. Intelligent Transportation Systems, September 2005, pp. 155–56.
6. Fan P, Mohamadian A, Nelson P, Haran J, John Dillenburg.: 'A novel direction-based clustering algorithm in vehicular ad hoc networks' In proc. of the transportation research board 86th annual meeting, Washington DC, United States, 2007.
7. Santos RA, Edwards A, Edwards Rand Seed L.: 'Performance evaluation of routing protocols in vehicular ad hoc networks', Int. Ad Hoc and Ubiquitous Computing 2005, pp. 80–91.
8. Fan P, Nelson P, Haran J, and Dillenburg J.: 'An improved compound clustering algorithm in vehicular ad-hoc networks' In proc. of the ninth int. conf. on applications of advanced technology in transportation, August 2006.
9. http://en.wikipedia.org/wiki/Intelligent_transportation_system,Wikipedia, 'Intelligent Transpotation System', 2008.
10. Perkins C, Belding-Royer E, Das E.: 'Ad-hoc on-demand distance vector (AODV) routing [EB/OL]', http://fags.orglrfcs356I.htmI2003–7, Accessed: 2 May, 2008.
11. Park Sanghyun, Shin Jayounq, Baek Senuqmin and Kim Sungchun.: 'AODV based Routing Protocol Considering Energy and Traffic in Ad-hoc networks', Proc. Int. Conf. on Wireless Networks, 2003, pp. 356–361.
12. 'AODV', http://moment.cs.ucsb.edu/AODV/aodv.html.
13. Blum J, Eskandarian A, Hoffman.L. 'Mobility management in IVC networks' proc. IEEE intelligent vehicles symposium, 2003.
14. Sergio Luis O. B. Correia, Joaquim Celestino Junior and Omar Cherkaoui.: 'Mobility-aware Ant Colony Optimization Routing for Vehicular Ad Hoc Networks', IEEE WCNC 2011–Network.

AgroKanti: Location-Aware Decision Support System for Forecasting of Pests and Diseases in Grapes

Archana Chougule, Vijay Kumar Jha and Debajyoti Mukhopadhyay

Abstract Grape is an important crop in Indian agriculture. There are many pests occurring on Grapes which cause huge yield loss to farmers. The grapes development is driven mainly by temperature and many pests have direct relation with temperature. We propose a decision support system named AgroKanti for managing pests on table grapes like powdery mildew and anthracnose. The decision support system is location based i.e. farmer is provided with details of pests considering current weather conditions at farmer's location. We support farmers with pest details like symptoms and management techniques for pests. We provide our system as an application on mobile phones. The knowledge base of pests is stored as ontology in OWL format. We have also developed a black box for agricultural experts where agricultural experts can generate pest ontology form text descriptions. We have used NLP techniques and AGROVOC library to extract pest details from text descriptions and generate ontology.

Keywords Decision support system · Ontology · Agriculture · Weather data extraction

A. Chougule (✉) · D. Mukhopadhyay
Maharashtra Institute of Technology, Pune, India
e-mail: chouguleab@gmail.com

D. Mukhopadhyay
e-mail: debajyoti.mukhopadhyay@gmail.com

V.K. Jha
Birla Institute of Technology, Mesra, Ranchi, India
e-mail: vkjha@bitmesra.ac.in

© Springer India 2016
S.C. Satapathy et al. (eds.), *Information Systems Design and Intelligent Applications*, Advances in Intelligent Systems and Computing 433,
DOI 10.1007/978-81-322-2755-7_70

677

1 Introduction

Agriculture is very important sector in Indian economy and grape is one of the major crops in Indian agriculture. There are grape experts in India who can help farmers to improve grape production. The formal representation of their knowledge using advanced technology can be directly used by decision support systems [1]. Expert knowledge through mobile application will be a great help to farmers. Many times farmers lose grape farms because of pests on grapes, as they do not know proper pest management techniques. The expert knowledge about grapes should be represented in ontology as it can be shared among diverse applications including Semantic web applications [2]. Procedural representation of expert knowledge about grape pests will help in building rich knowledge base and find new facts of grape pests. Knowledge base for AgroKanti is stored as ontology document. Inference rules and semantic reasoners like Bossam, Cyc, KAON2, Cwm, Drools, Flora-2, Jena and prova2 can be used to develop decision support system from generated pest ontology. We have used jena reasoner for AgroKanti.

AgroKanti provides pest management support for all stages of grape production as bud break, flowering and veraison. We have provided grape pest management support by providing details like reasons for pests, symptoms and treatments for pests. Compared to other resources of agricultural knowledge resources like internet, thesaurus and PDF documents; it becomes easier to provide desired specific information with AgroKanti knowledge base. Formal and specific representation of pest management knowledge and availability of the same on mobile phone helps farmers in easier understanding of expert knowledge. Farmers can have a look at treatment options available for particular a pest or disease on grapes.

As pests have tendency to develop resistance against controlling measures, it is very important to update farmers on new pest control measures for grapes. AgroKanti provides support for dynamic updating of pest knowledge base. It provides facility to update existing grape pest knowledge base at any point of time. We have used data mining techniques for automated construction of grape pest ontology from text descriptions of grape pests and update them dynamically [3].

Strength of AgroKanti is it provides support considering current weather conditions at farmer's location. There are strong relations between weather changes and insect pests and diseases of grapes [4]. We have studied these relations between weather conditions and grape pests and defined rules for pest management techniques accordingly. We extract weather details from meteorological websites of government of India.

Paper is organized as follows: We first provide work done by other people in developing agricultural expert system. We have then described architecture and implementation details of AgroKanti followed by performance evaluation. Conclusion and references are provided at end of the paper.

2 Related Work

Research papers have been published proposing decision support systems and expert systems for various crops by researchers all over world. We discuss some of such systems here.

Jinhui et al. [5] presented an online portal for agricultural ontology access. They have collected agricultural information from Web using distributed crawler. Collected information is used for generating OWL classes. The paper explains mapping for ontologies for multiple languages. The OWL classes extracted from information are mapped with OWL classes provided by AGROVOC and new merged OWL classes for are used as Knowledge base named as AOS. Jena APIs and Pellet inference engine is used for answering questions from farmers.

Tilva et al. [6] proposed weather based expert system for forecasting disease on corn crop. They have used fuzzy logic technique for developing inference engine of expert system. They have used temperature, humidity and leaf wetness duration as weather parameters for defining fuzzy rules to estimate plant disease. They have defined five classes for input and output member functions as very high, high, medium, low and very low.

A decision support system for management of Powdery Mildew in grapes is developed by Mundankar et al. [7]. They estimate disease risk by considering plant growth stage and weather condition. All the details about weather condition, field condition and plant growth stage are taken from end user through software interface. Expert system provides information regarding fungicide spray name and its dose for various field and weather conditions.

An agent oriented method for developing decision support system is adopted by Perini and Susi [8]. They described software development phases as early requirement analysis, late requirement analysis, architectural design and implementation for integrated production in agriculture. They listed various actors in agriculture production and showed their relationship in architectural design.

An expert system for the diagnosis of pests, diseases and disorders in Indian mango is proposed by Rajkishore Prasad et al. [9]. They described development of a rule-based expert system using ESTA; Expert System Shell for Text Animation. The system is based on answers to questions taken from farmers regarding disease symptoms.

An expert system for pest and disease management of Jamaican coffee is developed by Mansingh et al. [10] named as CPEST. It is built in wxCLIPS. Forward chaining is used as reasoning mechanism. They developed rule base containing 150 production rules. CPEST has three stages for solving problem as general data-gathering phase, diagnosis and possible treatments and integration of treatments.

A rule-based expert system to diagnose honeybee pests is described by Mahaman et al. [11] which can be used by beekeepers. It is implemented using EXSYS for Microsoft windows environment with backward chaining method Bange et al. [12] described a decision support system for pest management in

Australian cotton systems. It can be used on handheld devices to collect data required for pest management from different locations. An expert system for identification of pest diseases and weeds in olive crops is provided by Gonzalez-Andujar [13]. The knowledge base is created using interviewing technique and represented using IF-THEN rules. The knowledge base contains information for identification of 9 weed species, 14 insect species and 14 diseases.

Rebaudo and Dangles [14] developed an agent based model for integrated pest management coupling a pest model with a farmer behavior model. It is convinced in paper that passive IPM information diffusion is better than active diffusion. Potnikakos et al. [15] investigates effectiveness of location aware system of pest management for olive fruit fly. The described system uses information regarding olive fruit fly, meteorological conditions and spatiotemporal details of spraying areas. LAS have client-server architecture and it utilizes web services, geographic information system, expert system and multimedia technology.

An intelligent system for disease and pest diagnosis and control of tomatoes in greenhouses is proposed by Lopez-Morales et al. [16] named as JAPIEST. The system computes vapor pressure deficit to detect probable development of diseases on tomatoes. Graphical support is also provided with disease detection results.

3 Architecture and Implementation of AgroKanti

We have developed a decision support system for pest management of grapes which follows three tier architecture composed of the client layer, the application layer and the database layer. AgroKanti includes following elements (Fig. 1):

- Knowledge Base developed as ontology
- Weather data extractor
- Rule Base
- Inference Engine named as PestExpert
- User Interface

We have provided our system as an android based application to farmers. Knowledge base contains ontology having information about grape pests. These ontologies are created by extracting keywords from text descriptions of grape pests. Using natural language processing techniques [17] and AGROVOC [18] thesaurus provided by FAO, we try to find out most relevant words related to grape pests. AGROVOC is a large vocabulary of almost all areas of agriculture. It also provides support for multiple languages. We retrieve important keywords from grape pest descriptions by applying tokenization, stopping and stemming [3]. We have used Porter's stemming algorithm for stemming of retrieved keywords [19]. We have used open source data mining library in java: WEKA to apply TF-IDF algorithm [20] for finding most important keywords. These keywords are then verified with AGROVOC vocabulary and ranking to key words is provided accordingly. The

Fig. 1 AgroKanti system
architecture

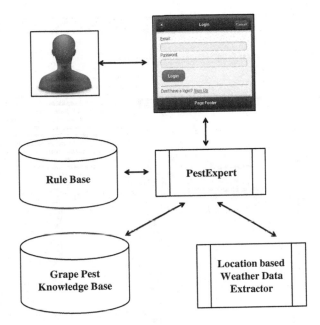

keywords are then used to mention classes, individuals and properties related to grape pests in pest ontology [21]. Web Ontology Language (OWL) is used to construct grape pest ontology. Classes, individuals, data properties, object properties and axioms are important parts of OWL document. In Grape pest ontology, major types of grape pests are mentioned as classes, subtypes of grape pests are mentioned as individuals and pest management details such as reasons, symptoms and management techniques are mentioned as data property values (Fig.2).

It is difficult for agricultural experts to represent knowledge in terms of ontologies. We have developed an application in java language which provides easy to use interface for agricultural experts. The application provides simple interface where grapes expert can add information like types of pests and diseases occurring on grapes and details like symptoms, reasons and remedy for grape pests. We have used Protégé APIs [22] to store this information in terms of ontology (Fig.3).

Location based Weather Extractor works as follows: For location based weather data extraction, we extract longitude and latitude of farmer's location using Location Manager Class from android library. From geographic coordinates we find out nearest weather station for farmer's location. Once we get nearest weather station of farmer where the grape farm is located; we then extract weather information from meteorological sites using jsoup library. Jsoup is a java library used to retrieve information from html pages using DOM application programming interface. We extract temperature, relative humidity and rainfall details for selected weather station.

Extracted weather information is then provided to PestExpert which is a rule based fuzzy inference engine. The relationship between weather conditions and

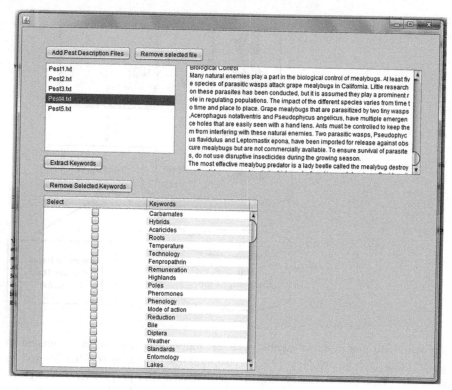

Fig. 2 Grape pest ontology generation from text

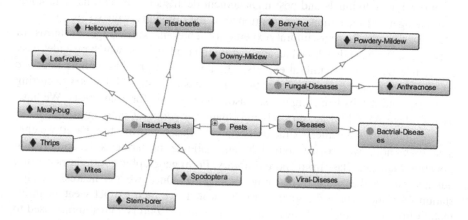

Fig. 3 Knowledge base for grape pests

```
<!-- http://www.semanticweb.org/himangipande/ontologies/2015/5/Grape-
Pests-Ontology#Powdery-Mildew -->

    <owl:NamedIndividual rdf:about="&Grape-Pests-Ontology;Powdery-
Mildew">
            <rdf:type rdf:resource="&Grape-Pests-Ontology;Fungal-
Diseases"/>
            <weather_con>T&gt; 70Â°F &amp; &lt;85Â°F</weather_con>
            <has_Symptom>Grayish-white growth on leaves and
berries.</has_Symptom>
            <Management>Spray of Sulfex (0.2%) or Wettasul (0.2%) or
Beyleton (0.1%) or Saprol (0.15%) at 10-15 days
intervals.</Management>
    </owl:NamedIndividual>
```

Fig. 4 Part of grape pests ontology

different pests on grapes is studied. For grapes, pests occur during different stages of growth. We consider eight stages of grape growth for generating rules as delay dormant, budbreak period, rapid shoot growth period, bloom to veraison period, veraison period, harvest period, postharvest period and dormant period [23]. Rules are generated considering these growth stages, weather conditions and expert knowledge and stored in rule base. PestExpert extracts corrective measures and suggestions using rule base for grape pests. These rules help to select correct measures to control pests in given weather conditions. Following are some examples of rules for grape pests (Fig. 4).

1. If T > 70 °F and < 85 °F then Pest ∼ **Powdery Mildew**
2. If T > 50 °F then Pest ∼ **Mealy bug**
3. If T > 68 °F and < 77 °F and WET = true then Pest ∼ **Downy Mildew**

DSS also provides suggestions based to prevent probable pests and yield losses which may occur in current weather conditions.

4 Performance Evaluation

We experimented the use of AgroKanti at 45 different locations of India. We listed pest management details provided by AgroKanti and compared it with actual pests and techniques used by farmers at these locations during same period. Figure 5 shows graph about forecasting of pests forecasted by AgroKanti and actual pests found at those locations for four pest types.

Fig. 5 Pest forecasting by AgroKanti

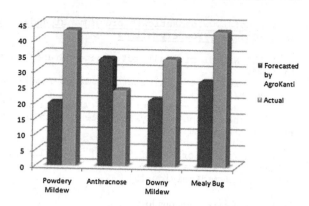

5 Conclusion

We have described a decision support system called as AgroKanti in this paper. We described generation of knowledge base for AgroKanti. As knowledge base is generated as ontology using natural language processing techniques; it can be reused by any other expert systems for grapes. Because there is strong relation between grape development, grape pest occurrences and weather conditions, we have provided innovative solution for weather based decision support. The system helps grape experts by providing very easy framework to generate pest ontology and to generate inference rules. We have given examples of rules to be used by fuzzy logic based inference engine of AgroKanti. The system helps grape growers by providing decision support for pest management considering real time weather conditions at farmer's location. As system is provided as android application it is easily available and accessible to farmers all over India. With developed system real time decision support is provided to farmers and an honest effort is taken to bridge gap between grape experts and grape growers. We expect it will help to minimize yield losses due to pests on grapes and increase profit to grape grower.

References

1. Devraj, Renu Jain: PulsExpert: An expert system for the diagnosis and control of diseases in pulse crops. Expert Systems with Applications Journal, Elsevier, vol. 38; 11463–11471 (2011).
2. Sukanta Sinha, RanaDattagupta, Debajyoti Mukhopadhyay: Designing an Ontology based Domain Specific Web Search Engine for Commonly used Products using RDF;CUBE 2012 International IT Conference, CUBE 2012 Proceedings, Pune, India; ACM Digital Library, USA; pp. 612–617; ISBN 978-1-4503-1185-4, (2012).
3. Olena Medelyan, ian H. Witten: Theasurus-based Index Term Extraction for Agricultural Documents; 2005/EFITA/WCCA joint congress on IT in Agriculture (2005).

4. Jagdev Sharma, Ajay Kumar Upadhyay: Effect of Climate Change on Grape and its Value-Added Products; Climate-Resilient Horticulture: Adaptation and Mitigation Strategies, Springer, India, pp. 67–73, (2013).

5. Xiong Jinhui, Yang Yong, Yang Zhifeng, Wang Shuya: An Online System for Agricultural Ontology Service. Third International Conference on Intelligent Networks and Intelligent Systems, 479–481, IEEEXplore (2010).

6. Vidita Tilva, Jignesh Patel, Chetan Bhatt: Weather Based Plant Disease Forecasting Using Fuzzy Logic; Nirma University International Conference on Engineering (NUICONE); IEEEXplore, (2012).

7. K Y Mundankar, S D Sawant, Indu S Sawant, J Sharma, P G Adsule: Knowledge Based Decision Support System for Management of Powdery Mildew Disease in Grapes; 3rd Indian International Conference on Artificial Intelligence (IICAI-07), pp. 1563–1571 (2007).

8. Anna Perini, Angelo Susi: Developing a decision support system for integrated production in agriculture; Environmental modeling and software 19; Elsevier journal; pp. 821–829, (2003).

9. Rajkishore Prasad, Kumar Rajeev Ranjan, A K Sinha: AMRAPALIKA: An expert system for the diagnosis of pests, diseases, and disorders in Indian mango; Knowledge-based systems 19; Elsevier journal; pp. 9–21, (2005).

10. Gunjan Mansingh, Han Reichgelt, Kweku-MuataOsei Bryson: CPEST: An expert system for the management of pests and diseases in the Jamaican coffee industry; Expert Systems with Applications 32; Elsevier journal; pp. 184–192, (2007).

11. B D Mahman, P Harizanis, I Filis, E. Antonopoulou, C P Yialouris, A B Sideridis: A diagnostic expert system for honeybee pests; Computer and Electronics in Agriculture; Elsevier journal; pp. 17– 31, (2002).

12. M P Bange, S A Deutscher, D Larsen, D Linsley, S Whiteside: A handheld decision support system to facilitate improved insect pest management in Australian cotton systems; Computers and Electronics in Agriculture 43; Elsevier journal; pp. 131–147, (2008).

13. J L Gonzalez-Andujar: Expert system for pests, diseases and weeds identification in olive crops; Expert Systems with Applications 36; Elsevier journal; pp. 3278–3283, (2009).

14. Francois Rebaudo, Olivier Dangles: An agent-based modeling framework for integrated pest management dissemination programs; Environmental modeling and software 45; Elsevier journal; pp. 141–149, (2013).

15. Costas M Pontikakos, Theodore A Tsiligiridis, Constantine P Yialouris, Dimitris C Kontodimas: Pest management control of olive fruit fly based on a location-aware agro-environmental system; Computer and Electronics in Agriculture; Elsevier journal; pp. 39–50, (2012).

16. V Lopez-Morales, O Lopez-Ortega, J Ramos-Fernandez, L B Munoz: JAPIEST: An integral intelligent system for the diagnosis and control of tomatoes diseases and pests in hydroponic greenhouses; Expert Systems with Applications 35; Elsevier journal; pp. 1506–1512, (2008).

17. OpenNLP, https://opennlp.apache.org/.

18. AGROVOC Thesaurus, http://aims.fao.org/agrovoc#.VF29AvmUc2U.

19. Porter: An algorithm for suffix stripping, Program, Vol. 14, no. 3, pp. 130–137 (1980).

20. TFIDF Algorithm, http://en.wikipedia.org/wiki/Tf-idf.

21. Jena Ontology API, http://jena.apache.org/documentation/ontology/.

22. Protégé, http://protege.stanford.edu/.

23. http:\\www.ipm.ucdavis.edu.

A Unified Modeling Language Model for Occurrence and Resolving of Cyber Crime

Singh Rashmi and Saxena Vipin

Abstract In the current scenario, distributed computing systems play significant role for accessing the various kinds of internet services. The different handheld devices like palmtop, laptop, mobile, etc. can be connected across the distributed network. People enjoy social networking websites, online purchasing websites, and online transaction websites in the daily routine life. On the other hand, hackers are regularly watching the activities of the people who are categorized as the authorized users connected across the globe. The present work is related to propose a model which is based upon the object-oriented technology for occurring of cyber crime across the distributed network. A well known Unified Modeling Language is used and one can easily write the code for implementation of model in any object-oriented programming language. After that a UML model is proposed for filing the FIR online against the cyber crime. The activities in the above procedure are represented by the UML activity diagram which is finally validated through the concept of finite state machine.

Keywords Cyber crime · UML · Activity diagram · FIR (first information report) · Finite state machine

1 Introduction

The Unified Modeling Language (UML) is widely used as a standard technique in software development and invented By Booch et al. [1, 2]. They characterized that how to show a problem in pictorial form through UML. There are various tools which have been produced to support UML model either static or dynamic model.

S. Rashmi (✉) · S. Vipin
Department of Computer Science, Babasaheb Bhimrao Ambedkar University
(A Central University), Vidya Vihar, Raebareli Road, Lucknow 226025, U.P., India
e-mail: rshmi08@gmail.com

S. Vipin
e-mail: vsax1@rediffmail.com

© Springer India 2016
S.C. Satapathy et al. (eds.), *Information Systems Design and Intelligent Applications*, Advances in Intelligent Systems and Computing 433,
DOI 10.1007/978-81-322-2755-7_71

687

Such UML tools translate any model into any programming language. UML includes a set of notations i.e. graphics notations which are used to create a model for easily understandable by anyone [3, 4]. Software engineers and researchers resolve the complex problem through UML by which they represent their problem in diagrammatic representation. Global decision reaction architecture built on the basis of requirement for the reaction after alert detection mechanisms in information system security and this security has been applied on telecom infrastructure system [5]. A model for security issues in distributed network, having features such as deployment of security strategy, cooperation of security components, automatic distribution, self-adaptive management function, etc. [6]. Cloud computing helps to remove high cost computing over distributed computing and minimize infrastructure for information technology based services and solutions. It provides a flexible and architecture accessible from anywhere through lightweight portable devices [7]. In network virtualization, virtualized infrastructure is also used to provide manifold independent networks over multiple framework providers [8]. Virtual networks managed by virtual network operator. A developed model that provides a structure to communities and can be used to purposive their level of alertness and to generate a strategy to upgrade their security perspective and magnify their possibility of auspiciously preventing and detecting from a cyber attack [9]. Cyber crime is typically occur when anyone accessing, modifying, destroying computer data without owner's permission. This unauthorized access can be committed against property, persons or government [10]. Cloud computing is used to circle components from technologies such as grid computing and autonomic computing into a new arrangement structure. This expeditious transformation around the cloud has stimulated concerns on a censorious issue for the victory of information security [11].

Cyber crime is usually mentioned as criminal actions using computer internet. What happens when cyber crime occurred in the real world and how people can protect and aware from occurrence of cyber crime [12]. In the modern scenario, day by day normal methods of cyber security become outmoded. They are getting failed in maintaining security [13]. Cloud computing provides a frame work for information technology based resolutions and favor that the industry and organizations uses. It also provides a flexible structure accessible through internet from all over the world using light weighted portable devices [14]. To calculated traffic congestion and standard of services during any attack over the network and how to provide network security under this situation [15]. In [16] the recent improvements to the potential of law as personal and public constructs are deployed for cloud association with crime and criminal activities. The research paper [17] reviewed that how to prevent the cybercrime influence from portable devices such as Smartphone, laptops, tablets etc. Models are described for permission based security and behavior-based detection for information security. In the present time countercyber attacks are mostly occurred in many countries due to cyber crime independently as an initial attack [18]. Cyber system must [19] also evolve to provide techniques for prevention and defense. Some experiments presents in this paper are blend cyber warfare with some approaches including planning and execution for deception in cyber defense.

The present work is related to the development of the model based on the object-oriented technique for identification of the cyber crime and filing the FIR against unauthorized users. One can develop the model in any programming language based on the object-oriented methodology because developed model is platform independent model. The proposed model is also validated by the use of concepts of Finite State Machine (FSM).

The purpose of proposed model is for identification of cyber crime which has been implemented and tested through the concept of software engineering. Different test cases have been generated for validation purpose on the proposed model and it is observed that the model is effective, reliable and robust.

2 UML Modeling for Occurrence and Filing of Cyber Crime

2.1 UML Class Model

UML class is a static representation of the problem which shows, how the problem is behaving or moving towards achievement of goal. The diagram is designed by the use of standard symbol available in Booch [3, 4]. In Fig. 1, the occurrence of cyber crime is represented through different classes. User is categorized as the authorized or unauthorized users. Association is shown between the two classes algorithm the representation of cardinality. Both kinds of users have internet connection and different web portals are grouped on the internet for the use of users. As represented in the class diagram, unauthorized user hacks the web portals multiple times by multiple unauthorized users. Hacking is controlled by hack class which is the type of cloning, steel card, steel data, steel bandwidth, login/password, etc. When the hacking occurs, then authorized users get information about the hacking. The different types of attributes and operations used to model the above diagram are recorded in the following Table 1.

2.2 UML Activity Model

The activity model shows the dynamic aspects of the problem. In the present work, an activity model is designed for occurrence of cyber crime. It connects the links from one activity to another activity controlled by an event. The different activities are summarized below in the following steps:

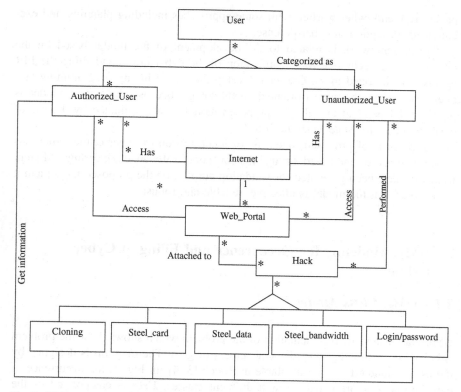

Fig. 1 UML class model for occurrence of cyber crime

Step 1   User applies for Internet Connection for surf the internet services;

Step 2   User categorized either authorized or unauthorized;

Step 3   User registered for internet connection, if user got connection then move to next step else user go to step 1;

Step 4   When user got connection for surfing net, user surfs the websites and access the data;

Step 5   According to step 2 user may be authorized or unauthorized who can access the websites;

Step 6   When unauthorized user hacks the data follow next step;

Step 7   Cyber crime occurs then it is reported to the user and moves to step 1;

The above steps are represented in the Fig. 2 which show the occurrence of cyber crime.

Table 1 Attributes and operations used for UML class model

Name of class	Attributes	Operations
User	User_id	Surf_webpages()
	User_name	Surf_apps()
	Mobile_number	Login()
	Nationality	Logout()
	Gender	
Authorized_User	Categorization_user	Mail_access()
	Address	Online_transaction()
	Date_of_birth	
	E-mail	
Unauthorized_User	Login_in_time	Steel_data()
	Login_out_time	Steel_password()
	Login_duration	Steel_card()
	Session_record_time	Unauthorized_login()
Internet	Connection_id	Access()
	Service_provider	Security()
	No._of_users Bandwidth	Surfing()
Web_Portal	Physical_location	Universal_login()
	Security_type	Facilitates_messaging
	Contact_information	Multi_channel_consistency()
	Business_information validation	Search()
Hack	Hacker_name	Access_unauthorized_data()
	Age	Hack_websites()
	Gender	Hack government_sites and data()
Cloning	Cloning_type	Credit_card_cloning()
	Cloning_device	Debit_card_cloning()
		Websites_cloning()
Steel_Card	Card_holder_name	Removing_funds()
	Expiry_date	Illegal_purchasing()
	Organization_name	Identity_theft()
	Card_number	
	Card_type	
Steel_Data	Type_of_data	Data_modification()
	Storage_device	Access_Data()
	Data_amount	
	Data_Address	
Steel_Bandwidth	Service_provider_name	Data_transmission()
	Bit_rate	Media_file_transmission()
	Capacity	Video_compression()
	City/State	

3 Validation of UML Activity Model Through Finite State Machine

Let us first explain the concept of Finite State Machine (FSM) which is a mathematical model of computation and is used to design logic circuits. A sequential logic unit takes an input and a current state to produce an output and new state. It can be represented using state transition table which shows current state, input state, new output state and the next state. It can also be represented using state transition diagram. It is defined by M and explained [20] as

$$M = \left(\sum, Q, \delta, q_0, F \right)$$

where
$\sum$ set of Inputs (Alphabets and symbols);
q_0 an initial state;
F final state;
δ transition between two states;
Q set of finite states;

On the basis of above definition of automata the Fig. 2 is converted into FSM by means of state and transition from one state to another state. The different states are recorded in the Table 2 and these are represented as (q_0, q_1, q_2, q_3, q_4, q_5 and q_6).

The two states let q_0 and q_1 are grouped through a transition event. The different transition events are given in Table 3.

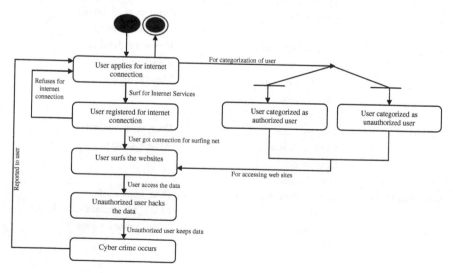

Fig. 2 UML activity model for occurrence of cyber crime

Table 2 Description of states selected from UML activity model

Name of State	Description of state
q_0	User applied for internet connection
q_1	User categorized as authorized user
q_2	User categorized as unauthorized user
q_3	User registered for internet connection
q_4	User surfs the websites
q_5	User hacks data
q_6	Cyber crime occur

Table 3 Description of events selected from UML activity model

Name of Input	Description of input
a	Categorization of user
b	Accessing websites
c	Surf for internet services
d	User got connection for surfing net
e	Reported to user that cyber crime occur
f	User access the data
g	User keeps data
h	User refuses for internet connection

From the definition of automata $\Sigma = \{a, b, c, d, e, f, h\}$ shows the set of input which are shown in the Table 3.

On the basis of above, a state transition diagram is designed which is represented in Fig. 3.

Above figure is used for validation purpose of UML activity model and different test cases are generated on the basis of transition table recorded in Table 4.

Valid Test Case 1 If unauthorized user hacks the data, cyber crime occurs and it is reported to the user.

Fig. 3 FSM representation from UML activity model

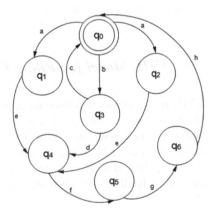

Table 4 Transition table

State	Event							
	a	b	c	d	e	f	g	h
q_0	q_1/q_2	q_3	–	–	–	–	–	–
q_1	–	–	–	–	q_4	–	–	–
q_2	–	–	–	–	q_4	–	–	–
q_3	–	–	q_0	q_4	–	–	–	–
q_4	–	–	–	–	–	q_5	–	–
q_5	–	–	–	–	–	–	q_6	–
q_6	–	–	–	–	–	–	–	q_0

$$\delta(q_0, a) \rightarrow q_1 \quad \Rightarrow \quad q_0 \rightarrow a \; q_1$$
$$\delta(q_1, e) \rightarrow q_4 \quad \Rightarrow \quad q_1 \rightarrow e \; q_4$$
$$\delta(q_1, f) \rightarrow q_5 \quad \Rightarrow \quad q_1 \rightarrow f \; q_5$$
$$\delta(q_5, g) \rightarrow q_6 \quad \Rightarrow \quad q_5 \rightarrow g \; q_6$$
$$\delta(q_6, h) \rightarrow q_0 \quad \Rightarrow \quad q_6 \rightarrow h \; q_0$$

After removing the non-terminals the string is $q_0 = aefghq_0 = aefgh$
Valid Test Case 2 The user is not registered for internet connection.

$$\delta(q_0, b) \rightarrow q_3 \quad \Rightarrow \quad q_0 \rightarrow b \; q_3$$
$$\delta(q_3, c) \rightarrow q_0 \quad \Rightarrow \quad q_3 \rightarrow c \; q_0$$

After removing the non-terminals the string is $q_0 = bcq_0 = bc$
Valid Test Case 3 If cyber crime occurs then it is reported to the user.

$$\delta(q_0, b) \rightarrow q_3 \quad \Rightarrow \quad q_0 \rightarrow b \; q_3$$
$$\delta(q_3, d) \rightarrow q_4 \quad \Rightarrow \quad q_3 \rightarrow d \; q_4$$
$$\delta(q_4, f) \rightarrow q_5 \quad \Rightarrow \quad q_4 \rightarrow f \; q_5$$
$$\delta(q_5, g) \rightarrow q_6 \quad \Rightarrow \quad q_5 \rightarrow g \; q_6$$
$$\delta(q_6, h) \rightarrow q_0 \quad \Rightarrow \quad q_6 \rightarrow h \; q_0$$

After removing the non-terminals the string is $q_0 = bdfghq_0 = bdfgh$.

3.1 UML Model for Filing Cyber FIR

UML model shows that how an authorized user is filing cyber FIR. The diagram shows that many authorized users have many internet connections. Police station and cyber cell both are connected with internet. Different police stations have different cyber cells. When an authorized user submitted cyber FIR to the police station, police station has cyber cell so the cyber cell performs enquiries and generate a feedback which is delivered to the authorized user (Fig. 4).

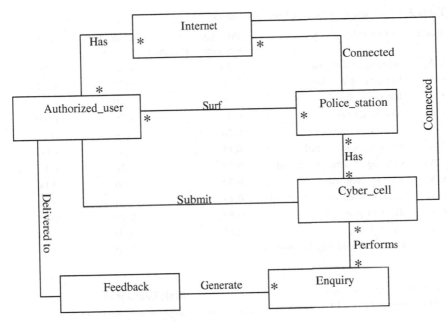

Fig. 4 UML model for filing cyber FIR

Risk analysis for occurrence of crime Risk is directly related to the loss due to cyber crime. In the present work percentage of loss due to cyber crime items has been evaluated. Let us define the two important factors associated to the risk analysis, these are given below:

(a) Probability of fault (CA_N)
(b) Cost (affected due to loss CA_N)

where CA_N are the items responsible for the cyber attack, then risk is computed by the following

$$R(CA_N) = P(CA_N) * C(CA_N)$$

The cyber attack algorithm for the computation of risks is recorded in Table 5.

The list of cyber attack is purely taken from the cyber crime cell and it consists of real data which is observed by grouping the 100 cyber cell complaints i.e. FIR. It is registered FIR either through online/offline mode and attacks are categorized through the unique code.

The decreasing sequence of losses is CA_{10}, CA_4, CA_2, CA_8, CA_1, CA_{12}, CA_7, CA_{11}, CA_5, CA_9, CA_6, and CA_3. From the Table 5 it is observed that the maximum loss is due to **Theft of Password** therefore, it should be resolved first to minimize the losses and the losses are minimized according to the said sequence of cyber attacks. A graphical view of computation of risk is also represented in Fig. 5.

Table 5 Calculated the risk based on cyber attack

Code	List of cyber attack	Probability of occurrence P (CA$_N$)	Cost affected C (CA$_N$)	R (CA$_N$)
CA$_1$	Stealing of database	0.20	0.80	0.16
CA$_2$	Hacking of websites	0.35	0.70	0.245
CA$_3$	Job scams/frauds	0.10	0.60	0.06
CA$_4$	Mobile crimes	0.45	0.70	0.315
CA$_5$	Antisocial activities	0.20	0.50	0.1
CA$_6$	Stealing of bandwidth	0.15	0.40	0.06
CA$_7$	Cloning of debit/credit card	0.20	0.70	0.14
CA$_8$	E-commerce fraud	0.30	0.60	0.18
CA$_9$	Unauthorized network access	0.15	0.55	0.0825
CA$_{10}$	Theft of password	0.50	0.80	0.4
CA$_{11}$	Identity theft	0.15	0.70	0.105
CA$_{12}$	Cyber blackmailing/harassment	0.25	0.60	0.15

Fig. 5 Risk evaluation on the basis of probability and factor

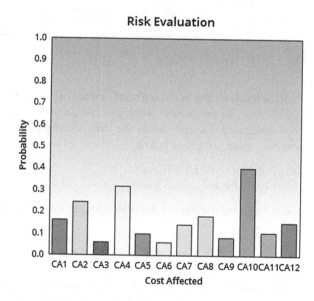

4 Concluding Remarks

From the above work, it is concluded that UML is a powerful modeling language for solution of the complex research problems. In the present work a UML model is proposed for the online FIR and computation of losses from the cyber attacks. The UML model is validated through FSM technique and various valid test cases have been generated for validation of proposed model. In the end, a technique for computation for risk analysis is proposed and finds the cyber attack having maximum risk analysis should be resolved first. The present paper can be extended further for method which can be suggested for minimization of losses like curve fitting method, optimization method, etc.

References

1. OMG, Unified Modeling Language Specification, http://www.omg.org (Accessed on 12th Feb. 2014), 1997.
2. OMG, Unified Modeling Language (UML)-Version1.5, OMG document formal/2003-3-01, (2003), Needham, MA.
3. Booch G., Rambaugh J., and Jacobson I.: The Unified Modeling Language User Guide, Twelfth Indian Reprint, Pearson Education, 2004.
4. Booch G., Rumbaugh J., and Jacobson I.: The Unified Modeling Language User Guide", China Machine Press, Beijing, 2006.
5. Feltus C., Khadraoui D. and Aubert J.: A security decision- reaction architecture for heterogeneous distributed network, published in Availability, Reliability and Security, publisher IEEE, pp. 1–8, Feb 2010.
6. Ping S. P.: An Improved Model of Distributed Network Information Security, published in Educational and Information Technology, publisher IEEE, Vol. 3, Sept 2010.
7. Subashini S., Kavitha V.: A Survey on Security Issues in Services in Delivery Models of Cloud Computing, Journal of Network and Computer Applications, Elsevier publication, Vol. 34, pp. 1–11, Jan 2011.
8. Goyette, R. Karmouch, A.: A Dynamic Model Building Process for Virtual Network Security Assessment, published in IEEE conference, pp. 482–487, Aug 2011.
9. White, Gregory B.: The Community Cyber Security Maturity Model" IEEE publication, pp. 173–178 Nov 2011.
10. Sekgwathe V. AND Talib M.: Cyber Crime Detection and Protection: Third World Still to Cope-Up, published by Springer, vol. 171, PP. 171–181, 2011.
11. Dimitrios Z. and Dimitrios L.: Addressing Cloud Computing Security Issues, Future Generation Computer System, Elsevier publication, Vol. 28, pp. 583–592, March 2012.
12. Zhang Y., Xiao Y. et al: A Survey of Cyber Crimes, published in Journal Security and Communication Networks, Vol. 5, pp. 422–437, April 2012.
13. Jain M., Vasavada J., Jain G. And Patel P.: Information Security and Auditing for Distributed Network, published in Instrumentation & Measurement, Sensor Network and Automation, publisher IEEE, Vol. 2, Aug 2012.
14. Bhadauria R. And Sanyal S.: Servey on Security Issues in Cloud Computing and Associated Mitigation Techniques, published by An International Journal of Computer Applications, Vol. 47, 2012.
15. Fang F., Xiaoyan L. And Jia W.: Network Security Situation Evaluation Method for Distributed Denial of Service, published in IMCCC, Publisher IEEE, pp. 16–21, Dec 2012.
16. Hooper C., Martine B. and Choo K.K.R.: Cloud Computing and its implications for cybercrime investigations in Australia, Published by Elsevier, 29, pp. 152–163, 2013.
17. Safavi S., Shukar Z. and Razali R.: Reviews on Cybercrime Affecting Portable Devices, Published by Elsevier, 11, pp. 650–657, 2013.
18. Kallberg J.: Aright to Cybercounter Strikes: The Risks of Legalizing Hack Back, in IT Professionals, Vol. 17, no. 1, pp. 30–35, Jan–Feb 2015.

19. Heckman K.E., Stech F.J., Schmoker B.S. and Thomas R.K.: Denial and Deception in Cyber Defense, in Computer, Vol. 48, no. 4, pp. 36–44, Apr. 2015.
20. Kumar N., Singh R. and Saxena V.: Modeling and Minimization of Cyber Attack through Optimization Techniques, An International Journal of Computer Application. 99(1), pp. 30–34, 2014.

Cross Lingual Information Retrieval (CLIR): Review of Tools, Challenges and Translation Approaches

Vijay Kumar Sharma and Namita Mittal

Abstract Today's Web spreads all over the world and world's communication over the internet leads to globalization and globalization makes it necessary to find information in any language. Since only one language is not recognized by all people across the world. Many people use their regional languages to express their needs and the language diversity becomes a great barrier. Cross Lingual Information Retrieval provides a solution for that barrier which allows a user to ask a query in native language and then to get the document in different language. This paper discusses the CLIR challenges, Query translation techniques and approaches for many Indian and foreign languages and briefly analyses the CLIR tools.

Keywords CLIR · Dictionary translation · Wikipedia translation · UNL · Corpora · Ontology · NER · Google translator · Homonymy · Polysemy

1 Introduction

Information Retrieval (IR) is a reasoning process that is used for storing, searching and retrieving the relevant information between a document and user needs. These tasks are not restricted to only Monolingual but also Multilingual. The documents and sentences in other languages are considered as unwanted "noise" in classical IR [1, 2]. CLIR deals with the situation where a user query and relevant documents are in different language and the language barrier becomes a serious issue for world communication. A CLIR approach includes a translation mechanism followed by mono lingual IR to overcome such language barriers. There are two types of translation namely query translation and documents translation. Query translation

V.K. Sharma (✉) · N. Mittal
Departmemt of Computer Science and Engineering, MNIT, Jaipur, India
e-mail: 2014rcp9541@mnit.ac.in

N. Mittal
e-mail: nmittal.cse@mnit.ac.in

© Springer India 2016
S.C. Satapathy et al. (eds.), *Information Systems Design and Intelligent Applications*, Advances in Intelligent Systems and Computing 433,
DOI 10.1007/978-81-322-2755-7_72

approaches are preferred due to a lot of computation time and space elapsed in document translation approaches [3]. Many workshops and Forums are acquainted to boost research in CLIR. Cross Language Evaluation Forum (CLEF) deals mainly with European languages since 2000. The NII Test Collection for IR System (NTCIR) workshop is planned for enhancing researches in Japanese and other Asian languages. First evaluation exercise by Forum for Information Retrieval Evaluation (FIRE) was completed in 2008 with three Indian languages Hindi, Bengali, Marathi. CLIA consortium includes 11 institutes of India for the project "Development of Cross Lingual Information Access system (CLIA)" funded by government of India. The objective of this project is to create a portal where user queries are responded in three possibilities such as responded in the query language, in Hindi and in English [2]. Literature Survey is discussed in Sect. 2. Issues and Challenges are discussed in Sect. 3. Various CLIR Approaches are discussed in Sect. 4. Section 5 includes Comparative Analysis and Discussion about CLIR translation technique and retrieval strategies. A brief analysis of CLIR tools also included in Sect. 5.

2 Literature Survey

Makin et al. were concluded that bilingual dictionary with cognate matching and transliteration achieves better performance. Parallel corpora and Machine Translation (MT) approaches are not well functioned. [4]. Pirkola et al. were experimented with English and Spanish languages and extract similar terms to develop transliteration rules [5]. Bajpai et al. were developed a prototype model where query was translated using any one technique including MT, dictionary based and corpora based. Word Sense Disambiguation (WSD) technique with Boolean, Vector space and Probabilistic model was used for IR [6]. Chen et al. were experimented with SMT and Parallel corpora for translation [7]. Jagarlamudi et al. were exploited statistical machine translation (SMT) system and transliteration technique for query translation. Language modeling algorithm was used for retrieving the relevant documents [8]. Chinnakotla et al. were used bilingual dictionary and rule based transliteration approach for query translation. Term-Term co-occurrence statistics were used for disambiguation [9]. Gupta et al. were used SMT and transliteration and the queries wise results was undergone mining and a new list of queries was created. Terrier open source[1] search engine was used for information retrieval [10]. Yu et al. were experimented with domain ontology knowledge method which is obtained from user queries and target documents [11]. Monti et al. were developed ontology based CLIR system. First linguistic pre-processing step was applied on source language query then transformation routines (Domain concept

---

[1]www.terrier.org.

mapping and RDF graph matching) and translation routines (Bilingual dictionary mapping and FSA/FSTs Development) were applied [12].

Chen-Yu et al. were used dictionary based approach and Wikipedia as a live dictionary for Out Of Vocabulary (OOV) terms. Further standard OKAPI BM25 algorithm was used for retrieval [13]. Sorg et al. were used Wikipedia as a knowledge resource for CLIR. Queries and documents both are converted to inter lingual concept space which is either Wikipedia article or categories. A bag-of-concept model was prepared then various vector based retrieval model and term weighting strategies experimented with the conjunction of Cross-Lingual Explicit Semantic Analysis (CL-ESA) [14]. Samantaray et al. were discussed concept based CLIR for agriculture domain. They were used Latent Semantic Analysis (LSA), Explicit Semantic Analysis (ESA) and Universal Networking language (UNL) and WordNet for CLIR and WSD [15]. Xiaoninge et al. were used Google translator due to high performance on named entity translation. Further Chinese character bigram was used as indexing unit, KL-divergence model was used for retrieval and pseudo feedback was used for improve average precision [16]. Zhang et al. were proposed search result based approach and appropriate translation was selected using inverse translation frequency (ITF) method that reduces the impact of the noisy symbols [17]. Pourmahmoud et al. were exploited phrase translation approach with bilingual dictionary and query expansion techniques were used to retrieve documents [18].

3 Issues and Challenges

Various issues and challenges are discussed in Table 1.

Table 1 List of CLIR issues and challenges

Issue and challenges	Homonymy	Polysemy	Word inflection	Phrase translation	Lack of resources	OOV Terms
Definition	Word having two or more different meaning	Word having multiple related meaning	Word may have different grammatical forms	Phrase gives different meaning then the words of phrase	Unavailability of resources for experimentation	Word which not found in dictionary. Like names, new term, technical terms
Example	"Left" means "opposite of right" or "past tense of leave"	"Ring" may be a wedding ring or boxing ring	Good, better, best are different forms of word "Good"	"Couch potato" used for someone who watches too much television	Dictionary, parallel corpora, MT system, character encoding	"H1N1 Malaysia" is a newly added term for influenza disease

Table 2 List of CLIR approaches with description

S. no	Approaches	Description and issues	References
1	Bi-lingual dictionary	Contains a list of source language words with their target language translations. Dictionary quality and coverage is an issue	[1, 2]
2	Corpora based	Corpora are the collection of natural language text in one or multiple languages. Parallel corpora are exactly the translation of each other sentence by sentence or word by word. Comparable corpora are not exactly the translation but cover same topic and contain equivalent vocabulary. Corpora based approach achieves better performance than the bi-lingual dictionary based approach, but these corpora are not available in all languages. In case of unavailability of corpora, it is very cumbersome and computationally expensive to construct parallel corpora of sufficient size	[1, 2]
3	Machine translation (MT)	MT tools used to translate queries into target documents language and target documents into source query language. MT tools save time in case of large text document but short documents are not translated correctly due to lack of context and syntactic structure for WSD. User queries are often short so MT system is not appropriate. MT system is computationally expensive for document translation. MT system is inefficient due to computation cost and unavailability	[20]
4	Transliteration	OOV terms are transliterated by either phonetic mapping or string matching techniques. Phonetic mapping is needed for the languages which have dissimilar alphabets. String matching techniques work best when the two languages having a shared common alphabet. Missing sound is an issue in phonetic mapping. Transliteration variant is an issue in string matching technique	[1, 4]
5	Co-occurrence method	Term-term co-occurrence method is used for translation disambiguation. Only a bilingual dictionary and a monolingual corpus are needed. Monolingual corpus of sufficient size is not available for a large set of languages and it is very cumbersome to create a monolingual corpus	[9, 21]
6	Ontology	An explicit specification for a conceptualization, the combination of ontological knowledge and its connection to the dictionaries gives a powerful approach for resolving CLIR problems	[11, 12]
7	Wikipedia	It is a Web-based, multilingual free content encyclopedia and written by volunteers from the whole world. There are total six million articles in 250 languages and still grow up. Wikipedia inter language link is defined between the same article in different language and it would be useful for translation disambiguation	[13, 14]

(continued)

Table 2 (continued)

S. no	Approaches	Description and issues	References
8	Google translation (GT)	GT is biased towards named entity and Terms in NTCIR topics are mostly name entities thats why Google translation may work well on NTCIR topics	[16]
9	Universal networking language (UNL)	In UNL, a sentence is parsed and a hyper-graph is constructed which having concepts (Universal words) as nodes and relations as arcs. A hyper-graph represents the set of binary relations between any two concepts	[15]
10	Web bases translation	The parallel and comparable web documents are also utilized for query translation and these documents are automatically discovered for different languages. In search result based approach, query terms are disambiguated by search result documents	[17, 22, 23]
11	Word sense disambiguation (WSD)	Appropriate sense of the word is identified. WSD mainly utilize four elements namely first is the word sense selection, second is the external knowledge source utilization, third is the context representation, fourth is the classification method selection	[24]
12	Named entity recognition (NER)	A natural language text is classified into predestined categories such as the person names, locations, organizations etc. State-of-the-art NER systems achieves near-human performance for English language	NER[1]
13	Lemmatization	Every word is simplified to its uninflected form or lemma. For example words "better" and "best" simplified in their uninflected form "good"	[2]

[1]http://en.wikipedia.org/wiki/Named_entity_recognition

4 CLIR Approaches

Various CLIR approaches are discussed in Table 2.

5 Comparative Analysis and Discussion

A comparative analysis of CLIR approaches is presented in Table 3.

Mean Average Precision (MAP) is the evaluation measure. MAP for a set of queries is the mean of the average precision score of each query and precision is the fraction of retrieved documents that are query relevant. Google translator is more

Table 3 Comparative analysis of CLIR approaches

Authors	Languages	Approaches	Datasets	Results (MAP)
Makin et al. [4]	H-TL	BD, CM, TR	BBC Hindi, NavBharat times website	0.2771 (JWS) 0.2449 (LCS)
Jagarlamudi et al. [8]	H-E	MT, PC, TR, LM	CLEF 2007	0.1994 (TD) 0.2156 (TDN)
Chinnakotla et al. [9]	H-E, M-E	BD, TR, COD	CLEF 2007	0.2336 (H (T)) 0.2952 (H (TD)) 0.2163 (M (T))
Chen-Yu et al. [13]	C, J and K	BD, WP, BM25	NTCIR-6	0.0992 (C-CJK-T) 0.0802 (C-CJK-D)
Yu et al. [11]	C-E	BD, HN, OL, COD	NTCIR-4	0.2652 (MITLAB-C-E)
Gupta et al. [10]	H-E	MT, TR, QM, Terrier System	FIRE 2010	0.3723 (BB2C retrieval model)
Sorg et al. [14]	E, G, F, and S	WP, BOC Model, CL-ESA, CAT-ESA, TREE-ESA	JRC-acquis (J) and Multext (M)	0.33 (M), 0.28 (J) (CLESA), 0.43 (M), 0.33 (J) (Cat-ESA), 0.46(M) and 0.31 (J) (Tree-ESA)
Xiaoning et al. [16]	C-E	GT, CCB, KL-Divergence and PF	NTCIR-7	0.3889
Chen et al. [7]	E, G, F, DT, I, S	L and H MT System, PC	CLEF 2003	0.3814 (F-G), 0.3446 (F-DT), 0.3859 (G-I), 0.4340 (I-S), 0.4694 (E-G), 0.4303 (E-S)
Zhang et al. [17]	E-C	SRWB and ITF	NTCIR-4	0.1582
Pourmahm-oud et al. [18]	P-E	BD, CT, QE, LM	Test collection prepared by themselves	0.3648 (without QE) 0.4337 (with QE)

BD bilingual dictionary, *CM* cognate matching, *TR* transliteration, *HN* HowNet, *PC* parallel corpora, *MT* machine translation, *LM* language modelling, *WP* wikipedia, *COD* co-occurrence distance, *OL* ontology, *QM* query mining, *QE* query expansion, *LSI* latent semantic indexing, *BOC* bag of concept, *GT* google translator, *CCB* chinese character bigram, *PF* pseudo feedback, *SRWB* search result web based approach, *ITF* inverse translation frequency, *CT* cohesion translation, *BT* back translation, *ER* entity recognition, *JWS* jaro winkler similarity, *LCS* longest common subsequence, *T* title, *D* description, *N* narration, foreign language (*E* English, *G* Germen, *DT*: Dutch, *I* Italian, *S* Spanish, *C* Chinese, *P* Persian, *F* Finnish, *J* Japanese, *K* Korean, *FR* French), Indian languages (*H* Hindi, *M* Marathi, *TL* Telugu)

effective due to biasing towards named entities and 0.3889 MAP achieved for English-Chinese [16]. Machine translation and Parallel corpora combinedly achieve better MAP that is 0.4694 for English-Germen [7] but lack of resources problem is there because a parallel corpora of enough size is not available for all languages.

Table 4 Comparative analysis of CLIR tools

S. no	Tools	Language supported	Translation technique	Functionality	Limitation
1	MULINEX	F, G and E	BD and BT	Interactive QD and QE, summaries and search results are translated on demand	Synonymy and Homonymy, User assisted query translation
2	KEIZAI	E, J and K	BD and PC	Interactive query translation along with English definition, target documents summary with English summary & document thumbnails visualization	Synonymy and homonymy, User assisted query translation
3	UCLIR	Arabic languages	BD and MT	Multi lingual query, interactive and non-interactive English query, Relevant retrieved document translated in English by word level (dictionary) or document level (MT), document thumbnails visualization	Non-interactive query approach include irrelevant translation, Interactive query approach is user assisted query translation
4	MIRACLE	English and other languages	BD	user can select or deselect some translation, query reformulation, automatic and user assisted query translation	Resources are not available, Homonymy and Synonymy
5	MULTILEX EXPLORER	Support multi lingual	WordNet and Web Search Engine	Exploring context of query, WSD, language selection, QE, automatic categorization, circle visualization	WordNet not available for all languages
6	MULTI SEARCHER	Support multi lingual	BD, PC, ER, Mutual information	User assisted disambiguation, Automatic translation disambiguation deal with the user's lack of knowledge in target language, Automatic Document categorization	Parallel Corpora not available for all languages

Mostly researcher used bilingual dictionary because it is available for all languages and also takes nominal computation cost. Bi-lingual dictionary with Cohesion translation and Query expansion achieves 0.4337 for Persian-English [18]. Wikipedia is used to identify OOV terms but Wikipedia with sufficient data is available for a limited number of languages. CLIR with Wikipedia achieves 0.46 MAP [14].

Ontology, WordNet, UNL and co-occurrence translation used for resolving term homonymy and polysemy issues. Dictionary coverage and quality, phrase translation, Homonymy, Polysemy and Lack of resources are major challenges for CLIR. Many comprehensive tools are cultivated to resolve the language barrier issue, such as MT tools and CLIR tools [19]. A brief study to the CLIR tools is summarized in the Table 4. All these tools uses bilingual dictionary because of nominal time computation. A common problem of user assisted query translation was tried to remove in MIRACLE, MULTI LEX EXPLORER and MULTI SEARCHER. Automatic query translation suffered by a problem of homonymy and polysemy.

6 Conclusion

CLIR enables searching documents via eternal diversity of languages across the world. It removes the linguistic gap and allows a user to submit a query in a language different than the target documents. A CLIR method includes a translation mechanism followed by monolingual retrieval. It is analyzed that query translation always efficient choice than document translation. In this paper, various CLIR issues and challenges and Query translation approaches with disambiguation are discussed. A comparative analysis of CLIR approaches is presented in Table 3. A CLIR approach with Bi-Lingual dictionary, Cohesion Translation, query expansion and Language Modeling achieves good MAP i.e. 0.4337. Another CLIR approach with Wikipedia, Bag of Concept and Cross language- Explicit Semantic analysis achieves better MAP i.e. 0.46. MT with parallel corpora CLIR approach achieves 0.4694 MAP. A brief analysis of CLIR tools is represented in Table 4. Dictionary Coverage and Quality, Unavailability of Parallel Corpora, Phrase Translation, Homonymy and Polysemy are concluded as major issues.

References

1. Nagarathinam A., Saraswathi S.: State of art: Cross Lingual Information Retrieval System for Indian Languages. In International Journal of computer application, Vol. 35, No. 13, pp. 15–21 (2006).
2. Nasharuddin N., Abdullah M.: Cross-lingual Information Retrieval State-of-the-Art. In Electronic Journal of Computer Science and Information Technology (eJCSIT), Vol. 2, No. 1, pp. 1–5 (2010).

3. Oard, D.W.: A Comparative Study of Query and Document Translation for Cross-language Information Retrieval. In Proceedings of the Third Conference of the Association for Machine Translation in the Americas on Machine Translation and the Information Soup., Springer-Verlag, pp. 472–483 (1998).

4. Makin R., Pandey N., Pingali P., Varma V.: Approximate String Matching Techniques for Effective CLIR. In International Workshop on Fuzzy Logic and Applications, Italy, Springer-Verlag, pp. 430–437 (2007).

5. Pirkola A., Toivonen J., Keskustalo H., Visala K., Jarvelin K.: Fuzzy translation of cross-lingual spelling variants. In: Proceedings of SIGIR'03, pp. 345–352 (2003).

6. Bajpai P., Verma P.: Cross Language Information Retrieval: In Indian Language Perspective. International Journal of Research in Engineering and Technology, Vol. 3, pp. 46–52 (2014).

7. Chen A., Gey F.C.: Combining Query Translation and Document Translation in Cross-Language Retrieval. In Comparative Evaluation of Multilingual Information Access Systems, Springer Berlin: Heidelberg, pp. 108–121 (2004).

8. Jagarlamudi J., Kumaran A.: Cross-Lingual Information Retrieval System for Indian Languages. In Advances in multilingual and multi modal information retrieval, pp. 80–87 (2008).

9. Chinnakotal M., Ranadive S., Dhamani O.P., Bhattacharyya P.: Hindi to English and Marathi to English Cross Language Information Retrieval Evaluation. In Advances in Multilingual and Multimodal Information Retrieval, springer-verlag, pp. 111–118 (2008).

10. Gupta S. Kumar, Sinha A., Jain M.: Cross Lingual Information Retrieval with SMT and Query Mining. In Advanced Computing: An International Journal (ACIJ), Vol.2, No.5, pp. 33–39 (2011).

11. Yu F., Zheng D., Zhao T., Li S., Yu H.: Chinese-English Cross-Lingual Information Retrieval based on Domain Ontology Knowledge. In International conference on Computational Intelligence and Security, Vol. 2, pp. 1460–1463 (2006).

12. Monti J., Monteleone M.: Natural Language Processing and Big Data An Ontology-Based Approach for Cross-Lingual Information Retrieval. In International Conference on Social Computing, pp. 725–731 (2013).

13. Chen-Yu S., Tien-Chien L., Shih-Hung W.: Using Wikipedia to Translate OOV Terms on MLIR. In Proceedings of NTCIR-6 Workshop Meeting, Tokyo, Japan, pp. 109–115 (2007).

14. Sorg P., Cimiano P.: Exploiting Wikipedia for Cross-Lingual and Multi-Lingual Information Retrieval. Elsevier, pp. 26–45 (2012).

15. Samantaray S. D.: An Intelligent Concept based Search Engine with Cross Linguility support. In 7th International Conference on Industrial Electronics and Applications, Singapore, pp-1441–1446 (2012).

16. Xiaoning H., Peidong W., Haoliang Q., Muyun Y., Guohua L., Yong X.: Using Google Translation in Cross-Lingual Information Retrieval, In Proceedings of NTCIR-7 Workshop Meeting, Tokyo, Japan, pp. 159–161 (2008).

17. Zhang J., Sun L. and Min J.: Using the Web Corpus to Translate the Queries in Cross-Lingual Information Retrieval. In Proceeding of NLP_KE, pp. 493–498 (2005).

18. Pourmahmoud S., Shamsfard M.: Semantic Cross-Lingual Information Retrieval. In International symposium on computer and information sciences, pp. 1–4 (2008).

19. Ahmed F., Nurnberger A.: Literature review of interactive cross language information retrieval tools. In The international Arab Journal of Information Technology, Vol. 9, No. 5, pp. 479–486 (2012).

20. Boretz, A., AppTek Launches Hybrid Machine Translation Software, in Speech Tag Online Magazine (2009).

21. Yuan, S., Yu S.: A new method for cross-language information retrieval by summing weights of graphs. In Fourth International Conference on Fuzzy Systems and Knowledge Discovery, IEEE Computer Society, pp. 326–330 (2007).

22. Nie, J., Simard M., Isabelle P., Durand R.: Cross-Language Information Retreval Based on Parallel Texts and Automatic Mining of Parallel Texts from the Web. In Proc. OfACM-SIGIR, pp. 74–81 (1999).
23. Lu W., Chien L., Lee H.: Anchor Text Mining for Translation of Web Queries: A Transitive Translation Approach. ACM Transactions on Information Systems 22(2), pp. 242–269 (2004).
24. Navigly R.: Word Sense Disambiguation: A Survey. ACM computing survey, Vol. 41, No. 2 (2009).

Minimizing the Cost of Losses Due to Cyber Attack Through B. B. (Branch and Bound) Technique

Narander Kumar and Priyanka Chaudhary

Abstract The advancement of computer and digitization of information system, cyber crime is now becoming one of the most significant challenges in our society. Threat of cyber crime is a growing danger to the industry, business and economic field that are influenced by the cyber criminals along with common person of our society. Since cyber crime is often an aspect of more complex criminological reigns such as money laundering, trafficking and cyber terrorism, the true damage caused through cyber crime to society that may be unknown. This paper presents Branch and Bound (B&B) technique to minimize the losses due to cyber crime. Branch and Bound is the effective technique to solve assignment problems. B&B is, however, an algorithmic technique, which provides the solution for each specific type of problem. There are numerous choice exist to solve each type of problem but Branch and bound (B&B) is the best way.

Keywords Cyber crime · Assignment problem · Branch and bound · Finite state machine · Cyber world

1 Introduction

In 21st century internet play an important role to utilize time and to improve for performance factor. Internet can be understood as it is a global network which is connected to the millions of computer and internet is world largest information system. Now a day's most of the information is stored in digital form in computer.

N. Kumar (✉) · P. Chaudhary
Department of Computer Science, B.B. Ambedkar University (A Central University),
Lucknow, U.P., India
e-mail: nk_iet@yahoo.co.in

P. Chaudhary
e-mail: cpriyanka22@gmail.com

709

© Springer India 2016
S.C. Satapathy et al. (eds.), *Information Systems Design and Intelligent Applications*, Advances in Intelligent Systems and Computing 433,
DOI 10.1007/978-81-322-2755-7_73

Information is digital form is very easy to stored, manipulate and access but the digitization of information is developed the cyber-crime. Cyber-crime is criminal activity which is done by internet, specifically illegal criminal trespass into the computer system or database of one and another which manipulate data or sabotage of equipment and data [1]. Now this time cyber crime is very crucial issues of all countries because most of the data are transferred through the internet even that government data also. Due to increasing number of cyber crime or online criminal activity, cyber space is unsafe for business world. Cyber space can be defined as *"the electronic world of computer network which facilitate the data transmission and exchange."* Due to the increasing use of internet or cyber space the cyber crimes are also increased. In this paper, we defined a technique using with branch and bound algorithm for reducing the cost of losses which are occurred by cyber crime. Branch and bound method was firstly defined by A.H. Land and A.G. Dog in 1960. It is a best way for solving a various type of assignment problem.

2 Review of Work

The Cyber crime is considered to be as a criminal activity that uses the computer network [2]. Identify theft is a another type of criminal activity in which unauthorized user can take more benefits of this system, public defamation, hacking, cyber stalking and any other type of social media sites, debit/credit card, child pornography and different types of cyber violation of copyright has been discussed in [3–7]. Preventing intellectual property theft is a priority for its criminal activity investigative program and focusing on theft of trade-marks and product infringements, such as counterfeit parts and other products that threaten safety has been discussed in FBI [8]. Cybercriminals activity is tending to attack through cyber-attack tools "dark markets" which is entrenched in online social media. A model based on probability for reducing cybercriminal networks from online social media has been discussed in [9]. Placement of detection nodes for distributed detection of different attacks in optimized manner, and reducing the number of these node, it minimize the cost of processing and more delays for identifying an attack in distributed network, has been given in [10]. Risk management in banking including field of measurement, risk identification, and assessment which reducing negative effects of financial risks as well as capital of a financial institutions has been discussed in [11]. An online sensor stream mining system to analyze situational behavior of humans in some specific vicinity and a real-time alert system to take countermeasures is discussed in [12]. The use of cyber-insurance products is good way for reducing the impact of financial, industry and economy losses from security infringements. Cyber-vulnerability assessment, and expected loss computation is studied in [13]. A FARE (Failure Analysis and Reliability Estimation), a framework for benchmarking reliability of cyber-physical systems is introduced in [14] to reduce the financial losses. To minimize any loss of information, a model interval-based data using Fuzzy Sets (FSs) is defined in which we transfer the interval based data into FS models, and it is also avoid to as much as possible assumptions about the distribution of the data has discussed in [15]. A worldwide problem of the banking industry is Cyber frauds and a malware author uses the Application Programming Interface (API) calls to perpetrate these crimes. To detecting for Malware based on API call sequences using with text and data mining, a static analysis method is defined in [16]. A case study, it is based on the sample of IBM Italian customers. The main objective of this authors is that it was make and validate robust models for handling a missing information, non-id data points and class unbalancedness using with a several classifiers and their subagged versions has discussed in [17]. Botnet Detection is becoming known as threat. It is related with the cyber crime prevention technique and it is also provide a distributed platform for different type of several criminal activities such as phishing, click fraud and malware dissemination has been presented in [18].

3 Formulation

The Branch and Bound methods are normally based on some relaxation of the ILP (Integer Linear Programming Model) model.

In the following optimization models, the variables x_{ij} are either excluded from the model or prevented by setting $c_{ij} = 4$.

ILP-model:

$$\text{Max} \sum_{i=1}^{n} \sum_{j=1}^{n} c_{ij}\, x_{ij} \quad \text{Subject to} \sum_{j=1}^{n} x_{ij} \quad \text{For every } i = 1\ldots, n$$

$$\sum_{i=1}^{n} x_{ij} \quad \text{For every } j = 1,\ldots, n$$

$$x_{ij} = 0 \text{ or } 1 \; i, j = 1,\ldots, n$$

4 Implementation

In the proposed paradigm, use of branch and bound method to find optimize solution. Here it leads to a tree of decision through which each branch represents a one possible way to continue to the loss from the current node. We evaluate the branches by finding at the lower bound of each current loss then continue with that branch i.e. the lowest bound. The algorithm stops when we found the possible valid solution and no any other node in the decision tree that has lower bound than we have found feasible solution (Tables 1 and 2).

Table 1 Different types of cyber crime

C1	C2	C3	C4	C5	C6	C7	C8	C9
Personal info hacked	Tax fraud	Web site hacked	Copied book matter	Audio music copied	Video music copied	Loan fraud	Cyber bully-ing	Credit/debit card info hacked

Table 2 Different type of person which are suffered by cyber crime

P1	Research scholar
P2	PG student
P3	Professional PG student
P4	UG student
P5	Diploma student
P6	Education department
P7	Government employee
P8	Private employee
P9	House women

We start by taking a node at lower bound, which are the minimum losses. In this method, take a minimum loss respective row of the data. If we take node first, that is suffered from 17 % of the cybercrime losses which is the minimum loss of this row. The same goes for second i.e. 33, third i.e. 25, 17, 100, 40, 33, 29, and 100 respectively. If we add all these values then we find that optimal solution is 17 + 33 + 25 + 17 + 100 + 40 + 33 + 29 + 100 = 394. This does not mean it is possible solution, it just only lowest possible solution that is guaranteed that is equal to or greater than to the 394. Now let we start make our decision tree with lowest bound that is 394.

With the help of the branch and bound method we minimize these cyber losses which represents by each cell in the Table 3.

Table 3 The matrix contains data in percentage of cyber loss due to the cyber crime on the basis of primary data collected

	C1	C2	C3	C4	C5	C6	C7	C8	C9
P1	50	33	50	67	67	17	33	100	33
P2	33	33	33	67	33	33	50	67	33
P3	50	–	50	100	50	25	25	75	50
P4	50	17	50	67	83	17	100	100	50
P5	100	–	–	100	100	–	100	100	100
P6	60	60	60	83	100	–	100	100	40
P7	83	50	83	83	83	–	83	83	33
p8	100	71	86	86	29	71	100	100	100
P9	100	100	100	100	100	–	100	100	100

Now we calculate lower bound of p1 ->1

$$P1C1 + P2C5 + P3C6 + P4C2 + P5C5 + P6C9 + P7C9 + P8C5 + P9C2$$
$$= 50 + 33 + 25 + 17 + 100 + 40 + 33 + 29 + 100$$
$$= 427$$

Now let us calculate all the lower bound values with the same procedure. After calculation we find p8 ->8 = 434 is optimal solution of minimize percentage of losses.

Primary data collection (Sample size) is 39. Therefore 434/39 = 11.12. So 11.12 % of minimize the losses percentage of the cyber crime through the assignment problem using Branch and Bound method.

5   Generation of Test Cases

We consider the concept of theory of automata to design the Finite State Machine (FSM) that is defined through M as:

$$M = (Q, \Sigma, \delta, q_0, F)$$

where
Q   Finite set of states;
Σ Finite set of input symbols (Alphabets and Numbers);
δ Transition between two states;
q_0 Initial state;
F   Final state;

From the above concept of the automata, a finite state machine diagram represented by the Fig. 1. This is based on the activity diagram represented in the Fig. 2.

There are seven states Q = {q_0, q_1, q_2, q_3, q_4, q_5, q_6} and these states are represented in the Table 4.

$\delta(q_0, a)$ represented the transition, where a is the set of inputs and inputs are considered as Σ = {a, b, c, d, e, f, g, h} and representation is recorded in the following Table 5.

On the basis of above, a transition table is given below:

$$\delta(q_0, a) \rightarrow q_1, \delta(q_1, b) \rightarrow q_2, \delta(q_2, c) \rightarrow q_0, \delta(q_2, d) \rightarrow q_3$$
$$\delta(q_3, e) \rightarrow q_4, \delta(q_4, f) \rightarrow q_5, \delta(q_5, g) \rightarrow q_6, \delta(q_6, h) \rightarrow q_0$$

By the use of above grammar different test cases are generated and explained below:

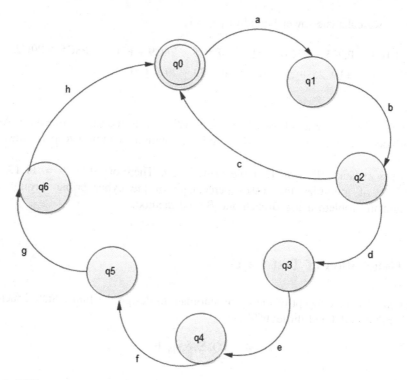

Fig. 1 FSM representation of activity diagram

Valid Test Case 1 *Cyber attacks losses are not optimized*
It is represented by

$$\delta(q_0, a) \to q_1 \Rightarrow q_0 \to a\ q_1; \delta(q_1, b) \to q_2 \Rightarrow q_1 \to b\ q_2$$
$$\delta(q_2, c) \to q_0 \Rightarrow q_2 \to c\ q_0$$

After changing the states or removing the non terminals, the string is given by

$$q_0 = abc\ q_0 = abc$$

This represents that the Cyber Attack losses are not optimized.
Valid Test Case 2 *Cyber attacks cost are optimized*
It is represented by

$$\delta(q_0, a) \to q_1 \Rightarrow q_0 \to a\ q_1; \delta(q_1, b) \to q_2 \Rightarrow q_1 \to b\ q_2$$
$$\delta(q_2, d) \to q_3 \Rightarrow q_2 \to d\ q_3; \delta(q_3, e) \to q_4 \Rightarrow q_2 \to e\ q_4$$
$$\delta(q_4, f) \to q_5 \Rightarrow q_2 \to f\ q_5; \delta(q_5, g) \to q_6 \Rightarrow q_2 \to g\ q_6$$
$$\delta(q_6, h) \to q_0 \Rightarrow q_2 \to h\ q$$

Fig. 2 UML activity diagram

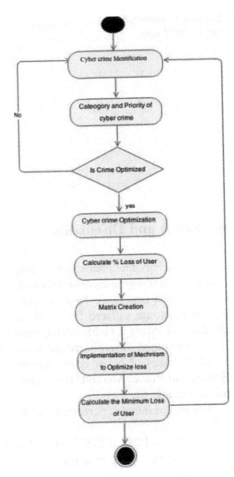

Table 4 Transition table

	a	b	c	d	e	f	g	H
q_0	q_1	–	–	–	–	–	–	–
q_1	–	q_2	–	–	–	–	–	–
q_2	–	–	q_0	q_3	–	–	–	–
q_3	–	–	–	–	q_4	–	–	–
q_4	–	–	–	–	–	q_5	–	–
q_5	–	–	–	–	–	–	q_6	–
q_6	–	–	–	–	–	–	–	q_0

After changing the states or removing the non terminals, the string is given by

$$q_0 = abdefghq_0 = abdefg$$

This represents that the cyber crime losses are optimized which is as per expectation.

Table 5 Representation of input symbol state

Name of input	Description input
a	List of cyber crime
b	Priority list of cyber crime
c	Not optimized list of cyber crime
d	Optimized list of cyber crime
e	List of cyber crime
f	Resultant matrix with cyber crime and losses
g	Final optimized matrix
h	Minimum loss result

6 Result and Discussion

No one can deny that internet can change our life, society and culture. Rapid growth of internet and digitization of information system, generate different type of cyber crime. In this paper, we have collect primary data through the different field of person which are suffered from different types of cyber crime. After collection of the data, we create a matrix and applied the assignment problem using Branch and Bound technique to minimize the losses caused by cyber crime. We can find that 11.12 % losses of cyber crime minimize through the assignment problem using Branch and Bound Method. If we apply assignment problem using the Hungarian method (Table 6).

The total minimum percentage of losses

$$= P1C6 + P2C1 + P3C7 + P4C2 + P5C4 + P6C3 + P7C9 + P8C5 + P9C8$$
$$= 17 + 33 + 25 + 17 + 100 + 60 + 33 + 29 + 100$$
$$= 414$$

Primary data collection (Sample size) is 39. Therefore 414/30 = 10.61. So 10.6 % of minimize the losses percentage of the cyber crime through the

Table 6 Resultant matrix after applying the Hungarian method

	C1	C2	C3	C4	C5	C6	C7	C8	C9
P1	17	3	17	34	34	[0]	0	67	20
P2	[0]	3	0	34	0	16	17	34	20
P3	25	M	25	75	25	16	[0]	50	45
P4	30	[0]	30	47	63	13	80	80	50
P5	0	M	M	[0]	0	M	0	0	20
P6	0	3	[0]	38	38	M	38	38	0
P7	30	50	30	30	30	M	30	30	[0]
p8	7	45	58	58	[0]	58	71	71	91
P9	0	3	0	0	0	M	0	[0]	4

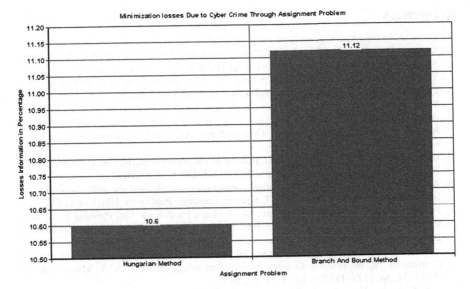

Fig. 3 Comparison of minimization losses due to the cyber crime through assignment problem using with Hungarian method and branch and bound method

assignment problem using Hungarian method. Then we find 10.60 % losses due to cyber crime. So we can say that Branch and Bound is better than Hungarian method for this particular problem. The performance of above both methods is given in the Fig. 3.

7 Conclusions

Cyber crime does have an extreme effect on the world. It affects each and every person there is no matter where they are belongs to which community or group. It is strange that the persons who break the secret into computers across the world only for enjoyment and entertainment. In order to deal with the cyber crime, society, the legal institutions and law enforcement authorities will also have to changes in their rules and regulations for cyber crime. All cyber crime is based on lack of awareness. Due to lack of information security awareness on the part of computer users, developers and administrators, many cyber vulnerabilities exists in present time. This is a duty of Government that they educate unaware persons about the dangerous areas of the cyber-world because prevention is better than cure. After applying an assignment problem using with branch and bound, we have find the minimize losses due to cyber crime in different areas as discussed in this paper. For the future work we can be applying different other types of optimization techniques to find minimize cost of Losses due to cyber crime.

References

1. Sumanjit Das, Tapaswini Nayak: Impact of Cyber crime: Issues and Challenge. International Journal of Engineering Sciences & Emerging Technologies. 6, 142–153 (2013).
2. Yanping Zhang, Yang Xiao, Kaveh Ghaboosi, Jingyuan Zhang, Hongmei Deng: A survey of cyber crimes. In Security and Communication Network, Vol. 5, pp. 422–437. John Wiley & Sons (2012).
3. Aideen Keane: Identity theft and privacy—consumer awareness in Ireland. In: International Journal of Networking and Virtual Organizations, pp. 620–633, Inderscience Publishers (2009).
4. Baca, M., Cosic, j., Cosic, Z.: Forensic analysis of social networks. In: Proceedings of the ITI 2013 35th International Conference, pp. 219–223, IEEE, Cavtat (2013).
5. Alison Ada.: Cyber stalking and Internet pornography: Gender and the gaze. In: Ethics and Information Technology. pp. 133–147, Kluwer Academic Publishers, Hingham (2002).
6. Jyoti Chhikara, Ritu Dahiya, Neha Garg, Monika Rani: Phishing & Anti-Phishing Techniques: Case Study. International Journal of Advanced Research in Computer Science and Software Engineering 3, 458–465 (2013).
7. Vukelic, B., Skaron, K: Cyber Crime and violation of copyright. In: 36th International Convention on Information Communication Technology Electronics & Microelectronics (MIPRO), pp. 1127–1130, IEEE, Opatija (2013).
8. Lau, R. Y. K., Yunqing Xia, Yunming Ye.: A Probabilistic Generative Model for Mining Cybercriminal Networks from Online Social Media. In: Computational Intelligence Magazine, Vol. 9 pp. 31–43. IEEE (2013).
9. Islam, M.H., Nadeem, K., Khan, and S.A.: Efficient placement of sensors for detection against distributed denial of service attack. In: International Conference on Innovation in information technology, pp. 653–657, IEEE, Al Ain (2008.
10. Ljiljanka Kvesi, Gordana Duki.: Risk Management and Business Credit Scoring. In: 34th International Conf. on Information Technology Interfaces proceeding of the ITI 2012, pp .47–50, IEEE, Cavtat (2012).
11. Bhole, L.M., Mahakud J.: Financial Institutions and Markets: Structure. Growth and Innovations. New Delhi, Tata McGraw-Hill Press (2012).
12. Ur Rehman, Z. Shaheen, M.: Situation-awareness and sensor stream mining for sustainable society. In: First Asian Himalayas International Conference on Internet 2009 AH-ICI, pp. 1–5, IEEE, Kathmandu (2009).
13. Arunbha Mukopadhyay, Sameer Chatterjee, Ambuj Mahanti, Samir, k. Sadhukhan: Cyber-risk decision models: To insure IT or not?. Decision Support Systems. 11–17 (2013).
14. Wu, L., Kaiser, G. FARE: A framework for benchmarking reliability of cyber-physical systems. In International conferences on Applications and Technology Conference (LISAT), pp. 1–6, IEEE, Long Island (2013).
15. Wagner, C., Miller, S., Garibaldi, J., Anderson, D. From Interval-Valued Data to General Type-2 Fuzzy Sets. Fuzzy Systems, IEEE Transactions, 1 (2014).
16. Sundar kumar, G.G. Ravi, V. Malware detection by text and data mining. In: International conference on Computational Intelligence and Computing Research (ICCIC), pp. 1–6, IEEE, Enathi (2013).
17. Paleologo, G., Elisseeff, A., Antonini, G.: Subagging for Credit Scoring Models. European Journal of Operational Research. 490–499 (2012).
18. Feily, M., Shahrestani, A., Ramadass, S.: A Survey of Botnet and Botnet Detection. In: 3rd International Conference on Emerging Security Information, Systems and Technologies 2009. SECURWARE '09. pp. 268–273, IEEE, Athens Glyfada (2009) 268–5.

Denoising Knee Joint Vibration Signals Using Variational Mode Decomposition

Aditya Sundar, Chinmay Das and Vivek Pahwa

Abstract Analysis of knee joint vibration (VAG) signals using signal processing, feature extraction and classification techniques has shown promise for the non-invasive diagnosis of knee joint disorders. However for such techniques to yield reliable results, the digitally acquired signals must be accurately denoised. This paper presents a novel method for denoising VAG signals using variational mode decomposition followed by wiener entropy thresholding and filtering. Standard metrics: mean squared error, mean absolute error, signal to noise ratio, peak signal to noise ratio and CPU consumption time have been calculated to assess the performance our method. Metric: normalized root mean squared error has also been evaluated to estimate the effectiveness of our method in denoising synthetic VAG signals containing additive white gaussian noise. The proposed method yielded a superior performance in denoising raw VAG signals in comparison to previous methods such as wavelet-soft thresholding, empirical mode decomposition-detrended fluctuation analysis and ensemble empirical mode decomposition-filtering. Our method also yielded better performance in denoising synthetic VAG signals in comparison to other methods like wavelet and wavelet packet-soft thresholding, wavelet-matching pursuit algorithm, empirical mode decomposition-detrended fluctuation analysis and ensemble empirical mode decomposition-filtering. The proposed method although computationally more complex, yields the most accurate denoising.

A. Sundar (✉) · C. Das · V. Pahwa
Department of Electrical, Electronics and Instrumentation, BITS, Pilani,
K.K. Birla Goa Campus, Goa, India
e-mail: aditsundar@gmail.com

C. Das
e-mail: chinmay39das@gmail.com

V. Pahwa
e-mail: viv.pahwa@gmail.com

© Springer India 2016
S.C. Satapathy et al. (eds.), *Information Systems Design and Intelligent Applications*, Advances in Intelligent Systems and Computing 433,
DOI 10.1007/978-81-322-2755-7_74

Keywords Vibroarthographic signals (VAG) · Variational mode decomposition (VMD) · Wiener entropy · Empirical mode decomposition (EMD) · Ensemble empirical mode decomposition (EEMD) · Detrended fluctuation analysis (DFA) · Mean squared error (MSE) · Signal to noise ratio (SNR) · Normalized root mean squared error (NRMS) · White gaussian noise (WGN)

1 Introduction

The knee is a hinge type synovial joint that connects the thigh and the lower part of the leg. This joint comprises of the fibula, patella, the extensive ligaments and two main muscle groups (i.e. quadriceps and hamstrings) [1]. The knee joint is a fibro-cartilaginous structure located between the tibia and the femur [1]. The knee joint can normally tolerate a moderate amount of stress without severe internal injury but cannot withstand rotational forces, commonly associated with athletic activities. Damage to the cartilage in the articular surface could cause rheumatic disorders such as osteoarthritis [2]. Non-invasive detection of knee joint disorders at an early age could provide information for physicians to undertake appropriate therapies to deter worsening of the disorder. Knee joint Vibroarthrographic signals (VAG) are recorded from the vibrations emitted from the mid patella during active movements such as flexion and extension. These signals have been used as tools for building real-time knee joint disorder diagnosis systems [3]. For such techniques to yield reliable results, the processed signals must be noise-free. However, real world signals are corrupted with noise. Typical sources of noise in VAG signals include power line interference, high frequency muscular contraction, base-line wandering and random artifacts introduced due to the sensor as well as other environmental factors [4]. Thus the signals must be accurately denoised, so that the extracted features reflect only the vital information contained in the signal. Several methods such as wavelet transform- soft thresholding [5], wavelet packet and matching pursuit algorithms [6], empirical mode decomposition-detrended fluctuation analysis (EMD-DFA) and ensemble empirical mode decomposition-filtering (EEMD-filtering) have been proposed for denoising these signals [7].

This paper presents a new method for denoising knee joint VAG signals using variational mode decomposition (VMD), followed by wiener entropy thresholding and filtering. Metrics: mean squared error (MSE), mean absolute error (MAE), signal to noise ratio (SNR), peak signal to noise ratio (PSNR) and CPU consumption time were used to evaluate the performance of this method. Using the proposed method 89 raw VAG signals were denoised and metrics are computed for all the aforementioned methods. The effectiveness of our method in removal of additive white gaussian noise (WGN) from VAG signals is also estimated. For this, WGN of SNR = 0 dB and SNR = 10 dB are added to the noise-free signal and these synthetic signals are denoised using the proposed method. Metrics normalized root mean squared error (NRMS) and CPU consumption time are calculated to assess the performance of different methods in denoising these signals. The proposed method yielded the most accurate denoising but with a slow response. However, since performing diagnosis of knee-joint disorders is not time-critical, our method can be adopted for real-time diagnosis systems.

2 Methodology

2.1 Database

The dataset used in this paper was compiled from the study conducted at University of Calagary, Canada [8]. In this study, the test subjects were asked to sit on a unmovable and rigid surface with their leg freely drooping. A miniature accelerometer was mounted on the mid-patella region and was held fixed with adhesive tape. The subjects were instructed to bend their leg over an angle ranging from 135° to 0° and back to 135°, within 4 s. The signals were then acquired using NI data acquisition boards (DAQ) and digitized with a 12 bit resolution. The signals were acquired at a sampling rate of 2 kHz and bandpass filtered between 10 Hz and 1 kHz. A total of 89 signals were acquired using this method, 51 signals from normal subjects and 38 signals from subjects with various disorders such as tear, anterior cruciate ligament and tibial chondromalacia injuries.

2.2 Variational Mode Decomposition (VMD)

Variational mode decomposition has been proposed by Dragomiretski and Zosso [9] to overcome the limitations of empirical mode decomposition like sensitivity to noise and sampling. VMD employs a non-recursive decomposition model, in which modes are extracted concurrently. The model searches for a number of modes and their respective center frequencies, so that the modes reproduce the input signal, while being smooth after demodulation into baseband. This closely corresponds to a narrow-bandprior in Fourier domain. The modes obtained using VMD are denoted as variation modes (VM), and the original signal can again be retrieved by summing all the modes. The variational model is efficiently optimized using an alternating direction method of multipliers approach. This method has shown better results in comparison to other decomposition methods like the EMD in decomposing real and artificial data. VMD has been used in previous works for denoising images [10] however, it has not been applied extensively for removal of noise from biomedical signals. The authors hence aim to explore the performance of the same in denoising knee joint VAG signals.

2.3 Wiener Entropy or Spectral Flatness

Wiener entropy or spectral flatness is a measure of the autocorrelation of an audio signal [8]. Wiener entropy can be used to estimate the noise-like component of a signal. Initially, the power spectrum of the signal is calculated using DFT. This is followed by calculating the ratio of the geometric mean and the arithmetic mean of the power spectrum. Let 'N' represent the length of the power spectrum and 'n' represent the value of a discrete sample of the power spectrum, where $n = 1, 2, 3, ..., N$. Wiener entropy is defined as:

$$\text{Wiener entropy} = \frac{\sqrt[n]{\prod_{n=1}^{N} H(n)}}{\frac{\sum_{n=1}^{N} H(n)}{N}} \tag{1}$$

2.4 Proposed Algorithm

The steps followed in denoising the signal using the proposed method are as described in the block diagram in Fig. 1.

In step 1, cascaded moving average filters have been used to remove noise from the main signal as performed by Wu [11]. In step 2, while performing VMD, the balancing parameter of the data-fidelity constraint (α), time-step of the dual ascent (τ), tolerance of convergence criterion were set to 1000, 0 and 10^{-7} respectively. The threshold value of Wiener, is chosen as 0.0001 based on the experiments conducted on the database. A 12 level decomposition was chosen as it yielded the most accurate noise removal. Figure 1 below summarizes the same using a block diagram.

Works by Rangayaan et al. [12] suggest that frequencies between 10 and 300 Hz characterize the VAG signals. Frequencies lower than 10 Hz comprise the base-line

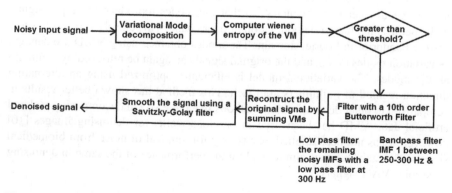

Fig. 1 Block diagram of the proposed denoising method

wander artifact or other low frequency artifacts and frequencies above 300 Hz present in the signal comprise the muscular contraction artifacts. Variation mode V1 contains the high frequency component of the VAG signal and is hence bandpass filtered between 250 and 300 Hz. All other modes that yielded a value entropy greater than the threshold were bandpass filtered between 10 and 300 Hz. In step 6 a 3rd order Savitzky-Golay filter with a frame size of 21 is used to filter the signal obtained after step 5. Figure 2 shows the signal obtained different steps of processing the signal and the total noise removed from the signal (base-line wander and other artifacts).

3 Evaluating the Performance of Denoising Algorithms

To assess the performance of denoising algorithms metrics: mean squared error, mean absolute error, signal to noise ratio, peak signal to noise ratio, normalized root mean squared error and CPU consumption time have been computed. Each of the metrics are described in brief.

3.1 Mean Squared Error (MSE)

MSE is a metric that is used to evaluate the accuracy of denoising. The lower the value of MSE, the closer is the denoised signal to the original, hence better denoising. Let $x(n)$ represent the noisy, raw VAG signal, $x'(n)$ represent the denoised signal and N represent the length of the signal. Let 'n' represent the sample number, where $n = 1, 2, 3, ..., N$. MSE can be defined as:

$$MSE = \frac{\sum_{n=1}^{N}(x(n) - x'(n))^2}{N} \tag{2}$$

3.2 Mean Absolute Error (MAE)

MAE, a metric similar to MSE is used to evaluate the accuracy of denoising. The lower the value of MAE, the better is the denoising. Using the aforementioned symbols, MAE can be defined as:

$$MAE = \frac{\sum_{n=1}^{N}|(x(n) - x'(n))|}{N} \tag{3}$$

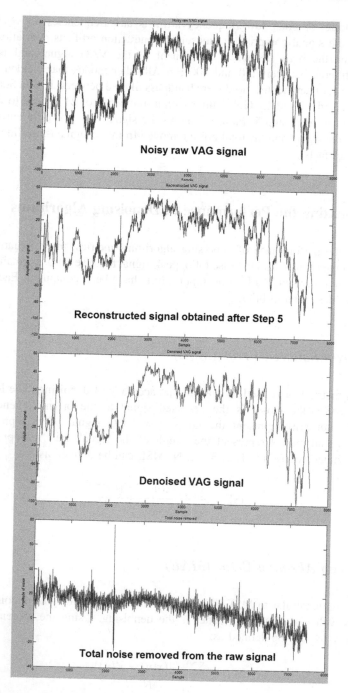

Fig. 2 Signal waveform obtained after processing using the proposed algorithm and the total noise removed from the signal

3.3 Signal to Noise Ratio (SNR)

SNR is a common metric used to estimate the performance denoising methods. SNR is inversely proportional to log (MSE). SNR can defined as:

$$SNR = 10 \log_{10} \frac{\sum_{n=1}^{N} x(n)^2}{\sum_{n=1}^{N} (x(n) - x'(n))^2} \qquad (4)$$

3.4 Peak Signal to Noise Ratio (PSNR)

PSNR is a metric similar to SNR. Similar to SNR, the higher the value of PSNR, the better is the denoising. PSNR is defined as:

$$PSNR = 20 \log_{10} \frac{\max(x(n))}{RMSE} \qquad (5)$$

where RMSE is the square root of MSE defined in Sect. 3.1.

3.5 Normalized Mean Squared Error (NRMS)

NRMS is a metric used to evaluate the performance of the denoising algorithm in retention the structural integrity post enhancement. Let $d(n)$ represent the signal obtained after removing the artificial noise added to $x'(n)$. NRMS is defined as:

$$NRMS = \sqrt{\frac{\sum_{n=1}^{N} (x'(n) - d(n))^2}{\sum_{n=1}^{N} (x'(n))^2}} \qquad (6)$$

The smaller the value of NRMS, the better the denoising. CPU consumption time is a measure used to measure the speed of the denoising algorithm. All computations presented in this paper have been performed using a CPU with an Intel core i7 processor operating at 2 GHz with a 6 GB RAM. All these metrics have been used to evaluate the performance of the denoising methods.

4 Evaluating the Performance of Denoising Algorithms

This section discusses the results obtained on denoising the VAG signals. 89 raw VAG signals: 51 normal and 38 abnormal signals have been denoised using 4 methods: empirical mode decomposition (EMD-DFA), wavelet-soft thresholding, EEMD-wiener entropy thresholding and the proposed method.

4.1 Denoising Raw VAG Signals

In the wavelet-soft thresholding method proposed by Rahangdale-Mittra [5], a 4th order coiflets wavelet has been used to decompose the signal into 3 levels. The wavelet sub-bands were then soft thresholded and reconstructed using wavelet-reconstruction. In the EMD-DFA method, the signal is first decomposed into 12 levels and detrended fluctuation analysis is performed to compute the fractal scaling index for each of the IMFs. IMFs that yielded a fractal scaling index value of less than 0.5 are then removed from the main signal. In the EEMD-Wiener entropy thresholding method, the signal is decomposed using ensemble empirical mode decomposition. This is followed by filtering the IMFs with wiener entropy greater than 0.0001. As IMF 1 contains the high-frequency component of the main signal, it is bandpass filtered between 250 and 300 Hz. All other noisy-IMFs are bandpass filtered between 10 and 300 Hz using a butterworth filter. The reconstructed signal is then filtered using a 3rd order Savitzky-Golay filter with a frame size of 41. Table 1 presents the average values of metrics obtained using each of the 4 methods.

From Table 1, it is observed that the proposed method yields the lowest average MSE, MAE values and the highest SNR, PSNR values. Hence our method yields the most accurate denoising. Our method however provides a much slower response in comparison to previously proposed methods due to the increased computational complexity involved in VMD.

4.2 Rejection of Artificial Noise

The White Gaussian Noise model (WGN) is commonly used in information theory to mimic the effects of stochastic or random processes. WGN has shown a strong ability to replicate the kind of noise contained in bio-signals [13]. The effectiveness of our method in removal of additive white Gaussian noise is evaluated using the metric NRMS. For this purpose WGN is added to the signals obtained after denoising the raw VAG signals using the proposed method. Gaussian noise of SNR = 0 dB and SNR = 10 dB are added to these signals and the obtained synthetic signals are denoised using 6 different methods: EMD-DFA, EEMD-filtering,

Table 1 Average value of metrics obtained on denoising 89 raw VAG signals, 51 normal (NO) and 38 abnormal (AB) signals

Metric	EMD-DFA		Wavelet-soft thresholding		EEMD-Wiener entropy thresholding		VMD-Wiener entropy thresholding	
	AB	NO	AB	NO	AB	NO	AB	NO
MSE	46.23	45.21	582.10	326.72	47.84	40.82	38.36	36.29
MAE	6.17	4.26	13.28	8.23	8.13	4.74	6.03	4.14
SNR	6.95 dB	6.32 dB	−4.2 dB	−3.73 dB	6.46 dB	5.81 dB	7.61 dB	6.54 dB
PSNR	21.90 dB	22.35 dB	10.69 dB	14.75 dB	17.88 dB	20.24 dB	21.79 dB	22.36 dB
CPU runtime (per signal)	4.82 s	4.65 s	0.4 s	0.41 s	5.88 s	5.92 s	30.2 s	28.2 s

Table 2 Average values of NRMS and CPU consumption time obtained by denoising 89 synthetic VAG signals using 5 different methods

Method	NRMS		Average CPU consumption time (s)	
	SNR = 0 dB	SNR = 10 dB	SNR = 0 dB	SNR = 10 dB
EMD-DFA	0.18	0.15	5 13	5.60
Matching pursuit	0.50	0.26	4.64	3.91
Wavelet packet	0.56	0.42	2.55	2.08
Wavelet soft-thresholding	0.70	0 75	1.09	0.97
EEMD-Wiener entropy thresholding	0.49	0.51	5.23	5.86
Proposed method	0.12	0.09	28.2	27.4

wavelet-matching pursuit, wavelet soft thresholding, wavelet packet- soft thresholding and the proposed method. A similar procedure has been carried out in previous works to evaluate the performance of denoising algorithms [14, 15]. In the wavelet-matching pursuit method proposed by Krishnan-Rangayyan [6] Gaussian functions were used for MP and the threshold for denoising was selected based on the value of the decay parameter. For wavelet-soft thresholding a 4th order symlet wavelet was chosen. In the wavelet packet denoising method, the best basis function was selected based on the Schur concavity cost function, and the noise-free signals are obtained by soft thresholding the obtained coefficients. Table 2 presents the average value of NRMS and CPU consumption time obtained by denoising the 89 synthetic VAG signals.

From Table 2, it can be observed that the proposed method yields the lowest average value of NRMS and hence shows the most accurate denoising, but again provides the slowest response.

5 Conclusion

In this paper a new method for denoising VAG signals using variational mode decomposition has been presented. Standard metrics: MSE, MAE, SNR, PSNR, NRMS and CPU consumption time have been computed to assess the performance of the proposed algorithm. The effectiveness of the algorithm in removal of noise from both raw VAG and synthetic VAG signals corrupted with additive white gaussian noise is also estimated. From our study it is clear that the proposed method yields better accuracy in denoising the VAG signals in comparison the previous methods, but with a slower response. However, since performing diagnosis of knee joint disorders is not time critical, our method is suitable for use in real world knee joint disorder screening systems.

References

1. J Umapathy K, Krishnan S, Modified local discriminant bases algorithm and its application in analysis of human knee joint vibration signals, IEEE Transactions on Biomedical Engineering, Volume 53, Issue 3, pp. 517–523, doi:10.1109/TBME.2005.869787.
2. V. Vigorita, B. Ghelman, D. Mintz, Orthopaedic Pathology, M - Medicine Series, Lippincott Williams and Wilkins, 2008.
3. G. McCoy, J. McCrea, D. Beverland, W. Kernohan, R. Mollan, Vibration arthrography as a diagnostic aid in diseases of the knee. a preliminary report, Journal of Bone and Joint Surgery, British Volume 69-B (2) (1987) pp. 288–293.
4. Akkan T, Senol Y, Capturing and analysis of knee-joint signals using acceleremoters, Proc 16th IEEE International Conference on Signal processing, Communication and Application, doi:10.1109/SIU.2008.4632614.
5. S.H.Rahangdale, A. K. Mittra, Vibroarthrographic Signals De-NoisingUsing Wavelet Subband Thresholding, International Journal of Engineering and Advanced Technolog, Volume: 3, Issue: 2, December 2013, ISSN: 2249–8958.
6. Krishnan S, Rangayyan R.M. Denoising knee joint vibration signals using adaptive time-frequency representations, IEEE Canadian Conference on Electrical and Computer Engineering, 1999, Volume 3, pp. 1495–1500, doi:10.1109/CCECE.1999.804930.
7. Jien-Chen Chen, Pi-Cheng Tung, Shih-Fong Huang, Shu-Wei Wu, Shih-Lin Lin and Kuo-Liang Tu, Extraction and screening of knee joint vibroarthographic signals using the empirical mode decomposition, International Journal of Innovative Computing, Information and Control, Volume 9, Number 6, 2013, ISSN: 1349–4198.
8. R. M. Rangayyan, Y. F. Wu, Screening of knee-joint vibroarthrographic signals using statistical parameters and radial basis functions., Medical & biological engineering & computing 46 (2008) 223–232. doi:10.1007/s11517-007-0278-7.
9. Dragomiretskiy K, Zosso D, Variational Mode Decomposition, IEEE Transactions on Signal Processing, Volume 62, Issue 3, Pg. 531–544, doi:10.1109/TSP.2013.2288675.
10. Lahmiri S, Boukadoum M, Biomedical image denoising using variational mode decomposition, IEEE Conference on Biomedical Circuits and Systems Conference (BioCAS), 2014, Pg. 340–343, doi:10.1109/BioCAS.2014.6981732.
11. Y. Wu, Knee Joint Vibroarthrographic Signal Processing and Analysis, Springer, 2014.
12. Rangayaan RM, Oloumi F, Wu Y, Cai S, Fractal analysis of knee-joint vibroarthographic signals via. power spectral analysis, Elsevier Biomedical signal processing and Control, Volume 8, Issue 1, 2013, pp. 23–29, doi:10.1016/j.bspc.2012.05.0.
13. Jit Muthuswamy, Biomedical Signal Analysis, Chapter 18, Standard Handbook of Biomedical Engineering and Design, 2004.
14. Rishendra Verma, Rini Mehrotra, Vikrant Bhateja, A New Morphological Filtering Algorithm for Pre-Processing of Electrocardiographic Signals, Lecture Notes in Electrical Engineering, Volume 1, pp. 193–201, (2013), doi:10.1007/978-81-322-0997-3_18.
15. Bhateja V, Urooj S, Verma R, Mehrotra R, A novel approach for suppression of powerline interference and impulse noise in ECG signals, Proceedings of IEEE International Conference on Multimedia, Signal Processing and Communication Technologies, pp. 103–107 (2013), doi:10.1109/MSPCT.2013.6782097.

Data Size Reduction and Maximization of the Network Lifetime over Wireless Sensor Network

Venu Madhav Kuthadi, Rajalakshmi Selvaraj and Tshilidzi Marwala

Abstract The main concept of this research is for increasing the network lifetime and decreases the data size over wireless sensor network. To perform this idea we proposed some novel technique which provides the reliable energy efficient routes and maximizing the network lifetime for finding the route that minimize the total energy for packet traversal. We also use the data compression model that reduce the size of data and joint balancing of nodes and optimize the dynamic compression for improving the lifetime of network. The data compression could be completed within some step, those are raw data could get broken in few branches and get compressed at distinct level of the compression, these compressed data could be decompressed at a certain level and again compressed with distinct level to forward directly or by using some base station. For transmitting the data to base station from the source node, every node has to be clustered and have to choose one cluster head in the group of every cluster, the CH (Cluster Head) is having the more energy in compared to the all other nodes. The CH (Cluster Head) is obtaining the entire message from the other neighbor's nodes and transmits it to the Base station. From source to destination data transmission, the nodes are searching shortest path that provide a high computation of complexity.

Keywords LEACH · Data compression · Dijkstra with AODV · Network lifetime · Huffman coder

V.M. Kuthadi (✉)
Department of AIS, University of Johannesburg, Johannesburg, South Africa
e-mail: vkuthadi@uj.ac.za

R. Selvaraj · T. Marwala
Faculty of Engineering and the Built Environment, University of Johannesburg, Johannesburg, South Africa
e-mail: selvarajr@biust.ac.bw

T. Marwala
e-mail: tmarwala@uj.ac.za

R. Selvaraj
Department of Computer Science, BIUST, Gaborone, Botswana

© Springer India 2016
S.C. Satapathy et al. (eds.), *Information Systems Design and Intelligent Applications*, Advances in Intelligent Systems and Computing 433,
DOI 10.1007/978-81-322-2755-7_75

1 Introduction

In general WSN (Wireless Sensor Network) contains a huge amount of sensor nodes. This paper defines wireless sensor nodes as a small device which respond to one or more stimuli, process and transmission of the information to a particular distance by utilizing the laser approaches and radio frequencies [1] Sensor networks normally senses physical phenomenon located near to incidence and convert these sensed measurement as signals which is processed to disclose the character of phenomena situated around the sensed area. There are various phenomenon types that could be Sensed. Acoustics, humidity, light, temperature, seismic activity, imaging, and some physical phenomenon that can cause transducer to reply. Sensor nodes contain the set of sensors, memory, processor, communication system, position locating system, mobilize and units of power. Wireless sensor network is collecting the data from targeted region and once it is collected then forward to base station and atmospheric processing node. Sensor node or a Base Station is either movable or immovable in nature. WSN (Wireless sensor Network) contains thousands of nodes deployed in homes, buildings, highways, cities to control and monitor the infrastructures atmosphere [2, 3].

In the research work [4] wireless sensor node which collects the information is commonly known as sink. These sinks collect information from the outer world by means of connecting through internet and these information's are used within limited period of time. The main problem of using WSN is its battery life which is limited and this is because of its size. Sensor nodes are with the limited resources and in small size like processor, battery size, and memory storage. Creation of these small sensor nodes and transmission of these sensed data through sensor nodes have made a way to implement an interesting application in new. The networks build a device known as WSNs. These WSNs allows the usage of small application widely. Some example of these network discussed in the paper are as follows, these networks are used in various fields, Once such field is the medical field. In medical field WSNs is used to check the conditions of the patient from home, it helps to take some emergency precaution and it saves time. Energy utilization is a main issue in the Wireless Sensor Network (WSN). As the Size of sensor node is small, it contains limited resources. In WSNs node battery life is one of the important factors because replacing of battery is too hard as it has some physical restrictions. For this reason many researchers are focusing on the power aware protocols algorithm structure. Mainly there are two major limitation which is been argued, they are less memory storage and Limited bandwidth process. These limitations are disabled by fabrication development technique. Therefore the limitation of energy will possibly solve soon as it has shown a slow process in the battery development [5, 6].

Wireless Sensor Networks Have the Following Characteristics

1. Sensor nodes are coordinated into single group known as cluster. CH performs the aggregation of data and the compressed data are received by Base Station.
2. WSN life time depends on the total number of time in which the 1st wireless sensor node is running out of the power.
3. Each sensor nodes communicate with base stations by multi-hop transmission or direct transmission since each node is immobile.
4. Wireless Sensor nodes is sensing the environmental condition at various locations at constant rate and base station receives data from sensor nodes.
5. Based on the distance, the Wireless sensor nodes adjust the wireless transmitter power [6].

There are more constraints that have a major effect on the WSNs lifetime. The important constraint of all is the lifetime of battery. A sensors node utilizes its batteries to process, sample, and data transmission. There are lots of studies that optimize the effective use of WSN dissipation of energy feature to avoid large consumption of energy in every protocol stack levels in WSNs (e.g. Hardware, communication, signal processing and various network protocols) [7]. It is exposed that to gain the productive and efficient WSN framework, entire nodes must utilize their energy evenly [7]. If not, few nodes which put out a large amount of data, consume the battery energies too early, and therefore there would be an observation hole in the network where some regions cannot be monitored. So, to avoid this problem an evenly dissipation of energy in network ought to be done.

Compression and data aggregation are key technique to make the energy efficiently in WSN operations. This Research works [8, 9] introduce the idea of tunable compression; which can tune the operation that is complexity calculation of data less-loss compression on the basis of energy accessibility. The approach discussed above comes towards the tools of compression like gzip where ten several levels of the compression ratios which are available. Nodes dissipate energy by means of data decompression and data compression; it is significant to decide the savings of energy which are gained by several compression methods. Data transmission Energy is the important parameter of WSN data communication. So, it is very important that traffic of data flow must be most effective one and data transmission energy must be utilizing it efficiently.

Structure of cluster can extend the sensor network lifetime through aggregate data of cluster head from the nodes over the cluster and sending it to base station. Sensor network deployed needs CFP (Cluster Formation Protocol) for partitioning the clusters from network and must have to select CH. This process uses two approaches called Cluster First Approach and Leader First Approach. Cluster first approach, 1st form the cluster and then the CH is selected. In leader first Approach CH is selected 1st and later formed the cluster [10]. Few clustering merits are: it decreases routing table size stored at specific nodes. Cluster Head can increase the individual and network sensors battery life. Cluster Head could perform the data collection in cluster and decreases the amount of redundant packets. Cluster Head reduce the energy rate using scheduled activities inside the clusters.

This research work is presenting the novel technique for increasing the network lifetime and the reduction in data size. The proposed technique also tries to overcome on the issue of replacement of battery, battery lifetime, shortest path selection for data transmission to destination from the source, access size of data etc.

The proposed technique like Huffman Coding is for the compression of data that reduce data size. The Modified LEACH algorithm is clustering the node and selects the cluster head, and the Dijkstra Algorithm with AODV protocol is providing an enhanced and efficient way to find the suitable and shortest path for the data transmission.

2 Literature Survey

The author proposes a new method on the basis of adaptive technique for preventing the problem of packets loss. The author presents an adaptive technique with acknowledgement scheme. This scheme is used to minimize a collision of packet and maximize the throughput of network. Network simulator-2 is used to build this acknowledgement concept. The author proposes a novel adaptive modulation method that helps for signal maximization and minimization of the error bit rate. The main issue in WSN is the nodes physical length. The author discuss that a node can communicate easily to its far node, and a node can't communicate to its close nodes as easily by means of some physical interference. Sensor network deployments create battery consumption and congestion. When sink receives the data from source then strength of signal became weak by some network interferences. Weakness of signal will affect the network reliability and performance. Because of this the author proposed a technique of adaptive modulation [11].

The author introduces the cross-layer design to develop the lifetime of network in WSN. This mechanism is combined with the advertisement transmission scheme for improving the small ALOHA slot on the basis of lifetime and throughput of WSN. In order to decrease the overhead transmission when compared with another medium access techniques, this cross layer design access method is preferred as it do not includes the information's of protocol with the data bit which are transferred. Result proves the cross-layer design combination and advertisement transmission scheme which maximize the network throughput up to 10 % and twice the lifetime of network so this method is better than slotted ALOHA WSN without advertisements [12].

The author presents a highest throughput on the analysis of wireless mesh backhaul network, which is possible to achieve by practical CSMA/CA (Carrier Sense Multiple Access with Collision Avoidance) MAC (Medium Access Control)

protocol. They route to MCF (multi commodity flow) for the formulated augmentation with the constraints of divergence graph, and an enhanced technique has
been taken over the collision of distributed CAMAC/CSMA. This type of overhead
is being ignored by the capacity of MCF that assumes the scheduling of impractical
centralized and the planning of the aggressive capacity in the result that is not
possible to achieve among the CA/CSMA. The author mentioned three most
important contributions:

1. The author proposed a generic technique of CA MAC/CSMA analysis with
   MCF to formulate the analysis capacity of optimal network, which is ready to
   generate the network throughput upper bound.
2. The author describes a novel idea of CSMA/CA theoretically studies and clique
   with the relationship of CSMA/CA area by throughput.
3. CSMA/CA clique is used like a tool; this tool derives a network throughput
   lower bound, and clique-based formulation of MCF is helping to achieve this
   throughput within CSMA/CA MAC protocol.

NS-2 simulation outcomes are presenting the lower and upper bound tightness;
these are newly developed and are compared with MCF formulation assumption of
centralized scheduling and slotted systems [13].

In this research work, a novel routing technique has been proposed for WSNs.
This technique extends the network lifetime by A-star technique and fuzzy technique combination [14]. The proposed technique helps to define the optimal path of
routing from the source to destination by using "loads of minimum traffic,
remaining battery life, and small amount of hops". This Proposed method is
demonstrated by balancing energy consumption and network lifetime maximization
[15].

The author has discussed in the research work [16], about sensor network
topology control and balance loads on the nodes of sensor, and it maximizes
network lifetime and network capacity. Distributed clustering is proposed by the
author for increasing lifetime of ad hoc network. Sensor node clustering in network
is most successful topology. Proposed technique doesn't have any virtual infrastructure about capacity of nodes and also it doesn't have any assumption about
multiple power availability levels in sensors nodes. The author presents HEED
protocol, this protocol selects cluster head based on node residual power hybrid and
secondary parameter like node proximity for their neighbors or degree of node.

The author of the research paper [11] has presented an algorithm for the WSN
(wireless sensor network) that is self-organized with huge amount of the static
nodes within the resources of energy. Some protocol in WSN is supporting slow
mobility through the node subset, sensor network formation, and efficient routing of
the energy to carry the processing function of cooperative signal over the nodes set.
The WSN (wireless sensor network) is being used for monitoring the health and
security, sampling of widespread environment and surveillance applications.

The author of the research work [6] has presented a discussion about the battery life in WSN (Wireless Sensor Networks), and how it could be efficiently managed, so the network lifetime would be increased. In the network of large scale, they consider about the multiple deployed sink nodes and a huge sensor node number for increasing the network manageability and the reduction over consumption of energy. In this research work, the author focuses on the problems of sink location most importantly in huge networks of WSN. There is more number of problems based on design criteria. The author concentrate to locate the sink nodes for the environment of sensors, If the author gives time constraint, that situation needs less operational time for sensor network.

2.1 Problem Formulation

The sensor node is a small electronic component so that it can be outfitted with limited energy only. In few situations we can't replace the energy resources. Battery life plays major role in the life of sensor node. Limited battery energy leads to failure of sensor node. If sensor node fails then it leads to network breakage and also we can't gather the data from specific area. In ad hoc sensor network and multi hocp sensor network, data origin plays double role or as data router. Some broken nodes will be a reason for the major change in topology and it needs packet rerouting and network reorganization. So we must give extra importance to the energy conservation. Most algorithms are proposed to save the energy and Clustering is a part of them. Here clusters are being formed as set of nodes by clustering. CHs are selected periodically and cluster member can communicate to their CHs. Cluster member receives the data from base station that is send through its CH. The Cluster Head can use multi clustering. CHs have to be rotated to make the energy balanced and to make an equal load over all the nodes. We could reduce the consumption of energy.

3 Proposed Work

3.1 Overview

The research work is proposing an enhanced technique for the problem over lifetime of the network. The low network lifetime is creating a mess over the transmission of data; the network lifetime totally based on the battery capacity. This paper is proposing an enhanced technique for increasing the network lifetime and as well for the reduction of data size. The reduction of data size is being reduced with the data compression technique; the Huffman Code has been used for reducing the size of data by compression of the data. Before compressing the data, all the node are being clustered by LEACH technique, the clustered node selects a CH over all the nodes based on the energy. Finally, data is being transmitted from source to base station depend upon the shortest path. The Dj-kstra Algorithm with the AODV protocol is providing an enhanced and efficient shortest path over the Wireless Sensor Network nodes.

3.2 Overall Architecture

See Fig. 1.

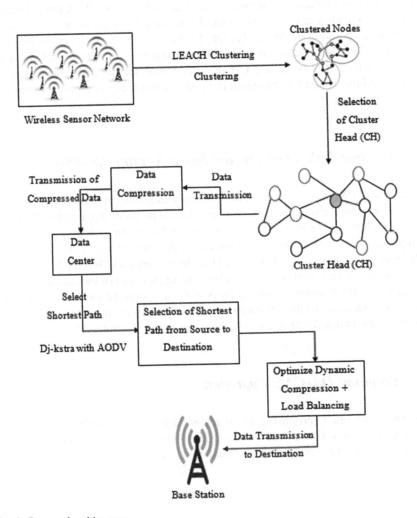

Fig. 1 Proposed architecture

3.3 Data Transmission

Once clusters are being formed and the schedule of the TDMA is fixed, the nodes starts to transmitting their data. Assume, some nodes are always there to send the data, they send it at the time of allocated transmission to CH (Cluster Head). This transmission needs only the minimum energy amount on the basis of strength received from the advertisement CH. Every non-cluster head radio could be switched off till transmission time of the allocated node, therefore it minimize the dissipation of energy. The receiver of the CH (Cluster Head) node must have to be on for receiving the data from clusters. Once every data is being received then cluster head is performing the functions of the optimization like aggregation of the data for functioning of another signal processing the compression of data into single signal. The composite signal is having a transmission of high-energy with base station on the big distance, and then sending to base station. CH (Cluster Head) is using CSMA with a code named as fixed spreading for sending these packets of data. It is an operation of the steady-state LEACH network. After certain period of time, a priori has been determined and the next round began with the determination of every node if it is becoming CH (Cluster Head) for the advertisement of the decision and this round for the remaining nodes that has been described in phase of the advertisement.

3.4 Lifetime Enhanced Cluster Based Routing in WSN

In the WSNs (Wireless Sensor Networks), data transmission is being constrained through the limited energy of battery nodes. Sensor nodes deploying densely would detect correlated data and redundant that may be the reason of energy wastage. For saving the energy and resources and for enhancing the lifetime of the network, the technique of clustering could be used [14]. In the network that based on cluster, Cluster Heads (CH's) performs the routing and aggregation of the data as shown in Fig. 1. The routing technique on the basis of cluster with the mechanism of power saving is being used for the enhancement of lifetime. Overlapped coverage area for deploying the nodes forming on the scheme of power saving.

3.5 Proposed DA-CAC Approach

This research work is proposing an enhanced Dijkstra's Algorithm and DA-CAC (Closest Adjacency Condition) for detection of the shortest path in the wireless sensor network. The promising mechanism is serving the requirement for the optimal path detection in the terms of consumption of energy and cost. It is also storing information about energy level of nodes for maintaining the communication in a continuous link. The approach has been started initially with discovery of route through the transmission of RREQ message to the neighbor nodes. After transmitting the message, wait for RREP within certain parameter and data transmission cost where every route could be identified. The cost of this approach is varying with the value of TTL fields and hop count. The nodes minimum counts are being selected from the hop counts for reaching at the destination and must have to keep the minimum value of the TTL. After the cost identification of all path storage information is performed through the source node to the routing table. The measurement of effective route should avoid some conditions even the nodes occur in the shortest path. If network is containing a structure on the basis of loop then the energy and performance would be degraded, then the need of the loop will be avoided. The proposed technique or mechanism measures about the occurrence of shortest path. If the entry of node is repeated in routing table, in that case it ought to be removed.

3.6 Data Compression

1. Selector: In this phase we select the new value with three existing values that are mentioned previously, these three reading values are considered as zero in initial stage.
2. Median Predictor: This phase predicts a median value from minimum and maximum values between three values that are previously selected and it computes the current value deviation through median value.
3. Huffman Coder: It calculates the Huffman code deviation from median predictor that provides result in the compressed data as shown in Fig. 2. The general concept of the Huffman coding is to map the alphabet for representing the alphabet, composing the bit sequences for variable size, symbols are easily occurring have little and small representation than the rarely occurred symbols. In this situation probabilities and R + 1 symbol are decreases by increasing values.

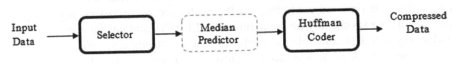

Fig. 2 Huffman code

Sensor node application is widely used for reporting the interested area status for users, and it's very useful to calculate the xi which is obtained by the sensor node. ri represents the binary form of the measured xi. Then R bits are provided by ADC, here R denotes the ADC resolution in the research work [9]. Compression technique predicts the middle value from three values which are mentioned previously for every new attainment of xi. And compression technique is used to compute the current value's deviation through median.

3.6.1 Dynamic Optimizations

Dynamics Optimization allows in situ constrains to empower and tunes the sensor node to adapt the changing needs of application and stimuli of environmental throughout the life time of sensor node. Dynamic Optimization is main process because the needs of application can be changed from time to time and based on environmental condition/stimuli which cannot predict the accuracy at the time of designing. In nature Some MAC layer, OS and optimization of routing are dynamic.

3.6.2 Frequency Scaling and Dynamic Voltage

Frequency Scaling and Dynamic Voltage adjusts sensor node processor voltage and frequency for optimizing the energy consumption. DV (Dynamic voltage) and Frequency scaling balance the performance to reduce the energy consumption using peak computation. It's very high in compared to the average throughput of application needs and sensor node that depends on the CMOS logic that contains a voltage based on high operating frequency. The authors demonstrate the system of DVFS will have a scheduler of voltage running in cycling with task scheduler of operating system and it provides the result of energy consumption efficiently.

3.7 LEACH and Its Descendant

Following techniques are utilized by LEACH for achieving the goals of design: adaptive cluster formation and self-configuring, randomized, local control for transferring the data and controlling by the media access for low energy and processing of data specifically on the application. LEACH contains number of rounds and every round is containing two phases that are steady and setup phase. The Setup phase provides the formation of cluster in suitable manner. In Steady phase data's are transmitted. LEACH reduces the intra-cluster and inter-cluster collision by CDMA MAC or TDMA. Formation of cluster depends upon number of properties like sensor types, numbers, range of communication and location. Sensor nodes are being in used for collecting the information of energy consumption and send to sink based on cluster head numbers and various techniques like radio range in various algorithms. This is because of the reduction in the consumption of energy with organized sensor node inside the clusters.

There are several form of the LEACH algorithm that is mentioned below LEACH-F, LEACH-C, LEACH-B, LEACH-ET, Energy–LEACH, TL-LEACH, MH-LEACH, ACHTH–LEACH, MELEACH-L, MS-LEACH, Trust-Based LEACH, LEACH-DCHS-CM, Re-Cluster-LEACH and etc.

The advantages of clustering algorithm are:

1. Decrease the size of routing table,
2. Decrease the exchanged message redundancy,
3. Decrease the consumption of energy, and
4. Prolong the life time of networks.

3.8 Cluster Head Selection

In networks cluster, cluster header performs routing and data aggregation. Routing method is cluster based, it is a mechanism of power saving, and then it is utilized for the enhancement of network lifetime. The area of overlapping coverage randomly organize the nodes basis scheme of power saving. We explain our proposed algorithm for selecting the CH (Cluster Head) for single-cluster network. Over Ns nodes, one node will be selected as cluster head; it is implicated to fuse the information from entire NCHs. The Model of single cluster network and algorithm design objectives is towards the increasing of time, meanwhile the initial node dies signify by NCHf, node of NCH is being on the farthest distance from cluster head, and distance amid of A and B, which is denoted by $d{:}A{:}B$, the power needed for NCHf and Cluster Header are shown below

$$(N_C = 1, \ \alpha = 1/N_S)$$

$$P_{CH} = E_{elec}^r (N_S - 1) + E_{DA}^r N_S + E_{elec}^r + E_{amp} d_{CH\cdot BS^r}^2$$

$$P_{NCHf} = E_{elec}^r + E_{amp} d_{CH\cdot NCHf^r}^2$$

3.9 Algorithm

3.9.1 Shortest Path Algorithm

Proposed DA-CAC Algorithm

```
Initialize AODV ();
Send the RREQ (Hop Counter, DestinationSeqNum, SourceSeqNum, TTL);
Receive RREP (Hop Counter, DestinationSeqNum, SourceSeqNum, TTL)
Cost_Analysis_ Route_Path ();
{
Max Cost= Max (Hop Counts + TTL Value);
Min Cost= Min (Hop Counts + TTL Value);
}
Costs for every Path==Store;
Comparison_Dijkstra's_Path (Path 1, Path 2 ...Path n);
If (Path==Shortest && Node==Twice)
{
Detection of Loop;
Delete path from list;
}
Gateway_Path_Count If (Path==Contains_Gateway_Node);
Then,
HopDistance (Closest Node to Gateway) // CAC (Closest Adjacency Condition)
Node
If (Node_Count_Shortest_Path>=2) // Convergence of dynamic route
Node of Shortest Path;
WakeUp== Closest Adjacency Condition (CAC) Node_Path_Elements;
If, low battery == Route Update // Check every node battery level
Else Perfect Route;
Suggested_CAC_Shortest_Path (Close to Gateway, low load, High Energy
Nodes, Loop free);
Shortest_Path for seeking the communication over node;
Routing table Updation;
Exit;
}
```

3.9.2 Data Compression Algorithm (Huffman Code)

```
encode (v, Table, prevArray[])
// v is known as present value and the prevArray [] containing three previous
values
i = prevArray[0]
j = prevArray[0]
k = prevArray[2]
p = minimum (l,m,n)
q = median (l,m,n)
r = maximum (l,m,n)
IF (p <= v <= r) THEN,
SET TO '0' the set_bit
SET the Previous bit set TO set_bit
IF (v >= q)
SET TO '0' the set_bit
SET Previous bit set TO << Pre_bit_set, set_bit >>
// Calculate Huffman Code
HC(diff(v,q))
SET Huffman Code TO HC(diff(v,q))
SET the final_code TO << Pre_bit_set, Huffman Code >>
ELSE
SET TO '1' the set_bit
SET Previous bit set TO << Pre_bit_set, set_bit >>
HC(diff(v,q))
SET hmc TO HC(diff(v,q))
SET the final_code TO << Pre_bit_set, Huffman Code >>
END IF
ELSE
SET the set_bit TO '0'
SET Previous bit set TO set_bit
IF (v > r)
SET the set_bit TO '0'
SET the Previous bit set TO << Pre_bit_set, set_bit >>
HC(diff(v,r))
SET hmc TO HC(diff(v,r))
SET the final_code TO << Pre_bit_set, Huffman Code >>
ELSE
SET the set_bit TO '1'
SET Previous bit set TO << Pre_bit_set, set_bit >>
HC(diff(v,r))
SET hmc TO HC(diff(v,r))
SET the final_code TO << Pre_bit_set, Huffman Code >>
END IF
END IF
RETURN to final_code
```

3.9.3 Modified LEACH Algorithm

1. Let, the Width and Height of the network is W, H respectively. The area of network is $A = W \times H$.
2. Let the N nodes within the energy of Ei at point of (Xi, Yi).
3. Let the Node 0 is Sink node at W/2, H/2.
4. The issue could be summarized for getting the connected graph $G = \{V,E\}$ from the source S, like E represents the Edge and V represent the Nodes, and L is the Lifetime, So maximize it. The lifetime is defined over the time t1, $Ei <= 0$, where the node I could be any sink node.
5. The Hello packet is broadcasting and another node know the position and energy of the respective nodes.
6. Generally, the sink nodes collects the data from sources, it is selecting the nodes with sufficient energy and maximum Neighbors as cluster heads.
7. Every CH (Cluster head) is informed about the cluster head.
8. Source is generating the RREQ packet.
9. A node forward RREQ packet when the node is cluster head.
10. The Route is being formed amid of the every sink source by cluster heads.
11. Data is being transmitted through source to sink.
12. Nodes loosing Energy as
13. $E_{loss} = E_{idle} + E_{receive} + E_{transmit}$,
14. $E = E - E_{loss}$, there is $E_{idle} = 1pJ/s$
15. $E_{transmit} = 3$ mJ/Packet (Length of the packet is 1024)
16. The $E_{receive} = 1$ mJ/Packet
17. If $E_i < 0$, at Transmission Time, then mark it as the lifetime of network.

4 Result and Discussion

This research study proposed a CHRS (Cluster based routing scheme) where based on the maximum coverage the CH is selected and the energy must be adequate to extend the communication. OMNeT++ is used to find the consumption of energy and to prolong life of WSN. The attribute of proposed technique measurement has conducted some simulator experiment. For the conducted experiment, following configuration is required.

(1) Windows 7, (2) Intel Pentium(R) (3) CPU G2020 and (4) Processer speed 2.90 GHz.

The Tables 1 and 2 is representing the simulation result over the 25 and 50 nodes. The simulation result is showing a sufficient and enhanced measurement over the nodes. The Packet size, Initial Energy, Packet Rate and lifetime of the network are increased.

Table 1 Simulation
parameters for 25 nodes

Area	500 * 500 m
Size of packet	512 Bytes
Nodes initial energy	100 mj
Rate of packet	500
Activity session numbers	14
Energy through MAC	0.003 mj/bit
Outside module energy	0.001 mj/bit

Table 2 Simulation
parameter for 50 nodes

Area	500 * 500 m
Size of packet	512 bytes
Nodes initial energy	1000 mj
Rate of packet	500
Activity session numbers	30
Energy through MAC	0.003 j/bit
Outside module energy	0.001 mj/bit

The Table 3 shows the measurement of the data compression technique, the measurement is showing the difference over the Original Size, Compressed Size, and Compressed Ratio.

The Table 4 is representing the various parameters that have been used in the simulation result. The process is showing the Area of Network, Rate of data, Highest transmission power, Range of Transmission, Packet size of Data, Size of Hello packet, Threshold of Battery death, and Sensing time of Channel.

The Fig. 3 proves that the LEACH algorithm reduces the uptake of energy because of less redundant neighbor number and transmission of redundant message. By the use of LEACH algorithm, the energy is being saved in the comparison with the case of CSA (K = 1) algorithm. The LEACH algorithm is using less energy in compared to any other technique.

The Fig. 4 shows the time slots of the Network lifetime, the modify LEACH algorithm is most enhanced and provide a good time slots over the network. The proposed algorithm is better than existing algorithm for increasing the lifetime of network.

Table 3 Data compression

	Sample 1	Sample 2	Sample 3
Orig_size	23,040 bits	23,040 bits	23,040 bits
Comp_size	7466 bits	7545 bits	7498 bits
Comp_ratio	67.60 %	67.25 %	67.46 %

Table 4 Values of various parameter used in simulations

Parameter	Value
Initial energy for every nodes	100 J
Area of network	$350 * 350$ m^2
Exponent of loss path (n)	3
Rate of data (r)	100 kbps
Transmitter circuit consumption of power (p_t)	100 mW
Highest power transmission (P_{max})	150 mW
Lowest power transmission (P_{min})	15 mW
Maximum# transmission in HBH system (Q_u)	7
Range of transmission (d_{max})	70 m
Packet size of data (L_d)	512 bytes
Size of ACK MAC packet (L_h)	240 bit
Size of ACK E2E packet (L_a)	96 byte
Size of Hello packet (L_{hello})	96 bytes
Threshold of battery death (B_a)	0
Highest collision probability (Pc_{max})	0.3
Sensing time of channel (T_{max})	50 μs
K_{idile}	0.2
K_{sense}	0.4
T_{hello}	10 s
T_{Tc}	20 s

4.1 Centralized and Localized

The Fig. 5 shows the relationship amid of the network life-time, threshold value of the network size, and desired level-k for the coverage for switching with the localized algorithm from centralized and also vice versa. The figure is also showing that the lifetime of the network is decreasing when k increasing.

4.2 Data Compression Technique

Figure 6 explains the comparison of existing data compression techniques with Huffman Code. The Huffman code is our proposed technique. This graph is showing the difference of existing and proposed techniques accuracy. The proposed technique is more accurate in compare to the other existing technique.

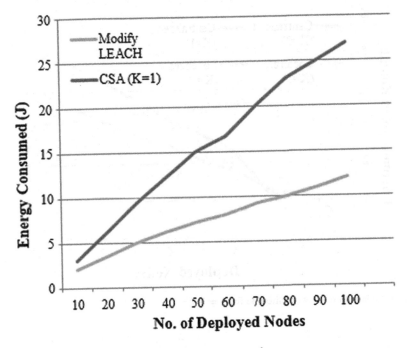

Fig. 3 Number of deployed nodes versus energy consumed

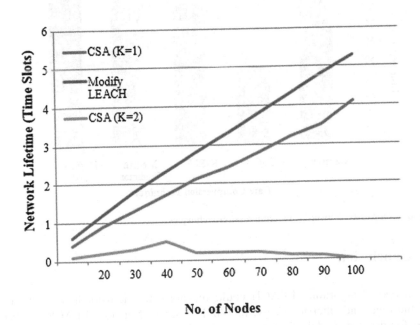

Fig. 4 Number of nodes versus network lifetime

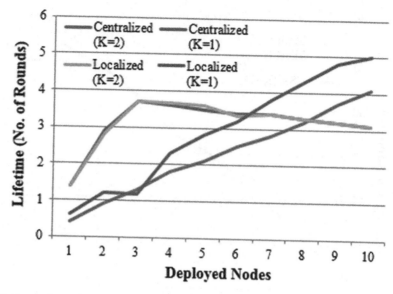

Fig. 5 Comparison of network lifetime for $k = 1$, $k = 2$

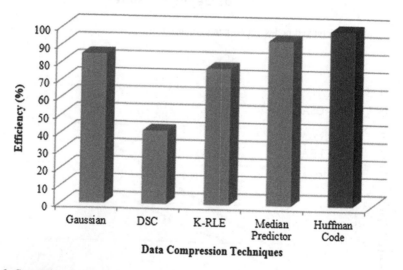

Fig. 6 Comparison between data compression techniques

5 Conclusion

Our proposed algorithm, LEACH could maximize the network lifetime through reliable route and efficient energy. In Wireless Sensor Network, LEACH has been used for detailed model of energy consumption in packet transfer. This research work is presenting the novel technique for increasing the lifetime of network and reducing the data size. For reducing the size of data, the Enhanced Huffman Code Data Compression Algorithm has been used. The compressed data is transmitted to base station through shortest path by using the Enhanced Dijkstra Algorithm with AODV protocol. This enhanced algorithm is seeking a communication among the other nodes in network that making a transmission of data towards source to destination. The modified LEACH algorithm clusters the node and selects cluster head based on the energy. Generally, when the sink nodes try to collect data from the source, it is selecting the maximum neighbor nodes and the cluster head with sufficient energy.

References

1. Akyildiz, I.F., Welilian Su, Yogesh, S., & Erdal, C.: A survey on sensor networks. IEEE Communications Magazine. 40(8), 103–114 (2002).
2. Al-Karaki, J.N., Kamal, A.E.: Routing Techniques in Wireless Sensor Networks: A Survey. IEEE Communications Magazine. 11, 6–28 (2004).
3. Kuthadi, V.M., Rajendra, C., & Selvaraj, R.: A study of security challenges in wireless sensor networks. JATIT 20 (1), 39–44 (2010).
4. Jakob Salzamann, Ralf Behnke, Dirk Timmermann.: Hex-MASCLE—Hexagon based Clustering with Self-Healing Abilities IEEE, 528–533 (2011).
5. Changsu Suh, Young-Bae Ko.: A Traffic Aware, Energy Efficient MAC Protocol for Wireless Sensor Networks. In: IEEE International Symposium on Circuits and Systems, South Korea, pp 2975–2978 (2005).
6. Basilis Mamalis, Damianos Gavalas, Charalampos Konstantopoulos & Grammati Pantziou.: Clustering in Wireless Sensor Networks. RFID and sensor Networks, 323–350 (2009).
7. K. Akkaya and M. Younis.: A Surrvey on routing protocols for wireless sensor networks. Ad Hoc Networks 3, 325–349 (2005).
8. Kuthadi, V.M., Selvaraj, R., & Marwala, T.: An Efficient Web Services framework for Secure Data Collection in Wireless Sensor Network. British Journal of Science 12(1), 18–31 (2015).
9. Y. Yu, B. Krishnamachari, and V. Prasanna. Data gathering with tunable compression in sensor networks. IEEE Transactions on Parallel and Distributed Systems, 19(2):276–287, 2008.
10. Kun Sun, Pai Peng, Peng Ning.: Secure Distributed Cluster Formation in Wireless Sensor Networks. In: 22nd Annual Computer Security Applications Conference, Florida, USA, pp 131–140 (2006).
11. Sunita.: Comprehensive Study of Applications of Wireless Sensor Network. International Journal of Advanced Research in Computer Science and Software Engineering 2 (11), 56–60 (2012).
12. Lucas D. P. Mendes, Rodrigues, J.J.P.C., Vasilakos, A.V, Liang Zhou.: Lifetime Analysis of a Slotted ALOHA-based Wireless Sensor Network using a Cross-layer Frame Rate Adaptation Scheme. In: IEEE International Conference on Communications, Kyoto, pp 1–5 (2011).
13. Yu Cheng, Hongkun Li, Peng-jun Wan, Xinbing Wang.: Wireless Mesh Network Capacity Achievable Over the CSMA/CA MAC. IEEE Transaction on Vehicular Technology.61 (7), 3151–3165 (2012).
14. Sohrabi, K., Gao, J., Ailawadhi, V., Pottie, G.J,: Protocols for Self-Organization of a Wireless Sensor Network. IEEE Personal Communication 7(5), 16–27 (2000).

15. Ramesh, K., Somasundaram, K.: A Comparative Study of Cluster-Head Selection Algorithms in Wireless Sensor Networks. International Journal of Computer Science & Engineering Survey 2(4), 153–164 (2011).
16. Priti, Sudhesh, K.: Comparision of Energy Efficient Clustering Protocols in wireless Sensor Networks. International Journal of Advanced Engineering and Global Technology 2(9), 962–969 (2014).

Author Index

© Springer India 2016
S.C. Satapathy et al. (eds.), *Information Systems Design and Intelligent
Applications*, Advances in Intelligent Systems and Computing 433,
DOI 10.1007/978-81-322-2755-7

Printed in the United States
By Bookmasters